# Marine Geochemistry

# Marine Geochemistry

**Emeritus Professor Roy Chester**
School of Environmental Sciences
The University of Liverpool
Liverpool, UK

**and**

**Professor Tim Jickells**
School of Environmental Sciences
University of East Anglia
Norwich, UK

**(W)WILEY-BLACKWELL**

A John Wiley & Sons, Ltd., Publication

This edition first published 2012 © 2012 by Roy Chester and Tim Jickells

Blackwell Publishing was acquired by John Wiley & Sons in February 2007. Blackwell's publishing program has been merged with Wiley's global Scientific, Technical and Medical business to form Wiley-Blackwell.

*Registered office:*  John Wiley & Sons, Ltd, The Atrium, Southern Gate, Chichester, West Sussex, PO19 8SQ, UK

*Editorial offices:*  9600 Garsington Road, Oxford, OX4 2DQ, UK
The Atrium, Southern Gate, Chichester, West Sussex, PO19 8SQ, UK
111 River Street, Hoboken, NJ 07030-5774, USA

For details of our global editorial offices, for customer services and for information about how to apply for permission to reuse the copyright material in this book please see our website at www.wiley.com/wiley-blackwell.

*Library of Congress Cataloging-in-Publication Data*

Chester, R. (Roy), 1936–
   Marine geochemistry / Roy Chester and Tim Jickells. – 3rd ed.
      p. cm.
   Includes bibliographical references and index.
   ISBN 978-1-118-34907-6 (cloth) – ISBN 978-1-4051-8734-3 (pbk.)  1. Chemical oceanography.  2. Marine sediments.  3. Geochemistry.  I. Jickells, T. D. (Tim D.)  II. Title.
   GC111.2.C47 2012
   551.46–dc23
                                     2012010712

A catalogue record for this book is available from the British Library.

Wiley also publishes its books in a variety of electronic formats. Some content that appears in print may not be available in electronic books.

Cover image: Blue ocean waves from underwater.©Solvod/Shutterstock.com
Cover design by: Simon Levy Associates

Set in 9 on 11.5 pt Sabon by Toppan Best-set Premedia Limited
Printed in Malaysia by Ho Printing (M) Sdn Bhd

1  2012

# Contents

COMPANION WEBSITE:
This book has a companion website:
www.wiley.com/go/chester/marinegeochemistry
with Figures and Tables from the book

This book is dedicated with great affection to

John Riley and Dennis Burton.

Two great pioneers in the field of Marine Chemistry

# Preface to the third edition

This edition of *Marine Geochemistry* has been created at a time when the role of the oceans in the Earth System is becoming ever more evident. The central role of ocean processes in climate change, and indeed in all aspects of global change, is increasingly important to all society. The scientific understanding of the role of the oceans and of how they function has developed sufficiently over recent years to justify a new edition of this book. The revisions incorporated in this new edition are the result of the collaboration between the two authors following Roy's retirement.

This edition has been updated to reflect recent advances in the field of marine geochemistry. In particular new insights into nutrient cycling and the carbon cycle have led to a large scale reorganisation of Chapters 8 and 9 compared to previous editions. The relatively recent recognition of the key role of iron as a nutrient is discussed in Chapters 9 and 11. In addition, a section on shelf seas has been added in Chapter 6 to draw together the new understanding of processes in these regions which are now evidently of considerable importance to the marine geochemical cycle, as well as being both of great societal value while also under considerable pressure from human activity.

We are grateful to our publishers for their patience and support, to Phil Judge for producing the new diagrams for this edition and to our respective institutions for allowing us the opportunity to develop this and previous editions. Our special thanks go to colleagues around the world who have published the science we attempt to summarize here.

Finally we would also like to thank Alison Chester and Sue Jickells for their help and support with this endeavour, and for so much more – including keeping us reasonably sane.

Roy Chester and Tim Jickells

# 1     Introduction

The fundamental question underlying marine geochemistry is, 'How do the oceans work as a chemical system?' At present, that question cannot be answered fully. The past four decades or so, however, have seen a number of 'quantum leaps' in our understanding of some aspects of marine geochemistry. Three principal factors have made these leaps possible:
1 advances in sampling and analytical techniques;
2 the development of theoretical concepts;
3 the setting up of large-scale international oceanographic programmes (e.g. DSDP, MANOP, HEBBLE, GEOSECS, TTO, VERTEX, JGOFS, SEAREX, WOCE), which have extended the marine geochemistry database to a global ocean scale.

## 1.1 Setting the background: a unified 'process-orientated' approach to marine geochemistry

Oceanography attracts scientists from a variety of disciplines, including chemistry, geology, physics, biology and meteorology. A knowledge of at least some aspects of marine geochemistry is an essential requirement for scientists from all these disciplines and for students who take courses in oceanography at any level. The present volume has been written, therefore, with the aim of bringing together the recent advances in marine geochemistry in a form that can be understood by all those scientists who use the oceans as a natural laboratory and not just by marine chemists themselves. Furthermore, the oceans are a key component of the Earth System, so an understanding of ocean geochemistry is central to understanding the functioning of the Earth as an integrated system (Lenton and Watson, 2011). One of the major challenges involved in doing this, however, is to provide a coherent global ocean framework within which marine geochemistry can be described in a manner that cannot only relate readily to the other oceanographic disciplines but also can accommodate future advances in the subject. To develop such a framework, it is necessary to explore some of the basic concepts that underlie marine geochemistry.

Geochemical balance calculations show that a number of elements that could not have come from the weathering of igneous rocks are present at the Earth's surface. It is now generally accepted that these elements, which are termed the excess volatiles, have originated from the degassing of the Earth's interior. The excess volatiles, which include H and O (combined as $H_2O$), C, Cl, N, S, B, Br and F, are especially abundant in the atmosphere and the oceans. It is believed, therefore, that both the atmosphere and the oceans were generated by the degassing of the Earth's interior. In terms of global cycling, Mackenzie (1975) suggested that sedimentary rocks are the product of a long-term titration of primary igneous-rock minerals by acids associated with the excess volatiles, a process that can be expressed as:

$$\text{primary igneous-rock minerals} + \text{excess volatiles}$$
$$\rightarrow \text{sedimentary rocks} + \text{oceans} + \text{atmosphere}$$

$$(1.1)$$

As this reaction proceeds, the seawater reservoir is continuously subjected to material fluxes, which are delivered along various pathways from external sources. The oceans therefore are a flux-dominated system. Seawater, however, is not a static reservoir in which the material has simply accumulated over geological time, otherwise it would have a very different composition from that which it has at present; for example, the material supplied over geological time far exceeds the amount now present in seawater. Further, the composition of seawater appears not to have changed markedly over very long periods of time; at least the last few million years and probably longer. Rather than acting as an accumulator, therefore, the flux-dominated seawater reservoir can be regarded as a reactor. Elements are intensively recycled within the vast oceans by biological and chemical processes, although the extent of this recycling and the associated lifetime of components of the chemical system within the oceans vary enormously. It is the nature of the reactions that take place within the reservoir, that is the manner in which it responds to the material fluxes, which defines the composition of seawater via an input → internal reactivity → output cycle. The system is ultimately balanced by the Earth's geological tectonic cycle that subducts ocean sediments into the Earth's interior and returns them to the land surface.

Traditionally, there have been two schools of thought on the overall nature of the processes that operate to control the composition of seawater.
1 In the equilibrium ocean concept, a state of chemical equilibrium is presumed to exist between seawater and sediments via reactions that are reversible in nature. Thus, if the supply of dissolved elements to seawater were to increase, or decrease, the equilibrium reactions would change in the appropriate direction to accommodate the fluctuations.
2 In the alternative steady-state ocean concept, it is assumed that the input of material to the system is balanced by its output, that is, the reactions involved proceed in one direction only. In this type of ocean, fluctuations in input magnitudes would simply result in changes in the rates of the removal reactions, and the concentrations of the reactants in seawater would be maintained.

At present, the generally held view supports the steady-state ocean concept. Whichever theory is accepted, however, it is apparent that the oceans must be treated as a unified input–output type of system, in which materials stored in the seawater, the sediment and the rock reservoirs interact, sometimes via recycling stages, to control the composition of seawater.

It is clear, therefore, that the first requirement necessary to address the question 'How do the oceans work as a chemical system?' is to treat the seawater, sediment and rock reservoirs as a unified system. It is also apparent that one of the keys to solving the question lies in understanding the nature of the chemical, geological and biological (biogeochemical) processes that control the composition of seawater and how these interact with the physical transport within the ocean system, as this is the reservoir through which the material fluxes flow in the input → internal reactivity → output cycle. In order to provide a *unified ocean* framework within which to describe the recent advances in marine geochemistry in terms of this cycle, it is therefore necessary to understand the nature and magnitude of the fluxes that deliver material to the oceans (the input stage), the reactive processes associated with the throughput of the material through the seawater reservoir (the internal reactivity stage), and the nature and magnitude of the fluxes that take the material out of seawater into the sinks (the output stage).

The material that flows through the system includes inorganic and organic components in both dissolved and particulate forms, and a wide variety of these components will be described in the text. In order to avoid falling into the trap of not being able to see the wood for the trees in the morass of data, however, it is essential to recognize the importance of the processes that affect constituents in the source-to-sink cycle. Rather than taking an element-by-element 'periodic table' approach to marine geochemistry, the treatment adopted in the present volume will involve a process-orientated approach, in which the emphasis will be placed on identifying the key processes that operate within the cycle. The treatment will include both natural and anthropogenic materials, but it is not the intention to offer a specialized overview of marine pollution. This treatment does not in any way underrate the importance of marine pollution. Rather, it is directed towards the concept that it is necessary first to understand the natural processes that control the chemistry of the ocean system, because it is largely these same processes that affect the cycles of the anthropogenic constituents.

Since the oceans were first formed, sediments have stored material, and thus have recorded changes in environmental conditions. The emphasis in the present volume, however, is largely on the role that the sediments play in controlling the chemistry of the oceans. The diagenetic changes that have the most immediate effect on the composition of seawater take place in the upper few metres of the sediment column. For this reason attention will be focused on these surface deposits and their role in biogeochemical cycles. The role played by sediments in recording palaeooceanographic change will be touched upon only briefly. It is, however, important to recognize that the oceans play a key role in the Earth System, a role that evolves over geological time, and the oceans also record the history of the evolution of the Earth System and its climate (e.g., Emerson and Hedges, 2008; Lenton and Watson, 2011).

In order to rationalize the process-orientated approach, special attention will be paid to a number of individual constituents, which can be used to elucidate certain key processes that play an important role in controlling the chemical composition of seawater. In selecting these process-orientated constituents it was necessary to recognize the flux-dominated nature of the seawater reservoir. The material fluxes that reach the oceans deliver both dissolved and particulate elements to seawater. It was pointed out above, however, that the amount of dissolved material in seawater is not simply the sum of the total amounts brought to the oceans over geological time. This was highlighted a long time ago by Forchhammer (1865) when he wrote:

> Thus the quantity of the different elements in seawater is not proportional to the quantity of elements which river water pours into the sea, but is inversely proportional to *the facility with which the elements are made insoluble by general chemical or organo-chemical actions in the sea. . .*

[our italics]. According to Goldberg (1963), this statement can be viewed as elegantly posing the theme of marine chemistry, and it is this 'facility with which the elements are made insoluble', and so are removed from the dissolved phase, which is central to our understanding of many of the factors that control the composition of seawater. This was highlighted more recently by Turekian (1977). In an influential geochemical paper, this author formally posed a question that had attracted the attention of marine geochemists for generations, and may be regarded as another expression of Forchhammer's statement, that is 'Why are the oceans so depleted in trace metals?' Turekian concluded that the answer lies in

the role played by particles in the sequestration of reactive elements during every stage in the transport cycle from source to marine sink.

Ultimately, therefore, it is the transfer of dissolved constituents to the particulate phase, and the subsequent sinking of the particulate material, that is responsible for the removal of the dissolved constituents from seawater to the sediment sink. The biological production and consumption of particles by the ocean microbial community and its predators is central to this process. It must be stressed, however, that although dissolved → particulate transformations are the driving force behind the removal of most elements to the sediment sink, the transformations themselves involve a wide variety of biogeochemical processes. For example, Emerson and Hedges (2008) and Stumm and Morgan (1996) identified a number of chemical reactions and physicochemical processes that are important in setting the chemical composition of natural waters at a fundamental physicochemical level. These processes included acid–base reactions, oxidation–reduction reactions, complexation reactions between metals and ligands, adsorption processes at interfaces, the precipitation and dissolution of solid phases, gas–solution processes, and the distribution of solutes between aqueous and non-aqueous phases. The manner in which reactions and processes such as these, and those specifically associated with biota, interact to control the composition of seawater will be considered throughout the text. For the moment, however, they can be grouped simply under the general term particulate ↔ dissolved reactivity. The particulate material itself is delivered to the sediment surface mainly via the down-column sinking of large-sized organic aggregates as part of the oceanic global carbon flux. Thus, within the seawater reservoir, reactive elements undergo a continuous series of dissolved ↔ particulate transformations, which are coupled with the transport of biologically formed particle aggregates to the sea bed. Turekian (1977) aptly termed this overall process the great particle conspiracy. In the flux-dominated ocean system the manner in which this conspiracy operates to clean up seawater is intimately related to the oceanic throughput of externally transported, and internally generated, particulate matter. Further, it is apparent that several important aspects of the manner in which this throughput cycle operates to control the inorganic and organic compositions of both the seawater reservoir and the sediment sink can be assessed in terms of the oceanic fates of reactive trace elements and organic carbon.

Many of the most important thrusts in marine geochemistry over the past few years have used tracers to identify the processes that drive the system, and to establish the rates at which they operate (Broecker and Peng, 1982). These tracers will be discussed at appropriate places in the text. The tracer approach, however, also has been adopted in a much broader sense in the present volume in that special attention will be paid to the trace elements and organic carbon in the source/input → internal reactivity → sink/output transport cycle. Both stable and radionuclide trace elements (e.g. the use of Th isotopes as a 'time clock' for both transport and process indicators) are especially rewarding for the study of reactivity within the various stages of the cycle, and organic carbon is a vital constituent with respect to the oceanic biomass, the down-column transport of material to the sediment sink and sediment diagenesis.

To interpret the source/input → internal reactivity → sink/output transport cycle in a coherent and systematic manner, a three-stage approach will be adopted, which follows the cycle in terms of a global journey. In Part I, the movements of both dissolved and particulate components will be tracked along a variety of transport pathways from their original sources to the point at which they cross the interfaces at the land–sea, air–sea and rock–sea boundaries. In Part II, the processes that affect the components within the seawater reservoir will be described. In Part III, the components will be followed as they are transferred out of seawater into the main sediment sink, and the nature of the sediments themselves will be described. The treatment, however, is concerned mainly with the role played by the sediments as marine sinks for material that has flowed through the seawater reservoir. In this context, it is the processes that take place in the upper few metres of the sediments that have the most immediate effect on the composition of seawater. For this reason attention will be restricted mainly to the uppermost sediment sections, and no attempt will be made to evaluate the status of the whole sediment column in the history of the oceans.

The steps involved in the three-stage global journey are illustrated schematically in Fig. 1.1. This is not meant to be an all-embracing representation of reservoir interchange in the ocean system, but is simply intended to offer a general framework within which to describe the global journey. By directing the journey in this way, the intention therefore is to treat the seawater, sediment and rock phases as integral parts of a unified ocean system.

In addition to the advantages of treating the oceans as a single system, the treatment adopted here is important in order to assess the status of the marine environment in terms of planetary geochemistry. For example, according to Hedges (1992) there is a complex interplay of biological, geological and chemical processes by which materials and energy are exchanged and reused at the Earth's surface. These interreacting processes, which are termed biogeochemical cycles, are concentrated at interfaces and modified by feedback mechanisms. The cycles operate on time-scales of microseconds to eons, and occur in domains that range in size from a living cell to the entire ocean–atmosphere system, and interfaces in the oceans play a vital role in the biogeochemical cycles of some elements. The chemistry of the vast oceans is ultimately profoundly shaped by their internal biological processes which are dominated by tiny organisms – microorganisms less than 1 mm in diameter. The carbon fixed from the atmosphere and transformed within the water column by these organisms affects the chemistry of the oceans and sustains most of the biological life within the oceans. The exchanges of $CO_2$ associated with these processes also play a critical role in the global carbon cycle and in the habitability of the whole planet.

The volume has been written for scientists of all disciplines. To contain the text within a reasonable length, a basic knowledge of chemistry, physics, biology and geology has been assumed and the fundamental principles in these subjects, which are readily available in other textbooks, have not been reiterated here. As the volume is deliberately designed with a multidisciplinary readership in mind, however, an attempt has been made to treat the more advanced chemical and physical concepts in a generally descriptive manner, with appropriate references being given to direct the reader to the original sources. One of the major aims of marine geochemistry in recent years has been to model natural systems on the basis

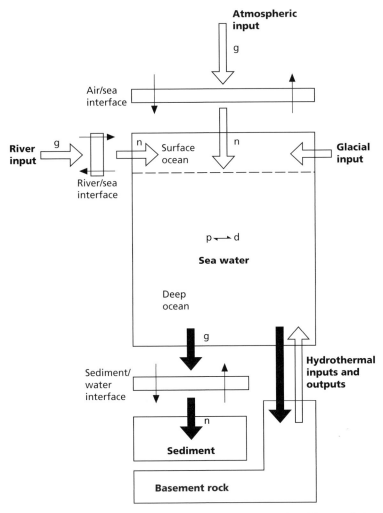

**Fig. 1.1** A schematic representation of the source/input → seawater internal reactivity → sink/output global journey. The large open arrows indicate transport from material sources, and the large filled arrows indicate transport into material sinks; relative flux magnitudes are not shown. The small arrows indicate only that the strengths of the fluxes can be changed as they cross the various interfaces in the system; thus, g and n represent gross and net inputs or outputs, respectively. Material is brought to the oceans in both particulate and dissolved forms, but is transferred into the major sediment sink mainly as particulate matter largely by biological processes. The removal of dissolved material to the sediment sink therefore usually requires its transformation to the particulate phase. This is shown by the p↔d term. The intention here, however, is simply to indicate that internal particulate–dissolved reactivity occurs within the seawater reservoir, and it must be stressed that a wide variety of chemical reactions and physicochemical processes are involved in setting the composition of the water phase: see text. For convenience, coastal zones are not shown.

of theoretical concepts. To follow this approach it is necessary to have a more detailed understanding of the theory involved, and for this reason a series of Worksheets have been included in the text. Some of these Worksheets are used to describe a number of basic geochemical concepts; for example, those underlying redox reactions and the diffusion of solutes in interstitial waters. In others, however, the emphasis is placed on modelling a variety of geochemical systems using, where possible, actual examples from literature sources; for example, the topics covered include a sorptive equilibrium model for the removal of trace metals in estuaries, a stagnant film model for the exchange of gases across the air–sea interface, and a variety of models designed to describe the interactions between solid and dissolved phases in sediment interstitial waters.

Overall, therefore, the intention is to provide a unifying framework, which has been designed to bring a state-of-the-art assessment of marine geochemistry to the knowledge of a variety of ocean scientists in such a way that allows future advances to be understood within a meaningful context.

## References

Broecker, W.S. and Peng, T.-H. (1982) *Tracers in the Sea.* Eldigio Press, New York: Columbia University.

Emerson, S.R. and Hedges, J.I. (2008) *Chemical Oceanography and the Marine Carbon Cycle*, Cambridge University Press.

Forchhammer, G. (1865) On the composition of sea water in the different parts of the ocean. *Philos. Trans. R. Soc. London*, **155**, 203–262.

Goldberg, E.D. (1963) The oceans as a chemical system, in *The Sea*, M.N. Hill (ed.), Vol. 2, 3–25. New York: John Wiley & Sons, Inc.

Hedges, J.I. (1992) Global biogeochemical cycles: progress and problems. *Mar. Chem.*, **39**, 67–93.

Lenton, T. and Watson, A. (2011) *Revolutions that Made the Earth*. Oxford University Press.

Mackenzie, F.T. (1975) Sedimentary cycling and the evolution of the sea water, in *Chemical Oceanography*, J.P. Riley and G. Skirrow (eds), Vol. 1, 309–364. London: Academic Press.

Stumm, W. and Morgan, J.J. (1996) *Aquatic Chemistry*, 3rd edn, New York: John Wiley & Sons, Inc.

Turekian, K.K. (1977) The fate of metals in the oceans. *Geochim. Cosmochim. Acta*, **41**, 1139–1144.

# Part I
# The Global Journey: Material Sources

# 2 The input of material to the ocean reservoir

The World Ocean may be regarded as a planetary dumping ground for material that originates in other geospheres, prior to its tectonic recycling, and to understand marine geochemistry it is necessary to evaluate the composition, flux rate and subsequent fate of the material that is delivered to the ocean reservoir.

## 2.1 The background

The major natural sources of the material that is injected into seawater are the continental crust, the oceanic crust and the atmosphere. Primary material is mobilized directly from the continental crust, mainly by low-temperature weathering processes and high-temperature volcanic activity. In addition, secondary (or pollutant) material is mobilized by a variety of anthropogenic 'weathering' processes, which often involve high temperatures. The various types of material released on the continents during both natural and anthropogenic processes include particulate, dissolved and gaseous phases, which are then moved around the surface of the planet by a number of transport pathways. The principal routes by which continentally mobilized material reaches the World Ocean are via river, atmospheric and glacial transport. The relative importance of these pathways, however, varies considerably in both space and time. Rivers and glaciers enter the oceans at particular locations and impact particularly coastal regions, while atmospheric transport disperses material more widely with fluxes decreasing away from source regions. Water in the form of ice can act as a major mechanism for the physical mobilization of material on the Earth's surface. The magnitude of the transport of this material depends on the prevailing climatic regime. At present, the Earth is in an interglacial period and large-scale ice sheets are confined to the Polar Regions. Even under these conditions, however, glacial processes are a major contributor of material to the oceans. For example, Raiswell *et al.* (2006) estimated that at present $\sim29 \times 10^{14}\,\mathrm{g\,yr^{-1}}$ of crustal products are delivered to the World Ocean by glacial transport, of which $\sim90\%$ is derived from Antarctica. Thus, ice transport is second only to fluvial run-off in the global supply of particulate material to the marine environment, although it is less important as a source of dissolved material because it is frozen and hence has reduced chemical weathering. The impacts of material fluxes associated with glaciers are seen predominantly in the high latitudes, particularly in the shelf seas of the Arctic and Southern Oceans.

Material also is supplied to the oceans from processes that affect the oceanic crust. These processes involve low-temperature weathering of the ocean basement rocks, mainly basalts, and high-temperature water–rock reactions associated with hydrothermal activity at spreading ridge centres. This hydrothermal activity, which can act as a source of some components and a sink for others, is now known to be of major importance in global geochemistry; for example, in terms of primary inputs it dominates the supply of dissolved manganese to the oceans. Although the extent to which this type of dissolved material is dispersed about the ocean is not yet clear, hydrothermal activity must still be regarded as a globally important mechanism for the supply of material to the seawater reservoir.

On a global scale, therefore, the main pathways by which material is brought to the oceans are:
1 river and glacial run-off, which delivers material to the surface ocean at the land–sea boundaries;

2 atmospheric deposition, which delivers material to all regions of the surface ocean;

3 hydrothermal activity, which delivers material to deep and intermediate waters above the sea floor.

The manner in which these principal pathways operate is described individually in the next three chapters, and this is followed by an attempt to estimate the relative magnitudes of the material fluxes associated with them.

# References

Raiswell, R., Tranter, M., Benning, L.G., Siegert, M., De'ath, R., Huybrechts, P. and Rayne, T. (2006) Contributions from glacially derived sediment to the global iron(oxyhyr)oxide cycle: implications for iron delivery to the oceans. *Geochim. Cosmochim. Acta*, 70, 2765–2780.

# 3     The transport of material to the oceans: the fluvial pathway

Much of the material mobilized during both natural crustal weathering and anthropogenic activities is dispersed by rivers, which transport the material towards the land–sea margins. In this sense, rivers may be regarded as the carriers of a wide variety of chemical signals to the World Ocean. The effect that these signals have on the chemistry of the ocean system may be assessed within the framework of three key questions (see e.g. Martin and Whitfield, 1983).

1 What is the quantity and chemical composition of the dissolved and particulate material carried by rivers?

2 What are the fates of these materials in the estuarine mixing zone?

3 What is the ultimate quantity and composition of the material that is exported from the estuarine zone and actually reaches the open ocean?

These questions will be addressed in this chapter, and in this way river-transported materials will be tracked on their journey from their source, across the estuarine (river–ocean) interface, through the coastal receiving zone and out into the open ocean. Chemical fluxes of some components have also been substantially modified by human impact and the nature and scale of this impact will be considered.

In addition to riverine inputs, glacial flows also contribute inputs to the oceans. Chemical weathering in glacial environments is similar or slower than rates in fluvial catchments (Anderson *et al.*, 1997). The largest glacial flows arise from Antarctica and Greenland, and hence inputs of anthropogenic materials from these systems are small compared to fluvial ones and organic matter will also be at low concentrations because of limited biological activity. Physi-cal weathering in glacial systems is very substantial (Raiswell *et al.*, 2006), but the chemical composition and behaviour of this material will be similar to that of fluvial particulate matter. Hence, the inputs of glaciers will not be treated separately but with river systems with differences noted where appropriate.

## 3.1 Chemical signals transported by rivers

### 3.1.1 Introduction

River water contains a large range of inorganic and organic components in both dissolved and particulate forms. A note of caution, however, must be introduced before any attempt is made to assess the strengths of the chemical signals carried by rivers, especially with respect to trace elements. In attempting to describe the processes involved in river transport, and the strengths of the signals they generate, great care must be taken to assess the validity of the databases used and, where available, 'modern' (i.e. post around 1975) trace-element data will be used in the present discussion of river-transported chemical signals, since some earlier data sets may include overestimates of concentrations due to contamination problems during the sampling and analysis which were not recognized at the time.

### 3.1.2 The sources of dissolved and particulate material found in river waters

Water reaches the river environment either directly from the atmosphere or indirectly from surface run-off, underground water circulation and the discharge

*Marine Geochemistry*, Third Edition. Roy Chester and Tim Jickells.
© 2012 by Roy Chester and Tim Jickells. Published 2012 by Blackwell Publishing Ltd.

of waste solutions. Freshwater reaches the ocean predominantly via rivers with about 5% of the total arising directly via groundwater. The fluxes of groundwater and its composition are less well known. This component can be locally very important, for instance on limestone islands where there is often no surface freshwater, but in terms of global fluxes, rivers dominate and will be the focus of attention here. The sources of the dissolved and particulate components that are found in the river water include rock weathering, the decomposition of organic material, wet and dry atmospheric deposition and, for some rivers, human activity induced discharges. The source strengths are controlled by a number of complex, often interrelated, environmental factors that operate in an individual river basin; these factors include rock lithology, relief, climate, the extent of vegetative cover and the magnitude of pollutant inputs.

The various factors that are involved in setting the composition of river water are considered in the following sections, and to do this it is convenient to use a framework in which the dissolved and the particulate components are considered separately. This distinction is usually based on simple filtration and hence dissolved components are defined as those passing through a filter usually with 0.2–1 μm pore diameter. This dissolved component therefore will include material that is in reality colloidal, that is, very small (<0.2 μm and >1 nm) particulate matter. Such material may include biological and mineral particles, and mixtures of the two. Colloids have a very high surface area relative to its mass. Since particle-water interactions usually involve interactions at surfaces, this colloidal component of the fraction operationally defined as dissolved may be particularly important (Gaillardet *et al.*, 2004)

### 3.1.3 Major and trace elements: the dissolved river signal

#### 3.1.3.1 Major elements

The average inorganic composition of rivers entering the principal oceans is given in Table 3.1, together with that of seawater (note nitrogen and phosphorus are considered later). From the average river and seawater compositions given in this table it can be seen that there are a number of differences between

**Table 3.1** Average major element composition of rivers $\mu mol\,l^{-1}$ except bicarbonate μ equivalents $l^{-1}$ (see Chapter 9) after Meybeck (2004) based on the summed fluxes in 680 individual river basins, seawater composition from Broecker and Peng (1982).

| | World average river composition | seawater |
|---|---|---|
| $Na^+$ | 240 | $4.7 \times 10^5$ |
| $K^+$ | 44 | $10 \times 10^3$ |
| $Ca^{2+}$ | 594 | $10 \times 10^3$ |
| $Mg^{2+}$ | 245 | $5.3 \times 10^4$ |
| $Cl^-$ | 167 | $5.5 \times 10^5$ |
| $SO_4^{2-}$ | 175 | $2.8 \times 10^4$ |
| $HCO_3^-$ | 798 | $2.3 \times 10^3$ |
| $SiO_2$ | 145 | 100 |

these two types of surface water. The most important of these is that in river water there is a general dominance of calcium and bicarbonate, whereas in seawater sodium and chloride are the principal dissolved components contributing to the total ionic, that is salt, content. The major element composition of river water, however, is much more variable than that of seawater, and some idea of the extent of this variability can be seen from the data in Table 3.1. Note dissolved Si is a major component but not dissociated at river water pH and is represented as $SiO_2$. Meybeck (2004) has ranked the global order of variability for the major dissolved constituents of river water as follows:

$Cl^- > SO_4^{2-} > Na^+ > Mg^{2+} > Ca^{2+} > SiO_2 > K^+ > HCO_3^-$.

The major factors that control these variations are discussed in the following sections.

There are a number of types of water on the Earth's surface, which can be distinguished from each other on the basis of both their total ionic content (salinity) and the mutual proportions in which their various ions are present (ionic ratios). Gibbs (1970) used variations in both parameters to identify a number of end-member surface waters. The cations that characterize the two principal water types are $Ca^{2+}$ for fresh water and $Na^+$ for highly saline waters, and Gibbs (1970) used variations in these two cations to establish compositional trends in world surface waters: see Fig. 3.1(a). He also demonstrated that the same general trends could be produced using variations in the principal anions in the two waters, that is $HCO_3^-$ for fresh water and

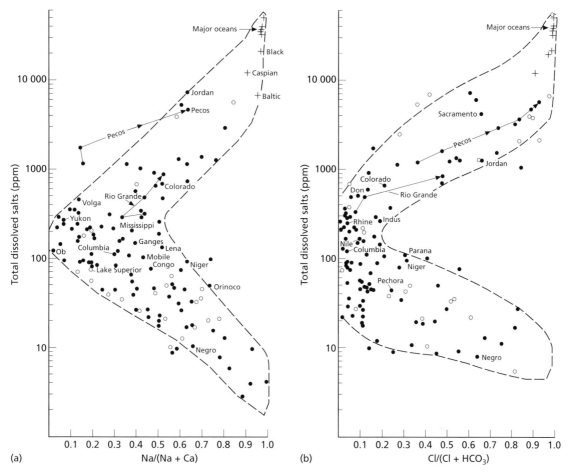

**Fig. 3.1** Processes controlling the composition of surface waters (from Gibbs, 1970). (a) Variations in the weight ratio Na/(Na + Ca) as a function of total dissolved salts. (b) Variations in the weight ratio Cl/(Cl + HCO$_3^-$) as a function of total dissolved salts. (c) Diagrammatic representation of the processes controlling end-member water compositions. See text for explanation. Note Gibbs uses 'total dissolved solids' in his diagram as a measure of all the dissolved ions in a sample, in more modern studies conductivity measurements are used and an approximate conversion to ionic strength and a total dissolved solids concentration of 200 mg l$^{-1}$ approximates to an ionic strength of about $2 \times 10^{-3}$ mol l$^{-1}$. Andrews *et al.* (2004).

Cl$^-$ for highly saline waters: see Fig. 3.1(b). By displaying the data in these two forms, Gibbs (1970) was able to produce a framework that could be used to characterize three end-member surface waters: see Fig. 3.1(c). These end-member waters were defined as follows:

**1** A precipitation- or rain-dominated end-member, in which the total ionic content is relatively very low, and the Na/(Na + Ca) and the Cl/(Cl + HCO$_3^-$) ratios are both relatively high. Conditions that favour the formation of this end-member are low weathering intensity and low rates of evaporation.

**2** A rock-dominated end-member, which is characterized by having an intermediate total ionic content and relatively low Na/(Na + Ca) and Cl/(Cl + HCO$_3^-$) ratios. This end-member is formed under conditions of high weathering intensity and low rates of evaporation.

**3** An evaporation–crystallization end-member, which has a relatively very high total ionic content and also

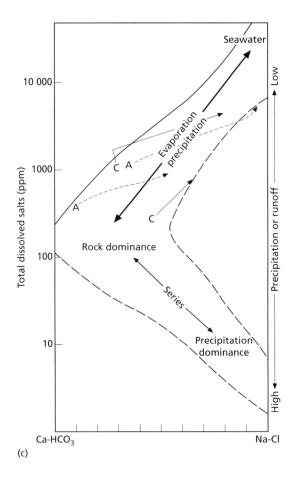

**Fig. 3.1** *Continued*

relatively high $Na/(Na + Ca)$ and $Cl/(Cl + HCO_3^-)$ ratios. Conditions that favour the formation of this end-member are high weathering intensity and high rates of evaporation.

Gibbs (1970) therefore was attempting to classify surface waters on the basis of the predominance of the principal external sources of the major ionic components, that is, precipitation and rock weathering, and the operation of internal processes, such as evaporation and precipitation. The 'Gibbs Diagram' essentially represents a hypothesis describing the large scale processes controlling river chemistry. This begins with the input of rainwater characterized by low ionic strength and a dominance of $Na^+$ and $Cl^-$ ions derived from the injection of seaspray into the atmosphere leading to deposition in rain, snow and so on, and freshwater whose composition is domi-

nated by that of the precipitation ('precipitation dominance'). As this rainwater percolates through soil and rock and begins to flow as river or ground-water, it will accumulate products of crustal weathering. Calcium carbonate is very abundant in the crust (globally ~20% of crust) and relatively soluble due to acid hydrolysis (Andrews *et al.*, 2004). Therefore the products of calcium carbonate dissolution ($Ca^{2+}$ and $HCO_3^-$) are appropriate tracers of weathering, although bicarbonate can be problematic because of its involvement in acid-base reactions with $CO_2$ and $CO_3^{2-}$ (see Chapter 9) on creating the 'rock dominance' regime. In addition, seaspray does contain some calcium from seawater itself, but this can be corrected for by assuming all chloride is from the atmosphere and the seaspray injected into the atmosphere has the same $Ca^{2+}/Cl^-$ ratio as seawater.

**Table 3.2** Average major element concentrations of rivers draining different catchment types; units, $\mu mol\, l^{-1}$ bicarbonate as $\mu$ equiv. $l^{-1}$ (data from Meybeck, 1981).

| Element | Rivers draining Canadian Shield | Mackenzie River; 'rock-dominated' end-member | Colorado River; 'evaporation–crystallization' end-member |
|---------|--------------------------------|--------------------------------------------|---------------------------------------------------------|
| $Na^+$ | 23 | 304 | 4000 |
| $K^+$ | 10 | 28 | 128 |
| $Ca^{2+}$ | 82 | 825 | 2075 |
| $Mg^{2+}$ | 29 | 433 | 1000 |
| $Cl^-$ | 53 | 251 | 2310 |
| $SO_4^{2-}$ | 19 | 361 | 2700 |
| $HCO_3^-$ | 165 | 1820 | 2213 |

The final component of the Gibbs diagram the 'evaporation/precipitation' regime represents freshwaters in hot arid environments where chemical precipitation takes place with calcium carbonate the first mineral to precipitate leading to a rise in the relative importance of sodium and chloride at the expense of calcium and bicarbonate.

The Gibbs diagram is of course very simplistic in its characterization of the complex processes regulating stream chemistry in terms of simple ratios. The approach breaks down in several situations which mostly fall into three categories.

1 If saline brines or evaporate minerals are important within a catchment, this will provide a source of NaCl in addition to atmospheric deposition. Examples of this situation include the Pecos River (Feth, 1971) and some tributaries of the Amazon draining parts of the Andes (Stallard and Edmond, 1983).

2 If weathering is dominated by rocks other than calcium carbonate. For example, the Rio Negro, a major tributary of the Amazon, drains highly weathered soils severely depleted in weatherable cations but containing some feldspars, resulting in low ionic strength waters where weathering is still the main source of cations.

3 If atmospheric deposition is not dominated by sodium chloride. In many parts of the world atmospheric deposition now contains a significant amount of sulfuric acid derived from emissions of combustion processes. This acts to increase the ionic strength without associated weathering products.

Despite these and other caveats, the Gibbs approach represents a useful simple first order description of river chemistry which Meybeck (2004) suggests is appropriate for 80% of the river systems he considered.

There is no doubt that there are considerable variations in the total ionic content of river waters. This can be illustrated with respect to a number of individual river types (see Table 3.2), and in a general way the variations can be related to the Gibbs classification.

1 Rivers with relatively small total ionic contents can be found:

(a) in catchments draining thoroughly leached areas of low relief where the rainfall is high, for example in some tropical regions of Africa and South America;

(b) in catchments that drain the crystalline shield rocks, for example those of Canada, Africa and Brazil, and rivers fed from glacial melt waters in arctic and Antarctic (e.g. 'dry valleys' region of Antarctica). As an example, very 'pure' waters, with total ionic contents of ~19 mg $l^{-1}$, are found on the Canadian Shield: see Table 3.2. It is waters such as these that will have their major ion composition most influenced by precipitation, even if they are not rain-dominated.

2 As rock weathering becomes increasingly more important, the total ionic content of the river water increases. The Mackenzie River, which drains sedimentary and crystalline formations, is an example of a river having a rock-dominated water type. The average total ionic content of the Mackenzie River water is ~200 mg $l^{-1}$, which is about an order of magnitude higher than that of the Canadian Shield rivers, and the concentration of $Ca^{2+}$ exceeds that of $Na^+$ by a factor of 4.7: see Table 3.2.

**3** Some river waters have relatively high total ionic contents and high Na/(Na + Ca) ratios. The Colorado River is an example of this type and has a total ionic content of ~700 mg l$^{-1}$ and a Na$^+$ concentration slightly in excess of Ca$^{2+}$: see Table 3.2. It is probable, however, that the major ion composition of this river has been influenced more by the input of saline underground waters draining brine formations than by evaporation–crystallization processes.

Meybeck (1981) took a global overview of the extent to which the three end-member waters are found on the Earth's surface. He concluded that the precipitation-dominated end-member (even if it exists at all) and the evaporation–crystallization (or evaporite) end-member together make up only around 2% only of the world's river waters, and that in fact ~98% of these surface waters are rock-dominated types.

Because the vast majority of the world's river waters belong to the rock-dominated category it is the extent to which the major rock-forming minerals are weathered, in other words the influence of the chemical composition of the source rocks, that is the principal factor controlling the concentrations of the major ions in the waters. This can be illustrated with respect to variations in the major ion composition of rivers that drain a number of different rock types: see Table 3.3. From this table it can be seen, for example, that sedimentary rocks release greater quantities of Ca$^{2+}$, Mg$^{2+}$, SO$_4^{2-}$ and HCO$_3^-$ than do crystalline rocks. Meybeck (1981) assessed the question of the chemical denudation rates of crustal rocks and concluded that:

**1** chemical denudation products originate principally (~90%) from sedimentary rocks, with about two-thirds of this coming from carbonate deposits;

**2** the relative rates at which the rocks are weathered follow the overall sequence evaporites >> carbonate rocks >> crystalline rocks, shales and sandstones.

These are general trends, however, and in practice the extent to which a crustal terrain is weathered depends on a complex of interrelated topographic and climatic factors.

**3** biogenic processes influence river chemistry, although these principally influence the behaviour of N, P and C.

It is apparent, therefore, that river waters can be characterized on the basis of their major ionic constituents. The total concentrations, and mutual proportions, of these constituents are regulated by a variety of interrelated parameters. Rock weathering, however, is the principal control on the dissolved major element chemistry of the vast majority of the world's rivers, with regional variations being controlled by the lithological character of the individual catchment. The dissolved solid loads transported by rivers are correlated with mean annual run-off (see e.g., Walling and Webb, 1987), and although the *concentrations* of dissolved solids decrease with increasing run-off, as a result of a dilution effect, the *flux* of dissolved solids increases.

#### 3.1.3.2 Trace elements

The sources that supply the major constituents to river waters (e.g., rock weathering, atmospheric deposition, pollution) also release trace elements into surface waters. Trace elements are components with dissolved concentrations of <1 mg l$^{-1}$, and many have concentrations orders of magnitude less than this. The measurement of trace elements is a demanding task in terms of sampling and analysis, but there is now a reasonable body of data and Table 3.4 is based

**Table 3.3** Major ion composition of rivers draining different rock types; units, µmol l$^{-1}$ bicarbonate as µ equiv. l$^{-1}$ (data from Meybeck, 1981).

| Element | Plutonic and highly metamorphic rocks | Volcanic rocks | Sedimentary rocks |
|---|---|---|---|
| Na$^+$ | Lithological influence displaced by oceanic influence | | |
| K$^+$ | 26 | 38 | 26 |
| Ca$^{2+}$ | 100 | 200 | 750 |
| Mg$^{2+}$ | 42 | 125 | 333 |
| Cl$^-$ | Lithological influence displaced by oceanic influence | | |
| SO$_4^{2-}$ | 20 | 60 | 250 |
| HCO$_3^-$ | 245 | 738 | 1640 |

**Table 3.4** Average concentrations of elements in upper crust, riverine particulate matter, dissolved in river water and seawater plus ratio of dissolved river concentration to crustal abundance. Note different units for ocean water. While most elements are listed, for some of the rarest elements there is insufficient data and these are omitted.

| Element | Average Upper Crust (AUC) $\mu mol\,g^{-1}$ | Average River particulate matter (RP) $\mu mol\,g^{-1}$ | Average River dissolved (ARD) $\mu mol\,l^{-1}$ | ARD/AUC | Average ocean $nmol\,l^{-1}$ | Deep Sea clay $\mu mol\,g^{-1}$ |
|---|---|---|---|---|---|---|
| Li | 3.5 | 3.6 | 0.27 | 0.07 | $26 \times 10^3$ | 6.5 |
| Be | 0.23 | – | $1 \times 10^{-3}$ | $4 \times 10^{-3}$ | 0.023 | |
| B | 1.6 | 6.5 | 0.9 | 0.56 | $0.42 \times 10^6$ | 20.3 |
| C inorganic | – | – | 798 | – | $2.25 \times 10^6$ | |
| N (nitrate) | | | 15* | | $30 \times 10^3$ | |
| Na | 520 | 310 | 240 | 0.46 | $0.47 \times 10^9$ | 870 |
| Mg | 620 | 490 | 245 | 0.40 | $53 \times 10^6$ | 109 |
| Al | 3000 | 3500 | 1.2 | $0.4 \times 10^{-3}$ | 1.1 | 3500 |
| Si | 11000 | 10000 | 145 | 0.013 | $100 \times 10^3$ | 10000 |
| P | 20 | 37 | 0.7* | 0.035 | 2000 | 45 |
| S | 1.9 | – | 175 | 92 | $28 \times 10^6$ | |
| Cl | 10.4 | – | 167 | 16 | $545 \times 10^6$ | |
| K | 590 | 510 | 44 | 0.07 | $10 \times 10^6$ | 718 |
| Ca | 640 | 530 | 594 | 0.93 | $10 \times 10^6$ | 250 |
| Sc | 0.31 | 0.40 | 0.027 | 0.09 | 0.015 | 0.44 |
| Ti | 80 | 117 | 0.010 | $0.12 \times 10^{-3}$ | 0.13 | 119 |
| V | 1.9 | 3.3 | $14 \times 10^{-3}$ | $7.4 \times 10^{-3}$ | 39 | 0.36 |
| Cr | 1.8 | 1.9 | $13 \times 10^{-3}$ | $7.2 \times 10^{-3}$ | 4.1 | 1.9 |
| Mn | 14 | 19 | 0.62 | 0.04 | 0.36 | 109 |
| Fe | 700 | 860 | 1.1 | $1.6 \times 10^{-3}$ | 0.54 | 1000 |
| Co | 0.29 | 0.34 | $2.5 \times 10^{-3}$ | $8.6 \times 10^{-3}$ | 0.02 | 0.9 |
| Ni | 0.8 | 1.5 | 0.0135 | $17 \times 10^{-3}$ | 8.2 | 3.4 |
| Cu | 0.44 | 1.6 | $23 \times 10^{-3}$ | $52 \times 10^{-3}$ | 2.4 | 3.1 |
| Zn | 1.0 | 5.4 | $9.2 \times 10^{-3}$ | $9.2 \times 10^{-3}$ | 5.3 | 1.8 |
| Ga | 0.25 | 0.36 | $0.43 \times 10^{-3}$ | $1.7 \times 10^{-3}$ | 0.017 | 0.23 |
| Ge | 0.019 | – | $0.09 \times 10^{-3}$ | $4.7 \times 10^{-3}$ | 0.076 | |
| As | 0.064 | 0.067 | $8.3 \times 10^{-3}$ | 0.134 | 16 | 0.17 |
| Se | 0.0011 | – | $0.89 \times 10^{-3}$ | 0.81 | 2 | |
| Br | 0.02 | 0.062 | | | $840 \times 10^3$ | 1.25 |
| Rb | 0.98 | 1.2 | 0.019 | 0.019 | 1400 | 1.3 |
| Sr | 3.65 | 1.7 | 0.69 | 0.19 | $90 \times 10^3$ | 2.8 |
| Y | 0.23 | – | $0.45 \times 10^{-3}$ | $1.9 \times 10^{-3}$ | 0.19 | 0.36 |
| Zr | 2.1 | – | $0.43 \times 10^{-3}$ | $0.2 \times 10^{-3}$ | 0.16 | |
| Nb | 0.13 | – | $0.018 \times 10^{-3}$ | $0.14 \times 10^{-3}$ | <0.05 | |
| Mo | 0.011 | 0.031 | 0.0044 | 0.4 | 104 | 0.08 |
| Cd | $0.8 \times 10^{-3}$ | – | $0.71 \times 10^{-3}$ | 0.9 | 0.62 | $2 \times 10^{-3}$ |
| Sb | $3.3 \times 10^{-3}$ | $20 \times 10^{-3}$ | $0.57 \times 10^{-3}$ | 0.17 | 1.6 | $6.5 \times 10^{-3}$ |
| Cs | $37 \times 10^{-3}$ | $45 \times 10^{-3}$ | $0.08 \times 10^{-3}$ | $2 \times 10^{-3}$ | 2.3 | 0.03 |
| Ba | 4.6 | 4.4 | 0.17 | 0.04 | 109 | 10.9 |
| La | 0.22 | 0.32 | $0.86 \times 10^{-3}$ | $3.9 \times 10^{-3}$ | 0.04 | 0.32 |
| Ce | 0.45 | 0.68 | $1.9 \times 10^{-3}$ | $4.2 \times 10^{-3}$ | $5 \times 10^{-3}$ | 0.7 |
| Pr | 0.05 | – | $0.28 \times 10^{-3}$ | $5.6 \times 10^{-3}$ | $5 \times 10^{-3}$ | 0.06 |
| Nd | 0.19 | 0.24 | $1 \times 10^{-3}$ | $5.2 \times 10^{-3}$ | $23 \times 10^{-3}$ | 0.28 |
| Sm | 0.03 | 0.05 | $0.24 \times 10^{-3}$ | $8 \times 10^{-3}$ | $3.8 \times 10^{-3}$ | 0.05 |
| Eu | 0.0066 | 0.01 | $0.64 \times 10^{-3}$ | $9.7 \times 10^{-3}$ | $1.1 \times 10^{-3}$ | $10 \times 10^{-3}$ |
| Gd | 0.025 | – | $0.25 \times 10^{-3}$ | $10 \times 10^{-3}$ | $5.7 \times 10^{-3}$ | 0.05 |
| Tb | $4.4 \times 10^{-3}$ | $6 \times 10^{-3}$ | $0.034 \times 10^{-3}$ | $7.7 \times 10^{-3}$ | $1.1 \times 10^{-3}$ | $6 \times 10^{-3}$ |
| Dy | 0.024 | – | $0.18 \times 10^{-3}$ | $7.5 \times 10^{-3}$ | $6.8 \times 10^{-3}$ | |

*Continued*

**Table 3.4** *Continued*

| Element | Average Upper Crust (AUC) $\mu mol\,g^{-1}$ | Average River particulate matter (RP) $\mu mol\,g^{-1}$ | Average River dissolved (ARD) $\mu mol\,l^{-1}$ | ARD/AUC | Average ocean $nmol\,l^{-1}$ | Deep Sea clay $\mu mol\,g^{-1}$ |
|---|---|---|---|---|---|---|
| Ho | $5 \times 10^{-3}$ | – | $0.043 \times 10^{-3}$ | $8.6 \times 10^{-3}$ | $2.2 \times 10^{-3}$ | $6 \times 10^{-3}$ |
| Er | 0.014 | – | $0.12 \times 10^{-3}$ | $8.6 \times 10^{-3}$ | $7.2 \times 10^{-3}$ | $16 \times 10^{-3}$ |
| Tm | $1.8 \times 0^{-3}$ | – | $0.019 \times 10^{-3}$ | $10 \times 10^{-3}$ | $1.2 \times 10^{-3}$ | $2 \times 10^{-3}$ |
| Yb | 0.011 | 0.02 | $0.098 \times 10^{-3}$ | $8.9 \times 10^{-3}$ | $6.9 \times 10^{-3}$ | 0.014 |
| Lu | $1.8 \times 10^{-3}$ | $2.8 \times 10^{-3}$ | $0.014 \times 10^{-3}$ | $7.8 \times 10^{-3}$ | $1.3 \times 10^{-3}$ | $3 \times 10^{-3}$ |
| Hf | 0.030 | 0.034 | 0.033 | $1.1 \times 10^{-3}$ | $19 \times 10^{-3}$ | 0.025 |
| Ta | $5 \times 10^{-3}$ | $6.9 \times 10^{-3}$ | $0.006 \times 10^{-3}$ | $1.2 \times 10^{-3}$ | | $5 \times 10^{-3}$ |
| W | 0.01 | – | $0.54 \times 10^{-3}$ | $54 \times 10^{-3}$ | 0.055 | |
| Re | $1.1 \times 10^{-6}$ | – | $2.1 \times 10^{-6}$ | 1.9 | 0.042 | |
| Pb | 0.082 | 0.72 | $0.38 \times 10^{-3}$ | $4.6 \times 10^{-3}$ | $13 \times 10^{-3}$ | 1 |
| Th | 0.045 | 0.06 | $0.18 \times 10^{-3}$ | $4 \times 10^{-3}$ | $0.09 \times 10^{-3}$ | 0.04 |
| U | 0.011 | 0.013 | $1.6 \times 10^{-3}$ | 0.145 | 13 | 2.9 |

Element for which there are no data and all gases are excluded.
AUC from Rudnick and Gao (2004), RP from Martin and Meybeck (1979), RDC from Gaillardet *et al.* (2004) except major ions (Na, K, Ca, Mg, Cl⁻, $SO_4^{2-}$, $HCO_3^-$ and dissolved Si) from Meybeck (2004).
Ocean average from Nozaki http://www.agu.org/eos_elec/97025e-table.html. Deep sea clay is from Martin and Whitfield (1983).
* See text, these fluxes are highly impacted by human activity.

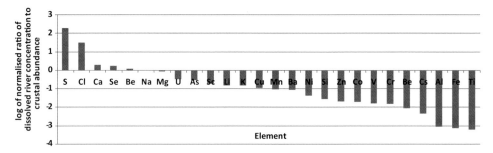

**Fig 3.2** Ratio of dissolved river concentrations of elements to their crustal abundances, normalized to sodium. This illustrates the decreasing mobility of elements in the weathering process from left to right based on Gaillardet *et al.* (2004). Note the log scale. Reprinted from Gaillardet *et al.* (2004) with permission from Elsevier.

on the thorough compilation presented by Gaillardet *et al.* (2004). The average concentrations span 10 orders of magnitude! We are currently a long way from a good understanding of the factors regulating trace metal concentrations in river waters but some trends are becoming clear. If the dissolved concentrations of the elements are compared to their average crustal abundances (Table 3.4) then a rough classification can be made which is illustrated for selected elements in Fig. 3.2. Based on this analysis Gaillardet *et al.* classify elements using the ratio of dissolved concentration to crustal average abundance, although in reality there is a continuous spectrum from very soluble to very insoluble elements.

The most mobile elements (ratios ≥1) are all present in natural waters as simple cations and anions and include chloride, calcium and molybdenum (present as the anion molybdate $MoO_4^{2-}$). Moderately mobile elements (ratios 0.1–1) include silicon, potassium, copper and nickel. Other elements including lead, aluminium and iron have very low dissolved concentrations compared to their crustal abundances. The patterns identified here are very similar to those seen in seawater and suggest an underlying fundamental geochemical control in all aquatic environments. We will discuss this issue in detail in Chapter 11 and show that this control reflects the charge and size of cations and their interaction with water. We will also see in Chapter 11 that many trace metals are not present as free cations in solution but rather are complexed often with organic matter, and the same is true of freshwater, although we are far from understanding the fundamental nature of these interaction, partly at least because we cannot currently characterize the organic matter. While in seawater, the pH is rather constant (see Chapter 9) this is not necessarily the case in freshwater where the buffering of pH by the carbonate system is rather variable, depending on the supply of carbonate from weathering. Decreases in pH tend to solubilize metals due to competition of the $H^+$ cation with the trace metals for adsorption sites on the particulate phase.

Beyond this fundamental geochemical solubility control, there are of course many additional processes operating leading to variability of the concentrations of any particular element between and within river systems. Clearly rock type has the potential to influence trace metal concentrations, with the most obvious example being enhancement of concentrations associated with mineral deposits. Anthropogenic activities including mining, but also many direct agricultural, industrial and urban inputs (and indirect ones such as through the atmosphere) can also alter the concentrations of trace metals. Examples of the interactions of these multiple factors can be seen in Table 3.5 which compares the global average concentrations with data from two river systems in the UK, one of which, the Trent, drains a large, diverse and heavily anthropogenically impacted region including the city of Birmingham and the other, the Swale, draining a rural low intensity upland agricultural catchment where

**Table 3.5** Average concentrations of some dissolved trace metals in rivers Swale (subject to mining legacy particularly for Pb and Zn) and Trent (subject to widespread and varied human impact) from Neal et al. (1996) compared to global average from Gaillardet et al. (2004) (all concentration µmol l$^{-1}$).

|    | Global Average | Trent | Swale |
|----|----------------|-------|-------|
| Al | 1.2 | 2.8 | 2.2 |
| Cr | 0.01 | 0.04 | 0.004 |
| Cu | 0.02 | 0.13 | 0.05 |
| Fe | 1.2 | 1.2 | 2.0 |
| Pb | $0.38 \times 10^{-3}$ | $5.8 \times 10^{-3}$ | $30 \times 10^{-3}$ |
| Mn | 0.6 | 0.65 | 0.13 |
| Ni | 0.01 | 0.24 | 0.04 |
| Zn | $6 \times 10^{-3}$ | 0.44 | 0.61 |

there is a legacy of lead and zinc mining with continued leaching from old spoil heaps.

### 3.1.4 Major and trace elements: the particulate river signal

In the present context, river particulate material (RPM) refers to solids carried in suspension in the water phase, that is, the suspended sediment load. River particulate material consists of a variety of components dispersed across a spectrum of particle sizes. These components include the following: primary aluminosilicate minerals, for example; feldspars, amphiboles, pyroxenes, micas; secondary aluminosilicates, for example the clay minerals; quartz; carbonates; hydrous oxides of Al, Fe and Mn; and various organic components. In addition to the discrete oxides and organic solids, many of the individual suspended particle surfaces are coated with hydrous Mn and Fe oxides and/or organic substances.

The mineral composition of RPM represents that of fairly homogenized soil material from the river basin, and as a result each river tends to have an individual RPM mineral signature. This was demonstrated by Konta (1985), who gave data on the distributions of crystalline minerals in RPM from 12 major rivers. The results of the study may be summarized as follows.

1 Clay minerals, or sheet silicates, were the dominant crystalline components of the RPM, although the distributions of the individual minerals differed. Mica–illite minerals were the principal sheet silicates present and were found in all the RPM samples. Kaolinite was typically found in higher concentrations in RPM from tropical river systems where weathering intensity is relatively high, for example the Niger and the Orinoco. Chlorite was found in highest concentrations in kaolinite-poor RPM, and tended to be absent in RPM from rivers in tropical or subtropical areas of intense chemical weathering. Montmorillonite was found only in RPM from some tropical and subtropical rivers.

2 Significant quantities of quartz were present in RPM from all the rivers except one.

3 Other crystalline minerals found as components of RPM included acid plagioclase, potassium feldspar and amphiboles.

4 Calcite and/or dolomite were reported in RPM from seven of the rivers, but it was not known if these minerals were detrital or secondary in nature.

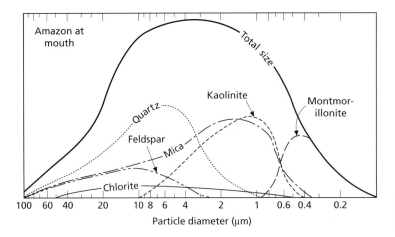

**Fig. 3.3** The size distribution of mineral phases transported by the Amazon River (from Gibbs, 1977).

The crystalline components of RPM therefore are dominated by the clay minerals, and the distribution of these minerals reflects that in the basin soils, which itself is a function of source-rock composition and weathering intensity. As a result, the clays in RPM have a general latitudinal dependence; for example kaolinite has its highest concentrations in RPM from tropical regions. This imposition of latitudinal control on the distribution of clay minerals in soils is used in Section 15.1 to trace the dispersion of continentally derived solids throughout the oceans. The mineral composition of RPM, however, also is dependent on particle size. For example, the size distribution of RPM transported by the Amazon is illustrated in Fig. 3.3, and demonstrates that, whereas quartz and feldspar are found mainly in the >2 μm diameter fraction, mica and the clay minerals, kaolinite and montmorillonite are concentrated in the <2 μm fraction. This mineral size fractionation has important consequences for the 'environmental reactivity' of elements transported via RPM, and this topic is considered below when elemental partitioning among the various components of the solids is discussed. For a 'first look', however, the chemical composition of RPM will be described in terms of the total samples.

Martin and Meybeck (1979) have provided an estimate of the global chemical composition of the total samples of river particulate material. This estimate was based on the determination of 49 elements in RPM from a total of 20 of the world's major rivers, and as there are considerably fewer problems involved in the analysis of particulate than of dissolved trace elements, it provides a generally reliable compositional database for RPM. Highly contaminated river systems were excluded from the study, although large scale global contamination by lead has probably affected the lead concentration in the RPM, and the findings therefore refer to RPM that has its composition controlled largely by natural processes. Under these conditions the RPM is derived mainly from the surface soil cover following the mechanical and chemical weathering of the surficial parent rocks. Martin and Meybeck (1979) were able to identify factors that lead to variability in the composition of RPM between rivers, but the most striking feature of the data in Table 3.4 is that, apart from the very soluble river components noted above, the average composition of riverine particulate matter approximates to that of average crustal material.

The variability for some of the major elements in the RPM can be related to climatic–weathering intensity conditions in the river catchments. Many tropical rivers have large areas in their drainage basins in which the rate of mechanical erosion is generally low, and the RPM originates mainly from highly developed soil material that has undergone chemical weathering, that is transport-limited regimes. The RPM in this type of river is enriched in those elements that generally are relatively insoluble during chemical weathering, for example Al, Ti and Fe, and is depleted in the more soluble elements, which are leached into the weathering solutions, for example Na and Ca. In contrast, many temperate and arctic rivers have drainage basins in which mechanical erosion can greatly exceed chemical weathering, that is, weathering-limited regimes. As a result, the parent material of the RPM in these rivers is either original rock debris or poorly weathered soils. Relative to the world average RPM, that found in temperate and arctic rivers tends to be depleted in Al, Ti and Fe, and enriched in Na and Ca, and its overall composition is closer to that of fresh rock than the RPM from tropical rivers.

It is apparent, therefore, that there is a considerable variability in the concentrations of some elements in RPM from different rivers and also of course glaciers, even when the material is derived from mainly natural sources. Nonetheless, the data for RPM given in Table 3.4 do offer an indication of chemical composition of the crust-derived solid material that is brought to the ocean margins by river transport. As we will see later, the actual amount of RPM transported to the oceans by rivers varies greatly depending on the catchment characteristics.

Up to this point we have described the elemental chemistry of RPM in terms of total sample composition. It was pointed out above, however, that RPM consists of a variety of individual components, which are present in a range of particle sizes. The elements in suspended particulates, and also in deposited sediments, are partitioned between these individual host components, some of which can bind them more strongly than others. In this respect, it is important to make a fundamental distinction between two genetically different element–host associations:

1 elements associated with the crystalline mineral matrix (the detrital or residual fraction), which are in an environmentally immobile form and largely have their concentrations fixed at the site of weathering;

2 elements associated with non-crystalline material (the non-detrital or non-residual fraction), which are in environmentally mobile forms and have their concentrations modified by dissolved $\leftrightarrow$ particulate reactivity.

A number of elemental associations are usually identified within the non-detrital fraction itself, for example an exchangeable fraction, a carbonate-associated fraction, a metal oxide-associated fraction, an organic-matter-associated fraction. A variety of techniques have been used to establish the parti-

tioning of elements among these host components, and one of the most common involves the sequential leaching of the primary sample with a number of reagents that are designed progressively to take into solution elements associated with the individual hosts. Such sequential leaching techniques are open to a number of severe criticisms (see e.g., Chester, 1988), not the least of which is that the host fractions identified are operationally defined in terms of the technique used, and so are not necessarily analogues of natural binding fractions. However, when the various constraints on their use are recognized, sequential leaching techniques can provide important data on the manner in which elements are partitioned in materials such as suspended river particulates, sediments and aerosols.

A number of studies have been carried out on the partitioning of elements in RPM. Outstanding among these are the pioneer investigations reported by Gibbs (1973; 1977) on suspended particulates from the Amazon and Yukon rivers. These two rivers were selected because they are large, relatively unpolluted systems representative of the major rivers that have their catchments in tropical (Amazon) and subarctic (Yukon) land masses and drain a wide variety of rock types. In both studies, Gibbs gave data on the partitioning of a number of elements in river-transported phases. To do this, he distinguished between: elements in *solution* (i), and those associated with *particulates* in an ion-exchange phase (ii), a precipitated metallic oxide coating phase (iii), an organic matter phase (iv), and a crystalline (detrital or residual) phase (v). A summary of the partitioning data reported by Gibbs (1973) is given in Table 3.6. A number of general conclusions can be drawn from these data.

1 The fraction of the total amounts of the metals transported in solution ranges from <1% for Fe to ~17% for Mn, but solution transport is not dominant for any of the trace metals.

2 The partitioning signatures of Fe, Mn, Ni, Cu, Co and Cr are similar in RPM from both the tropical and the subarctic river systems.

3 Cu and Cr are transported mainly in the crystalline particulate phase.

4 Mn is transported principally in association with the precipitated metal oxide particulate phase.

5 Most of the total amounts of Fe, Ni and Co are partitioned between the precipitated metal oxide and crystalline particulate phases.

**Table 3.6** The percentage partitioning of elements in RPM (data from Gibbs, 1973).

| River | Transport phase | Element | | | | | |
|---|---|---|---|---|---|---|---|
| | | Fe | Ni | Co | Cr | Cu | Mn |
| Amazon | Solution | 0.7 | 2.7 | 1.6 | 10.4 | 6.9 | 17.3 |
| | Ion exchange | 0.02 | 2.7 | 8.0 | 3.5 | 4.9 | 0.7 |
| | Metal oxide coating | 47.2 | 44.1 | 27.3 | 2.9 | 8.1 | 50 |
| | Organic matter | 6.5 | 12.7 | 19.3 | 7.6 | 5.8 | 4.7 |
| | Crystalline matrix | 45.5 | 37.7 | 43.9 | 75.6 | 74.3 | 27.2 |
| Yukon | Solution | 0.05 | 2.2 | 1.7 | 12.6 | 3.3 | 10.1 |
| | Ion exchange | 0.01 | 3.1 | 4.7 | 2.3 | 2.3 | 0.5 |
| | Metal oxide coating | 40.6 | 47.8 | 29.2 | 7.2 | 3.8 | 45.7 |
| | Organic matter | 11.0 | 16.0 | 12.9 | 13.2 | 3.3 | 6.6 |
| | Crystalline matrix | 48.2 | 31.0 | 51.4 | 64.5 | 87.3 | 37.1 |

6 The non-crystalline carrier phases represent the particulate metal fraction which is most readily available to biota, and the most important of these carriers are the metal oxide coatings.

7 There is a particle-size–concentration relationship for a number of elements in the RPM, with concentrations of Mn, Fe, Co, Ni and Cu all increasing dramatically with decreasing particle size.

It may be concluded, therefore, that in river systems that receive their supply of particulate trace elements mainly from natural sources; crystalline, metal oxide and organic host fractions are the principal metal carriers. It is the non-crystalline phases that are the most readily environmentally available, and the proportions of particulate trace metals associated with these phases generally increase in river systems that receive inputs from pollutant sources. The various total element, solid-state partitioning, mineral and particle size data for RPM can be combined to develop the general concept that rivers transport two different types of suspended solids:

1 a trace-element-poor, large-sized (≥2 μm diameter) fraction that consists mainly of crustal minerals such as quartz and the feldspars, a large proportion of the trace elements in this fraction being in crystalline solids and generally environmentally immobile;

2 a trace-element-rich, small-sized (≤2 μm diameter), surface-active fraction that consists largely of clay minerals, organic matter and iron and manganese oxide surface coatings, a large proportion of the trace elements in this fraction being associated with non-crystalline carriers (e.g., oxide coatings and organic phases) and environmentally available.

It must be stressed that there is no sharp division between these two fractions, but the distinction between them is extremely important both (i) from the point of view of differential transport, that is the small-sized material can be carried for a longer period in suspension, and (ii) because it is the small-sized, non-crystalline, environmentally available trace elements that undergo the dissolved ↔ particulate equilibria that play such an important role in controlling the river → estuarine → coastal sea → open ocean transport–deposition cycles of many elements and in regulating the composition of seawater. The non-residual elements can undergo changes between the various solid-state carrier phases themselves. All of these particulate ↔ dissolved and solid-state changes are sensitive to variations in environmental parameters (e.g., the concentrations of particulate material and complexing ligands, redox potential, pH), so that during the global transportation cycle the mobile surface-associated elements can undergo considerable speciation migration.

### 3.1.5 Organic matter and nutrients

#### 3.1.5.1 Dissolved organic carbon (DOC)

DOC represents a wide range of material whose composition is not well known – as discussed below. For some purposes an estimate of total organic carbon is required and can be determined using an oxidation technique to convert all the DOC to $CO_2$. The DOC in river waters originates mainly from three general sources:

1 phytosynthesis associated with fluvial production (autochthonous, mainly low molecular weight labile material);

2 the leaching of soils where the organic material is derived from plant and animal material via microbial activity (allochthonous, mainly high molecular weight refractory material);

3 anthropogenic inputs.

DOC concentrations in rivers span an enormous range from values for rivers in high rainfall relatively poor soil regions such as Northern Canada having low concentrations (e.g., Yukon 0.4 µmol l$^{-1}$, Degens et al., 1991) to tropical rivers draining peat soils (so called black rivers) with very high concentrations (e.g., river Dumai in Indonesia 5000 µmol l$^{-1}$, Alkhatib et al., 2007).

A number of authors have found a climate-related pattern in DOC concentrations in rivers of which Table 3.7 is an example. Thus, the lowest values are found for rivers draining glacial and arid environments and the highest for those draining tropical wetland regions. Perdue and Ritchie (2004) suggest a global average DOC concentration of 400–480 µmol l$^{-1}$.

In non-polluted river systems the DOC pool thus contains organic matter synthesized and degraded in both the terrestrial (allochthonous) and aqueous (autochthonous) environments. Perdue and Ritchie (2004) suggest the allochtonous fraction dominates. The DOM (dissolved organic matter) is poorly characterized chemically but is highly degraded material, generally rich in aromatic groups, low in nitrogen (average composition ~50% C, 5% H, 40%O, 1.5% N) and with an average molecular weight of ~1000 kDa (Perdue and Ritchie, 2004; Degens et al., 1991). According to Ertel et al. (1986) between ~40 and ~80% of fluvial DOC consists of combined humic substances, which generally are considered to be refractory material that can escape degradation in the fluvial–estuarine environment and so reach the ocean. Ertel et al. (1986) gave data on the humic and fulvic acid components of the DOC humic fraction in the Amazon River system. Both these components contain lignin (a phenolic polymer unique to vascular plants) and appear to be formed from the same allochthonous source material, but they differ in the extent to which they have suffered biodegradation in the soil, with fulvic acids being more oxidized than humic acids. In the Amazon system the fulvic acids do not undergo reactions with suspended particles and behave in a conservative manner. In contrast, humic acids can be adsorbed on to particle surfaces and also can undergo flocculation in the estuarine environment (see Sections 3.2.5 and 3.2.6). In view of this, Ertel et al. (1986) suggested that humic acids may not contribute carbon, or lignin, to the oceans, and that it is fulvic acids that represent the major proportion of the DOC input to seawater. The autochthonous DOC includes potentially very labile (metabolizable) biochemical material, such as proteins and carbohydrates, and less labile components, such as lipids and pigments; much of this labile material is likely to be degraded in the river or estuarine system and so will not escape into the open ocean (see also the refractory–labile classification of POC below).

The DOC in rivers represents a source of carbon to the ocean, although only a small and variable amount is readily biodegradable (e.g., Wiegner et al., 2006). This DOC also plays other important roles including absorbing light and complexing trace metals.

### 3.1.5.2 Particulate organic carbon (POC)

The filtration of water samples obviously allows the distinction between DOC and POC to be readily made, but it needs to be recognized that the distinction between these phases can change: for example, as POC is degraded by bacteria to DOC or as DOC material is adsorbed to particles or taken up by bacteria. POC concentrations vary greatly and depend in large measure on river hydraulics and the ability of water flow to keep material in suspension. As we shall see later this means that Asian rivers draining

**Table 3.7** Riverine DOC concentrations (µmol l$^{-1}$) in a number of climatic zones (Meybeck, 1988).

| | |
|---|---|
| Tundra | 167 |
| Taiga | 583 |
| Temperate | 333 |
| Wet tropical | 666 |
| Dry tropical | 250 |
| Semi arid | 83 |

rapidly eroding soils dominate the global flux. Particulate organic carbon concentrations in a number of world rivers are lie in the range 40–1500 $\mu mol\,l^{-1}$, although higher values can be found in those rivers draining marshes and those that receive relatively large inputs of sewage and industrial wastes. Perdue and Ritchie (2004) suggest a global average POC concentration of 377 $\mu mol\,l^{-1}$. Fluvial POC consists of living (i.e. bacteria and plankton) and non-living (detritus) fractions. In general, there is a decrease in fluvial POC with logarithmically increasing concentrations of total suspended material (TSM), which results from a reduction in primary productivity as a result of decreased light penetration arising from the presence of high suspended matter loads, and a dilution with mineral matter. It is necessary to distinguish between refractory (non-metabolizable) and labile (metabolizable, or degradable) fractions of river-transported POC. This is an important distinction because, although the refractory fractions will survive microbial attack, the labile fractions are an important food source for organisms and can be lost within rivers, estuaries and the sea. Degens *et al.* (1991) suggest on average 35% of POC is labile but this figure will represent an average over a very large range. The remainder of the fluvial POC, which is highly degraded, or refractory, in character, can escape to the oceans and so represents a significant source of organic carbon to marine sediments.

In the discussion to date the enormous variability in concentrations of DOC and POC between rivers has been emphasized. There is also variability with time reflecting changes in water flow, inputs, weathering and biological productivity and hence average concentrations need to be used with some caution. Degens *et al.* (1991) present an interesting compilation which despite the obvious limitations imposed by data availability, emphasizes the large scale regional differences in both carbon transport and the forms of carbon by deriving average fluxes from different continents. In Table 3.8 the ratios of DIC/DOC and DOC/POC are presented to illustrate this, reflecting large scale differences in rock type and flow regimes.

### 3.1.5.3  The nutrients

The key plant nutrients nitrate, phosphates and silicate in river water are each derived from different sources.

**Table 3.8** Variations in carbon transport to the oceans from different continents. (based on Degens *et al.*, 1991).

| Continent | Total Flow km³ yr⁻¹ | DIC/DOC | POC/DOC |
|---|---|---|---|
| South America | 11039 | 1.3 | 0.5 |
| North America | 5840 | 3.6 | 0.4 |
| Africa | 3409 | 0.4 | 0.3 |
| Asia | 12205 | 1.7 | 1.4 |

*Nitrate.* Nitrate ($NO_3^-$), which originates mainly from soil leaching, terrestrial run-off (including that from fertilized soils) and waste inputs, is the most abundant stable inorganic species of nitrogen in well-oxygenated waters, but dissolved organic nitrogen may dominate in humid tropical and subarctic rivers. River water also contains particulate nitrogen, which is mainly biological in origin. In pristine river systems nitrate concentrations may be of the order of 10–20 $\mu mol\,l^{-1}$ (Jickells *et al.*, 2000) and in some remote streams in southern Chile concentrations as low as 0.5 $\mu mol\,l^{-1}$ have been observed (Perakis and Hedin, 2002). In the latter rivers DON is the dominant form of dissolved nitrogen with an average concentration of about 4 $\mu mol\,l^{-1}$. Concentrations in river systems with very high inputs from agricultural and industrial sources now have very much higher concentrations for example >300 $\mu mol\,l^{-1}$ in the Rhine (Fig. 3.4). Galloway (2004) suggests riverine fluxes of nitrogen have increased by four fold from 1890 to 1990, and even the 1890 situation does not represent a pristine global situation. Even in highly nitrate enriched river systems, it should be noted that denitrification within the river system (organic matter oxidation in the absence of oxygen using nitrate as an electron acceptor see Chapters 9 and 14) removes substantial amounts of nitrate. For example Donner *et al.* (2004) estimate 22% of nitrate inputs to the Mississippi are removed in this way, and Seitzinger *et al.* (2002) suggest a value of 40% for a range of US east coast rivers. This loss mechanism takes place mainly in reducing sediments in river backwaters, and hence river management, such as straightening can reduce the effectiveness of this loss process.

*Phosphates.* Phosphorus is present in river waters in dissolved and particulate forms. The dissolved phosphorus is mainly orthophosphate (principal species

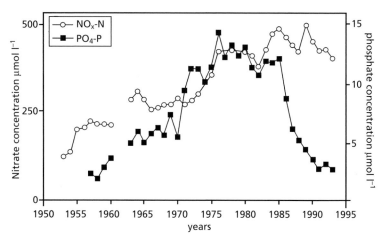

**Fig. 3.4** Changes in dissolved phosphate (PO$_4$-P) and nitrate (NO$_x$-N) concentrations in the Rhine 1950 to 1995 (modified from Van Dijk *et al.*, 1996).

HPO$_4^{2-}$), together with dissolved organic phosphate and, in polluted systems, polyphosphate. According to Meybeck (1982) the global average river-water concentration of total dissolved phosphorus is 0.9 µmol l$^{-1}$, and that of total particulate phosphorus is 17 µmol l$^{-1}$, of which 60% is in an inorganic form and the rest in an organic form. The sources of dissolved phosphate (PO$_4^{3-}$) in river waters include the weathering of crustal minerals (e.g. aluminium orthophosphate, apatite) and anthropogenic inputs (e.g. from the oxidation of urban and agricultural sewage and the breakdown of polyphosphates used in detergents). Dissolved phosphorus is removed during biological production, and is often considered to be the limiting nutrient in river systems. In addition, however, the concentrations of dissolved inorganic phosphorus in river waters are affected significantly by chemical processes involved in mineral–water equilibria, for example, those involving adsorption on to phases such as clay minerals and ferric hydroxides. This adsorption means that phosphorus added as fertilizer is more effectively retained within soils than nitrate in fertilizer, and the input of phosphorus associated with sewage discharges, which bypass such soil based adsorption, is particularly important. Dissolved inorganic phosphorus concentrations in uncontaminated rivers are often <1 µmol l$^{-1}$, while in rivers subject to large scale contamination concentrations can be 10 times this

value (e.g., Prastka *et al.*, 1998; Fig. 3.4). Falkowski *et al.* (2000) suggest a four-fold enhancement of the global P cycle due to human activity. Reduction in phosphorus concentrations in sewage discharges as a result of regulations to limit phosphorus concentrations in detergents, and active phosphorus removal strategies within sewage works, have led to substantial reductions in dissolved inorganic phosphorus concentrations in rivers in recent years (Fig. 3.4). *Silicate*. Dissolved reactive silicate is present in river waters almost exclusively as silicic acid (H$_4$SiO$_4$), and is derived mainly from the weathering of silicate and aluminosilicate minerals. Silicon is also present in river water in a variety of particulate forms, which include inorganic minerals (e.g. quartz, aluminosilicates) and biological material (e.g. the opaline skeletons of diatoms). Dissolved silicon is a major constituent of river water, making up ~10% of the total dissolved solids, and its global average concentration has been estimated to be 145 µmol l$^{-1}$ (Meybeck, 2004). Unlike nitrogen and phosphorus, anthropogenic sources play a relatively minor role in the supply of dissolved silicon to rivers. However, it has been argued that the increased nitrogen and phosphorus loadings and the management of rivers, particularly by damming, has lead to increased primary production within rivers which has acted to remove dissolved reactive silicate and hence anthropogenic activity leads to lower fluxes and there is evidence that this is changing patterns of primary production particularly in the Black sea and the Gulf of Mexico (Conley *et al.*, 1993; Humborg *et al.*, 1997; Turner and Rabalais, 1994).

### 3.1.6 Chemical signals carried by rivers: summary

1 Rivers transport large quantities of both dissolved and particulate components, and in this sense they may be regarded as carriers of a wide variety of chemical signals to the land–sea margins. Glaciers also carry large amounts of particulate material.

2 The dissolved major element composition of river water is controlled mainly by the chemical composition of the source rock that is weathered in the catchment region. These relatively simple rock–water chemistry relationships, however, do not have the same degree of control on the dissolved trace element composition of river water, which appears to be strongly influenced by chemical constraints within the aqueous system itself; for example particulate–dissolved equilibria that involve both inorganic and biological particles, and which are influenced by factors such as pH and the concentrations of complexing ligands. Anthropogenic inputs also can influence the concentrations of trace metals in some river systems.

3 The chemical composition of RPM from different rivers systems shows considerable variation, some of which may result from climate-induced weathering intensity differences in catchment regions. Crystalline (residual), metal oxide and organic host fractions are the principal particulate trace-metal carrier phases in rivers that receive their trace elements mainly from natural sources.

4 There is a catchment-related pattern in the fluvial transport of POC and DOC by rivers, the highest concentrations being found in rivers draining tropical regions and the lowest in those flowing over glacial and arid environments.

We have now tracked the transport of fluvial material to the river–ocean interface. Before reaching either the coastal receiving zone or the open regions of the sea itself, however, the river-transported material must pass through the estuarine environment. This environment can act as a filter, with the result that the fluvial signals can be severely modified before they are finally exported from the continents.

The nature of these estuarine modifications is considered in the next section.

## 3.2 The modification of river-transported signals at the land–sea interface: estuaries

### 3.2.1 Introduction

Fairbridge (1980) defined an estuary as 'an inlet of the sea reaching into a river valley as far as the upper limit of the tidal rise'. This definition allows three distinct estuarine sections to be distinguished:

1 a marine (or lower) estuary, which is in connection with the sea;

2 a middle estuary, which is subject to strong seawater–freshwater mixing;

3 an upper (or fluvial) estuary, which is characterized by freshwater inputs but which is subjected to daily tidal action.

Estuaries are therefore zones in which seawater is mixed with, and diluted by, fresh water. These two types of water have different compositions (see Table 3.1), making estuaries very complex environments in which the boundary conditions are extremely variable in both space and time. As a result, river-transported signals are subjected to a variety of physical, chemical and biological processes in the estuarine mixing zone. From this point of view, estuaries can be thought of as acting as filters of the river-transported chemical signals, which often can emerge from the mixing zone in a form that is considerably modified with respect to that which entered the system (see e.g. Schink, 1981). This concept of the estuarine filter is based on the fact that the mixing of the two very different end-member waters will result in the setting up of strong physicochemical gradients in an environment that is subjected to continuous variations in the supply of both matter and energy. It is these gradients that are the driving force behind the chemical processes acting within this filter.

In the present section an attempt will be made to understand how estuaries work as chemical, physical and biological filters. Before doing this, however, it is necessary to draw attention to three important points.

1 The estuarine filter is selective in the manner in which it acts on different elements; for example,

some dissolved river species are simply diluted in an estuary and then carried out to sea, whereas others undergo reactions that lead to their addition to, or removal from, the dissolved phase.

2 The effects of the filter can vary widely from one estuary to another, so that it is difficult to identify common global estuarine processes.

3 It is necessary to take into account the status of an estuary before any attempt is made to extrapolate its dynamics on to an ocean-wide scale.

In many cases estuaries act as traps for riverine material, particularly sediments as we shall see in Chapter 6. However, for a few really large estuaries such as the Amazon, estuarine mixing takes place on the shelf and even in waters of oceanic depths, and hence transformations may not necessarily lead to trapping within the estuary itself. Because of their environmental significance as material traps, estuaries require careful management, and as a result they have been the subject of considerable scientific interest over the past two or three decades. Much of this interest, however, has inevitably been directed towards relatively small urban estuarine systems, which often are highly perturbed by anthropogenic activities. Investigations of this type have provided invaluable insights into estuarine processes, but many of the estuaries themselves have little relevance for ocean flux studies on a natural global-ocean scale. Attempts have been made to study the chemical dynamics of major estuaries, such as those of the Amazon, the Zaire and the Changjiang (Yangtze) rivers, and it is the processes operating on the river signals in these large systems that will have a major influence on the chemistry of the oceans. In attempting to assess the importance of estuarine processes on fluvial signals it is necessary therefore to distinguish between such globally relevant estuaries and those that have a much more limited local effect. It must be stressed, however, that an approach which concentrates only on large river–estuarine systems has also important limitations. For example, Holland (1978) has pointed out that the combined run-off of the 20 largest rivers in the world accounts for only ~30% of the total global river run-off; further, these rivers drain the wettest areas of the globe. Thus, although processes operating in globally relevant estuaries may provide a better understanding of the effects that the estuarine filter has on the fluxes of material that reach the oceans, the fluxes themselves will be biased and may not represent truly global values.

### 3.2.2 The estuarine filter: the behaviour of elements in the estuarine mixing zone

The estuarine filter operates on dissolved and particulate material that flows through the system, and it can both modify and trap fluvially transported components within the mixing zone. The modification of the signals takes place via a number of chemical and biological processes that involve dissolved ↔ particulate speciation changes. The equilibria involved can go in either direction, that is particulate material can act either as a source of dissolved components, which are released into solution, or as a sink for dissolved components, which are removed from solution. As all the water in an estuary is eventually flushed out, usually on a time-scale of days or weeks, it is only the sediment that can act as an internal (i.e. estuarine) sink for elements that are brought into the system by river run-off. The sediments are not a static reservoir within the estuarine system, but are in fact subjected to a variety of physical, biological and chemical processes (e.g. bioturbation, diagenesis), which can result in the recycling of deposited components back into the water compartment. These recycling processes include:

1 the chemically driven diffusion of components from interstitial waters;

2 the physically driven flushing out of interstitial waters into the overlying water column;

3 the tidal resuspension of surface sediments, and sometimes their transfer from one part of an estuary to another.

The sediments themselves play a significant biogeochemical role in estuaries, and according to Bewers and Yeats (1980) they therefore can be regarded as acting as a third end-member in estuarine mixing processes, that is in addition to river water and seawater.

The physicochemical processes that control the estuarine filter therefore must be considered to operate in terms of a framework involving particulate–dissolved recycling associated with three estuarine end-members, that is; river water, seawater (which may consist of more than one component end-member) and sediments. This is illustrated in a very simplified diagrammatic form in Fig. 3.5 in terms of

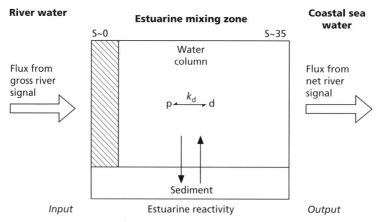

**River water**

**Estuarine mixing zone**

**Coastal sea water**

S~0                    S~35

Water column

Flux from gross river signal

Flux from net river signal

p $\xleftarrow{k_d}$ d

Sediment

*Input*          Estuarine reactivity          *Output*

**Fig. 3.5** Simplified schematic representation of the modification of a river-transported signal in the estuarine environment. p $\xleftarrow{k_d}$ d indicates particulate–dissolved reactivity associated with physical, chemical and biological processes in the estuarine mixing zone. In natural waters the equilibrated partitioning of an element between dissolved and particulate phases can be described by a conditional partition coefficient $k_d$ where $k_d = X/C$, where $X$ is the concentration of the exchangeable element on the particulate phase and $C$ is the concentration of the element in the dissolved phase: see Worksheet 3.1. ↑↓ Indicates two-way exchange of components between the water and sediment phases. □ Indicates the low-salinity zone of enhanced particulate–dissolved reactivity. For a discussion of gross and net river fluxes, see Section 6.1. S is salinity. Reprinted from Gaillardet *et al.*, (2003) with permission from Elsevier.

the estuarine modification of the river signal. Within this framework the estuarine filter operates on the fluvial flux as it flows through the system. The filter does not, however, affect all fluvially transported chemical signals, and for this reason it is necessary as a first step to establish whether or not estuarine reactivity has actually taken place.

### 3.2.3 The identification of estuarine reactivity: the 'mixing graph' approach

One of the principal processes that modify a river-transported chemical signal in an estuary is the physical mixing of fresh and saline waters of markedly different compositions along salinity (and other property) gradients. In the absence of any biogeochemical processes (reactivity) that lead to the addition or removal of a component, the physical mixing of the end-member waters would result in a linear relationship between the concentrations of a component and the proportions in which the two waters have undergone mixing (the salinity gradient), providing, that is, that the compositions of the end-members remain constant over a time approximating to the estuarine flushing time, and that there are no other sources or sinks for the components. This physical mixing relationship offers a useful baseline for assessing the effects that reactive biogeochemical processes have on the distribution of a component in an estuary. The technique used most commonly for this utilizes mixing graphs or mixing diagrams. In these diagrams, the concentrations of a component in a suite of samples (usually including the end-member waters) is plotted against a conservative index of mixing, that is a component whose concentrations in estuarine waters are controlled only by physical mixing. Salinity (a measure of the ionic strength of waters, see Chapter 7 for a rigorous definition) is the most widely used conservative index of mixing, although other parameters (e.g. chlorinity) also can be used for this purpose.

Mixing diagrams have been applied most commonly to dissolved components, and the theoretical relationships involved are illustrated in Fig. 3.6(a). If the distribution of a dissolved component is controlled only by physical mixing processes its concentrations in a suite of estuarine waters along a salinity gradient will tend to fall on a straight line, the theoretical dilution line (TDL), which joins the concentrations of the two end-members of the mixing series;

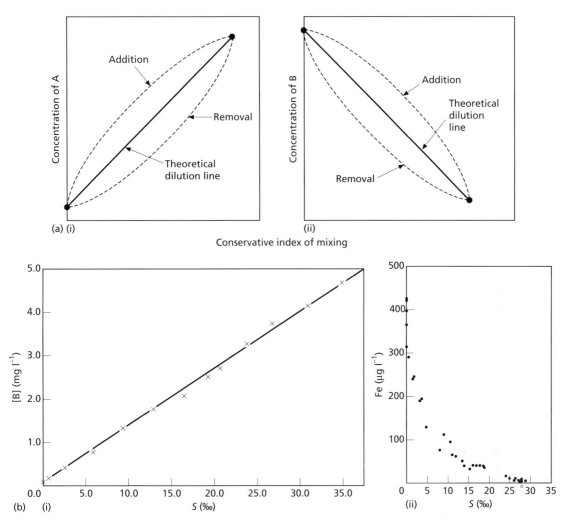

**Fig. 3.6** The behaviour of dissolved elements in estuaries. (a) Idealized estuarine mixing graphs for (i) a component with a concentration higher in seawater than in river water, and (ii) a component with a concentration higher in river water than in seawater (from Liss, 1976). (b) Estuarine mixing graphs and estuarine profiles for a series of representative dissolved elements. (i) Conservative behaviour of dissolved B in the Tamar estuary (from Liddicoat *et al.*, 1983). (ii) Non-conservative behaviour of dissolved Fe in the Beaulieu estuary showing removal at low salinity (from Holliday and Liss, 1976). (iii) Non-conservative behaviour of dissolved Ba in the Changjiang estuary showing addition to solution in the mid-salinity range (modified from Edmond *et al.*, 1985). (iv) Non-conservative behaviour of dissolved Cu in the Savannah estuary showing additions at both low and high salinities (from Windom *et al.*, 1983). Closed and open circles indicate data for different surveys. Broken curve indicates general trend of mixing curve only. (v) Non-conservative behaviour of dissolved Mn in the Tamar estuary showing both removal (low salinity) and addition (mid-range salinities) (modified from Knox *et al.*, 1981). Δ = Mn(II), o = total Mn; open symbols denote surface concentrations; closed symbols denote bottom concentrations. Reprinted with permission from Elsevier.

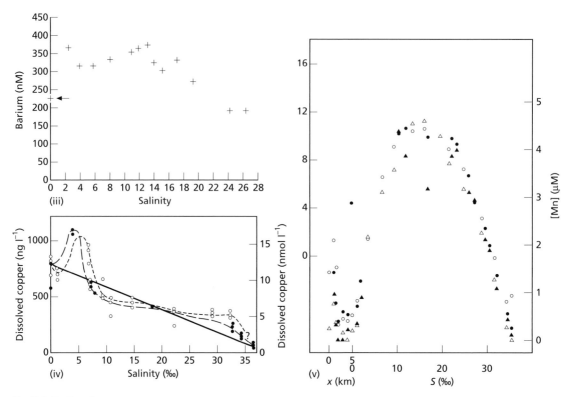

**Fig. 3.6** *Continued*

under these conditions its behaviour is described as being conservative or non-reactive. In contrast, when the component is involved in estuarine reactions that result in its addition to, or its loss from, the dissolved state its concentrations will deviate from the TDL, and its behaviour is termed non-conservative or reactive. For such a component the dissolved concentration data will lie above the TDL if it is added to solution, and below the TDL if it is removed from solution: see Fig. 3.6(a). Non-conservative behaviour therefore induces curvature in an estuarine mixing graph as concentrations deviate from the TDL. This curvature may be restricted to certain ranges of salinity, and thus can allow a geochemist to identify the specific estuarine zone in which the reactions take place.

The mixing diagram is a relatively simple concept for describing the behaviour of a dissolved component during estuarine mixing, but in practice it suffers from a number of fundamental problems. For example, although it is generally easy to identify

conservative behaviour, deviations from the TDL are more difficult to interpret because they may result from factors other than estuarine reactivity. Two particularly important complicating factors in using mixing diagrams are inputs into the estuary from more than one freshwater source (e.g. more than one river or a river and a sewage works), and temporal variability in inputs which might for example arise from changes in freshwater flow. For the latter reason it is preferable if possible to conduct estuarine surveys under conditions of relatively constant river flow.

There have been various attempts to rationalize the interpretation of estuarine mixing graphs. For example, Boyle *et al.* (1974) derived a mathematical relationship for the variation of the flux of a dissolved component with salinity in an estuary. In doing this they developed a general model for mixing processes between river and ocean water in which definitive criteria were established for the identification of non-conservative behaviour among dissolved components. In this model it was assumed that the concentration of a dissolved component is a continuous, single-valued function of salinity, and the following relationship was developed for the variation of the flux of the dissolved component with salinity:

$$\frac{dQ_c}{dS} = -Q_w(S - S_r)\frac{d^2C}{dS^2} \qquad (3.1)$$

where $Q_w$ is the flux of the river water, $Q_c$ is the flux of the dissolved component transported by the river water, $S_r$ is the salinity of the fluvial end-member, $S$ is the salinity at a given isohaline surface and $C$ is the concentration of the dissolved component at the same isohaline surface.

For conservative behaviour there is no gain or loss of the dissolved component during mixing, and hence

$$\frac{dQ_c}{dS} = 0 = \frac{d^2C}{dS^2} \qquad (3.2)$$

and the plot of the concentration of the dissolved component against salinity will be a straight line.

For non-conservative behaviour, the second derivative $d^2C/dS^2$ will not be equal to zero, and the plot of the concentration of the dissolved component against salinity will be a curve described by Equation 3.1.

This explicit formulation of the mixing process, that is of $C(S)$, shows that over straight-line segments of the mixing curve simple two-end-member dilution processes are taking place, and that to establish non-conservative behaviour curvature must be demonstrated. By applying the model to various examples of the estuarine behaviour of dissolved components, the authors were able to demonstrate that previous examples of non-conservative 'curvature' behaviour could, in fact, be shown to fit to straight-line segments on the mixing curve. This was interpreted as resulting from the introduction of a third end-member component, that is in addition to the river-water and seawater end-members, into the mixing processes. This third end-member can be a tributary stream. More commonly, however, it is coastal or shelf water of intermediate salinity. For example, it was demonstrated that in the Mississippi estuary there is mixing between three end-members: Mississippi river water, shelf water of intermediate salinity and open Gulf water of high salinity. In situations such as this, the mixing between river water and

seawater must therefore be interpreted in terms of two end-member seawaters, each of which can produce different straight-line relationships, which when combined can, under some circumstances, appear as curvature on the mixing diagram. The 'straight-line segment' approach to mixing diagrams outlined by Boyle *et al.* (1974) therefore offers a mathematical model for the interpretation of estuarine mixing processes.

Unless strict constraints are imposed, the mixing graph procedure can be both insensitive and imprecise. None the less, it has been used as an important first step in assessing the direction, if any, in which the estuarine filter has affected a river-transported signal, and in this respect a number of different estuarine behaviour patterns can be identified.

### 3.2.3.1 Conservative behaviour

Some elements appear to behave in a conservative manner in all estuarine situations. These include the major dissolved components that contribute to the salinity of seawater, for example $Na^+$, $K^+$, $Ca^{2+}$ and $SO_4^{2-}$. Boron, another major element, also seems to behave conservatively in many estuaries (see e.g. Fanning and Maynard, 1978). Liddicoat *et al.* (1983) demonstrated this for the Tamar estuary (UK), and concluded that the dissolved boron in the estuarine waters was derived almost entirely from seawater. The mixing graph for boron in the Tamar is illustrated in Fig. 3.6(b(i)), and provides an excellent example of conservative behaviour during estuarine mixing.

### 3.2.3.2 Non-conservative behaviour

Mixing graphs that demonstrate that a component behaves in a non-conservative manner in estuaries can have a variety of forms, which can indicate the gain or loss of the dissolved component, and also can sometimes identify the zone in which the estuarine reactivity has taken place. A number of different types of estuarine mixing graphs are described below.
1 Removal of dissolved components. For these components estuarine concentrations fall below the TDL. An example of this type of mixing graph is illustrated for iron in Fig. 3.6(b(ii)), from which it is evident that the removal of dissolved iron has taken place largely in the low-salinity region.

2 Addition of dissolved components. For these components estuarine concentrations lie above the TDL. An example of this is provided by the behaviour of Ba in the Changjiang (Yangtze) estuary: see Fig. 3.6(b(iii)). A more complex mixing graph showing the addition of dissolved components has been reported by Windom *et al.* (1983) for the behaviour of copper in the Savannah estuary (US). This graph is illustrated in Fig. 3.6(b(iv)) and indicates that there are additions to the fluvial Cu at both low (≤5‰) and high (≥20‰) salinities.

3 Combined addition and removal. This type of estuarine profile has been identified for dissolved manganese in the Tamar estuary (UK) by Knox *et al.* (1981). The estuarine profile is illustrated in Fig. 3.5(b(v)), and shows that dissolved Mn is removed from solution at low salinities, but is added to solution at mid-range salinities (see also Section 3.2.7.2).

It is apparent, therefore, that if sufficient constraints are applied mixing graphs can be used to identify: (i) whether or not a dissolved component has undergone estuarine reactivity; (ii) the direction, that is gain or loss, in which the dissolved component has been affected; and (iii) the general estuarine region (e.g. low-, mid-range, or high-salinity zones) over which the reactivity has been most effective. The mixing graph, however, suffers from one very fundamental limitation, that is, it does not provide information on *why* the component behaves in the way it does or on the *nature* of biogeochemical reactions that have caused the estuarine reactivity. In order to know how the estuarine filter operates it is therefore necessary to understand the nature of the reactive processes that occur in the mixing zone.

Biogeochemical reactivity in natural waters is controlled by a number of physicochemical parameters, which include pH, redox potential, salinity, and the concentrations of complexing ligands, nutrients, organic components and particulate matter. All these parameters undergo major variations in estuaries, as a result of which a variety of dissolved ↔ particulate transformations are generated in the mixing zone. These transformations are driven by physical, chemical and biological factors, and include: (i) sorption at the surfaces of suspended particles; (ii) precipitation; (iii) flocculation–aggregation; and (iv) uptake via biological processes. In general, the extent to which the transformations occur depends on the nature and concentrations of both the particulate and the dissolved components in the mixing zone.

### 3.2.4 Estuarine particulate matter (EPM)

#### 3.2.4.1 The classification of EPM

On the basis of the general scheme proposed by Salomons and Forstner (1984), the following classes of estuarine particulates can be identified.

1 River-transported or fluvial particulates. These solids, which are transferred across the river–estuarine boundary, include crustal weathering products (e.g. quartz, clay minerals), precipitated oxyhydroxides (principally those of iron and manganese), terrestrial organic components (e.g. plant remains, humic materials) and a variety of pollutants (e.g. fly ash, sewage).

2 Atmospherically transported particulates. The transfer of material across the atmosphere–estuarine boundary can be important in some estuaries. The atmospheric components involved in this transfer include crustal weathering products and pollutants such as fly ash.

3 Ocean-transported particulates. Particulate matter transferred across the ocean–estuarine boundary includes biogenous components of marine origin (e.g. skeletal debris, organic matter) and inorganic components (e.g. those originating in coastal sediments or formed in the marine water column).

4 Estuarine-generated particulates. This type of particulate material has an internal source and includes inorganic and organic flocculants and precipitates, and both living and non-living particulate organic matter. Of the various processes that lead to the formation of estuarine particulates,

    (a) flocculation,

    (b) precipitation and

    (c) the biological production of organic matter

are especially important.

Flocculation is a process that causes smaller particles (colloids or semi-colloids) to increase in size and form larger units thereby increasing their likelihood of sedimentation, and has been described in detail by Potsma (1967) and Drever (1982). In estuaries, where the mixing of saline and fresh water leads to an increase in ionic strength, flocculation affects both organic and inorganic components; these include river-transported clay mineral suspensions, colloidal species of iron and dissolved organics (such as humic material). The aggregation of particles into larger sizes also can take place via biological mediation; for example through the production of faecal pellets by filter-feeding organisms (see Chapter 10).

Various types of precipitation processes occur in estuaries; of these, heterogeneous precipitation in the presence of particle clouds (e.g. in turbidity maxima) is especially important in the removal of dissolved Mn, and other metals, from solution.

The estuarine biomass is formed as a result of primary production, which is related to factors such as the supply of nutrients and the turbidity of the waters. In some turbid waters light penetration into the water column is so limited as to essentially preclude any significant photosynthesis, while in other systems it can be important. According to the estimate made by Williams (1981), the global internal estuarine photosynthetic production is $\sim 0.43 \times 10^{14}$ mol C yr$^{-1}$, which is $\sim 1.5\%$ of the total marine production (Section 9.1). Rivers transport $\sim 0.3 \times 10^{14}$ mol C yr$^{-1}$ as DOC + POC into estuaries (Perdue and Ritchie, 2004), which is the same order of magnitude as that resulting from primary production. According to Reuther (1981), however, much of this imported carbon is refractory, that is resistant to microbial attack, and so does not enter into recycling within the estuarine system (see Section 3.1.5.2).

#### 3.2.4.2 The distribution of EPM

The physical processes that control the distribution of EPM involve water circulation patterns, gravitational settling and sediment deposition and resuspension. All these processes, together with primary production, combine to set up the particle regime in an estuary.

Estuarine circulation patterns exert a fundamental influence on the processes that control the distribution of EPM. A number of characteristically different types of estuary can be distinguished on the basis of water circulation patterns (see Fig. 3.7), and these are described below in relation to the manner in which they constrain estuarine particle regimes, and therefore influence estuarine reactivity.

The most commonly occurring estuaries are of the positive type, that is loss from evaporation at the surface is less than the input of fresh water from rivers. The basic factor that determines the type of circulation in an estuary of this type is the part played by tidal currents (inflow of seawater) in relation to river flow (inflow of fresh water), and in a general way the type of circulation developed can be related to the dominance of one of these two water flow regimes and the influence of vertical mixing driven by tides.

In some estuaries the river flow is dominant. Under these conditions the less dense river water forms an upper layer over the more dense saline water, and the estuary becomes stratified. There are a number of types of stratified estuaries.

*Salt-wedge estuary.* When the circulation is almost completely dominated by river flow, a salt-wedge estuary can be formed. As the name implies, the salt water in this type of estuary extends into the river as a wedge under the freshwater outflow. There is also a steep density gradient between the fresh water and saline water, which prevents mixing between the two water layers until the river flow attains a critical velocity.

*Highly stratified estuary.* A second type of well developed vertical stratification, but one in which the saline layer is not confined to a wedge shape, can be formed in some estuaries. These estuaries are termed highly stratified, or are referred to as having a two-layer flow. Like the salt-wedge type, the river flow in a highly stratified estuary is large relative to the tidal flow and there is still a two-layer stratification. Now, however, the velocity of the seaward-flowing river water is sufficient to cause internal waves to break at the saline–freshwater interface. This results in the mixing of saline water from below into the upper freshwater layer, resulting in the salinity of the upper water layer increasing as it moves seawards.

Both Meade (1972), and later Potsma (1980), have pointed out that the distribution of suspended particulate matter in an estuary is largely controlled by the dynamics of the water circulation pattern. According to these authors, most of the suspended particulate material in both salt-wedge and highly stratified estuaries is of fluvial origin and is transported seawards in the upper water layer. The suspended matter does not thereafter enter the estuarine cycle proper.

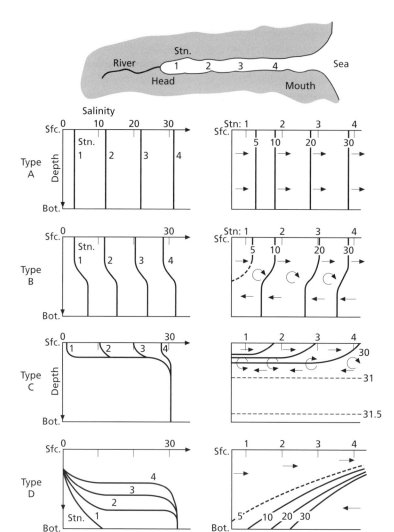

**Fig. 3.7** Salinity–depth profiles and longitudinal salinity sections in estuaries displaying different water circulation patterns (from Aston, 1978; after Pickard and Emery, 1982). Type A, a vertically mixed estuary; type B, a slightly stratified estuary; type C, a highly stratified estuary; type D, a salt-wedge estuary. Right column numbers 5, 10, 20 and 30 represent the salinity. Reprinted with permission from Elsevier.

*Partially stratified estuary.* As tidal forces become more important, vertical mixing will increase. In a partially stratified (or partially mixed) estuary, there is still a two-layer structure developed, with fresh water at the surface overlying saline water at depth. In this situation, however, there is vertical mixing between the mainly inflowing bottom water and the mainly outflowing surface water. This mixing takes place through eddy diffusion, which, unlike entrainment, is a two-way process, that is salt water is mixed upwards and fresh water is mixed downwards, with the result that a layer of intermediate salinity is formed. In this type of estuary there are also longitudinal variations in salinity in both the upper and lower layers, and undiluted fresh water is found only at the head of the estuary.

In the partially stratified estuary, the lower-water landward flow is strong enough to move suspended material up the estuary to the head of the salt intrusion. This may include both fluvial material that has settled from the upper layer and marine material that has been transported landwards. This material may settle and then resuspend as tidal flows change over the tidal cycle. A turbidity maximum, or particle cloud, may be developed at the head of the salt intrusion in the region where suspended solids transported from both up- and downstream directions can be trapped and regularly resuspended. Such a turbidity maximum, which is a region of concentration of suspended material, is an important site for dissolved ↔ particulate interactions. In some estuaries very high concentrations (up to several hundred g of sediment per l) have been reported near the bed. This so-called fluid mud, which is built up of material sinking from the overlying turbidity maximum during neap tides, can undergo resuspension and restricted transport on spring tides.

*Vertically well-mixed estuary.* Under certain estuarine conditions tidal flow can become completely dominant. For example, in some estuaries, which usually have a small cross-section, a strong tidal flow can exceed the river flow and the velocity shear on the bottom may be of sufficient magnitude for turbulence to mix the water column completely. These estuaries are termed homogeneous or fully mixed estuaries and, because there is no appreciable river flow, the suspended particulates are concentrated in the nearshore region. There is some doubt, however, as to whether *completely* homogeneous estuaries actually exist in nature.

It is apparent, therefore, that the nature and strength of the circulation patterns in an estuary will exert a strong control on the suspended particle regime and can lead, in the case of the partially stratified type, to the development of zones of relatively high suspended solid concentrations (turbidity maxima), which are major sites for physical, chemical and biological reactions between particulate and dissolved species. It is in these zones, therefore, that the estuarine filter is especially active. Estuarine circulation also affects the residence time of the water in estuaries. This varies from a few days to a few months, and according to Duinker (1986) it increases with the increasing extent of vertical mixing; thus, residence times generally increase in the sequence salt-wedge < stratified < partially mixed < well-mixed estuarine types, although obviously the size of an estuary in relation to exchange flows is also important in determining residence times.

## 3.2.5 The concentration and nature of dissolved material in estuaries

Concentrations of many components are higher in river water than seawater and hence concentrations tend to decrease through estuaries, although this is not true of some of the major ions in seawater. The latter ions tend to be unreactive in ocean waters and also in estuaries, leading to conservative mixing. For the ions with higher concentrations in rivers, more complex estuarine behaviour can in some cases be observed. The nature or speciation of the dissolved components is also affected by the water mixing. It is the speciation of an element between particulate, colloidal and dissolved (e.g. ion pairs, and both inorganic and organic complexes) forms, rather than its total concentration, that controls its environmental reactivity. Competitive complexing between the principal inorganic ligands (e.g. $Cl^-$, $SO_4^{2-}$, $CO_3^{2-}$, $OH^-$), the organic (e.g. humic material) ligands and particulate matter is the main factor that controls the nature of the inorganic species in natural waters. Complexation processes are considered in more detail in Section 11.6, with respect to seawater. Elemental speciation, however, also is important in the estuarine mixing zone.

From the point of view of the behaviour of both dissolved and particulate elements during estuarine mixing, three of the most important physicochemical differences between the river water and seawater end-members are those associated with the following parameters:

**1** ionic strength (salinity), which varies from zero in the river end-member to 0.7 M in the marine end-member;

**2** ionic composition and the concentrations of complexing ligands: for example, in river water the most important ions are calcium and bicarbonate, whereas seawater is dominated by sodium and chloride;

**3** pH: river waters can be either acidic or alkaline, with pH values that are usually in the range 5–8. Thus, in some estuaries the pH of the river end-member may be similar to that of the seawater end-member (average pH ~8.2). River waters, however, usually are more acidic than seawater, with the result that under many estuarine conditions there is a pH gradient, which increases with increasing salinity. Changes in parameters such as those identified above will affect the speciation of elements as the two end-member waters undergo estuarine mixing.

pH and $P_e$ often are considered to be the master variables in natural waters (see Chapter 14). $P_e$ is a convenient way of expressing the equilibrium redox potential, which is an important parameter in estuarine chemistry because in estuaries elements can be exposed to reducing conditions in both the water column and sediment interstitial waters, and under some estuarine conditions sulfide complexing also must be taken into account in speciation modelling. Even with a knowledge of the master variables, however, there are many uncertainties involved in understanding the speciation of elements in any natural waters, and these are magnified when the mixed-water estuarine system is considered. For a detailed treatment of the fundamental concepts involved in chemical speciation the reader is referred to Ure and Davidson (2002), and for estuarine waters to the review presented by Dyrssen and Wedborg (1980).

At this stage, it is sufficient to understand that during the mixing of fresh and saline waters there is competition for dissolved trace elements between various complexing agents, such as those described above, and between these agents and the suspended particulate matter that is present in the resulting estuarine 'soup'.

### 3.2.6 Dissolved ↔ particulate interactions in estuaries

The concentrations of both complexing agents and particulate matter undergo large variations during estuarine mixing. It is only particulate matter, however, that can be trapped in the mixing zone and this is coupled to dissolved material via particulate ↔ dissolved equilibria.

Processes that are involved in the addition or removal of dissolved components in the estuarine 'soup' include the following.

1 Flocculation, adsorption, precipitation and biological uptake, which result in the removal of components from the dissolved phase and their transfer to the particulate phase. It is these processes that can result in the retention of components in estuaries via particulate trapping in sediments.

2 Desorption from particle surfaces and the breakdown of organics, which result in the addition of components to the dissolved phase. This will result in the flushing out of the components if they remain in the dissolved phase.

3 Complexation and chelation reactions with inorganic and organic ligands, which stabilize components in the dissolved phase. These components will also be flushed out of an estuary with the water mass.

Theoretically air–sea gas exchange is also able to add or remove components, although in practice this is only important for a very small number of compounds such as some sulfur compounds, $N_2O$ and $CO_2$.

The manner in which these different types of processes can be linked to (i) the kind and concentration of EPM, and (ii) the speciation of dissolved components, can be described in terms of a number of experimental approaches that have been used to model estuarine chemistry.

Laboratory studies designed to investigate chemical reactions in estuarine waters tend to be of two types:

1 those involving the mixing of end-member waters and the characterization of the reaction products formed;

2 those attempting to study individual reaction processes, such as adsorption–desorption, by experimentally varying the controlling parameters (e.g. pH, ionic strength).

A number of laboratory studies on the chemical reactions taking place in estuaries have involved the artificial mixing of both filtered and non-filtered samples of the fresh and saline end-member waters (see e.g. Coonley *et al.*, 1971; Duinker and Notling, 1977). This often is referred to as the product approach, and perhaps the most significant advances in our understanding of estuarine chemistry to emerge from this approach are those which have been reported by Sholkovitz and his co-workers.

The Sholkovitz model (Sholkovitz, 1978) was an attempt to make a quantitative prediction of the reactivity of trace metals in the estuarine mixing zone. The model is based on the following assumptions.

1 A fraction of the dissolved trace metals in river water exist as colloids in association with colloidal forms of humic acids and hydrous iron oxides.

2 In the estuarine mixing zone the removal of the metals takes place via the flocculation of these colloids and/or their subsequent adsorption on to humic acids and hydrous iron oxide flocs.

3 The extent to which this removal takes place in the estuarine soup is determined by competition for the trace metals by seawater anions, humic acids and hydrous iron oxides, in the presence of seawater cations.

The Sholkovitz model was then tested in a variety of experiments using the product approach resulting from the mixing of river-water and seawater end-members. One of the most significant findings to emerge from the product-approach experiments was the importance of the role played by the rapidly floc-culating humic fraction in the formation of metal humates. The importance of the flocculation of humics and hydrous iron oxides was demonstrated by Sholkovitz and Copland (1981), who carried out an investigation into the fates of dissolved Fe, Cu, Ni, Cd, Co and humic acids following the artificial mixing of river and seawater. Although they restricted their experimental work to the filtered water of a single organic-rich river, the River Luce (UK), they were able to demonstrate the effects that resulted from the flocculation of fluvial-transported colloids by the seawater cations encountered on estuarine mixing. In this process, $Ca^{2+}$ was found to be the main coagulating, that is colloid destabilizing, agent. On the basis of the River Luce data, the authors concluded that the extent of this flocculation removal (expressed as a percentage of the dissolved concen-tration) varied from *large* (e.g. Fe ~80%, humic acids ~60%, Cu ~40%), to *small* (e.g. Ni and Cd ~15%), to essentially *nothing* (e.g. Co and Mn ~3%). However, Sholkovitz and Copland (1981) pointed out that their findings, based on the organic-rich river water, should not be applied *a priori* to estuar-ies in general. For example, Hoyle *et al.* (1984) showed that dissolved rare-earth elements (REE) were flocculated in association with Fe–organic-matter colloids when water from the River Luce was mixed with seawater; however, REE removal did not occur when the river end-member used in the experi-ments was organic-poor in character. In addition to the cationic destabilization of fluvially transported colloidal humic substances, organic complexes can stabilize some dissolved elements in solution; for example, Waslenchuk and Windom (1978) con-cluded that the conservative transport of dissolved As in estuaries from the southeastern US results from the stabilization of the element by organic complexes (see also Section 11.6.2).

In practice, estuarine dissolved ↔ particulate reac-tions depend on a number of variables. Several of these were considered by Salomons (1980) in a series of interlinked laboratory experiments in which the influence of pH, chlorinity and the concentration of suspended material on the adsorption of Zn and Cd was assessed under estuarine conditions. Three

important conclusions regarding the nature of the competition for trace metals in the estuarine mixing zone can be drawn from this study.

**1** The adsorption of both metals increased with increasing pH over the experimental range (pH 7.0–8.5).

**2** The adsorption of Cd, and to a lesser extent that of Zn, decreased with increasing chlorinity, probably as a result of competition from the chloride ion for the complexation of the metals, thus keeping them in solution.

**3** The adsorption of both elements increased with increasing turbidity, that is with an increase in the concentration of suspended matter. This apparently occurred to the extent that the suspended matter was able to compete effectively with chloride ions for metal complexation.

Extrapolated to real estuarine situations the latter two findings mean that the effectiveness of suspended particulate matter for the capture of dissolved Cd and Zn decreases as salinity increases, but that this effect can be overridden when the concentrations of the suspended particulates are relatively high, for example in the presence of a turbidity maximum, which can act as a zone of enhanced adsorption (Salomons and Forstner, 1984). A model designed to describe adsorption in the presence of a turbidity maximum in the estuarine environment is summar-ized in Worksheet 3.1.

The studies described previously have focused largely on the removal of trace elements from solu-tion on to particulate matter. In contrast, other labo-ratory experiments have shown that some elements can be added to the dissolved state via desorptive release from particulate matter during estuarine mixing (see e.g. Kharkar *et al.*, 1968; Van der Weijden *et al.*, 1977). This desorptive release, however, does not apparently affect all elements. For example, Li *et al.* (1984) added radiotracer spikes to experiments in which river and seawater were mixed and con-cluded that Co, Mn, Cs, Cd, Zn and Ba will be desorbed from river suspended particulate material on estuarine mixing, whereas Fe, Sn, Bi, Ce and Hg will undergo removal by adsorption.

The findings deduced from laboratory studies can be used to interpret estuarine mixing diagrams. This can be illustrated with respect to the study reported by Windom *et al.* (1983), who carried out an inves-tigation of the estuarine behaviour of copper in the

**Worksheet 3.1: A simple sorptive equilibrium model for the removal of trace metals in a low-salinity turbid region of an estuary (after Morris, 1986)**

A variable fraction of the river influx of many dissolved trace metals is removed from solution in the low-salinity, high-turbidity, estuarine mixing zone. The equilibrated partitioning of a trace metal between the dissolved and particulate phases in natural waters can be described by the term $K_d = X/C$, where $K_d$ is a conditional partition coefficient, $X$ is the concentration (w/v) of exchangeable metal on the particulate phase, and $C$ is the dissolved metal concentration (w/v). To set up his sorptive equilibrium model, Morris (1986) made the following assumptions: (i) the diffusion of solutes from the estuary to the very low-salinity region (<0.2‰) is minor so that the river can be considered to be the only significant source of dissolved metal; and (ii) the salinity related changes in the conditional partition coefficient are negligible. The problems involved in describing the removal of the river influx of dissolved trace metals in the turbid low-salinity estuarine mixing zone therefore can be simplified to a consideration of the change in sorptive equilibrium, at constant $K_d$, induced by adding suspended particulates to the river water. Under these conditions, the mass balance for the conservation of a trace metal, for unit volume of river water, can be written:

$$C_R + X_R P_R + X_S P_S = C + X(P_R + P_S) \qquad \text{(WS3.1.1)}$$

where $C_R$ and $C$ are the equilibrated dissolved metal concentrations (w/v) in the river water and the turbidized water, respectively; $X_R$ and $X$ are the equilibrated exchangeable metal concentrations (w/w) on the particles in the river water and the turbidized water, respectively; $X_S$ is the exchangeable metal concentration (w/w) on the added particulate load; $P_R$ is the suspended particulated load (w/w) carried by the river and $P_S$ is the additional suspended particulate load (w/v). Substituting $X_R = K_d C_R$ and $X = K_d C$ into Equation WS3.1.1, and introducing a term $\alpha = X_S/X$, yields:

$$C/C_R = (1 + K_d P_R)/[1 + K_d P_S(1 - \alpha)] \qquad \text{(WS3.1.2)}$$

Equation WS3.1.2 therefore predicts the change in dissolved metal concentration induced by adding particles to river water as a function of the particle load, the added particulate load, and $\alpha$ (the fraction of the exchangeable metal on the added particles relative to ultimate equilibration). For sorptive removal to occur, the added particles must have a lower concentration of exchangeable metal than the particles in the turbidized water, that is, $\alpha < 1$.

Morris (1986) considered how the predicted concentration ratio ($C/C_R$) within the zone of removal varies with $K_d$, $P_R$ and $\alpha$. The relationship for $K_d$ is illustrated in Fig. WS3.1.1, and shows how the predicted concentration ratio ($C/C_R$) varies with $K_d$ for additions of suspended particulate load ($P_S$) ranging from 0 to 1000 mg l$^{-1}$, with a representative depletion factor of $\alpha = 0.90$, and a river suspended particulate load ($P_R$) = 5 mg l$^{-1}$. The curves illustrate that at constant $P_R$ and $\alpha$, the degree of removal for any $P_S$ value increases (i.e. the ratio $C/C_R$ decreases) with increasing $K_d$, the most sensitive changes being in the $K_d$ range $1 \times 10^3$ to $1 \times 10^6$ mg l$^{-1}$. The extent of removal is also sensitive to changes in $P_R$ and $\alpha$. Morris (1986) then used data from the Tamar and plotted the concentration ratio ($C/C_R$) against the resuspended particulate load ($P_S$) for Cu, Zn and Ni. The results are illustrated in Fig. WS3.1.2, and show that for Cu and Zn the data correspond with the form of the relationship predicted by the model.

From the equilibrium sorption model it may be predicted that for elements with $K_d$ values higher than ~$1 \times 10^3$ there will be significant removal in the turbid waters. Data in the literature

*continued*

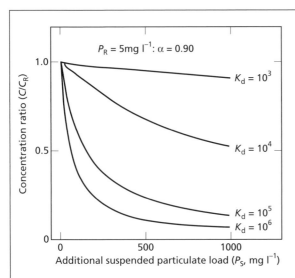

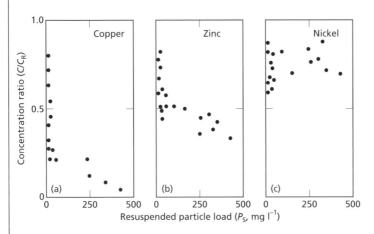

**Fig. WS3.1.1** The influence of the partition coefficient ($K_d$) on the sorptive uptake of fluvial dissolved constituents in the very-low-salinity, high-turbidity estuarine region.

**Fig. WS3.1.2** Variations in the concentration ratio ($C/C_R$) for dissolved (a) Cu, (b) Zn and (c) Ni with changes in suspended particulate load ($P_s$) in the very-low-salinity, high-turbidity zone in the Tamar estuary (UK) (from Morris, 1986). Reprinted with permission from Elsevier.

indicate that $K_d$ values in river water generally can exceed this value, and Morris (1986) therefore concluded that sorptive removal is potentially an important feature of the behaviour of trace metals in moderately to highly turbid estuaries.

Savannah and Ogeechee estuaries (USA). The field survey data showed that Cu behaved in a non-conservative manner, with additions to the dissolved fluvial load being found in both estuaries. The additions occurred at both low (<5‰) and high (>20‰) salinities, with a possible removal taking place at intermediate salinities (see Fig. 3.6biv). The addition of Cu to the dissolved phase at the two ends of the estuarine salinity range implied that the element was being released from suspended particulate matter and/or bottom sediments. In order to distinguish between these two potential release mechanisms the authors carried out a series of experiments in which Ogeechee river water was mixed with mid-shelf seawater in various proportions before filtration and analysis of the samples. This series of experiments, in which only the water and its suspended particulate matter were used, thus eliminated any effects that would have arisen from the release of Cu from bottom sediments. The results revealed that, following an initial release of Cu at low salinities, the element behaved in an essentially conservative manner over the salinity range ~8 to ~34‰. The authors concluded, therefore, that the addition of Cu to solution at salinities <5‰ was a result of its release from suspended particulate matter, but that the addition at salinities >20‰ was a result of its mobilization from bottom or resuspended sediment.

It may be concluded that mixing graphs can be used to establish whether or not a dissolved component has suffered reactivity in the estuarine mixing zone, and laboratory experiments can be used to interpret the processes involved; although it must be remembered that such experiments never fully reproduce the complex estuarine situation. Despite this constraint, laboratory experiments have led to at least a first-order understanding of some of the processes, for example competition for dissolved components from complexing ligands and particulate material, which control particulate ↔ dissolved equilibria in the estuarine 'soup'. Perhaps the single most important finding to emerge from all this work, however, is that both the mixing graphs and the laboratory experiments have shown that some elements can, under different conditions, behave differently i n the mixing zone. This reinforces the concept that for many elements there is no such thing as a global estuarine behaviour pattern. For this reason it is necessary to look at the behaviour of components in a number of individual estuaries. With this in mind, an attempt will be made in the following section to summarize the behaviour patterns of some components in the estuarine mixing zone, and to relate these to both non-biological and biological controls; in doing this attention will be focused, where possible, on the large globally important estuarine systems. The components considered represent examples of types of behaviour, and also include some of the more geochemically important elements.

### 3.2.7 The behaviour of individual components in the estuarine mixing zone

#### 3.2.7.1 Total particulate material

At the present day much of the particulate material that enters estuaries is trapped within the system. For example, only ~8% of the particulate matter entering the upper zone of the Scheldt estuary is exported to the North Sea (Duinker et al., 1979); more than 90% of the RPM carried by the Mississippi is deposited in the delta (Trefry and Presley, 1976); more than 95% of the suspended solids in the Amazon settle out within the river mouth (Milliman et al., 1975); over the entire St Lawrence system (i.e. upper estuary, lower estuary, gulf) ~90% of the particulate input is held back (Bewers and Yeats, 1977). Overall, therefore, it would appear that ~90% of RPM reaching the land–sea margins is retained in estuaries. Glaciers do not necessarily discharge into the oceans via estuaries in the same way as rivers do, but much of the particulate matter from glaciers is still believed to be trapped on the continental shelf, although in the case of iron the leakage beyond the continental shelf may still be important for ocean productivity (Raiswell et al., 2006).

#### 3.2.7.2 Iron and manganese

Iron and manganese are extremely important elements in aquatic geochemical processes because their various oxide and hydroxide species act as scavengers for a variety of trace metals. The processes that control the estuarine chemistries of the two elements are different, however.

*Iron.* Iron is present in natural waters in a continuum of species ranging through true solution, colloidal suspension and particulate forms. Most determinations of dissolved iron given in the literature refer to species that have passed through a 0.45 µm filter, and this definition will be used here.

Dissolved iron in river water is present mainly as hydrous Fe(III) oxides, which are stabilized in colloidal dispersion by high-molecular-weight humic acids (see e.g. Boyle *et al.*, 1977; Sholkovitz *et al.*, 1978; Hunter, 1983). Humic material is affected by speciation changes on the mixing of fresh and saline waters, and the dominant mechanism for the removal of iron to the particulate state appears to be the flocculation of mixed iron-oxide–humic-matter colloids of fluvial origin, which undergo electrostatic and chemical destabilization during estuarine mixing, partly as a result of neutralization of colloid charges by marine cations. The coagulative removal of iron therefore takes place at low salinities, and a typical estuarine mixing graph for iron is given in Fig. 3.6(bii).

Figueres *et al.* (1978) summarized the behaviour of iron in estuaries and found that in 15 out of the 16 systems considered, fluvial dissolved iron was removed during estuarine mixing; expressed in terms of the fluvial input, the removal ranged from ~70 to ~95%. This pattern has been confirmed with respect to the major estuaries, for example those of the Amazon, the Zaire and the Changjiang, which will have the greatest influence on oceanic chemistry. It may be concluded, therefore, that iron is one of the few elements for which a global behaviour pattern, that is, removal at low salinities, has been demonstrated almost unambiguously for the estuarine environment. However, Shiller and Boyle (1991) found dissolved iron to behave essentially conservatively in the Mississippi plume region, and they ascribe the different behaviour seen in this system to low riverine dissolved iron concentrations in this alkaline river.

*Manganese.* In contrast to iron, the distribution patterns for manganese in estuaries are variable and both gains and losses of dissolved manganese, and even conservative behaviour, have been reported in the literature. These variations can be related to the aqueous chemistry of the element.

Manganese is a redox-sensitive element that is biogeochemically reactive in the aqueous environment, and it readily undergoes transformations between the dissolved and particulate phases in response to physicochemical changes. Much of the dissolved manganese supplied by rivers is in the reduced and more soluble Mn(II) state, and this can undergo oxidative conversion to the much less soluble and hence predominantly particulate Mn(IV) as the physicochemical parameters change on the mixing of fresh and saline waters (see also Chapters 11 and 14). The oxidation of Mn(II) is autocatalytic, the product being a solid manganese dioxide phase with a composition that depends on the reaction conditions, that is, it is heterogeneous. The process is pH-dependent and also proceeds faster in the presence of a particulate phase that is able to adsorb the Mn(II), for example, in the presence of a turbidity maximum. In estuarine waters, therefore, it is redox-driven processes involving dissolved ↔ particulate reactions that play the dominant role in the chemistry of manganese. These redox reactions also continue after the deposition of Mn(IV) particulates in the bottom sediments. Many estuarine sediments are reducing at depth so that the Mn(IV) solids can become reduced during diagenesis, thus releasing Mn(II) into the interstitial waters and setting up a concentration gradient with respect to the overlying water column, which usually contains smaller concentrations of dissolved manganese. Dissolved Mn migrates up this gradient and, if the upper sediments are reducing, it may escape into the overlying water column, either directly or during the tidal resuspension of bottom material. If the upper sediments are oxic, much of the dissolved Mn will be reprecipitated as Mn(IV) hydrous oxides. However, Mn in this form also can enter the water column during resuspension processes, and this has been demonstrated for the St Lawrence by Sundby *et al.* (1981). These authors showed that the manganese enrichment in the sediment surface layers was not uniform, but occurred mostly on discrete, small-sized (0.5–4 µm), Mn-enriched particles. These particles were also found in the water column, including the upper waters, and it was suggested that they can become caught up in the estuarine circulation pattern and, as a consequence of their small size, some may escape into the open ocean. This, therefore, provides a potential mechanism for the leaking of particulate manganese from the estuarine system.

Under certain estuarine conditions manganese can behave in a conservative manner. This was demonstrated for the Beaulieu estuary (UK) by Holliday and Liss (1976). Here, there was no evidence for the removal of dissolved manganese from solution, a situation that may be attributed to the low concentration of suspended solids and the rapid freshwater replacement time in the estuary (Mantoura and Morris, 1983). In many estuaries, however, dissolved manganese shows pronounced non-conservative behaviour patterns. These patterns can be of different types, but that reported by Knox *et al.* (1981) for the Tamar (UK) will serve to illustrate the behaviour of the element in an estuary where a turbidity maximum is present and where the bottom sediments are in a reduced condition. The principal features of the dissolved Mn profile in this estuary are (i) a decrease in the concentrations of dissolved manganese at low salinities; and (ii) a broad increase at higher, mid-estuarine, salinities (see Fig. 3.6bv). Essentially, this profile can be interpreted in terms of a fluvial input of Mn(II), which is strongly removed from solution by the oxidative precipitation of Mn(IV) at the turbidity maximum that is present in the low-salinity region of the Tamar estuary. This turbidity maximum removes a large quantity of the fluvially transported Mn(II) before it can enter the estuary proper, and in order to account for the mid-estuarine Mn(II) increase, the authors proposed that in this region interstitial waters from anoxic sediments had been swept into the overlying water column. Under these conditions, therefore, the redox-driven production of Mn(II) took place in the sediment reservoir. Different behaviour patterns of dissolved Mn, in which the production of Mn(II) takes place in the water column, have been reported for other estuaries, for example, those of the Rhine and Scheldt (Duinker *et al.*, 1979).

It may be concluded, therefore, that profiles of dissolved manganese in estuarine surface waters are controlled by the fluvial input of dissolved Mn(II), the oxidative conversion of the dissolved Mn(II) to particulate Mn(IV), and the reduction and resolubilization of Mn(IV) in either the water column or the sediment compartments. However, owing to inter-estuarine variations in the parameters that control the redox-driven cycling of manganese, the element does not appear to exhibit a global estuarine mixing pattern. In many estuaries it appears that dissolved Mn is *removed* from solution, but this is not a universal feature, especially in large river–estuarine systems. For example, Edmond *et al.* (1985) reported that in the Changjiang estuary dissolved manganese appears to *increase* up to a salinity of ~12‰, beyond which it behaves conservatively.

### 3.2.7.3  Barium

There is now a considerable body of evidence in the literature which suggests that there may be a common pattern in the behaviour of barium during estuarine mixing. This involves the desorption of the element from particulate matter at low salinities, often followed by conservative behaviour across the rest of the mixing zone (see e.g., Fig. 3.6biii). Significantly, this estuarine production of barium has been reported for large river–estuarine systems, such as those of the Amazon (Boyle, 1976), the Mississippi (Hanor and Chan, 1977), the Zaire (Edmond *et al.*, 1978) and the Changjiang (Edmond *et al.*, 1985); in the latter system the fluvial flux of dissolved barium was increased by 25% following the estuarine production of the element. The release of Ba is at least in part related to ion exchange reactions with increasing concentrations of major ions forcing the release of Ba originally adsorbed to riverine particulate matter as salinity increases during estuarine mixing, Coffey *et al.* (1997).

### 3.2.7.4  Boron

A number of authors have suggested that boron can be removed from solution during estuarine mixing (see e.g. Liss and Pointon, 1973). Fine-grained suspended material, especially illite-type clays, are thought to adsorb the boron from solutions, a process that is enhanced as salinity increases. Fanning and Maynard (1978), however, reported data showing that boron behaves conservatively in both the Zaire and Madalena river plumes (see also Fig. 3.6bi), thus casting doubt on the concept that the inorganic adsorption of the element on to suspended solids at the freshwater–salinewater interface is a globally significant geochemical process.

### 3.2.7.5  Aluminium

Several studies carried out on the behaviour of dissolved aluminium in small river–estuarine systems have indicated removal of the element at low salinities, either via flocculation, sorption on to resuspended sediment particles or precipitation with silica (see e.g. Hosokawa *et al.*, 1970; Hydes and Liss, 1977; Sholkovitz, 1978; Macklin and Aller, 1984). A consensus of the data in the literature therefore appears to indicate that the Al delivered to the oceans by fluvial sources may be largely removed in the estuarine and coastal zones before it can reach the open ocean (see e.g. Orians and Bruland, 1985), the removal apparently varying between ~20 and >50% of the fluvial flux. It must be stressed, however, that few data exist on the behaviour of dissolved Al in the estuaries of large rivers. Van Bennekon and Jager (1978) carried out a preliminary study in the Zaire plume and reported that in this tropical system, which is characterized by low suspended solids, high dissolved organic matter loads and a short residence time of the water in the low-salinity region, there was a production of dissolved aluminium with maximum values at a salinity of ~5‰. The authors concluded, however, that it is not possible to predict a global estuarine behaviour pattern for dissolved aluminium because variations in the concentrations and type of suspended solids, the amount and type of dissolved organics and the residence time of the water in different salinity regions must all be taken into account when assessing the fate of the element.

### 3.2.7.6 The nutrients

A variety of behaviour patterns have been reported for nitrogen, phosphorus and silicon in estuaries (see e.g. GESAMP, 1987). Nutrients are carried by rivers in both dissolved and particulate forms, and their estuarine reactivity can be brought about by chemical (e.g. adsorption or desorption) and biological (e.g. photosynthetic production) processes which can result in trapping within estuaries or transformation of their chemical form (e.g. Kaul and Froelich, 1984). In the absence of these processes, the nutrients can behave conservatively, with their distributions being controlled by the physical mixing of the river-water and seawater end-members.

Several studies have reported Si removal within estuaries, but these appear to all arise from biological uptake, some of which may be subsequently regenerated.

Nitrogen is delivered to estuaries as nitrate (and small amounts of nitrite), ammonium, particulate nitrogen and dissolved organic nitrogen; the latter is considered in the next section. Particulate nitrogen will tend to sediment within the estuary, along with other riverine particulate matter, but it may be converted within the sediments to soluble nitrogen components by bacterial action. Ammonium (and nitrite) are usually minor nitrogen components in most estuaries unless there are low oxygen concentrations or particularly large inputs of the reduced N components, such as from sewage discharges. Hence the main issue to consider is the behaviour of nitrate. As an anionic species, nitrate is not strongly affected by particle water reactions, since most particulate matter is itself negatively charged and consequently reacts more with cations. Nitrogen can, however, be removed by biological uptake and subsequent burial and also by denitrification in sediments or the water column, a process linked to the oxidation of organic carbon in low oxygen environments (Chapter 9). In estuaries where these processes are of minor importance, nitrate is often found to behave conservatively with dilution of high riverine nitrate by lower nitrate seawater. In rivers with extremely low nutrient concentrations, such as those draining lowland peat swamps, conservative mixing of low nitrate and silicon river water with higher concentration offshore waters has been reported (Alkhatib *et al.*, 2007).

Phosphorus behaviour in estuaries appears to be very complex, reflecting the quite strong tendency of dissolved phosphorus to adhere to particulate matter. Some researchers have reported release of dissolved phosphorus from particles within estuaries and others reported active removal. Using a model similar to that in Worksheet 3.1, Prastka *et al.* (1998) argued that in relatively pristine rivers, release of phosphorus from riverine particles might be expected, but the effect of increasing riverine dissolved phosphorus concentrations is to perturb the particle water pseudo equilibrium and encourage estuarine transformation of dissolved phosphorus to particulate phases.

A number of studies have been carried out on the behaviour of the nutrients in globally important river systems and these can be illustrated with reference to the Amazon, the Changjiang and the Zaire estuarine systems.

*The Amazon estuary.* Data reported for the Amazon plume during a *summer* diatom bloom (Edmond *et al.*, 1981) showed that photosynthetic activity commenced when suspended particulates had decreased to ≤10 mg l$^{-1}$, which occurred at a salinity of ~7‰, thus making sufficient light available for primary production to be initiated. The diatom bloom occurred over the salinity range ~7–15‰, and here there was a complete depletion of dissolved nitrate and phosphate and a 25% depletion in dissolved silica. The underlying salt wedge became enriched in nutrients remineralized from sinking planktonic material, and mass balance calculations showed the following trends.

1 Almost all the phosphorus was remineralized in the salt wedge, or at the sediment surface. Thus, phosphate was little affected by estuarine processes, because planktonic uptake and remineralization were in approximate balance.

2 Only a minor part of the silica removed in production was remineralized, and ~20% of the total fluvially transported silica was transferred to the sediment sink as diatom tests. In a subsequent paper, however, DeMaster *et al.* (1983) estimated that only ~4% of the fluvial silica flux accumulates on the Amazon shelf as biogenic opal. It therefore would appear that between ~80% and ~95% of the fluvial silica escapes the Amazon plume.

3 The remineralization of fixed nitrogen was only partial, with ~50% being transformed into forms other than nitrate or nitrite. Thus, only about half of the fixed nitrogen escaped the estuarine zone.

The finding that fluvially transported phosphate and silica essentially escaped from the Amazon plume under summer diatom bloom conditions was in agreement with the model predictions made by Kaul and Froelich (1984). Nutrient dynamics, however, can vary seasonally. In the *winter* months under conditions of very much reduced biological activity in the Amazon plume, phosphate concentrations increased at low salinities, probably as a result of desorption from particulate material (a chemical effect resulting in an additional fluvial source of dissolved phosphate), but both nitrate and silica were essentially conservative over the salinity range ~3 to ~20.

*The Changjiang estuary.* In this estuary high concentrations of suspended particulates are maintained up to a salinity of ≥20‰, and as a result summer plankton blooms are found only on the inner shelf, at salinities greater than this. Edmond *et al.* (1985)

have provided data on the seasonal behaviour of nutrients in the Changjiang estuary.

1 In the *summer* months both silica and nitrate behaved conservatively out to salinities of ~20, but showed some depletion beyond this in the presence of plankton blooms, thus indicating biological removal. In *winter*, however, when there was no significant photosynthetic activity, both nutrients behaved in a conservative manner over the entire salinity range of the mixing zone. This was an important conclusion with regard to the estuarine behaviour of silica, because it showed that despite very high concentrations of suspended matter there was no significant removal by chemical processes associated with the particulates.

2 In *summer* phosphate was almost completely depleted at salinities >18‰. In contrast, in the *winter* months, there was an increase in dissolved phosphate at low salinities (<5‰), indicating a desorption of phosphate from particulate material: a situation similar to that found in the Amazon plume.

*The Zaire estuary.* The nutrient relationships in this estuary are somewhat different from those in the Amazon and Changjiang systems, because photosynthetic activity in the Zaire does not occur at salinities less than ~30‰, and even at values above this the water is still relatively turbid and *in situ* production remains low (Cadee, 1978). Data on the nutrient dynamics in the Zaire estuary have been provided by Van Bennekon *et al.* (1978).

1 Silica in the estuary is conservative out to salinities at which phytoplankton blooms occur, again indicating a lack of interaction with particulate matter, beyond which the concentrations decrease as a result of biological uptake.

2 Phosphate profiles showed maximum concentrations at salinities of ~10‰, owing partly to desorption from suspended particulate matter, followed by uptake at higher salinities as a result of biological removal.

3 Nitrate had a profile which was similar to that of phosphate, with highest concentrations at low salinities, but the addition to the fluvial source of nitrate mainly arose from the mixing of subsurface seawater, with higher nitrate concentrations, into the surface waters.

Nutrients are carried by fluvial transport in both particulate and dissolved forms. When sufficient dissolved nutrients are available, biological uptake is controlled by the availability of light. Turbidity has a major influence on light penetration and the distribution of suspended material appears to be the controlling variable on the onset of large-scale biological productivity in estuaries (see e.g. Milliman and Boyle, 1975). Once biological activity is initiated nutrients are removed from solution and the sinking of planktonic debris, containing nutrients in a particulate form, can lead to one of two general situations.

1 The remineralization of the nutrients in the saline layer and their subsequent transport out to sea where, when the flow is debouched into open shelves, they can lead to an increase within subsurface waters and can sustain coastal productivity following upwelling.

2 The trapping of the nutrients in the bottom sediment.

Although the picture is by no means clear yet, there is evidence that some nutrients may exhibit global estuarine patterns, and these can be summarized as follows.

1 In relatively pristine systems the fluvial supply of dissolved phosphate appears to be increased at low salinities via desorption from particulate matter. Phosphate is removed during photosynthetic production, but much of this is regenerated back into the water column from the sinking biota. As an overall result, estuarine processes can enhance the fluvial supply of dissolved phosphate to the oceans.

2 Silica is removed during phytoplankton blooms but much of this is regenerated from sinking biota. In the absence of biological activity silica appears to behave conservatively and is not removed on to suspended particulates; the overall result is that the fluvial silica flux largely escapes estuaries and is delivered intact to the oceans.

3 The situation for nitrate is less clear, but it would seem that in some estuaries a considerable fraction of fluvial nitrate may be retained within the system.

In polluted estuaries enriched in nitrate and phosphate, however, the nutrient dynamics may be different from those in the large, relatively unpolluted estuaries. Jickells *et al.* (2000) considered the way nitrate fluxes have changed with time in the Humber estuary, the largest English estuary. They showed that in the prehistoric condition the estuary was a net sink for nitrate, trapping riverine and coastal sourced

nitrate. Since that time riverine nitrate concentrations have increased substantially due to human activity and the intertidal area of the estuary has been subject to extensive reclamation for agricultural and urban land use with more than 90% of the intertidal area lost. The combined effect of both of these human induced changes has been to alter the estuary from a nutrient sink to a source to the adjacent North Sea with essentially conservative behaviour of nitrate. This pattern of human induced change with increased inputs and removal of the estuarine filter has been repeated in many other estuaries around the world.

### 3.2.7.7 Organic carbon

*Particulate organic carbon (POC).* In attempting to assess the fate of fluvial POC (and DOC) in the estuarine and coastal zones it is necessary to distinguish between a labile and a refractory fraction (see Section 9.2). The total fluvial flux of POC has been estimated to be $173 \times 10^{12}\,g\,yr^{-1}$ (Perdue and Ritchie, 2004) and according to Ittekkot (1988) ~35% of this is labile and may undergo oxidative destruction in estuaries and the sea. The remaining 65% is refractory and can escape to the ocean environment to be accumulated in sediments there or retained in estuarine sediments.

*Dissolved organic carbon (DOC).* Data for the behaviour of DOC during estuarine mixing are somewhat contradictory. Evidence is available from both laboratory experiments and field surveys, which suggests that at least some fractions of fluvial DOC can be removed in estuaries and coastal waters by processes such as flocculation, adsorption on to particulates and precipitation (see e.g. Sieburth and Jensen, 1968; Menzel, 1974; Schultz and Calder, 1976; Sholkovitz, 1978; Hunter and Liss, 1979). These removal mechanisms would lead to the nonconservative behaviour of DOC at the land–sea margins. However, this was challenged by Mantoura and Woodward (1983). These authors reported data derived from a large-scale, two-and-a-half-year survey of the distribution, variability and chemical behaviour of DOC in the Severn estuary and Bristol Channel (UK), a system selected because of its relatively long water residence times (100–200 days) and its particle-rich waters, both of which would favour the in situ detection of any of the DOC removal processes. The survey revealed, however, that even under these conditions DOC behaved in a conservative manner in the estuary. Other recent studies (e.g. Spencer *et al.*, 2007; Alkhatib *et al.*, 2007) also detect complex and variable DOC behaviour in estuaries at different times, but in rivers with limited anthropogenic perturbation it does seem that much of the bulk DOC can pass through the estuary, although this may disguise significant changes in its chemical composition reflecting selective removal or inputs of individual components.

### 3.2.7.8 Trace metals

There is considerable variability in the data for the behaviour of a number of trace metals in estuaries. Both conservative and non-conservative behaviour patterns have been reported for individual dissolved trace metals in different estuarine systems, and these patterns are obviously dependent on local conditions. For example, there is considerable evidence that trace metals such as Zn, Cu, Cd, Pb and Ni *can* exhibit non-conservative patterns during estuarine mixing, and Campbell and Yeats (1984) showed that ~50% of the dissolved Cr was removed at low salinities in the St Lawrence estuary; at salinities >5‰ the element behaved conservatively. Danielsson *et al.* (1983), however, gave data on the distributions of dissolved Fe, Pb, Cd, Cu, Ni and Zn in the Gota River estuary (Sweden), a relatively unpolluted salt-wedge system, and demonstrated that the processes which remove dissolved trace metals from solution did not operate very effectively in the system. The result of this was that, apart from iron, the metals behaved in an essentially conservative manner during estuarine mixing. Balls (1985) used salinity versus concentration plots to demonstrate the conservative behaviour of Cu and Cd in the Humber estuary (UK). Shiller and Boyle (1991) demonstrated essentially conservative behaviour of Cu, Ni, Mo, Zn, and Cr while Cd showed desorption from particles during estuarine mixing which has been attributed to very effective complexation of $Cd^{2+}$ by chloride ions during estuarine mixing. It is apparent, therefore, that the non-conservative behaviour of some trace metals is not a universal feature of the estuarine environment. Further, even when non-conservative behaviour has been identified for trace metals it may

involve either the loss of, or the production of, dissolved species. From the point of view of the influence that estuarine reactivity has on the trace metal chemistry of the World Ocean, it is therefore perhaps most sensible to concentrate on the behaviour of trace metals in the large estuarine systems. Although data on this topic are still relatively scarce, it is worthwhile summarizing the findings of a number of studies.

Boyle *et al.* (1982) concluded that Cu, Cd and Ni are usually unreactive during the mixing of river and seawater in the Amazon plume: a low organic and high particulate system. In contrast, Edmond *et al.* (1985) reported that in the Changjiang estuary Ni was desorbed from particulate matter at low salinities, then behaved conservatively at values above 8‰; however, Cu and Be behaved in a conservative manner over the entire mixing zone in the same estuary. Moore and Burton (1978) described the distribution of dissolved Cu in the Zaire estuary, and although trends in the data were not very clear there was a suggestion that Cu was desorbed at intermediate salinities. Both Boyle *et al.* (1982) and Edmond *et al.* (1985) have provided evidence that Cd undergoes desorption at low salinities in the Amazon and Changjiang estuaries, respectively.

Thus, it appears that, with the exception of iron, which is almost universally removed from solution (see previously), there are few common trends in the conservative/non-conservative behaviour patterns of trace metals in estuaries although large scale systematic removal is not widely seen. When non-conservative behaviour does occur, however, low-salinity desorption, that is, gain to solution, followed by essentially conservative behaviour appears to be an important process in the behaviour of Ba, Ni and Cd in major estuarine systems.

### 3.2.8 The estuarine modification of river-transported signals: summary

1 Estuaries, which are located at the land–sea margins, exhibit large property gradients and can be regarded as acting as filters of river-transported chemical signals. This filter acts mainly via dissolved–particulate reactivity. This reactivity can be mediated by both physicochemical and biological processes, and is dependent on a range of interrelated factors, which include: the physical regime of an estuary, the

residence time of the water, primary production, pH, differences in composition between the end-member waters, and the concentrations of suspended particulate material, organic and inorganic ligands and nutrients. The effects of the filter, however, can vary widely from one estuary to another, and it is difficult to identify global estuarine processes.

2 The estuarine filter is also selective in the manner in which it acts on individual elements. Some dissolved elements are simply diluted on the mixing of river and seawater (conservative behaviour), whereas others undergo estuarine dissolved–particulate reactions, which lead to their addition to, or removal from, the dissolved phase (non-conservative behaviour).

3 Particulate elements that are added to the dissolved phase (e.g. by desorption, biological degradation) increase the fluvial flux and can be carried out of the estuarine environment during tidal flushing. In contrast, elements that are added to the particulate phase (e.g. by adsorption, flocculation, biological uptake) can be trapped in the bottom sediments; however, processes such as nutrient regeneration, sediment resuspension and diagenetic release followed by the flushing out of interstitial waters can recycle these elements back into the water phase.

4 The dissolved–particulate reactivity can take place over a number of estuarine regions. (a) Physicochemically mediated particulate–dissolved reactions are especially intense at the low-salinity region where the initial mixing of the fresh and saline end-member waters takes place. Here, processes such as flocculation (e.g. Fe) and particle adsorption–precipitation (e.g. Mn) are active in the removal of elements from solution, the latter being enhanced in the presence of a turbidity maximum. (b) Primary production, which involves the biologically mediated generation of particulate matter and the removal of nutrients from solution, is controlled largely by the availability of nutrients and the turbidity of the waters, and tends to reach a maximum at mid- and high-salinity ranges where the concentrations of particulate matter (turbidity) decrease allowing light for photosynthesis and where nutrients are present in sufficient quantity to initiate production. Changes in nutrient inputs due to human activity have undoubtedly increased the potential for photosynthetic growth in estuaries where light conditions allow, and other changes in estuarine management, such as land reclamation have also altered the functioning of estuaries.

5 Around 90% of the particulate matter transported into estuaries via the fluvial flux is trapped in the estuarine environment under present-day conditions.
6 With the exception of iron, which is removed at low salinities, few dissolved elements exhibit a global estuarine behaviour pattern. For example, dissolved Cu has been shown to exhibit either conservative or non-conservative behaviour (including both addition to and removal from solution) in different estuaries. Evidence suggests, however, that desorption at low salinities, followed by conservative behaviour over the rest of the mixing zone, is an important process in the behaviour of Ba, Ni and Cd in *major* estuarine systems.

# References

Alkhatib, M., Jennerjahn, T.C. and Samiaji, J. (2007) Biogeochemistry of the Dumai River estuary, Indonesia, a tropical black-water river. *Limnol. Oceanogr.*, **52**, 2410–2417.

Anderson, S.P., Drever, J.I. and Humphrey, N.F. (1997) Chemical weathering in glacial environments. *Geology*, **25**, 399–402.

Andrews, J.E., Brimblecombe, P., Jickells, T.D., Liss, P.S. and Reid, B.J. (2004) *An Introduction to Environmental Chemistry*. Blackwell Publishing, Oxford.

Aston, S.R. (1978) Estuarine chemistry, in *Chemical Oceanography*, J.P. Riley and R. Chester (eds), Vol. 7, 361–440. London: Academic Press.

Balls, P.W. (1985) Copper, lead and cadmium in coastal waters of the western North Sea. *Mar. Chem.*, **15**, 363–378.

Bewers, J.M. and Yeats, P.A. (1977) Oceanic residence times of trace metals. *Nature*, **268**, 595–598.

Bewers, J.M. and Yeats, P.A. (1980) Behaviour of trace metals during estuarine mixing, in *River Inputs to Ocean Systems*, J.-M. Martin, J.D. Burton and D. Eisma (eds), 103–115. Paris: UNEP/UNESCO.

Boyle, E.A. (1976) *The Marine Geochemistry of Trace Metals*. Thesis, MIT-WHOI, Cambridge, MA.

Boyle, E.A., Collier, R., Dengler, A.T., Edmond, J.M., Ng, A.C. and Stallard, R.F. (1974) On the chemical mass balance in estuaries. *Geochim. Cosmochim. Acta*, **38**, 1719–1728.

Boyle, E.A., Edmond, J.M. and Sholkovitz, E.R. (1977) The mechanism of iron removal in estuaries. *Geochim. Cosmochim. Acta*, **41**, 1313–1324.

Boyle, E.A., Huested, S.S. and Grant, B. (1982) The chemical mass balance of the Amazon plume – II. Copper, nickel and cadmium. *Deep-Sea Res. A*, **29**, 1355–1364.

Broecker, W.S. and Peng, T.-H. (1982) Tracers in the Sea. Lamont-Doherty Geological Observatory.

Cadee, G.C. (1978). Primary production and chlorophyll in the Zaire River, estuary and plume. *Neth. J. Sea Res.*, **12**, 368–381.

Campbell, J.A. and Yeats, P.A. (1984) Dissolved chromium in the St. Lawrence estuary. *Estuarine Coastal Shelf Sci.*, **19**, 513–522.

Chester, R. (1988) The storage of metals in sediments, in *Workshop on Metals and Metalloids in the Hydrosphere*, Bochum 1987, UNESCO International Hydrological Programme, 81–110.

Coffey, M., Dehairs, F., Collette, O., Luther, G., Church, T. and Jickells, T. (1997) The behaviour of dissolved barium in estuaries. *Est. Coastal Shelf Sci.*, **45**, 113–121.

Coonley, L.S., Baker, E.B. and Holland, H.D. (1971) Iron in the Mullica River and Great Bay, New Jersey. *Chem. Geol.*, **7**, 51–64.

Conley, D.J., Schleske, C.L. and Stoermer, E.F. (1993) Modification of the biogeochemical cycle of silica with eutrophication. *Mar. Ecol. Progr. Ser.*, **101**, 179–192.

Danielsson, L.G., Magnusson, B., Westerlund, S. and Zhang, K. (1983) Trace metals in the Gota River estuary. *Estuarine Coastal Shelf Sci.*, **17**, 73–85.

Degens, E.T., Kempe, S., Richey, J.E. (1991) Summary: Biogeochemistry of major world rivers, in *Biogeochemistry of World Rivers*, E.T. Degens, S. Kempe and J.E. Richey (eds). SCOPE. New York: John Wiley & Sons, Inc.

DeMaster, D.J., Knapp, G.B. and Nittrover, C.A. (1983) Biological uptake and accumulation of silica on the Amazon continental shelf. *Geochim. Cosmochim. Acta*, **47**, 1713–1723.

Donner, S.D., Kucharik, C.J. and Foley, J.A. (2004) Impact of changing land use practices on nitrate export by the Mississippi river. *Global Biogeochem. Cycl.*, **18**, GB1028, doi:10.1029/2003GB002093.

Drever, J.L. (1982) *The Geochemistry of Natural Waters*. Englewood Cliffs, NJ: Prentice-Hall.

Duinker, J.C. (1986) Formation and transformation of element species in estuaries, in *The Importance of Chemical 'Speciation' in Environmental Processes*, M. Bernhard, F.T. Brinckman and P.J. Sadler (eds), 365–384. Berlin: Springer-Verlag.

Duinker, J.C. and Kramer, C.J.M. (1977) An experimental study on the speciation of dissolved zinc, cadmium, lead and copper in Rhine River and North Sea water, by differential pulsed anodic stripping voltammetry. *Mar. Chem.*, **5**, 207–228.

Duinker, J.C., Wollast, R. and Billen, G. (1979) Behaviour of manganese in the Rhine and Scheldt estuaries. II. Geochemical cycling. *Estuarine Coastal Shelf Sci.*, **9**, 727–738.

Dyrssen, D. and Wedborg, M. (1980) Major and minor elements, chemical speciation in estuarine waters, in *Chemistry and Biochemistry of Estuaries*, E. Olausson and I. Cato (eds), 71–120. New York: John Wiley & Sons, Inc.

Edmond, J.M., Boyle, E.A., Drummond, D., Grant, B. and Mislick, T. (1978) Desorption of barium in the plume of the Zaire (Congo) River. *Neth. J. Sea Res.*, **12**, 324–328.

Edmond, J.M., Boyle, E.A., Grant, B. and Stallard, R.F. (1981) Chemical mass balance in the Amazon plume I: the nutrients. *Deep-Sea Res. A*, **28**, 1339–1374.

Edmond, J.M., Spivack, A., Grant, B.C., Ming-Hui, H., Zexiam, C., Sung, C. and Xiushau, Z. (1985) Chemical dynamics of the Changjiang estuary. *Cont. Shelf Res.*, **4**, 17–36.

Ertel, J.R., Hedges, J.I., Devol, A.H., Richey, J.E. and Ribeiro, M. (1986) Dissolved humic substances of the Amazon River system. *Limnol. Oceanogr.*, **31**, 739–754.

Fairbridge, R.W. (1980) The estuary: its identification and geodynamic cycle. In *Chemistry and Biochemistry of Estuaries*, E. Olausson and I. Cato (eds), 1–36. New York: John Wiley & Sons, Inc.

Falkowski, P., Scholes, R.J., Boyle, E., *et al.* (2000) The global carbon cycle: A test of our knowledge of earth as a system. *Science*, **290**, 291–296.

Fanning, K.A. and Maynard, V.I. (1978) Dissolved boron and nutrients in the mixing plumes of major tropical rivers. *Neth. J. Sea Res.*, **12**, 345–354.

Feth, J.H. (1971) Mechanisms controlling world water chemistry: evaporation–crystallization processes. *Science*, **172**, 870–872.

Figueres, G., Martin, J.M. and Meybeck, M. (1978) Iron behaviour in the Zaire estuary. *Neth. J. Sea Res.*, **12**, 329–337.

Gaillardet, J., Viers, J. and Dupré (2004) Trace elements in Rivers Surface and Ground Water, Weathering and Soils (ed. J.J. Drever), Vol. 5: *Treatise on Geochemistry*, H.D. Holland and K.K. Turekian (eds), pp.225–272. Oxford: Elsevier-Pergamon.

Galloway, J.N. (2004) The Global Nitrogen Cycle in Biogeochemistry. W.H. Schlesinger (ed.) Vol. 8: *Treatise on Geochemistry*, H.D. Holland and K.K. Turekian (eds), pp.557–583. Oxford: Elsevier-Pergamon.

GESAMP (1987) *Land/Sea Boundary Flux of Contaminants from Rivers*. Paris: UNESCO.

Gibbs, R.J. (1970) Mechanisms controlling world river water chemistry. *Science*, **170**, 1088–1090.

Gibbs, R.J. (1973) Mechanisms of trace metal transport in rivers. *Science*, **180**, 71–73.

Gibbs, R.J. (1977) Transport phases of transition metals in the Amazon and Yukon rivers. *Bull. Geol. Soc. Am.*, **88**, 829–843.

Hanor, J.S. and Chan, L.A. (1977) Non-conservative behaviour of barium during mixing of Mississippi River and Gulf of Mexico waters. *Earth Planet. Sci. Lett.*, **37**, 242–250.

Holland, H.D. (1978) *The Chemistry of the Atmosphere and Oceans*. New York: John Wiley & Sons, Inc.

Holliday, L.M. and Liss, P.S. (1976) The behaviour of dissolved iron, manganese and zinc in the Beaulieu Estuary. *Estuarine Coastal Mar. Sci.*, **4**, 349–353.

Hosokawa, I.O., Ohshima, F. and Kondo, N. (1970) On the concentration of the dissolved chemical elements in the estuary of the Chikugogawa River. *J. Oceanogr. Soc. Jap.*, **26**, 1–5.

Hoyle, J., Elderfield, H., Gledhill, A. and Grieves, M. (1984) The behaviour of the rare earth elements during mixing of river and sea water. *Geochim. Cosmochim. Acta*, **48**, 148–149.

Humborg, C., Ittekot, V., Cociasu, A. and Bodungen, B.V. (1997) Effects of Danube River dam on Black Sea biogeochemistry and ecosystem structure. *Nature*, **386**, 385–388.

Hunter, K.A. (1983) On the estuarine mixing of dissolved substances in relation to colloid stability and surface properties. *Geochim. Cosmochim. Acta*, **47**, 467–473.

Hunter, K.A. and Liss, P.S. (1979) The surface charge of suspended particles in estuarine and coastal waters. *Nature*, **282**, 823–825.

Hydes, D.J. and Liss, P.S. (1977) The behaviour of dissolved Al in estuarine and coastal waters. *Estuarine Coastal Shelf Sci.*, **5**, 755–769.

Ittekkot, V. (1988) Global trends in the nature of organic matter in river suspensions. *Nature*, **332**, 436–438.

Jickells, T., Andrews, J., Samways, G., Sanders, R., Malcolm, S., Sivyer, D., Parker, R., Nedwell, D., Trimmer, M. and Ridway, J. (2000) Nutrient fluxes through the Humber estuary, Past, Present and Future. *Ambio*, **29**, 130–135.

Kaul, L.W. and Froelich, P.N. (1984) Modelling estuarine nutrient geochemistry in a simple system. *Geochim. Cosmochim. Acta*, **48**, 1417–1433.

Kharkar, D.P., Turekian, K.K. and Bertine, K.K. (1968) Stream supply of dissolved Ag, Mo, Sb, Se, Cr, Co, Rb, and Cs to the oceans. *Geochim. Cosmochim. Acta*, **32**, 285–298.

Knox, S., Turner, D.R., Dickson, A.G., Liddicoat, M.I., Whitfield, M. and Butler, E.I. (1981) Statistical analysis of estuarine profiles: application to manganese and ammonium in the Tamar estuary. *Estuarine Coastal Shelf Sci.*, **13**, 357–371.

Konta, J. (1985) Mineralogy and chemical maturity of suspended matter in major rivers sampled under the SCOPE/UNEP Project, in *Transport of Carbon and Minerals in Major World Rivers, Part 3*, E.T. Degens, S. Kempe and R. Herrera (eds), pp.569–592. Hamberg: Mitt. Geol.-Palont. Inst. Univ. Hamburg, SCOPE/UNEP, Sonderband 58.

Li, Y.-H., Burkhard, L. and Teroaka, H. (1984) Desorption and coagulation of trace elements during estuarine mixing. *Geochim. Cosmochim. Acta*, **48**, 1879–1884.

Liddicoat, M.I., Turner, D.R. and Whitfield, M. (1983) Conservative behaviour of boron in the Tamar Estuary. *Estuarine Coastal Shelf Sci.*, **17**, 467–472.

Liss, P.S. (1976) Conservative and non-conservative behaviour of dissolved constituents during estuarine mixing, in *Estuarine chemistry*, J.D. Burton and P.S. Liss (eds), pp.93–130. London: Academic Press.

Liss, P.S. and Pointon, M.J. (1973) Removal of dissolved boron and silicon during estuarine mixing of sea and river waters. *Geochim. Cosmochim. Acta*, **37**, 1493–1498.

Macklin, J.E. and Aller, R.C. (1984) Dissolved Al in sediments and waters of the East China Sea: implications for authigenic mineral formation. *Geochim. Cosmochim. Acta*, **48**, 281–298.

Mantoura, R.F.C. and Morris, A.W. (1983) Measurement of chemical distributions and processes, in *Practical Procedures for Estuarine Studies*, A.W. Morris (ed.), pp.101–138. Plymouth: IMER.

Mantoura, R.F.C. and Woodward, E.M.S. (1983) Conservative behaviour of riverine dissolved organic carbon in the Severn Estuary: chemical and geochemical implications. *Geochim. Cosmochim. Acta*, **47**, 1293–1309.

Martin, J.-M. and Meybeck, M. (1979) Elemental mass balance of material carried by major world rivers. *Mar. Chem.*, **7**, 173–206.

Martin, J.-M. and Whitfield, M. (1983) The significance of the river input of chemical elements to the ocean, in *Trace Metals in Sea Water*, C.S. Wong, E. Boyle, K.W. Bruland, J.D. Burton and E.D. Goldberg (eds), pp.265–296. New York: Plenum.

Meade, R.H. (1972) Transport and deposition of sediments in estuaries. *Geol. Soc. Am. Mem.*, **133**, 91–120.

Menzel, D.W. (1974) Primary productivity, dissolved and particulate organic matter, and the sites of oxidation of organic matter, in *The Sea*, E.D. Goldberg (ed.), Vol. 5, pp.659–678. New York: John Wiley & Sons, Inc.

Meybeck, M. (1981) Pathways of major elements from land to ocean through rivers, in *River Inputs to Ocean Systems*, J.-M. Martin, J.D. Burton and D. Eisma (eds), pp.18–30. Paris: UNEP/UNESCO.

Meybeck, M. (1982) Carbon, nitrogen and phosphorus transport by world rivers. *Am. J. Sci.*, **282**, 401–450.

Meybeck, M. (1988) How to establish and use world budgets of riverine material, in *Physical and Chemical Weathering in Geochemical Cycles*, A. Lerman and M. Meybeck (eds), pp.247–272. Dordecht: Kluwer Academic Publishers.

Meybeck, M. (2004) Global occurrence of major elements in rivers. Surface and Ground Water, Weathering and Soils (ed. J.J. Drever) Vol. 5 *Treatise on Geochemistry*, H.D. Holland and K.K. Turekian (eds), pp.207–223. Oxford: Elsevier-Pergamon.

Milliman, J.D. and Boyle, E.A. (1975) Biological uptake of dissolved silica in the Amazon River estuary. *Science*, **189**, 995–957.

Milliman, J.D., Summerhayes, C.P. and Barreto, H.T. (1975) Oceanography and suspended matter of the

Amazon River, February–March 1973. *J. Sediment. Petrol.*, **45**, 189–206.

Moore, R.M. and Burton, J.D. (1978) Dissolved copper in the Zaire estuary. *Neth. J. Sea Res.*, **12**, 355–357.

Morris, A.W. (1986) Removal of trace metals in the very low salinity region of the Tamar Estuary, England. *Sci. Total Environ.*, **49**, 297–304.

Neal, C., Smith, C.J., Jeffery, H.A., Jarvie, H.P. and Robson, A.J. (1996) Trace element concentrations in the major rivers entering the Humber estuary, NE England. *J. Hydrology*, **182**, 37–64.

Orians, K.J. and Bruland, K.W. (1985) Dissolved aluminium in the central North Pacific. *Nature*, **316**, 427–429.

Perakis, S.S. and Hedin, L.O. (2002) Nitrogen loss from unpolluted South American forests mainly via dissolved organic compounds. *Nature*, **415**, 416–419.

Perdue, E.M. and Ritchie, J.D. (2004) Dissolved organic matter in freshwater, in Surface and Ground Water, Weathering and Soils, J.J. Drever (ed.) Vol. 5 *Treatise on Geochemistry*, H.D. Holland and K.K. Turekian (eds), pp.273–318. Oxford: Elsevier-Pergamon.

Pickard, G.L. and Emery, W.J. (1982) *Descriptive Physical Oceanography*. Oxford: Pergamon Press.

Potsma, H. (1967) Sediment transport and sedimentation in the estuarine environment, in *Estuaries*, G.H. Lauff (ed.), 158–159. American Association for the Advancement of Science, Publication No. 83.

Potsma, H. (1980) Sediment transport and sedimentation, in *Chemistry and Biochemistry of Estuaries*, E. Olausson and I. Cato (eds), 153–186. New York: John Wiley & Sons, Inc.

Prastka, K., Sanders, R. and Jickells, T. (1998) Has the role of estuaries as sources or sinks of dissolved inorganic phosphorus changed over time? Results of a Kd study. *Mar. Pollut. Bull.*, **36**, 718–728.

Raiswell, R., Tranter, M., Benning, L.G., Siegert, M., De'ath, R., Huybrechts, P. and Rayne, T. (2006) Contributions from glacially derived sediment to the global iron(oxyhyr)oxide cycle: implications for iron delivery to the oceans. *Geochim. Cosmochim. Acta*, **70**, 2765–2780.

Reuther, J.H. (1981) Chemical interactions involving the biosphere and fluxes of organic material in estuaries, in *River Inputs to Ocean Systems*, J.-M. Martin, J.D. Burton and D. Eisma (eds), 239–242. Paris: UNEP/UNESCO.

Rudnick, R.L. and Gao, S. (2004) Composition of the continental crust, in The Crust, R.L. Rudnick (ed.) Vol. 3 *Treatise on Geochemistry*, H.D. Holland and K.K. Turekian (eds), pp.1–64 Oxford: Elsevier-Pergamon.

Salomons, W. (1980) Adsorption processes and hydrodynamic conditions in estuaries. *Environ. Technol. Lett.*, **1**, 356–365.

Salomons, W. and Forstner, U. (1984) *Metals in the Hydrosphere*. Berlin: Springer-Verlag.

Schultz, D.J. and Calder, J.A. (1976) Organic $^{13}C/^{12}C$ variations in estuarine sediments. *Geochim. Cosmochim. Acta*, **40**, 381–385.

Seitzinger, S.B., Styles, R.V., Boyer, E.W., Alexander, R.B., Billen, G., Howarth, R.W., Mayer, B. and Breemen, N.V. (2002) Nitrogen retention in rivers: model development and application of watersheds in the northeastern USA. *Biogeochemistry*, **57/58**, 199–237.

Shiller, A.M. and Boyle, E.A. (1991) Trace elements in the Mississippi outflow region: behaviour at high discharge. *Geochim. Cosmochim. Acta*, **55**, 3241–3251.

Schink, D. (1981) Behaviour of chemical species during estuarine mixing, in *River Inputs to Ocean Systems*, J.-M. Martin, J.D. Butron and D. Eisma (eds), pp.101–102. Paris: UNEP/UNESO.

Sholkovitz, E.R. (1978) The flocculation of dissolved Fe, Mn, Al, Cu, Ni, Co and Cd during estuarine mixing. *Earth Planet. Sci. Lett.*, **41**, 77–86.

Sholkovitz, E.R. and Copland, D. (1981) The coagulation, solubility and adsorption properties of Fe, Mn, Cu, Ni, Cd, Co and humic acids in river water. *Geochim. Cosmochim. Acta*, **45**, 181–189.

Sholkovitz, E.R., Boyle, E.A. and Price, N.B. (1978) The removal of dissolved humic acids and iron during estuarine mixing. *Earth Planet. Sci. Lett.*, **40**, 130–136.

Sieburth, J.M. and Jensen, A. (1968) Studies on algal substances in the sea. I. Gelbstoff (humic materials) in terrestrial and marine waters. *J. Exp. Mar. Biol. Ecol.*, **2**, 174–180.

Spencer, R.G.M., Ahad, J.M.E., Baker, A., Cowie, G.L., Ganeshram, R., Upstill-Goddard, R.C. and Uher, G. (2007) The estuarine mixing behaviour of peatland derived organic carbon and its relationship to chromophoric dissolved organic matter in tow North Sea estuaries (UK). *Est. Coastal Shelf Sci.*, **74**, 131–144.

Stallard, R.F. and Edmond, J.M. (1983) Geochemistry of the Amazon 2. The influence of geology and weathering environment on the dissolved load. *J. Geophys. Res.*, **88**, 9671–9688.

Sundby, B., Silverburg, N. and Chesselet, R. (1981) Pathways of manganese in an open estuarine system. *Geochim. Cosmochim. Acta*, **45**, 293–307.

Trefry, J.H. and Presely, B.J. (1976) Heavy metal transport from the Mississippi River to the Gulf of Mexico, in *Marine Pollutant Transport*, Windom, H.L. and Duce, R.A. (eds), pp.39–76. Lexington, MA: Lexington Books.

Turner, R.E. and Rabalais, N.N. (1994) Coastal eutrophication near the Mississippi river delta. *Nature*, **368**, 619–621.

Ure, A.M. and Davidson, C.M. (2002) *Chemical Speciation in the Environment*, 2nd edn. Oxford: Blackwell Publishing Ltd.

Van Bennekon, A.J. and Jager, J.E. (1978) Dissolved aluminium in the Zaire River plume. *Neth. J. Sea Res.*, **12**, 358–367.

Van Bennekon, A.J., Berger, G.W., Helder, W. and De Vries, R.T.P. (1978) Nutrient distribution in the Zaire Estuary and river plume. *Neth. J. Sea Res.*, **12**, 296–323.

Van der Weijden, C.H., Arnoldus, M.J.H.L. and Meurs, C.J. (1977) Desorption of metals from suspended material in the Rhine estuary. *Neth. J. Sea Res.*, **11**, 130–145.

Van Dijk, G.M., Stålnacke, P., Grimvall, A., Tonderski, A., Sundblad, K. and Schäfer, K. (1996) Long-term trends in nitrogen and phosphorus concentrations in the lower river Rhine. *Arch. Hydrobiol. Suppl.*, **113**, 99–109.

Walling, D.E. and Webb, B.W. (1987) Material transport by the world's rivers: evolving perspectives, in *Water for the Future: Hydrology in Perspective*, pp.313–329. Wallingford: International Association of Hydrological Sciences, Publication no. 164.

Waslenchuk, D.C. and Windom, H.L. (1978) Factors controlling the estuarine chemistry of arsenic. *Estuarine Coastal Shelf Sci.*, **7**, 455–462.

Wiegner, T.N., Seitzinger, S.P., Glibert, P.M. and Bronk, D.A. (2006) Bioavailability of dissolved organic nitrogen and carbon from nine rivers in the eastern United States. *Aq. Microbial Ecol.*, **43**, 277–287.

Williams, P.J. (1981) Primary productivity and heterotrophic activity in estuaries, in *River Inputs to Ocean Systems*, J.-M. Martin, J.D. Burton and D. Eisma (eds), pp.243–249. Paris: UNEP/UNESCO.

Windom, H.L., Wallace, G., Smith, R., Dudek, N., Maeda, M., Dulmage, R. and Storti, F. (1983) Behaviour of copper in southeastern United States estuaries. *Mar. Chem.*, **12**, 183–194.

# 4     The transport of material to the oceans: the atmospheric pathway

Material is transported within the atmosphere to the oceans as particles and gases. In this chapter, the focus is primarily on aerosols while the air-sea exchange of long lived atmospheric gases is considered in Chapter 8. The trophosphere is a reservoir in which particles have a relatively short residence time, usually in the order of days for those with diameters in the range ~0.1 to ~10 μm, before removal from the atmosphere by wet or dry deposition processes which are considered in more detail in Chapter 6. Despite these short residence times, rapid transport rates associated with the global wind systems still allow such particles to be transported hundreds of kilometres every day. The particles carried in the marine atmosphere have a different history from those transported to the oceans via river run-off, one of the most important differences being that they do not undergo trapping, or modification, in the estuarine *filter* at the land-sea margins. At the present time, estuaries act as an effective trap for river-transported solids, holding back ~90% of river particulate material (RPM) (see Section 6.1.2). Potentially, therefore, the atmosphere is perhaps the most important pathway for the long-range transport of particulate material directly to open-ocean areas. This has become increasingly apparent over the past few decades. For example, Delany *et al*. (1967) concluded that the land-derived material in equatorial North Atlantic deep-sea sediments deposited to the east of and on the Mid-Atlantic Ridge has been derived wholly from wind transport. Such long-range transport, spanning several thousand kilometres, has also been found over the Pacific Ocean, and Blank *et al*. (1985) estimated that almost all of the non-biogenic material in deep-sea sediments in the central North Pacific is essentially aeolian in origin. In addition to supplying material to sediments, atmospheric deposition can have a pronounced effect on the biogeochemistry of the oceanic mixed layer by supplying nutrients and trace components. Atmospheric material, however, has to be transferred across an interface before it is introduced into the ocean system; this is the air-sea interface (or microlayer), which is involved in the exchange of particulate and gaseous phases between seawater and the atmosphere. The atmospheric transport of material also directly affects climate via, for instance, the reflection and absorption of solar radiation (IPCC, 2007).

## 4.1 Material transported via the atmosphere: the marine aerosol

### 4.1.1 Introduction

A suspension of solid and liquid material in a gaseous medium usually is referred to as an aerosol. Prospero *et al*. (1983) defined a number of aerosol types on the basis of their compositions and sources, and a summary of their classification is given in Table 4.1. It is important to stress at this stage that the components making up the world aerosol originate from two different types of processes: (i) the direct formation of particles (e.g. during crustal weathering, sea-salt generation, particulate volcanic emissions); and (ii) the indirect formation of particles in the atmosphere itself by chemical reactions and by the condensation of gases and vapours. A generalized relationship between the processes responsible for the generation of aerosol particles and their size spectra is illustrated

*Marine Geochemistry*, Third Edition. Roy Chester and Tim Jickells.
© 2012 by Roy Chester and Tim Jickells. Published 2012 by Blackwell Publishing Ltd.

in Fig. 4.1. In this figure, the particles are divided into two broad groups, fine particles (diameter, $d < 2\,\mu m$) and coarse particles ($d > 2\,\mu m$), and there are three size maxima, two in the fine class and one in the coarse class. The maxima in the *fine* mode relate to two particle populations: (i) those in the

Aitken nuclei range; and (ii) those in the accumulation range. Aitken nuclei originate predominantly from combustion processes via gas to particle conversions within the atmosphere, which include anthropogenic, volcanic emissions and biomass burning, and also from some non-combustion processes. Particles in the accumulation mode are thought to result primarily from the coagulation of Aitken nuclei into larger aggregates. In contrast, particles in the *coarse* mode have been formed by mechanical action, which can involve both low-temperature processes (e.g. the generation of crustal dusts and sea-salts) and high-temperature processes (e.g. the formation of some industrial fly-ash).

The particle size of the marine aerosol has been described by Junge (1972), who identified five classes of size-related components in a North Atlantic aerosol. These are (i) particles with diameters $>40\,\mu m$, (ii) sea-spray particles, (iii) mineral dust particles, (iv)

**Table 4.1** The classification of aerosols on the basis of their composition or sources (based on Prospero *et al.*, 1983).

| Natural aerosols | Anthropogenic aerosols |
|---|---|
| Sea-spray residues | Direct anthropogenic particle |
| Windblown mineral dust | emissions |
| Volcanic effluvia | Products from the conversion |
| Biogenic materials | of anthropogenic gases |
| Smoke from the burning of | |
| land biota | |
| Natural gas-to-particle | |
| conversion products | |

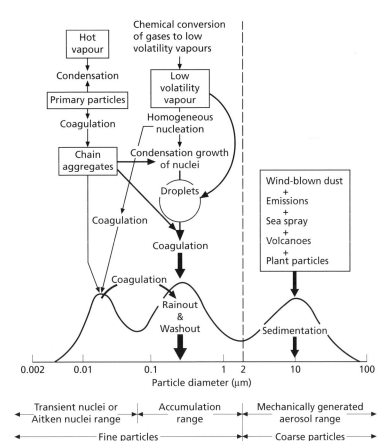

**Fig. 4.1** Schematic representation of the processes involved in the generation and removal of atmospheric particles (from Whitby, 1977). Reprinted with permission from Elsevier.

tropospheric background particles and (v) particles with diameters <0.06 μm. The nature of the tropospheric background material has been the subject of much speculation, but it is generally thought that an aerosol of fairly uniform composition, with an inorganic component dominated by sulfate, is present in over ~85% of the troposphere. If the very small particles ($d < 0.06$ μm) are excluded, the marine aerosol therefore can be regarded as being composed of the tropospheric background material, upon which is superimposed continental dust (the mineral aerosol) and seaspray (the sea-salt aerosol).

Close to the source, the composition of a specific aerosol component, such as mineral dust, will be related closely to that of the parent material. During their residence time in the atmosphere, however, both coarse and fine particles can undergo physical and chemical modifications, and the character of an aerosol will change with increasing distance from the source. For example, the concentration falls off, the particle size distribution is altered and the chemical composition is modified and tends to become more uniform. One result of aerosol transport is that the larger particles are removed more rapidly than small particles, so that aerosols over coastal regions will differ from those over distant open-ocean areas, which represent an integration of material from many sources (see e.g. Maring and Duce, 1990). The residence times of aerosols varies from hours or less for really large particles to many tens of days for the fine mode aerosols (Raes *et al.*, 2000). Such residence times are much shorter than air mass mixing times within a hemisphere, or the even longer mixing times between hemispheres and hence there are strong gradients in aerosol concentrations in the Earth's atmosphere.

Aerosols therefore may be regarded as the end-product of a complex series of processes and are not static components, but are best regarded in terms of a dynamic aerosol continuum (Prospero *et al.*, 1983). The physical and chemical composition of the marine aerosol therefore is extremely variable in both space and time, and the aerosol characteristics are governed by a combination of processes that are involved in the generation, conversion, transport and removal of particles. In the following sections the marine aerosol will therefore be discussed in terms of this general 'generation–conversion–transport–removal' framework.

## 4.1.2 The transport of aerosols within the troposphere

Most of the continentally derived material (both natural and anthropogenic) that contributes to the world aerosol is injected initially into the planetary boundary layer of the atmosphere. This is the layer in which the direct influence of the underlying surface is felt, and it has a height of ~1000 m to ~1500 m over the land and ~300 m to ~600 m over the sea (Hasse, 1983). The upper surface of the boundary layer is defined by an inversion, which inhibits the transfer of material to the upper atmosphere (Prospero, 1981). According to Prospero (1981), the primary transport path by which material generated close to the continents, (i.e. within tens to hundreds of kilometres) reaches the sea surface may be via this marine boundary layer. While most emission and transport over the oceans takes place within the marine boundary layer, large scale dust storms and volcanic emissions can inject material high into the atmosphere above the boundary layer to heights of several kilometres where transport rates are faster and removal rates slower (e.g. Jickells *et al.*, 2005). It must also be remembered that, because most collections of the marine aerosol are made in the boundary layer, they are probably not representative of the concentrations at high levels in the troposphere.

The transport of aerosols within the troposphere takes place via the major wind systems, which operate on a global scale. For simplicity it is convenient to describe atmospheric circulation in terms of a meridional three-cell model. This of course only represents long-term average circulations (climate) while shorter term weather produces considerable day to day variability around this average transport. In this model the circulation in each hemisphere can be related to the presence of three cells of alternating belts of easterly and westerly zonal winds, the cells being separated by pressure zones. This is illustrated in Fig. 4.2(a) with respect to the Northern Hemisphere. The three pressure zones are:

**1** the equatorial low-pressure belt or trough, originally termed the 'Doldrums' but now referred to as the Inter-Tropical Convergence Zone (ITCZ);
**2** the subtropical high-pressure belt (~30°N): an area of dry sinking air characterized by calms and often called the 'Horse Latitudes';
**3** the low-pressure belt at ~60°N.

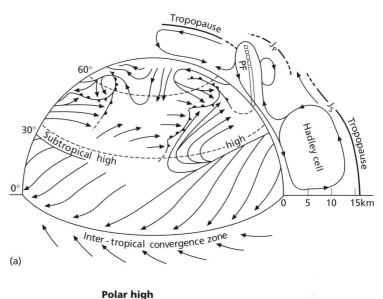

(a)

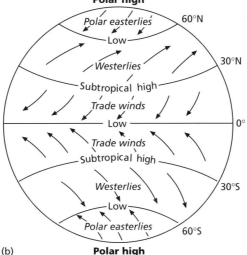

(b)

**Fig. 4.2** Circulation in the atmosphere. (a) Schematic representation of the general circulation in the atmosphere at the surface and a meridional cross-section (from Hasse, 1983; after Defant and Defant, 1958). $J_P$ and $J_S$ indicate the average positions of the Polar Front (PF) and the Subtropical Jet Streams. (b) Schematic representation of the surface wind distribution (from Iribarne and Cho, 1980). With kind permission from Springer Science + business Media B.V.

The main features of the circulation lying between these zones are illustrated in Fig. 4.2(b) and are described in very general terms in the following.

**1** In the Hadley cell operating between the equatorial low-pressure and subtropical high-pressure belts, the prevailing zonal winds are north-easterly or south-easterly, and are termed the trades: the Northeast Trades in the Northern Hemisphere and the Southeast Trades in the Southern Hemisphere. These trades are directionally quite stable and have relatively constant velocities, in the range ~16 to ~32 km h$^{-1}$. The trades converge at the ITCZ. In some regions (e.g. India, Southeast Asia and China) the trade-wind pattern can be modified by the development of a monsoon system. Here there is seasonal reversal of wind direction in which transequatorial westerlies replace the regular trade-wind easterlies. These monsoons can result in the transfer of air across the Equator and so can initiate the interhemispheric exchange of aerosol components, which is otherwise rather limited.

**2** The circulation in the temperate or mid-latitude Ferrel cell, which lies between the subtropical high-pressure belt at ~30°N and the low-pressure belt at ~60°N, has a prevailing zonal wind that is westerly in direction. The westerlies, however, are not as directionally stable as the trades and exhibit considerable variability in both space and time. Further, they have a greater force than the trades, and are sometimes referred to as the 'upper westerlies' because their force increases with altitude, to reach a maximum at ~12 km. Cores of high-speed winds are embedded in the westerlies, the two main types being the Subtropical Jet Stream and the Polar Front Jet Stream, both of which are important for the transport of aerosols in the mid-latitudes.

**3** A third or polar cell is found to the north of the low-pressure belt located at ~60°N. Here, the winds are the polar easterlies, although the atmospheric circulation system is extremely complex.

Aerosols are dispersed about the surface of the planet by these various zonal wind systems and by transfer between them. Such transfer, however, tends to be inhibited by the various pressure belts. Despite this, the inter-hemispheric transfer of aerosols can take place; for example, when the ITCZ is shifted across the Equator, or when it is absent during the development of some monsoon systems.

### 4.1.3 Sources of material to the marine atmosphere

The particulate and gaseous material supplied to the atmosphere can originate from either natural or anthropogenic sources, and when considering the elemental chemistry of both of these it is useful to distinguish between *low*-temperature and *high*-temperature generation processes (see Section 4.1.1 and Fig. 4.1). This is important because the forms in which the elements are present in aerosols, that is. their speciation, can be strongly dependent on the temperature at which they were released from their parent hosts (see Section 4.2.1.3) in part because of the different size distribution of aerosols formed from primary particulate sources (generally larger) and gas-to-particle conversions (finer) as discussed earlier. Some of the more important sources of material to the world atmosphere are listed below.

#### 4.1.3.1 Natural sources

*The Earth's crust.* The Earth's surface can supply both particulate and gaseous components to the atmosphere. The generation of particulate crustal material involves a low-temperature mechanical mobilization of surface deposits by wind erosion, and these products are an important component of many marine aerosols.

*The oceans.* Like the crust, the surface of the ocean can supply material to the atmosphere in both particulate and gaseous forms. Particulate material is formed during the low-temperature generation of sea-salts by mechanical action. Volatilization from the sea surface, in the form of a sea-to-air gaseous flux, also can be a source of some volatile components (see Section 4.1.4.3 and Chapter 8).

*Volcanic activity.* Volcanoes can release particulate material, for example ash, together with gaseous phases formed from high-temperature volatilization processes. These gaseous phases may undergo condensation reactions, which can result in the enrichment of the particulate material in some volatile elements.

*The biosphere.* The supply of material to the atmosphere from the biosphere occurs through the high-temperature burning of vegetation (phytomass), by the entrainment into the atmosphere of small biological particles and organisms (Jaenicke *et al.*, 2007) and by the emission of particulate and vapour phases from plant surfaces and soils. Phytomass burning includes natural forest and grassland fires and biomass burning for agriculture. Phytomass burning is probably more significant at low latitudes, and it has been estimated that >80% of the annual phytomass burn occurs in the 'developing' countries (see e.g. Suman, 1986). Over the time scale of the last 2000 years, the amount of biomass burning, as recorded by charcoal in sediments, has been closely linked to climate variability. However, over the last few hundred years this linkage has been broken by human intervention which initially increased biomass burning in the period from 1750 to the early twentieth century associated with increasing land use, followed by a decrease as land use change decreased and active management of fires became important until the latter part of the twentieth century when growing populations have driven further land clearance (Marlon *et al.*, 2008). Material released during phytomass burning includes $CO_2$, $CH_4$ and particulates (Marlon *et al.*, 2008); an important component of the latter is charcoal (elemental carbon): see for example Cachier *et al.* (1989) and Buat-Menard *et al.* (1989).

*Outer space.* Extraterrestrial sources provide a small, but interesting, supply of material to the atmosphere. This extraterrestrial material includes micrometeorites (cosmic spherules) and a number of cosmic-ray-produced radioactive or stable nuclides (see Section 15.4).

### 4.1.3.2 Anthropogenic sources

There are a wide variety of anthropogenic processes that release both particulate and gaseous material into the atmosphere. These include fossil fuel burning, mining and the processing of ores, waste incineration, the production of chemicals, agricultural utilization, and numerous other industrial and social activities.

### 4.1.3.3 Source strengths

Particle emission source strengths of some of the processes listed previously are given in Table 4.2. It should be noted that these are global estimates and have very substantial uncertainties. Furthermore these emissions change with time. Emissions of several components will have been very much lower in the pre-industrial era and more recently, emission controls have decreased the emissions of some components. Overall, natural particle production processes dominate, with sea-salt and dust production of similar magnitude. However, in the case of fine mode aerosol formed from gas to particle conversions, anthropogenic emissions, particularly of sulfate and nitrate precursors, now rival natural emissions.

The distributions of these components are considered in the following sections. Although sulfate, sea-salt and mineral dust can dominate the marine aerosol, they can sometimes however, be present

**Table 4.2** Global emissions of major aerosol components after Seinfeld and Pandis (1998).

| Source | Strength ($10^{12}$ g yr$^{-1}$) |
| --- | --- |
| Sea-salt | 1300 |
| Dust | 1700* |
| Carbon | 130[#] |
| Sulfate | 340[$] |
| Nitrate | 80[£] |
| Total particles | 3600[!] |

*~90% from deserts, 2% volcanoes, remainder industrial.
#~8% soot, ~40% biological debris, remainder from oxidation of volatile organic carbon (VOC), mostly natural terpenes (~85% of total VOC) rather than anthropogenic (~15%).
$~45% from natural precursor emissions.
£~60% anthropogenic.
!(85% natural).

in intimate mixtures within individual aerosol particles rather than as individual components (see e.g. Andreae *et al.*, 1986).

### 4.1.4 The principal components of the marine aerosol

The largest global sources of marine aerosols are the sea surface (sea-salt), the Earth's crust (mineral dust) and anthropogenic processes (mainly sulfates). Sea-salts are emitted from all of the surface ocean, although the main emissions are from the windiest regions, particularly the southern Ocean (Chin *et al.*, 2009), but the mineral dust and sulfate sources are concentrated in specific belts. The principal sources of mineral dust are the large deserts, and anthropogenic aerosols (essentially sulfates) originate from the major industrial regions. Both of these sources are located predominantly in the Northern Hemisphere and hence impacts of these land based emissions are seen primarily in the Northern Hemisphere oceans (Fig. 4.3a, e.g. Chin *et al.*, 2009). The Atlantic Ocean is much smaller and narrower than the Pacific and is surrounded by major emission sources particularly the Sahara desert and the industrial regions of North America and Europe. Hence the North Atlantic and adjacent seas such as the Mediterranean receive proportionately more atmospheric deposition than the Pacific overall, although there are strong gradients

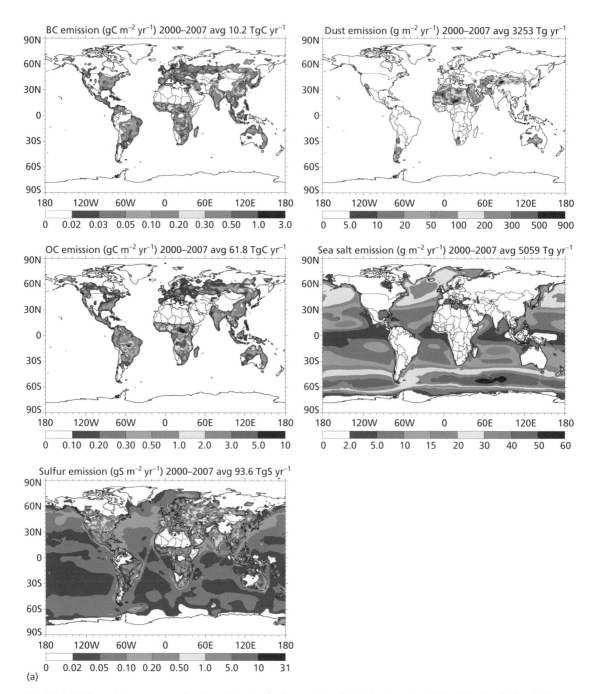

**Fig. 4.3** (a) Major emission source regions for sulfur g S m$^{-2}$ yr$^{-1}$, organic and black (soot) carbon (OC + BC) g C m$^{-2}$ yr$^{-1}$, soil dust g m$^{-2}$ yr$^{-1}$ and seasalt g m$^{-2}$ yr$^{-1}$, for year 1990, based on Chin *et al.* (2009). (b) Global total aerosol distribution based on satellite imagery. (See Colour Plate 1)

Optical depth All, All, Blue, Annual 2006 F15_0031
Summarizes L2 AS_AEROSOL, RegBestEstimateSpectralOptDepth field F12_0022, 0.5 deg res

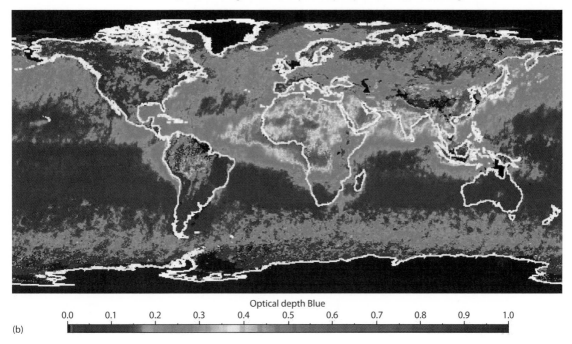

Optical depth Blue

0.0    0.1    0.2    0.3    0.4    0.5    0.6    0.7    0.8    0.9    1.0

(b)

**Fig. 4.3** *Continued*

within both areas. The Northern Indian Ocean receives large inputs from the adjacent desert and populous/industrial regions, but in this region the impact of the monsoon makes these inputs highly seasonal (Tindale and Pease, 1999). The South Atlantic can be significantly impacted by emissions form biomass burning in Africa and South America (Chin *et al.*, 2009), but otherwise the inputs of terrestrial derived atmospheric inputs to Southern Hemisphere Oceans, and particularly the Southern Ocean itself, are low compared to the Northern Hemisphere Oceans. The Arctic Ocean is remote from dust sources regions but strongly impacted by industrial emissions leading to the Arctic haze phenomenon (e.g. Garrett and Verzella, 2008).

The overall global pattern of aerosol is shown in Fig. 4.3(b). Fig. 4.3 is derived from a model which incorporates both ground based observational data and information from satellite observations. The latter in particular have transformed our understanding of global atmospheric transport over recent decades. The figure illustrates the impacts of the relatively localized emissions from industrial regions

and deserts, while the impact of seasalt emissions are more dispersed.

### 4.1.4.1 The mineral aerosol over the World Ocean

There is considerable interest in dust transport and deposition to the oceans for a number of reasons. Mineral aerosol is a significant component of the global aerosol with resultant impacts on global climate via its reflectance of solar radiation. The atmospheric dust deposited in deep sea sediments (and ice cores) represents an important record of climate change. It is also now clear that atmospheric dust represents an important source of nutrients, particularly iron, to surface ocean waters (see Chapters 9 and 11).

High-temperature volcanic activity can generate particles that form part of the mineral aerosol. This volcanic activity is a sporadic source, but at the time of large-scale eruptions, very large quantities of material can be injected into the atmosphere and can affect world climate. Apart from these periods of large eruptions, low temperature sources of dust from arid regions dominate the input of mineral aerosol to the atmosphere.

Particulate material in the form of continental dust is mobilized into the atmosphere mainly by wind erosion. The process is strongly dependent on the nature of the surface cover in the source (or catchment) area, which itself is dependent on the prevailing geological, weathering and general climatic regimes. In regions having loose surface deposits, for example desert and arid land areas, there is a readily available reservoir of particulate material that is susceptible to wind erosion and transport during dust storm events. In contrast to the conditions found in arid regions, surface covers of forest, grassland and snow or ice will considerably reduce the production rates of atmospheric dusts. Most mineral dust present in the atmosphere is produced from surface soils. Dried out lake beds appear to be a particularly important source of mineral dust within desert regions (Prospero *et al.*, 2002; Jickells *et al.*, 2005). Dust storms in these areas are highly episodic, and the resultant inputs of dust to the oceans can vary dramatically from day to day. Long term measurements at Barbados, thousands of kilometres downwind of the Sahara, have demonstrated the seasonality of transport and also the interannual variation which

is closely tied to aridity within the Sahara (Prospero and Lamb, 2003). On still longer time scales, it is clear that dust fluxes to the oceans have varied with long-term changes in climate, particularly increasing during glacial periods when the climate globally was drier and winds were stronger (Rea, 1994; Jickells *et al.*, 2005). Thus, the dust cycle is tied to climate and at the same time can influence climate itself directly via reflecting sunlight, and indirectly by stimulating ocean productivity and hence uptake of $CO_2$ (Jickells *et al.*, 2005; also see Chapters 8 and 11).

Mahowald *et al.* (2005) estimated the magnitude of the inputs to the global tropospheric dust cycle and a summary of the data is given in Table 4.3. Modelled estimates of global dust deposition are shown in Fig. 4.4.

**Table 4.3** The global tropospheric dust cycle. Dust deposition to major ocean basins (Mahowald *et al.*, 2005).

| Ocean Basin | Dust deposition $10^{12}\,g\,yr^{-1}$ |
|---|---|
| North Pacific | 72 |
| South Pacific | 29 |
| North Atlantic | 202 |
| South Atlantic | 17 |
| Indian Ocean | 118 |

Average dust deposition (g m² per yr)

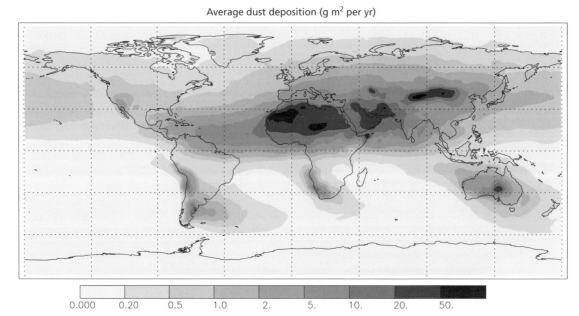

Fig. 4.4 The global deposition of mineral aerosol (based on Jickells *et al.*, 2005; and provided by Natalie Mahowald).

It is clear from this table and figure that the Northern Hemisphere has a higher tropospheric dust burden than the Southern Hemisphere, mainly as a result of the larger area of land mass in the northern latitudes. Also the Sahara Desert plays an extremely important role in the Northern Hemisphere dust cycle. The net result is that the tropical North Atlantic region receives a particularly high dust loading while the Southern Ocean receives a dust loading that is in the order of 100 times or more lower, with a dust envelope over the North Atlantic under laying the Trade Winds.

The major SEAREX programme in the Pacific Ocean used long term land based records of atmospheric sampling to reveal much about long range atmospheric transport, including the importance of Asian dust sources in the period February to June. The results from this study, illustrated in Fig. 4.5, show clearly the way concentrations decrease away from this source throughout the North Pacific, although the seasonality of dust concentrations is similar at all sites with concentration maxima throughout during the February to June period associated with Asian dust seasons (Prospero *et al.*, 1989). Concentrations were lower for stations in the tropical South Pacific but increased somewhat further south at Norfolk Island under the influence of emissions from Australian deserts.

The dominant feature in the distribution of mineral dust over the Pacific Ocean is the seasonal input of dust 'pulses' from Asian sources to the North Pacific. The effect of these dust 'pulses' can be seen in concentration versus frequency diagrams for the mineral aerosol collected at a number of the Pacific network stations which illustrate the size distribution of the aerosol measured at these stations. A set of these diagrams is illustrated in Fig. 4.5(b). From this figure it can be seen that in the North Pacific the data for Midway and Enewetak exhibit biomodal distributions, the lower concentration mode representing the background mineral aerosol and the upper mode representing perturbations from aerosols associated with the Asian dust storms; the percentages of the samples falling within each mode are shown in Fig. 4.5(b), from which it can be seen that at Enewetak ~75% of the mineral dust lies in the upper mode. In contrast, the data for Fanning Island, close to the Equator in the North Pacific, and American Samoa, in the South Pacific, are essentially unimodal and

may be regarded as being background aerosol. At Norfolk Island, which receives inputs from the Australian mainland, the aerosol belongs mainly to the background mode (85%), but there is evidence of dust events (15% of the total inputs) perturbing this background.

*The Mediterranean.* The Mediterranean Sea represents an interesting transitional environment in terms of atmospheric deposition. It is confined to a narrow latitudinal band, and from the point of view of aerosol inputs it is of special interest because it lies between large source regions for both anthropogenic and mineral dust generation (see Fig. 4.3). As a result the Mediterranean Sea has contrasting aerosol-generation regimes on its opposite shores. On the northern shore it is bordered by nations with economies ranging from highly industrial to agricultural, and air masses transported into the Mediterranean atmosphere from the north have often crossed part of the European 'pollution belt', which provides a continuous supply of anthropogenic 'background' aerosol. In contrast, it is bordered on its southern shore by the North African (Saharan) desert belt, which acts as a large reservoir for the supply of mineral dusts. These dusts are delivered in a seasonal pattern, mainly in the form of sporadic, that is. non-continuous, dust 'pulses', which perturb the anthropogenic 'background' aerosol. Mineral dusts also can be supplied to the east of the region from the deserts of the Middle East. In addition to the European anthropogenic source and the Saharan and Middle Eastern desert sources, there are inputs of particulate material to the Mediterranean atmosphere from the Atlantic Ocean to the west, from the local sea surface and from volcanic activity in the region. Aerosols are transported into the Mediterranean atmosphere from all these sources, and the inputs vary seasonally. For example, there is maximum transport of African dust to central and eastern basins during the spring, and to the western and central basins during the summer: patterns that follow the distribution of Mediterranean cyclones. As a result the concentrations of mineral dust can exhibit wide concentration variations in the Mediterranean atmosphere. For example, Chester *et al.* (1984b) showed that incursions of Saharan dust could be identified in the lower troposphere over the Tyrrhenian Sea. The mineral dust concentrations in

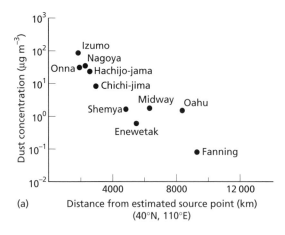

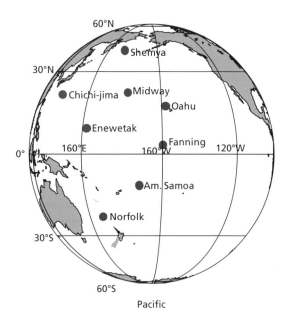

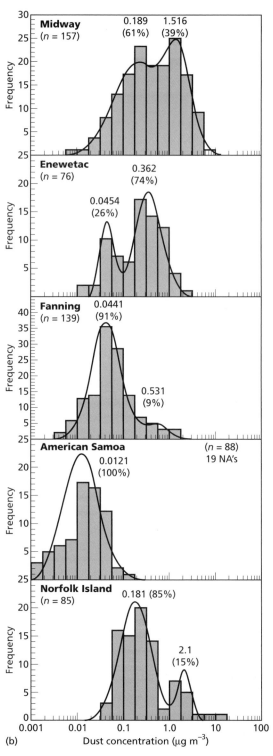

**Fig. 4.5** The distribution of mineral aerosol over the Pacific Ocean (from Prospero *et al.*, 1989). (a) The concentration of mineral aerosol over the North Pacific plotted as a function of distance from the estimated source point in Asia. (b) Frequency distributions of mineral aerosol concentrations over the Pacific Ocean. The mineral aerosols for the Midway and Enewetak Island sites in the North Pacific show a bimodal distribution, the lower mode representing the background aerosol and the upper mode representing dust pulses from

Asian desert sources. In contrast, the aerosol frequency distributions for the Fanning Island site, which is close to the Equator, and the American Samoa site in the South Pacific, are essentially unimodal and represent background aerosol only. The locations of the island sites are shown in insert figure. Note Izumo, Nagoya, Onno and Hachijo-jama are on or close to the coast of Japan and are not marked here. Reprinted with permission from Elsevier.

these pulses (average, $25\,\mu g\,m^{-3}$ of air) were an order of magnitude higher than those in the 'European background' air (average, $1.4\,\mu g\,m^{-3}$ of air). Pulses of this kind are intermittent, however, and make it difficult to identify an average mineral aerosol loading for the Mediterranean.

Evidence is now becoming available that long-term patterns in dust transport can be linked to large-scale variability in the atmosphere. In this context, the study reported by Moulin *et al.* (1997) provides a major landmark in our knowledge of the factors controlling the movement of atmospheric dust. These authors used daily satellite observations of airborne dusts to obtain an 11-year regional-scale analysis of dust transport out of Africa to the Atlantic Ocean and the Mediterranean Sea. They were able to show that seasonal variability in dust transport to both marine regions could be explained by synoptic meteorology. The most important conclusion of the study, however, concerned the variations in dust transport. Despite differences in the source regions and transport processes, these interannual variations were similar over both the Atlantic and the Mediterranean, which suggested that large-scale forces are constraining the dust export to both regions. Precipitation is one of the most efficient processes that can induce interannual variability in the transport of African dust, and Moulin *et al.* (1997) combined changes in precipitation patterns with those in atmospheric circulation to identify the constraints on dust transport out of Africa by linking them to the North Atlantic Oscillation (NAO). The NAO is a standing oscillation, or 'see-saw', between pressure differences in the Azores High and the Iceland Low pressure centres, and produces major changes in the meteorological regime. Winters with a high NAO index induce a northward shift of the North Atlantic westerlies, which provide much of the moisture to northern Africa and Europe, and results in drier conditions over southern Europe, the Mediterranean Sea and northern Africa. During years with a low NAO index, precipitation is likely to be greater over the Mediterranean Sea and parts of northern Africa, thus limiting both the uptake and transport of dust. Moulin *et al.* (1997) found that the NAO index covaries with desert dust transport for both the Mediterranean and the Atlantic, and were thus able to establish that there is a large-scale climatic control on dust export from Africa, which is effected through changes in precipitation and atmospheric circulation over the regions of dust mobilization and transport. It is probable also that the NAO 'see-saw' will affect the transport and distribution of other aerosols, such as anthropogenic sulfates.

### 4.1.4.2 The sea-salt aerosol over the World Ocean

The sea surface provides a vast reservoir for the generation of aerosols, comparable in size to dust emissions (Table 4.2). According to Berg and Winchester (1978), the bursting of bubbles produced by the trapping of air in surface water by breaking waves, or whitecaps, is the chief mechanism for the formation of particulate matter in the marine atmosphere, and the mechanism has been described in detail by Blanchard (1983). Sea spray produced by the direct shearing of droplets from wave crests, however, may play a role in the generation of sea-salts (see e.g. Lai and Shemdin, 1974; Koga, 1981).

Sea-salts have particle sizes ranging from $<0.2\,\mu m$ to $>200\,\mu m$ in diameter, with a distinct maximum in number distribution below a diameter of $2\,\mu m$. More than 90% of the sea-salt aerosol mass, however, is located in giant particles having a median mass diameter (MMD) between $2-20\,\mu m$, and McDonald *et al.* (1982) have shown that large and giant particles dominate the deposition of sea-salts, even though they are present in the air in only relatively low numbers.

The concentration of sea-salt in the marine atmosphere varies with both altitude (as larger aerosol particles deposit before they can be transported to altitude) and windspeed, with sea-salt production increasing as a cubic function of wind speed (Chin *et al.*, 2009) The overall distribution of the sea-salt aerosol therefore differs from that of the mineral aerosol in that the highest sea-salt concentrations occur at high latitudes in areas of relatively strong winds as in Fig. 4.6, whereas the highest mineral aerosol concentrations are found in arid low-latitude areas: see Fig. 4.4. In addition the high-latitude regions of both hemispheres usually have higher atmospheric sea-salt concentrations than do the low latitudes. The strong uniform surface winds in the Southern Hemisphere at high latitudes result in high, and relatively constant, atmospheric sea-salt concentrations in both winter and summer periods.

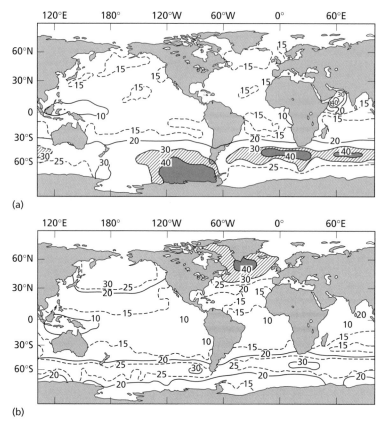

**Fig. 4.6** Sea-salt concentrations over the World Ocean (from Erickson *et al.*, 1986); the isopleths are lines of constant sea-salt concentration in $\mu g \, m^{-3}$ of air. (a) Sea-salt concentrations during the boreal summer (June–August). (b) Sea-salt concentrations during the boreal winter (December–February).

By contrast the high-latitude winds in the Northern Hemisphere vary seasonally, and the atmospheric sea-salt concentrations exhibit a difference of a factor of three between winter and summer periods.

### 4.1.4.3 Sulfate, nitrate and ammonium: the main acid base aerosol components

Sulfates, such as $(NH_4)_2SO_4$, are an important constituent of the tropospheric background aerosol. Sulfates from both natural and pollutant sources can be transported into the marine atmosphere from the continents. In addition, sulfates can be generated from the sea surface, either directly during the formation of sea-salts or indirectly from gas-to-particle conversion reactions. The $SO_4^{2-}:Na^+$ ratio in seawater (see Chapter 3) has been used to estimate the sea-salt-associated sulfate in marine aerosols; thus, for ratios in excess of the seawater value, the non-sea-salt sulfate component is usually referred to by convention as excess sulfate.

This excess sulfate is primarily found in the fine mode aerosol and derives from the oxidation of $SO_2$ gas. This produces sulfuric acid, contributing substantially to the acidity of rainwater. The sulfuric acid exists as an aerosol because of its low vapour pressure, and it can also react with other aerosols such as seasalt or dust. Alternatively the sulfuric acid can react with ammonia gas, the main atmospheric alkali species, to form ammonium sulfate. Ammonia is emitted from waste decomposition on land and to a lesser extent from the oceans with total emissions of about $64 \times 10^{12} \, g \, N \, yr^{-1}$ with more than 80% of it anthropogenic (Duce *et al.*, 2008).

Total emissions of $SO_2$ are estimated at between 90 and $125 \times 10^{12}$ g S yr$^{-1}$ (IPCC, 2007). The main source of this $SO_2$ is now fossil fuel (72%) with an additional small contribution from biomass burning (2%) and 7% from volcanoes (IPCC, 2007). All these sources are based on land. Clearly all will vary from year to year, for instance depending on the intensity of volcanic emissions, from both the low level fumarole emissions and the particularly episodic large explosions. The anthropogenic sulfur emissions also vary, having been close to zero in pre-industrial times, but even more recently there are quite large changes. For instance the IPCC (2007) report suggests a decrease in global S emissions from 10–25% between 1980 and 2000 as a result of changes in emission patterns with strong declines in Europe and North America offset by increases in the Asian region.

The additional source of $SO_2$ is from the oxidation of the gas dimethyl sulfide (DMS) which is emitted by marine phytoplankton. Algae make the compound dimethylsulfoniopropionate (DMSP) within their cells, possibly to provide osmoregulation and/or as antioxidant. The amounts produced vary with physiological condition and between species. This DMSP is released into the surface ocean from the phytoplankton cells, probably by cell lysis, and it is then degraded to DMS, a small part of which then escapes to the atmosphere, the remainder being consumed by bacteria in the water column (Simó, 2001). DMS is oxidized in the atmosphere to sulfate predominantly and a second product methane sulfonic acid (MSA). It has been suggested that the emission of DMS and the resultant formation of aerosol sulfate which can reflect sunlight, can potentially contribute to a climate feedback system where changes in solar radiation could alter phytoplankton productivity and DMS emissions which could then alter the reflectivity of the atmosphere and hence climate – the so called CLAW hypothesis (Charlson et al., 1987). There is some evidence that this DMS is linked to aerosol formation (Ayers et al., 1991), and that the aerosol produced could have a significant effect on climate (Gunson et al., 2006), but there are many uncertainties still in our understanding of the DMS cycle and its link to climate and the CLAW hypothesis currently remains just that, a hypothesis.

There are therefore three major sources of S emissions to the atmosphere that form sulfate aerosol; anthropogenic $SO_2$, sea-salt sulfate and DMS, with additional volcanic sources. Sea-salt sulfates will tend to be in the coarse mode aerosol because they are formed with sea-salt, while the other sources will predominantly form fine mode sulfate aerosol after oxidation. Identifying the relative significance of these sources can be done by using tracers such as sodium for sea-salt, perhaps MSA for DMS and anthropogenic tracers for anthropogenic S. An additional tool for obtaining such source discrimination is the measurement of S isotopes since the isotopic signatures of industrial and DMS sulfur are substantially different (e.g. Wadleigh, 2004).

The other major acidifying emission to the atmosphere is from the emission of the oxides of nitrogen (NO and $NO_2$) produced during high temperature combustion mainly (>70%) with additional sources from soil emissions and lightening. Total emissions are estimated to be $52 \times 10^{12}$ g N yr$^{-1}$. These oxides are further oxidized to nitric acid which can react reversibly with ammonia to form ammonium nitrate, or with sodium chloride to form sodium nitrate. The latter reaction dominates in the marine environment and this means that although the nitrate starts out as a gas, it is found mainly in the coarse mode aerosol with the seasalt, unlike most other components formed by gas to particle conversions. The deposition of nitrate and ammonium represents a significant nitrogen source to the oceans (Duce et al., 2008) as we shall see in Chapter 6.

#### 4.1.4.4 Organic carbon in the marine aerosol

The atmosphere is a primary pathway for the transport of some types of organic substances to the oceans. The organic components in the atmosphere may be divided into particulate organic matter (POM), for which the organic carbon fraction is termed POC, and vapour-phase organic material (VOM), for which the organic carbon fraction is referred to as VOC. The principal sources for both POM and VOM in the atmosphere are vegetation, soils, the marine and freshwater biomass, biomass burning and a variety of anthropogenic activities; to these must be added the in situ production of some organic substances within the atmosphere itself. The sinks for all these organics can be related to the mechanisms that remove them from the atmosphere. For VOM, four removal mechanisms are important:

conversion to POC, dry deposition, wet deposition and transformation to inorganic gaseous products. The removal of POM is dominated by wet and dry deposition. In addition, the oxidation of POM to inorganic gaseous products, for example, $CO_2$, can result in its loss from the atmosphere.

The distributions and types of organic substances in the atmosphere have been described by Duce *et al.* (1983b) and Peltzer and Gagosian (1989).

*Particulate organic matter (POM).* Particulate organic aerosols can be produced in two ways, either directly as particles or as a result of gas-to-particle conversions, and the resultant POM can be divided into non-viable and viable species.

The POC in the marine atmosphere is a complex mixture derived from a large variety of organic compounds, but most data for non-viable POM was confined originally to POC. From the study reported by Buat-Menard *et al.* (1989), concentrations of the carbonaceous aerosol in the marine atmosphere appear to vary over the range ~0.05 to ~1.20 $\mu g\,C\,m^{-3}$ of air, with values being significantly higher in the Northern (~0.4 to ~1.2 $\mu g\,C\,m^{-3}$ of air) than in the Southern Hemisphere (~0.05 to ~0.3 $\mu g\,C\,m^{-3}$ of air), suggesting that the Northern Hemisphere POC is dominated by continental sources. Buat-Menard *et al.* (1989) related the sources of POC in the marine atmosphere to particle size and reported that in the Northern Hemisphere marine-derived carbon (i.e. that on large particles, $d \geq 3\,\mu m$ associated with sea-spray) had a mean concentration of 0.07 $\mu g\,C\,m^{-3}$ of air, whereas the continentally derived POC (i.e. that on small particles formed mainly from gas to particle conversions, $d \leq 1\,\mu m$) had a mean concentration of 0.45 $\mu g\,C\,m^{-3}$ of air; for the Southern Hemisphere the concentrations were 0.07 and 0.06 $\mu g\,C\,m^{-3}$ of air for the marine-derived and continentally derived POC, respectively. The carbon isotopic signature of the atmospheric POC also suggests a continental source from combustion processes particularly over the Northern Hemisphere. In contrast, the atmospheric POC in the Southern Hemisphere is dominated by natural continental emissions. Recently O'Dowd *et al.*, (2004) have suggested that marine derived organic matter may make a more important contribution to the aerosol during periods of enhanced water column productivity.

Clearly, therefore, two features are apparent in the distribution of POC in the marine atmosphere: (i) concentrations are higher in the Northern than in the Southern Hemisphere; and (ii) small-sized continentally derived material having an anthropogenic source dominates the atmospheric POC in the Northern Hemisphere, whereas marine-derived and natural continentally derived material contribute roughly the same amounts to the POC in the Southern Hemisphere.

In Table 4.2 the total emission of organic carbon to the atmosphere is estimated at $130 \times 10^{12}\,g$ C $yr^{-1}$, resulting in atmospheric carbon inputs through the atmosphere to the oceans that are comparable in size to riverine inputs, but very small compared to internal carbon production by photosynthesis within the oceans. Of the atmospheric carbon input, ~85% is natural, divided almost equally between primary biogenic particles and secondary material formed by the oxidation of volatile compounds released by vegetation such as isoprene ($C_5H_8$). Soot or black carbon and aerosol formed from oxidation of anthropogenic volatile organic compounds each contribute about $10 \times 10^{12}\,g\,C\,yr^{-1}$, predominantly to the fine mode aerosol (Seinfeld and Pandis, 1998). In addition to providing a modest input of organic carbon, the organic aerosol component also provides a significant source of organic nitrogen to the oceans, although the source of this material is still uncertain (Duce *et al.*, 2008)

In addition to total particulate organic carbon, various authors have provided data on the distributions of individual organic compounds in the marine atmosphere. The numbers of compounds are of course very large and therefore, different researchers have considered different groups of compounds reflecting their specific research priorities.

Much of the earlier work concentrated on dusts collected from the North Atlantic Northeast Trades (see e.g., the data compiled by Simoneit, 1978). For example, Simoneit *et al.* (1977) showed that higher plant wax is present in dusts from these Northeast Trades and demonstrated the potential of aeolian transport to supply terrestrial lipids to the marine environment. More recently Conte and Weber (2002) showed that the seasonal changes in leaf wax isotopic composition during the growing season is faithfully recorded in samples collected in the central Atlantic, testifying to the efficiency of aerosol transport.

Large-scale programmes were needed, however, to confirm the role of the atmosphere in the transport of organic material to the oceans. An outstanding example of such a programme was SEAREX, and considerable advances have been made to our understanding of the long-range transport of organic material to the sea surface as a result of data obtained from this programme. As part of SEAREX, a variety of organic source markers, selected to provide information on marine versus terrestrial sources, were determined in aerosols collected over the Pacific Ocean. The characteristic organic fractions of background aerosols from continental and coastal regions are dominated by a variety of compound classes of biogenic lipids, and a number of these source markers were determined in aerosols and rain samples from stations in the SEAREX network.

Peltzer and Gagosian (1989) provided SEAREX data on the following: $n$-alkanes, $C_{17}$–$C_{40}$ (terrestrial sources); fatty alcohols, $C_{21}$–$C_{36}$ (terrestrial source) and $C_{13}$–$C_{20}$ (marine source), and $C_{13}$–$C_{18}$ fatty acid salts (marine). Some of the principal findings arising from the study are summarized in the following.

**1** Organic carbon accounted for around 10% of the total aerosol in the remote marine atmosphere. Less than ~1% of this organic material has been identified as individual compounds, yet it has provided valuable insights into the sources, transport paths, transformation mechanisms and processes controlling the fluxes of organic material to the sea surface.

**2** Over the Pacific, the concentrations of the terrestrially derived and the marine-derived compounds varied independently.

**3** Terrestrially derived compounds. The $n$-alkanes and the $C_{21}$–$C_{36}$ fatty alcohols were the most abundant of these compounds, which is consistent with their predominance in the epicuticular waxes of vascular plants. In contrast, the long-chain fatty acids, which are only minor constituents of the plant waxes, were present at lower concentrations in the aerosols. There were considerable variations in the concentrations of the terrestrially derived lipids both at individual sites and between sites. However, there was an overall trend for the concentrations to be highest in mid-latitudes and lowest in the tropics. Seasonal changes in waxes at sampling sites were also consistent with seasonal changes in atmospheric transport routes to the sites, such as discussed earlier for dust.

**4** Marine-derived compounds. The $C_{13}$–$C_{18}$ fatty acid salts were the most abundant of these compounds at all the sites, with the $C_{13}$–$C_{20}$ fatty alcohols being typically the least abundant of all the lipid compounds.

**5** The SEAREX study also highlighted the importance of transformations that affect organic material in the marine atmosphere. For example, unsaturated fatty acids, which are characteristic of marine organisms, were absent from the atmospheric samples; however, their photochemical oxidation products were identified as a major class of organic compounds in both aerosols and rain water. There was also evidence that wax coatings on some aerosols protect them to some extent from atmospheric degradation.

**6** The major fluxes of the atmospherically transported organic material to the sea surface resulted from wet rather than dry deposition processes.

The viable POM species in aerosols include fungi, bacteria, pollen, algae, yeasts, moulds, mycoplasma, viruses, phages, protozoa and nematodes, and a number of these have been identified in marine air. For example, Delany *et al.* (1967) noted the presence of various marine organisms, freshwater diatoms and fungi in aerosols collected at Barbados (North Atlantic) in the path of the Northeast Trades. Folger (1970) reported the presence of phytoliths, freshwater diatoms, fungi, insect scales and plant tissue in the atmosphere over the North Atlantic. Various insects also have been found in marine air samples, and the atmosphere offers a pathway by which some plant and animal species can colonize remote islands. More recently Jaenicke *et al.*, (2007) emphasized the importance of biological particles, both living and dead, to the overall aerosol mass burden.

*Vapour-phase organic matter (VOM).* Vapour-phase organics have varying degrees of chemical reactivity, and their atmospheric lifetimes therefore vary over a wide time-scale. Some more complex VOM material is oxidized rapidly to form aerosol organic carbon (see above), while some is transformed to gaseous reactants and ultimately carbon dioxide. Methane is the most abundant atmospheric hydrocarbon, originating naturally from wetlands, with inputs now augmented by about a factor of 2 by emissions from coal, gas and hydrocarbon industries, and agricultural and waste disposal sources In addition to methane, which can remain in the atmosphere for between 4–7 years, a large range of vapour-phase organic compounds, which have considerably shorter lifetimes (~1–100 days), are present in marine air. For a detailed inventory of methane and these non-methane vapour-phase organics see Seinfeld and Pandis (1998).

In addition to natural organic material, the atmosphere is important in the transport of pollutant organics to the oceans; these include DDT residues and PCBs (see e.g. Goldberg, 1975). (DDT is a complex mixture containing mainly *p,p*'-dichlorodiphenyltrichloroethane and PCBs are polychlorinated biphenyls. They were used as insecticides.) Usage of these notoriously harmful compounds have been greatly reduced but other potentially harmful organic compounds, including polycyclic aromatic hydrocarbons (PAHs) released from combustion, sources remain of concern. In addition, patterns of industrial and agricultural organic chemical emissions change continuously in response to changing patterns of human activity and the residues of the use of these chemicals are often to be found in the atmosphere. Air–sea exchange is a critical link in this transport process, and differences in chemical properties of individual heavy synthetic organics can affect the mechanism by which they are transferred to the ocean surface and their reactivity in seawater: see Sections 6.2.3 and 6.4.3. For a review of this topic, and for comprehensive data on synthetic vapour-phase organics, the reader is referred to Duce *et al.* (1983b); Atlas and Giam (1986) and Seinfeld and Pandis (1998). Many of these compounds have intermediate volatility that mean they exchange continuously between the gas and solid (aerosol or soil) phase as temperatures change and for some this can lead to accumulation in high latitudes, low temperature ecosystems (Wania and Mackay, 1993).

## 4.2 The chemistry of the marine aerosol

### 4.2.1 Elemental composition

As we shall see in Chapter 6, atmospheric fluxes of many elements represent an important component

**Table 4.4** The concentration ranges of some elements in the marine aerosol.

| Element | Estimated concentration range* ($ng\,m^{-3}$ of air) |
|---------|------------------------------------------------------|
| Al      | $1–10^4$                                             |
| Fe      | $1–10^4$                                             |
| Mn      | $0.1–10^2$                                            |
| Cu      | $0.1–10^1$                                            |
| Zn      | $0.1–10^2$                                            |
| Pb      | $0.1–10^2$                                            |

* Approximate values only.

of the global cycles of these elements and hence there is considerable interest in the sources and cycling of these elements within the atmosphere.

The concentrations of many particulate elements in the marine atmosphere vary over several orders of magnitude (see Table 4.4), and at any specific location the concentrations are dependent on a number of factors. These include: (i) the efficiency with which the host components containing the element are mobilized in the air; (ii) the time the components spend in the atmosphere and the distance they are transported through it, that is. the extent to which the components are 'aged'; and (iii) the processes by which the components are removed from the air.

According to Berg and Winchester (1978) the composition of the marine aerosol resembles that expected from the mixing of finely divided materials from large-scale sources. The main sources of material to the atmosphere have been described in Section 4.1.3, and a source–control relationship offers a convenient framework within which to describe the chemistry of the marine aerosol.

One of the most common methods of relating an element in an aerosol to its source is by using a source indicator, or marker, that is derived predominantly from one specific source. In order to assess the enrichment of an element relative to a source it is common practice to define the excess, that is. non-source fraction, in terms of an enrichment factor (EF), which is calculated with respect to an equation of the type:

$$EF_{source} = (E/I)_{air} / (E/I)_{source} \qquad (4.1)$$

in which $(E/I)_{air}$ is the ratio of concentrations of an element $E$ and the indicator element $I$ in the aerosol, and $(E/I)_{source}$ is the ratio of their concentrations in the source material.

The three most important sources for the supply of particulate matter to the marine atmosphere are: (i) the Earth's crust (the crustal aerosol), (ii) the ocean surface (the sea-salt aerosol), both of which involve natural low-temperature generation processes, and (iii) a variety of usually high-temperature anthropogenic processes (the enriched aerosol).

### 4.2.1.1 The crustal aerosol

Aluminium is the indicator element most commonly used for the crustal source because of it is the most abundant metal within the crust and its atmospheric cycle has been little altered by human impact. Although there are problems involved in selecting a composition for the source (or precursor) material, that of the average crustal rock is frequently used for the calculation of $EF_{crust}$ values, according to the equation:

$$EF_{crust} = (E/Al)_{air} / (E/Al)_{crust} \qquad (4.2)$$

in which $(E/Al)_{air}$ is the ratio of the concentrations of an element $E$ and Al in the aerosol, and $(E/Al)_{crust}$ is the ratio of their concentrations in average crustal rocks (see Chapter 3).

Because of the various constraints involved in the calculation, $EF_{crust}$ values should be treated only as order-of-magnitude indicators of the crustal source. Thus, values close to unity are taken as an indication that an element has a mainly crustal origin, and those >10 are considered to indicate that a substantial portion of the element has a non-crustal origin.

Rahn (1976) and Rahn et al. (1979) have tabulated the $EF_{crust}$ values for some 70 elements in over 100 samples from the world aerosol, and a geometric mean for the whole population gives a first approximation of the most typical value for each element. These geometric means are plotted in Fig. 4.7, from which it can be seen that over half the elements have $EF_{crust}$ values that range between 1–10, indicating that they are present in the aerosol in roughly crustal proportions. These are termed the crustal or non-enriched elements (NEEs). The remaining elements have $EF_{crust}$ values in the range 10 to ~$5 \times 10^3$, and their concentrations in the aerosol are not crust-controlled. These are referred to as the enriched or anomalously enriched elements (AEEs).

The NEEs will almost always retain their character in all aerosols. It is important, however, to stress that the degree to which an AEE is actually enriched can vary considerably as the relative proportions of the various components in an aerosol change. Data to illustrate this are given in Table 4.5, which lists the $EF_{crust}$ values for several elements from a variety of marine locations. These data show, for example, that $EF_{crust}$ values of the AEEs Cu, Pb and Zn can vary between <10 and >100. Thus, under certain conditions the $EF_{crust}$ values of the AEEs can indicate that they are present in crustal proportions. Some of the factors that control the source strengths, and so the $EF_{crust}$ values, of both the NEEs and the AEEs in the marine aerosol can be illustrated with respect to the distributions of Fe and Cu in the Atlantic atmosphere. The average Al, Cu and Fe concentrations, and $EF_{crust}$ values for Fe and Cu, from a number of aerosol populations sampled on a north–south Atlantic transport are listed in Table 4.6. Iron is a crustal element, and the Fe $EF_{crust}$ values are all <10 and exhibit little variation between the populations. In contrast, Cu is an AEE in the world aerosol, but Cu $EF_{crust}$ values in the Atlantic aerosol vary considerably. In most of the populations they are >10. However, the average Cu $EF_{crust}$ values fall in the aerosol collected over the Straits of Gibraltar and reach a minimum of around unity in the Northeast Trades population, which was sampled off the coast of West Africa, where pulses of Saharan dust-laden air are common: see Section 4.1.4.1. It is apparent, therefore, that although Cu is classified as an AEE in the world aerosol (average Cu $EF_{crust}$ value ~100), under some conditions crustal material can become the dominant source of the element and can completely mask the non-crustal components. It may be concluded, therefore, that although the atmospheric concentrations of the NEEs (e.g. Fe) vary considerably, their $EF_{crust}$ values remain relatively constant. In contrast, both the concentrations and $EF_{crust}$ values of the AEE vary, the latter being lowered by increases in the input of crustal material if this is not accompanied by inputs of AEEs.

### 4.2.1.2 The sea-salt aerosol

In order to assess the importance of the sea surface as a source for the marine aerosol, Na can be used as the marine indicator element and the precursor

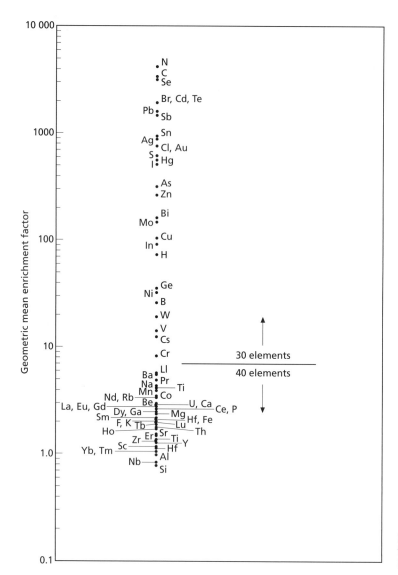

**Fig. 4.7** Geometric mean $EF_{crust}$ values for elements in the world aerosol (from Rahn et al., 1979).

source composition is assumed to be that of bulk seawater. The $EF_{sea}$ is then calculated according to the equation

$$EF_{sea} = (E/Na)_{air}/(E/Na)_{sea\ water} \qquad (4.3)$$

in which $(E/Na)_{air}$ is the ratio of the concentrations of an element $E$ and Na in the aerosol, and $(E/Na)_{sea\ water}$ is the ratio of their concentrations in bulk seawater.

A selection of $EF_{sea}$ values for aerosols for a number of marine regions is given in Table 4.7. The regions are ranked on the basis of their distance from the major continental trace metal sources, and so represent increasingly more pristine marine environments. The elements in the table can be divided into three general groups.

*Group 1* contains K and Mg. For these two elements there is little intersite variation in their $EF_{sea}$ values, which are all around unity and indicate that the elements have a predominantly oceanic source at all locations.

**Table 4.5** $EF_{crust}$ values for some anomalously enriched crustal elements in marine aerosols.

| | $EF_{crust}$ | | | |
|---|---|---|---|---|
| Element | North Atlantic, westerlies* | Bermuda (North Atlantic)[†] | Enewetak (North Pacific)[‡] | North Atlantic, Northeast Trades[§] |
| Cu | 120 | 9.6 | 3.2 | 1.2 |
| Zn | 110 | 26 | 14 | 3.8 |
| Pb | 2200 | 170 | 40 | 9.1 |
| Cd | 730 | 570 | 130 | 9.4 |

*Data from Duce et al. (1975). [†]Data from Duce et al. (1976). [‡]Data from Duce et al. (1983a). [§]Data from Murphy (1985).

**Table 4.6** Average concentrations and $EF_{crust}$ values for Al, Fe and Cu in Atlantic aerosols (data from Murphy, 1985).

| | Element | | | | | |
|---|---|---|---|---|---|---|
| | Al | Fe | | Cu | | |
| Oceanic region | Concentration ($ng\,m^{-3}$ of air) | Concentration ($ng\,m^{-3}$ of air) | $EF_{crust}$ | Concentration ($ng\,m^{-3}$ of air) | $EF_{crust}$ | |
| North Atlantic westerlies | 48 | 36 | 1.2 | 1.0 | 30 | |
|   Straits of Gibraltar | 827 | 626 | 1.1 | 2.4 | 8.2 | |
|   Northeast Trades | 5925 | 3865 | 1.0 | 4.5 | 1.2 | |
| South Atlantic | | | | | | |
|   Southeast Trades | 17 | 13 | 1.2 | 0.30 | 44 | |
|   westerlies | 2.7 | 2.6 | 1.8 | 0.29 | 225 | |

**Table 4.7** $EF_{sea}$ values for a number of elements in the marine aerosol*.

| | $EF_{sea}$ | | |
|---|---|---|---|
| Element | North Atlantic | North Pacific, Hawaii | North Pacific, Enewetak |
| K | 1.3 | 1.1 | 1.3 |
| Mg | 0.9 | 1.0 | 1.1 |
| Al | $1 \times 10^6$ | $3 \times 10^5$ | $1 \times 10^5$ |
| Co | $8 \times 10^3$ | $<1 \times 10^3$ | $1 \times 10^3$ |
| Cu | $6 \times 10^4$ | $7 \times 10^3$ | $3 \times 10^3$ |
| Fe | $4 \times 10^7$ | $6 \times 10^5$ | $3 \times 10^4$ |
| Mn | $4 \times 10^5$ | $8 \times 10^3$ | $2 \times 10^4$ |
| Pb | $5 \times 10^5$ | $5 \times 10^5$ | $3 \times 10^4$ |
| V | $9 \times 10^3$ | $4 \times 10^2$ | $2 \times 10^2$ |
| Zn | $2 \times 10^6$ | $6 \times 10^5$ | $1 \times 10^5$ |
| Sc | $1 \times 10^5$ | – | $4 \times 10^4$ |

| | UK coastal aerosol | North Atlantic, westerlies | Atlantic, Northeast Trades | Atlantic, Southeast Trades | South Atlantic, westerlies |
|---|---|---|---|---|---|
| Mo | 116 | 20 | 13 | 9.5 | 1.4 |

*Data for Mo from Chester et al. (1984a); data for all other elements from Weisel et al. (1984).

*Group* 2 contains Mo. This element has $EF_{sea}$ values that range up to ~100 close to the continents but fall to around unity in the remote South Atlantic westerlies. This indicates that in remote areas bulk seawater can be a significant source for this element (see e.g. Chester *et al.*, 1984a).

*Group* 3 contains Al, Fe, Mn, Co, V, Cu, Pb and Zn. For these elements there is a progressive decrease in their $EF_{sea}$ values towards the more pristine oceanic environments. However, even at the remote sites the elements are still enriched, by factors ranging up to 600 000, relative to bulk seawater.

From their $EF_{sea}$ values, therefore, it would appear that the ocean is not a significant source for the elements in Group 3 above, even in areas remote from the influence of other sources. The calculation of the $EF_{sea}$ values, however, presents problems that are considerably more complex than those associated with $EF_{crust}$ values. These problems involve the selection of a composition for the oceanic precursor material, and the most serious of them arises in response to the nature of the processes involved in the generation of sea-salts. During the bursting of bubbles, which is the principal mechanism in sea-salt formation, part of the sea-surface microlayer can be skimmed off to be incorporated into the salt particles. This microlayer can contain elevated concentrations of trace metals. Some studies have suggested that bubble bursting through the microlayer can significantly enrich the resulting aerosol in trace metals. However, Hunter (1997) has argued that this enrichment process is predominantly one involving bubble capture of particulate matter from seawater and the microlayer, rather than enrichments of dissolved metals from these sources, and this conclusion is consistent with field and laboratory studies of the enrichment of radionuclides in aerosols generated from seawater (Walker *et al.*, 1986).

### 4.2.1.3 The enriched aerosol

There are a number of processes that can supply the AEE to the atmosphere, and many of these are associated with some form of high-temperature processes (see Section 4.1.1). Unlike the low-temperature processes involved in the production of crustal and oceanic particulate material, however, the high-temperature processes do not, in general, have readily identifiable indicator elements that can be determined on a routine basis. Despite this, attempts have been made to identify the anthropogenic source, and element to element ratios can be used for this purpose. For example, Rahn (1981) used 'non-crustal Mn/non-crustal V' ratios as a tracer of large-scale sources of pollution in Arctic aerosols, and Rahn and Lowenthal (1984) used a seven 'element:element' ratio system to demonstrate that regional elemental tracers offer a way of determining the long-range sources of pollution in the atmosphere. Data for particulate anthropogenic trace metal emissions have also been used to establish metal sources, especially in coastal aerosols. For example, several authors have used metal/metal ratios from the data set provided by Pacyna *et al.* (1984) for the composition of anthropogenic emissions from Europe, to establish the sources of metals to the North Sea atmosphere (see e.g. Schneider, 1987; Yaaqub *et al.*, 1991; Chester *et al.*, 1994). In general, however, it is not the usual practice to calculate enrichment factors for high-temperature sources because the patterns of enrichment at any one location are likely to depend on local patterns of industrial activity. Although the presence of elements associated with these sources is indicated by the fraction that is in excess of those accounted for by the $EF_{crust}$ and $EF_{sea}$ values, other approaches must be used for their actual identification. Two such approaches applicable to the marine aerosol will be discussed here; these involve the particle-size distribution and the solid-state speciation of elements in aerosols.

*The particle-size distribution of elements in the marine aerosol.* The particle size of an element in the atmosphere is controlled by the manner in which it is incorporated into its host component, which, in turn, is a function of its source. The overall particle-size–source relationships have been illustrated in Fig. 4.1, from which two general populations can be identified. The coarse particles are in the mechanically generated range and have diameters >~2 μm. These include the crustal and sea-salt components, and during their formation these particles acquire their elements directly from the precursor material. In contrast, the fine particles are in the Aitken nuclei or accumulation range and have diameters <~2 μm. Those in the accumulation range are formed by processes such as condensation and gas-to-particle conversions, and elements associated with them can be acquired from the atmosphere itself. For example, during high-temperature processes (such as volcanic activity, fossil-fuel combustion, waste incineration and the processing of ore materials) some elements can be volatilized from the parent material in a vapour phase. They can be removed from this vapour phase via condensation and gas-to-particle processes during which the elements may be adsorbed on to the surfaces of the ambient aerosols. The processes are size-dependent because, in general, small particle condensation nuclei will have a large ratio of surface area to volume (see e.g., Rahn, 1976). As a result, many of the more volatile elements released during high-temperature processes become associated with small particles in the accumulation range, that is. with diameters $<\sim$1–0.1 μm: see Fig. 4.1.

Because of the different processes involved in the generation of their host particles, elements having a crustal or an oceanic source should be present in association with larger ($<\sim$1 μm diameter) aerosols, whereas those having a high-temperature source should be found on smaller ($>\sim$1 μm diameter) aerosols, and hence EF values for these metals are often much greater in the fine mode aerosol compared to the coarse mode. There are also differences in the particle-size distributions of the mechanically generated crustal and sea-salt aerosols. The particle-size distribution of an element in an aerosol population should therefore offer an insight into its source. This has been demonstrated in a number of studies. For example, Fig. 4.8 illustrates the mass particle-size distributions of:

1 $^{210}$Pb which is produced from the decay of radon gas within the atmosphere, which can be used as an example of an element that has a mass predominantly associated with submicrometre-sized particles typical of many high-temperature-generated pollutant-derived elements produced after atmospheric gas-to-particle conversion;

2 Al, which is typical of crust-derived elements;

3 Na, which is typical of sea-salt-derived elements.

From data such as these, three general conclusions can be drawn regarding the particle size distribution of elements in the marine aerosol.

1 Sea-salt-associated elements have most of their total mass on particles with MMDs in the range $\sim$3–7 μm; these elements include Na, Ca, Mg and K.

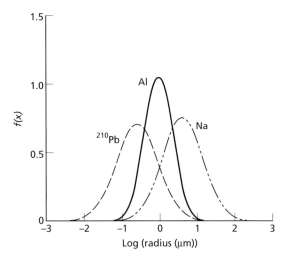

**Fig. 4.8** Mass size distributions of $^{210}$Pb, Al and Na in aerosols from the marine atmosphere (from Arimoto and Duce, 1986).

2 Crust-derived elements have most of their total mass on particles with MMDs in the range $\sim$1–3 μm; these elements include the NEEs Al, Sc, Fe, Co, V, Cs, Ce, Rb, Eu, Hf and Th.

3 Elements associated with high-temperature sources (mainly anthropogenic) have most of their total mass on particles with MMDs $< \sim$0.5 μm; these elements include the AEEs Pb, Zn, Cu, Cd and Sb.

It may be concluded, therefore, that in an aerosol population the particle-size distributions of elements between the three size classes, together with EF data for the individual size classes, can be used to establish the predominance of their anthropogenic, crustal and sea-salt sources.

*The solid-state speciation of elements in the marine aerosol.* The partitioning of elements among the components of an aerosol, that is. their solid-state speciation, can reveal information on their sources. This can be illustrated with respect to the study reported by Chester *et al.* (1989), in which a sequential leaching technique (i.e. the sequential use of stronger liquid chemical reagents to dissolved metals from samples) was used to establish the solid-state speciation of a series of trace metals in anthropogenic and crustal 'end-member' aerosols. The technique separated three trace-metal binding associations: (i) an exchangeable association; (ii) an

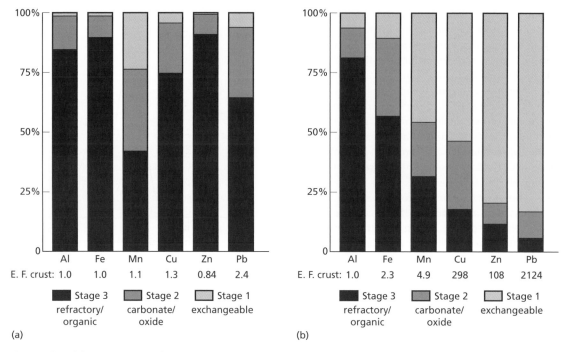

**Fig. 4.9** The solid-state speciation of elements in aerosols: (a) Crustal 'end-member' aerosol (SCAP, Saharan Crustal Aerosol Population), (b) anthropogenic 'end-member' aerosol (LUAP, Liverpool Urban Aerosol Population) (after Chester *et al.*, 1989).

oxide and carbonate association; and (iii) a refractory and organic association. The results are illustrated in Fig. 4.9, and may be summarized as follows.

1 Al and Fe are generally refractory in both aerosol 'end-members', that is greater than ~80% of the total Al ($\Sigma$Al) and greater than ~60% of the $\Sigma$Fe are in refractory associations.

2 Mn is speciated between all three binding associations in both 'end-members', with no single association being especially dominant.

3 Cu, Zn and Pb exhibit very different speciation signatures in two 'end-member' aerosols. In the crustal 'end-member' aerosol, ~65% of the $\Sigma$Pb, ~75% of the $\Sigma$Cu and ~90% of the $\Sigma$Zn are in *refractory* associations. In contrast, in the anthropogenic 'end-member' ~50% of the $\Sigma$Cu and ~90% of the $\Sigma$Zn and the $\Sigma$Pb are in exchangeable associations. The differences between the solid-state speciation signatures of Cu, Zn and Pb are entirely consistent with their different sources to the atmosphere for the two 'end-member' aerosols. Crustal dust is generated during low-temperature weathering–mobilization processes. In contrast, high-temperature processes

are often involved in the emission of anthropogenic trace metals, such as Cu, Zn and Pb, to the atmosphere. During these high-temperature processes, the metals can be released from the parent material into the vapour phase from which subsequently they can be taken up by small-sized ambient particles in a weakly held exchangeable surface association (see particle-size discussion above).

The exchangeable fraction of the metals in the aerosol is most likely to be solubilized upon deposition to seawater and so become available to ocean biota. Sequential leaching can provide valuable information to assess this bioavailability and in general, crustal lattice bound metals are much less easily dissolved than metals associated with the weakly bound and exchangeable fractions. However, even within the lattice bound metal fraction, elements will dissolve to some extent and in the case of iron, even this modest dissolution is significant as a source of iron to the remote ocean regions (Jickells *et al.*, 2005) as we shall discuss further in Chapter 11. Iron dissolution from aerosols has therefore been extensively studied and is now known to vary systematically from <1% in high dust regions to >10% in very low dust regions, probably reflecting increasing iron solubility from finer dust particles and possibly an additional contribution from atmospheric chemical reactions during long range transport (Baker and Jickells, 2006).

In addition to the various chemical and particle-size indicators described above, which can be used to assess the general sources of elements to the atmosphere, the actual source regions supplying material to an individual air mass often can be determined using air mass back-trajectory techniques. These involve back-tracking the atmospheric path followed by an air mass in order to identify the potential aerosol source regions it has crossed during at least some part of its history, and they have been used in both short-range (see e.g. Chester *et al.*, 1994) and long-range (see e.g. Merrill, 1989) aerosol transport studies.

## 4.2.2 Elemental source strengths

There have been a number of attempts to estimate the global source strengths of elements to the atmosphere. There are large uncertainties involved in making these estimates and they should be treated at best as no more than approximations. Nonetheless, a number of overall trends in elemental atmospheric source strengths can be identified from the published estimates.

Lantzy and Mackenzie (1979) compiled data on the natural and anthropogenic global source of elements released into the atmosphere, and derived an atmospheric interference factor (AIF) calculated according to the equation:

$$AIF = (E_a/E_n) \times 100 \tag{4.4}$$

in which $E_a$ is the total anthropogenic emission of an element $E$, and $E_n$ is its total natural emission. Thus, an AIF of 100 indicates that the anthropogenic flux of an element is equal to its natural flux. Table 4.8 is based on a more recent compilation of anthropogenic and natural emissions and is used here to derive a revised AIF. The scale of anthropogenic perturbations are less than in the Lantzy and Mackenzie estimate due mainly to lower anthropogenic emission estimates, reflecting in part better data and in the case of some elements such as lead (see below) real reductions in anthropogenic emissions. The data for the global atmospheric source strengths, and the AIF values, for the various elements are listed in Table 4.8. The elements fall into three main groups.

1 Mn has an AIF value <10 and for this element natural fluxes dominate the supply to the atmosphere.
2 As, Sb, Cr, Cu, Hg, Mo and Se have AIF values in the range 30–100 and hence suffer significant atmospheric perturbation.

**Table 4.8** A comparison of global anthropogenic and natural emissions of some trace metals to the atmosphere for the 1990s as $10^9\,g\,yr^{-1}$ and the atmospheric interference factor (AIF) as the ratio of the two emissions to provide an index of human perturbations of the system (based on Pacyna and Pacyna, 2001).

| Trace Metal | Anthropogenic Emissions | Natural Emissions | AIF (Anthropogenic/Natural) × 100 |
|---|---|---|---|
| As | 5 | 12 | 42 |
| Cd | 3 | 1.3 | 230 |
| Cr | 15 | 44 | 34 |
| Cu | 26 | 28 | 93 |
| Hg | 2.2 | 2.5 | 88 |
| Mn | 11 | 317 | 3 |
| Mo | 2.6 | 3.0 | 87 |
| Ni | 95 | 30 | 320 |
| Pb | 119 | 12 | 1000 |
| Sb | 1.6 | 2.4 | 67 |
| Se | 4.6 | 9.3 | 49 |
| V | 240 | 28 | 860 |
| Zn | 57 | 45 | 130 |

3 Cd, Ni, Pb, V and Zn have AIF values >100 and hence their atmospheric cycles are very significantly disturbed by human activity.

It should be remembered that these are global averages and that most anthropogenic emissions take place in the Northern Hemisphere where effects will be greatest (Robinson and Robbins, 1971).

A number of other authors have made estimates of the global source strengths of trace metals to the atmosphere. Those provided by Weisel *et al.* (1984) emphasized the potential importance of seaspray emissions which as noted above are not now thought to be particularly important. Nriagu (1989) emphasized the importance of biogenic sources, some of which, such as biomass burning and biogenic particles, are probably rather important, but still very uncertain. More recently, Pyle and Mather (2003) have re-evaluated the importance of volcanic mercury emissions and suggested these have previously been underestimated (and possibly those of some other metals), particularly from persistent fumarole emissions, which provide a rather constant emission augmented by the periodic larger scale eruptions Mercury is unique amongst the trace metals in being found predominantly in the atmosphere in the gas phase as $Hg^0$ and as a result has a much longer atmospheric residence time (~1 year) than most other metals (Lohman *et al.*, 2008).

As noted already, volcanic emission vary with time; with episodic eruption enhancing the more steady fumarole emissions. Anthropogenic emissions also vary with time. Clearly in the pre-industrial era emissions were much less than today and they subsequently increased with this activity and with changing extent and distribution of industrial activity. Coal burning in power stations for example is a much bigger source of trace metal emissions than gas burning. Large coal combustion sources represent local emissions of relatively large amounts of particles, although the trace metal content of individual particles may not be particularly high. In contrast emissions from a metal smelter may be small in terms of overall mass of particles, but the particles emitted may well be rather heavily enriched in trace metals associated with the smelting, for example Zn from a Zn smelter (Nriagu and Davidson, 1986).

A particular example of changes in emission is that of lead. Tetraethyl lead was widely used from about 1930–1990 as a fuel additive in cars before its health hazards and incompatibility with catalytic converters (required to reduce traffic nitrous oxide emissions which contribute to ozone production) led to its phasing out in the 1990s. Lead is a relatively volatile metal so it is readily vaporized during fuel burning with the vapour then condensing onto fine aerosol, making it available for long range atmospheric transport. The history of lead (and other trace metals emitted to the atmosphere) is recorded in ice cores (Boutron *et al.*, 1991; Rosman *et al.* 1993; Candelone *et al.*, 1995), sediments (Renberg *et al.*, 2000; see also Fig. 4.10) and corals (Kelly *et al.*, 2009; see

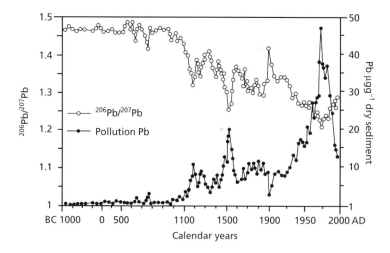

**Fig. 4.10** The history of atmospheric lead pollution as recorded in a sediment core from a remote Swedish lake. Increases in lead concentrations are evident throughout much of human history from Roman times onward with a particularly strong increase during the period centred around 1975. The changes in concentrations are accompanied by changes in the lead isotopic composition ($^{206}Pb/^{207}Pb$) indicating a change in lead source as well as flux (Renberg *et al.*, 2000).

Chapter 11). Arctic ice cores, for example, show concentrations increased from a few pg Pb per $g^{-1}$ of ice 500 years ago to several hundred pg $g^{-1}$ in the 1960s before decreasing to close to pre-industrial values by the late 1990s. In Chapter 11 we will consider how this lead emission history has propagated through the ocean water column. Lead mined from different rock sources has an isotopic composition that depends on the geological age of the source region, and thus measurement of the isotopic composition of the lead can provide additional information about the source region of the ores. In Arctic ice cores, this has been used to estimate the changing relative importance of European and North American emissions (Rosman *et al.*, 1993). Concerns over global lead contamination have led to studies of the long term history of lead pollution (e.g. Fig. 4.10) which have revealed a rich history which shows small increases in lead levels, even in remote regions, associated with smelting activity during the height of the Roman Empire, followed by a decline in concentration into the Middle Ages and a subsequent increase in concentrations with the industrial revolution followed by the major increase in levels associated with lead use in car fuels and a more recent decline (Renberg *et al.*, 2000).

### 4.2.3 Geographical variations in the elemental composition of the marine aerosol

It was pointed out in Section 4.2.1 that the concentrations of many elements in the marine atmosphere vary over several orders of magnitude from one region to another. The reasons for these variations can now be assessed in the light of the factors discussed above. The overall composition of the marine aerosol is controlled by the extent to which components derived from the various sources are mixed together in the atmosphere. A considerable amount of data are now available on the concentrations of elements in marine aerosols from a wide variety of environments, and a compilation of some of the more recent data is given in Table 4.9. In this table the marine locations are arranged in an increasing order of remoteness from the primary, that is, non-oceanic, sources. The principal trends in atmospheric concentrations that emerge from this data set can be summarized as follows.

1 The highest concentrations of the AEEs (as illustrated by Cu, Pb and Zn) are found over coastal seas that are relatively close to the continental anthropogenic sources.

2 The highest concentrations of NEEs (as illustrated by Al) are found over regions close to the continental arid land and desert sources.

3 The concentrations of both the AEE and the NEE decrease with remoteness from the continental sources, and for those regions for which sufficient data are available the general rank order is: coastal seas > North Atlantic > North Pacific ~ tropical Indian > South Pacific.

4 The $EF_{crust}$ values are highest in aerosols from the coastal seas, but those for the most remote aerosols for example, the Tropical South Pacific aerosol are higher than those for the less remote Tropical North Pacific aerosol, which suggests that the residence times of the small-sized AEE-containing anthropogenic particles are longer than those of the larger NEE-containing mineral particles, leading to an increase in the relative proportions of the enriched metals during long range transport.

The various data in Table 4.9 offer an indication of the concentrations of elements in the marine atmosphere. These elements are available for deposition at the sea surface. Before entering the ocean system proper, however, they have to cross the air–sea interface, and the characteristics of this important transition zone are described in the next section.

## 4.3 Material transported via the atmosphere: the air–sea interface and the sea-surface microlayer

The air-sea interface is the site of a very specialized marine environment, the sea-surface *microlayer*, at which a number of other geochemically important processes occur.

Surface-active organic materials are found on the surfaces of all natural water bodies, including the oceans. These organic materials, which are usually of a biological origin but can also include anthropogenic substances (e.g. petroleum products), sometimes manifest themselves as visible slicks in calm weather, but some sort of interface persists even in stormy conditions. This interface, or microlayer, forms a distinct ecosystem and is an extremely important feature of the ocean reservoir. The thickness of

**Table 4.9** Concentrations (1) in ng m$^{-3}$ and Enrichment Factors (Ef) values (2) of selected trace metals in marine aerosols from some sites selected to show global gradients.

| | Al | | Cu | | Zn | | Pb | | |
|---|---|---|---|---|---|---|---|---|---|
| | 1 | 2 | 1 | 2 | 1 | 2 | 1 | 2 | Ref |
| North Sea | 294 | 1.0 | 6.3 | 30.5 | 41 | 164 | 34.5 | 781 | a |
| Black Sea | 540 | 1.0 | – | – | 46 | 100 | 60 | 741 | b |
| N. Atlantic Trade Winds | 5925 | 1.0 | 4.5 | 1.1 | 16 | 3.2 | 6.9 | 7.7 | c |
| N Arabian Sea | 1227 | 1.0 | 2.6 | 7.2 | 10 | 18 | 4.3 | 27 | d |
| Tropical North Atlantic | 160 | 1.0 | 0.79 | 7.4 | 4.4 | 32 | 9.9 | 407 | e |
| South Atlantic Westerlies | 2.7 | 1.0 | 0.29 | 161 | 1.8 | 784 | 0.97 | 2364 | f |
| Tropical Indian Ocean | 11 | 1.0 | 0.08 | 10 | 0.1 | 13 | 0.17 | 158 | g |
| Tropical North Pacific | 21 | 1.0 | 0.045 | 3.2 | 0.17 | 38 | 0.12 | 38 | h |
| Tropical South Pacific | 0.72 | 1.0 | 0.01 | 27 | 0.07 | 114 | 0.02 | 146 | i |

a Chester *et al.* (1994), b Hacisalihoglu *et al.* (1992), c Murphy (1985), d Chester *et al.* (1991), e Buat-Menard and Chesselet (1979), f Murphy (1985), g Chester *et al.* (1991), h Duce *et al.* (1983a), i Arimoto *et al.* (1987).

the microlayer has been variously reported to extend from that of a monomolecular layer to several hundred micrometres, but because it is notoriously difficult to sample it is usually defined operationally in terms of the device used to collect it (e.g. Hunter, 1997). The microlayer is the site across which the atmosphere–ocean system interacts, that is, where the sea 'breathes', and it has unique chemical, physical and biological properties, which are very different from those of the underlying seawater although our knowledge of this interface is still rather limited (Liss and Duce, 1997). In particular, the microlayer is enriched in hydrophobic and particulate organic material, particularly lipids and possibly some contaminants such as PAHs (Hunter, 1997). The existence of this organically enriched microlayer affects the behaviour of waves and the passage of material into and out of seawater, both particulate and gaseous. The complexity of the processes that can influence trace metals during exchange through the microlayer between the atmosphere and the surface ocean is illustrated in Fig. 4.11. The microlayer is also subject to higher light levels than deeper waters with the associated possibility of increased photochemical activity. The microlayer also has a distinct biological community associated with it (Liss and Duce, 1997). On the basis of the brief discussion outlined above it therefore may be concluded that the microlayer is a complex and reactive environment that forms the interface separating the sea from the atmosphere and it may modify the transport of fluxes both into and out of the oceans, particularly for those compounds that interact with the organics concentrated in the microlayer.

## 4.4 The atmospheric pathway: summary

1 A particulate 'aerosol veil' is present over all the oceans, but the concentrations of material in the veil vary over several orders of magnitude, ranging from greater than ~$10^3$ µg m$^{-3}$ of air off desert and arid regions to less than ~$10^{-3}$ µg m$^{-3}$ of air in remote oceanic areas.

2 The principal sources of material to the marine aerosol veil are the Earth's crust (the crustal aerosol), the sea surface (the sea-salt aerosol) and a variety of mainly high-temperature anthropogenic processes (the enriched, or anthropogenic, aerosol).

3 There is a general tendency for the sources of both crustal and anthropogenic material to be concentrated in the Northern Hemisphere, where they occur in specific latitudinal belts.

4 The sources of anthropogenic material include episodes of 'high pollution' but, in general, anthropogenic components are supplied to the atmosphere on a more or less continuous basis and form part of the 'background' aerosol.

5 The global mineral dust source strength is about $1700 \times 10^{12}$ g yr$^{-1}$; however, unlike anthropogenic

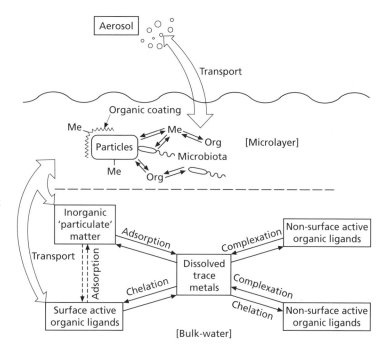

**Fig. 4.11** Schematic representation of the alternatives for the fates of trace metals at the air–sea interface (from Lion and Leckie, 1981). The trace metals may be thought of as competing with each other, and with the major marine cations, for adsorption sites and for available complexing ligands. At equilibrium, the results of interactions of this type will reflect the differing aqueous chemistries (including speciation) of the various trace metals.

material, much of the dust from the major desert sources is delivered to the marine atmosphere in the form of intermittent, that is, non-continuous, pulses.

6 The particulate elements in the marine aerosol have characteristic particle-size spectra constrained by their sources; sea-salt-associated elements have most of their total mass on particles with MMDs in the range ~3–7 μm, crust-derived elements have most of their total mass on particles with MMDs in the range ~1–3 μm, and anthropogenically derived elements have most of their total mass on particles with MMDs < 0.5 μm.

7 The concentrations of many particulate elements in the marine aerosol vary over several orders of magnitude, the highest concentrations being found close to the continental (crustal and anthropogenic) sources and the lowest in remote pristine regions.

8 Atmospherically transported particulates have to cross the air–sea interface before entering the bulk ocean. This interface is the site of the sea-surface microlayer, which is an organic-rich, particle-rich and trace-metal-rich zone, and is one of the most reactive environments in the oceans.

# References

Andreae, M.O., Carlson, R.J., Bruynseels, F., Storm, H., van Grieken, R. and Maenhaut, W. (1986) Internal mixtures of sea salt, silicates, excess sulfate in marine aerosols. *Science*, **232**, 1620–1623.

Arimoto, R. and Duce, R.A. (1986) Dry deposition and the air–sea exchange of trace elements. *J. Geophys. Res.*, **91**, 2787–2792.

Arimoto, R., Duce, R.A., Ray, B.J., Hewitt, A.D. and Williams, J. (1987) Trace elements in the atmosphere of American Samoa: concentrations and deposition to the tropical South Pacific. *J. Geophys. Res.*, **92**, 8465–8479.

Atlas, S. and Giam, C.S. (1986) Sea–air exchange of high molecular weight synthetic organic compounds, in *The Role of Air–Sea Exchange in Geochemical Cycling*, P. Buat-Menard (ed.), pp.295–329. Dordrecht: Reidel.

Ayers, G.P., Ivey, J.P. and Gillett, R.W. (1991) Coherence between seasonal cycles of dimethyl sulphide, methanesulphonate and sulphate in marine air. *Nature*, **349**, 404–406.

Baker, A.R. and Jickells, T.D. (2006) Mineral particle size as a control on aerosol iron solubility, *Geophys. Res. Lett.*, **33**, L17608, doi: 10.1029/2006GL026557.

Berg, W.W. and Winchester, J.W. (1978) Aerosol chemistry of the marine atmosphere, in *Chemical Oceanography*, J.P. Riley and R. Chester (eds), Vol. 7, 173–231. London: Academic Press.

Blanchard, D.C. (1983) The production, distribution, and bacterial enrichment of the sea-salt aerosol, in *Air–Sea Exchange of Gases and Particles*, P.S. Liss and W.G.N. Slin (eds), pp.407–454. Dordrecht: Reidel.

Blank, M., Leinen, M. and Prospero, J.M. (1985) Major Asian aeolian inputs indicated by the mineralogy of aerosols and sediments in the western North Pacific. *Nature*, **314**, 84–86.

Boutron, C.F., Görland, U., Candelone, J.-P., Bolshov, M.A. and Delmas, R.J. (1991) Decrease in anthropogenic lead, cadmium and zinc in Greenland snows since the late 1960s. *Nature*, **353**, 153–156.

Buat-Menard, P. and Chesselet, R. (1979) Variable influence of the atmospheric flux on the trace metal chemistry of oceanic suspended matter. *Earth Planet. Sci. Lett.*, **42**, 399–411.

Buat-Menard, P., Cachier, H. and Chesselet, R. (1989) Sources of particulate carbon to the marine atmosphere, in *Chemical Oceanography*, J.P. Riley, R. Chester and R.A. Duce (eds), Vol. 10, pp.251–279. London: Academic Press.

Cachier, H., Bremond, M-P. and Buat-Menard, P. (1989) Carbonaceous aerosols from different tropical biomass burning sources. *Nature*, **340**, 371–373.

Candelone, J.-P., Hong, S., Pellone, C. and Boutron, C.F. (1995) Post Industrial Revolution changes in large-scale atmospheric pollution of the northern Hemisphere by heavy metals documented in central Greenland snow and ice. *J. Geophys. Res.*, **100**, 16605–16616.

Charlson, R.J., Lovelock, J.E., Andreae, M.O. and Warren, S.G. (1987) Oceanic phytoplankton, atmospheric sulfur, cloud albedo and climate. *Nature*, **326**, 655–661.

Chester, R., Sharples, E.J. and Murphy, K.J.T. (1984a) The distribution of particulate Mo in the Atlantic aerosol. *Oceanol. Acta*, **7**, 441–450.

Chester, R., Sharples, E.J., Sanders, G.S. and Saydam, A.C. (1984b) Saharan dust incursion over the Tyrrhenian Sea. *Atmos. Environ.*, **18**, 929–935.

Chester, R., Murphy, K.J.T. and Lin, F.J. (1989) A three stage sequential leaching scheme for the characterization of the sources and environmental mobility of trace metals in the marine aerosol. *Environ. Technol. Lett.*, **10**, 887–900.

Chester, R., Berry, A.S. and Murphy, K.J.T. (1991) The distributions of particulate atmospheric trace metals and mineral aerosols over the Indian Ocean. *Mar. Chem.*, **34**, 261–290.

Chester, R., Bradshaw, G.F. and Corcoran, P.A. (1994) Trace metal chemistry of the North Sea particulate aerosol: concentrations, sources and sea water fates. *Atmos. Environ.*, **28**, 2873–2883.

Chin, M., Diehl, T., Dubrovik, O., Eck, T.F., Holben, B.N., Sinyuk, A. and Streets, D.G. (2009) Light absorption by pollution, dust and biomass burning aerosols: a global model study and evaluation with AERONET measurements. *Ann. Geophys.*, **27**, 3439–3464.

Conte, M. and Weber, J.C. (2002) Plant biomarkers in aerosols record isotopic discrimination of terrestrial photosynthesis. *Nature*, **417**, 639–641.

Defant, A. and Defant, F. (1958) *Physikalische Dynamik der Atmosphäre*. Frankfurt: Akademie Verlagsgesellschaft.

Delany, A.C., Delany, A.C., Parkin, D.W., Griffin, J.J., G. Goldberg and Reinmann, B.E. (1967) Airborne dust collected at Barbados. *Geochim. Cosmochim. Acta*, **31**, 885–909.

Duce, R.A., Hoffman, G.L., Ray, B.J. *et al.* (1976) Trace metals in the marine atmosphere: sources and fluxes, in *Marine Pollutant Transfer*, H.L. Windom and R.A. Duce (eds), pp.77–120. Lexington, MA: Lexington Books.

Duce, R.A., Hoffman, G.L. and Zoller, W. (1975) Atmospheric trace metals at remote Northern and Southern hemisphere sites: pollution or natural? *Science*, **187**, 59–61.

Duce, R.A., Arimoto, R., Ray, B.J., Unni, C.K. and Harder, P.J. (1983a) Atmospheric trace metals at Enewetak Atoll: I. Concentrations, sources and temporal variability. *J. Geophys. Res.*, **88**, 5321–5342.

Duce, R.A., Mohnen, V.A., Zimmerman, P.R. *et al.* (1983b) Organic material in the global troposphere. *Rev. Geophys. Space Phys.*, **21**, 921–952.

Duce, R.A., LaRoche, J., Altieri, K. *et al.* (2008) Impacts of atmospheric anthropogenic nitrogen on eth open ocean. *Science*, **320**, 893–897.

Erickson, D., Merrill, J.I. and Duce, R.A. (1986) Seasonal estimates of global atmospheric sea-salt distributions. *J. Geophys. Res.*, **91**, 1067–1072.

Folger, D.W. (1970) Wind transport of land-derived mineral, biogenic and industrial matter over the North Atlantic. *Deep-Sea Res.*, **17**, 337–352.

Garrett, T.J. and Verzella, L.L. (2008) An evolving history of arctic aerosols. *Bull. Am. Met. Soc.*, **89**, 299–302.

Goldberg, E.D. (1975) Marine pollution, in *Chemical Oceanography*, J.P. Riley and G. Skirrow (eds), Vol. 3, pp.39–89. London: Academic Press.

Gunson, J.R., Spall, S.A., Anderson, T.R., Jones, A., Totterdell, I.J. and Woodage, M.J. (2006) Climate sensitivity to ocean dimethylsulphide emissions. *Geophys. Res. Lett.*, **33**, L07701, doi:10.1029/2005GL024982.

Hacisalihoglu, G., Eliyakut, F., Olmez, I., Balkas, T.I. and Tuncel, G. (1992) Chemical composition of particles in the Black Sea atmosphere. *Atmos. Environ.*, **26**, 3207–3218.

Hasse, L. (1983) Introductory meteorology and fluid mechanics, in *Air–Sea Exchange of Gases and Particles*, P.S. Liss and W.G.N. Slin (eds), pp.1–51. Dordrecht: Reidel.

Hunter, K.A. (1997) Chemistry of the sea-surface microlayer, in *The Sea Surface and Global Change*, P.S. Liss and R.A. Duce (eds), pp.287–319. Cambridge: Cambridge University Press.

Iribarne, J.V. and Cho, H.-R. (1980) *Atmospheric Physics*. Dordrecht: Reidel.

IPCC (2007) *Climate Change 2007: The Physical Science Basis*. Contribution of working group 1 to the Fourth Assessment Report of the Intergovernmental Panel on Climate Change, Solomons, S. *et al.* (eds). Cambridge: Cambridge University Press.

Jaenicke, R., Matthia-Maser, S. and Gruber, S. (2007) Omnipresence of biological material in the atmosphere. *Environ. Chem.*, 4, 217–220.

Jickells, T.D., An, Z.S., Andersen, K.K. *et al.* (2005) Global iron connections between desert dust, ocean biogeochemistry and climate. *Science*, 308, 67–71.

Junge, C.E. (1972) Our knowledge of the physico-chemistry of aerosols in the undisturbed marine environment. *J. Geophys. Res.*, 77, 5183–5200.

Kelly, A.E., Reuer, M.K., Goodkin, N.F. and Boyle, E.A. (2009) Lead concentrations and isotopes in corals and water near Bermuda 1780–2000. *Earth Planet. Sci. Lett.*, 283, 93–100.

Koga, M. (1981) Direct production of droplets from breaking wind waves – its observation by a multicolored overlapping exposure photography technique. *Tellus*, 33, 552–563.

Lai, R.J. and Shemdin, O.H. (1974) Laboratory study of the generation of spray over water. *J. Geophys. Res.*, 79, 3055–3063.

Lantzy, R.J. and Mackenzie, F.T. (1979) Atmospheric trace metals: global cycles and assessment of man's impact. *Geochim. Cosmochim. Acta*, 43, 511–525.

Lion, L.W. and Leckie, J.O. (1981) Chemical speciation of trace metals at the air–sea interface: the application of an equilibrium model. *Environ. Geol.*, 3, 293–314.

Liss, P.S. and Duce, R.A. (1997) *The Sea Surface and Global Change*. Cambridge: Cambridge University Press.

Lohman, K., Seigneur, C., Gustin, M. and Lindberg, S. (2008) Sensitivity of the global atmospheric cycle of mercury to emissions. *Applied Geochem.*, 23, 454–466.

McDonald, R.L., Unni, C.K. and Duce, R.A. (1982) Estimation of atmospheric sea salt dry deposition: wind speed and particle size dependence. *J. Geophys. Res.*, 87, 1246–1250.

Mahowald, N.M., Baker, A.R., Bergametti, G., Brooks, N., Duce, R.A., Jickells, T.D., Kubilay, N., Prospero, J.M. and Tegen, I. (2005) Atmospheric global dust cycle and iron inputs to the ocean. *Global Biogeochem. Cycl.*, 19 GB4025, doi 10.1029/2004GB002402.

Maring, H.B. and Duce, R.A. (1990) The impact of atmospheric aerosols on trace metal chemistry in open surface sea water. 3. Lead. *J. Geophys. Res.*, 95, 5341–5347.

Marlon, J.R., Bartein, P.J., Carcailet, C., Gavin, D.G., Harrison, S.P., Higuera, P.E., Joos, F., Power, M.J. and Prentice, I.C. (2008) Climate and human influence on global biomass burning over the past two millennia. *Nature Geosciences*, 1, 697–702.

Merrill, J.T. (1989) Atmospheric long-range transport to the Pacific Ocean, in *Chemical Oceanography*, J.P. Riley, R. Chester and R.A. Duce (eds), Vol. 10, pp.15–20. London: Academic Press.

Moulin, C., Lamber, C.E., Dulac, F. and Dayan, U. (1997) Control of atmospheric export of dust from North Africa by the North Atlantic Oscillation. *Nature*, 387, 691–694.

Murphy, K.J.T. (1985) *The trace metal chemistry of the Atlantic aerosol*. PhD thesis, University of Liverpool.

Nriagu, O.N. (1989) Natural versus anthropogenic emissions of trace metals to the atmosphere, in *Control and Fate of Atmospheric Trace Metals*, J.M. Pacyna and B. Ottar (eds), pp.3–13. Dordrecht: Kluwer Academic Publishers.

Nriagu, J.O. and Davidson, C.I. (1986) *Toxic Metals in the Atmosphere*. New York: John Wiley & Sons, Inc.

O'Dowd, C.D., Facchini, M.C., Cavalli, F., Ceburnis, D., Mircea, M., Decesari, S., Fuzzi, S., Yoon, Y.J. and Putaud, J-P. (2004) Biogenically driven organic contribution to marine aerosol. *Nature*, 431, 676–680.

Pacyna, J.M., Semb, A. and Hanson, J.E. (1984) Emissions and long-range transport of trace elements in Europe. *Tellus*, 36B, 163–178.

Pacyna, J.M. and Pacyna, E.G. (2001) An assessment of global and regional emissions of trace metals to the atmosphere from anthropogenic sources worldwide. *Environmental Reviews*, 9, 269–298.

Peltzer, E.T. and Gagosian, R.B. (1989) Oceanic geochemistry of aerosols over the Pacific Ocean, in *Chemical Oceanography*, J.P. Riley and R. Chester (eds), Vol. 10, pp.282–338. London, Academic Press.

Prospero, J.M. (1981) Eolian transport to the World Ocean, in *The Sea*, C. Emiliani (ed.), Vol. 7, pp.801–874. New York: John Wiley & Sons, Inc.

Prospero, J.M., Charlson, R.W., Mohnen, V. *et al.* (1983) The atmospheric aerosol system: an overview. *Rev. Geophys. Space Phys.*, 21, 1607–1629.

Prospero, J.M., Uematsu, M. and Savoie, D.L. (1989) Mineral aerosol transport to the Pacific Ocean, in *Chemical Oceanography*, J.P. Riley, R. Chester and R.A. Duce (eds), Vol. 10, pp.137–218. London: Academic Press.

Prospero, J.M., Ginoux, P., Torres, O., Nicholson, S.E. and Gill, T.E. (2002) Environmental characterization of global sources of atmospheric soil dust identified with the Nimbus 7 total ozone mapping spectrometer (TOMS) absorbing aerosol products. *Rev Geophys.*, 40, 2.1–2.31.

Prospero, J.M. and Lamb, P.J. (2003) African droughts and dust transport to the Caribbean: climate change implications. *Nature*, 302, 1024–1027.

Pyle, D.M. and Mather, T.A. (2003) The importance of volcanic emissions for the global atmospheric mercury cycle. *Atmos. Environ.*, 37, 5115–5124.

Raes, F., Dingenen, R.V., Vignati, E., Wilson, J., Putaud, J.-P., Seinfeld, J.H. and Adams, P. (2000) Formation and cycling of aerosols in the global troposphere. *Atmos. Environ.*, 34, 4215–4240.

Rahn, K.A. (1976) *The Chemical Composition of the Atmospheric Aerosol*. Kingston, RI: Technical Report, Graduate School of Oceanography, University of Rhode Island.

Rahn, K.A. (1981) The Mn/V ratio as a tracer of large-scale sources of pollution aerosol for the Arctic. *Atmos. Environ.*, 15, 1547–1564.

Rahn, K.A. and Lowenthal, D.H. (1984) Elemental tracers of distant regional pollution aerosols. *Science*, 223, 132–139.

Rahn, K.A., Borys, R.D., Shaw, G.E., Schutz, L. and Jaenicke, R. (1979) Long-range impact of desert aerosol on atmospheric chemistry: two examples, in *Saharan Dust*, C. Morales (ed.), pp.243–266. New York: John Wiley & Sons, Ltd.

Rea, D.K. (1994) The paleoclimatic record provided by eolian deposition in the deepsea: The geological history of wind. *Rev. Geophys.*, 32, 159–193.

Renberg, I., Brännvall, M.-J., Bindler, R. and Emteryd, O. (2000) Atmospheric lead pollution history during four millennia (2000BC to 2000AD) in Sweden. *Ambio*, 29, 150–156.

Robinson, E. and Robbins, R.D. (1971) *Emissions, Concentrations, and Fate of Particulate Atmospheric Pollutants*. Washington, DC: Publication 4067, American Petroleum Institute.

Rosman, K.J.R. and Chisolm, W., Boutron, C.F., Candelone, J.P. and Görlach, U. (1993) Isotopic evidence for the source of lead in Greenland snows since the late 1960s. *Nature*, 362, 333–335.

Schneider, B. (1987) Source characterization for atmospheric trace metals over Kiel Bight. *Atmos. Environ.*, 21, 1275–1283.

Seinfeld, J.H. and Pandis, S.N. (1998) *Atmospheric Chemistry and Physics*. New York: John Wiley & Sons, Inc.

Simó, R. (2001) Production of atmospheric sulfur by oceanic plankton: biogeochemical, ecological and evolutionary links. *Trends in Ecology and Evolution*, 16, 287–294.

Simoneit, B.R.T. (1978) The organic chemistry of marine sediments, in *Chemical Oceanography*, J.P. Riley and R. Chester (eds), Vol. 7, pp.233–311. London: Academic Press.

Simoneit, B.R.T., Chester, R. and Eglinton, G. (1977) Biogenic lipids in particulates from the lower atmosphere over the eastern Atlantic. *Nature*, 267, 682–685.

Suman, D.O. (1986) Charcoal production from agricultural burning in Central Panama and its deposition in the sediments of the Gulf of Panama. *Environ. Conservation*, 13, 51–60.

Tindale, N.W. and Pease, P.P. (1999) Aerosols over the Arabian Sea: Atmospheric transport pathways and concentrations of dust. *Deep-Sea Res. II*, 1577–1595.

Walker, M.I. Mackay, W.A., Pettenden, N.J. and Liss, P.S. (1986) Actinide enrichment in marine aerosols. *Nature*, 323, 141–143.

Wadleigh, M.A. (2004) Sulphur isotopic composition of aerosols over the western North Atlantic Ocean Can. *J. Fish. Aq. Sci.*, 61, 817–825.

Wania, F. and Mackay, D. (1993) Global fractionation and cold condensation of low volatility organochlorine compounds in polar regions. *Ambio*, 22, 10–18.

Weisel, C.P., Duce, R.A., Fasching, J.L. and Heaton, R.W. (1984) Estimates of the transport of trace metals from the ocean to the atmosphere. *J. Geophys. Res.*, 89, 11607–11618.

Whitby, K.T. (1977) The physical characteristics of sulphur aerosols. *Atmos. Environ.*, 12, 135–159.

Yaaqub, R.R., Davies, T.D., Jickells, T.D. and Miller, J.M. (1991) Trace elements in daily collected aerosols at a site in southeast England. *Atmos. Environ.*, 25, 985–996.

# 5    The transport of material to the oceans: the hydrothermal pathway

Figure 5.1 shows a schematic representation of ocean crust around mid ocean ridges. At mid ocean ridges (MORs) ocean plates move apart and fresh basalt is emplaced. As the newly formed crust cools and subsides the basalt is covered by ocean sediments (see Chapters 13–16) consisting largely of insoluble desert dust and biogenic debris. The upper part of the igneous oceanic basement (Layer 2 of the oceanic crust), which underlies the sediment column (Layer 1 of the oceanic crust), is composed predominantly of basaltic lavas and their intrusive equivalents. These basalts are by far the commonest rock type found on the sea bed, and the chemical compositions of a number of marine basalts are given in Table 5.1. The presence of hot basalts (at about 1100°C) creates a hydrothermal circulation as seawater percolates down through the fractured ocean crust toward the heat source, and in the process becomes warmed and then rises back toward the seabed. These basaltic lavas can interact with seawater over a wide range of temperatures and time scales. In very general terms, the reactions can be classified into three types:

1 those involving hydrothermal activity, either at high temperature at depth in the crust at the centres of sea-floor spreading, or at intermediate temperatures on the ridge flanks;

2 those associated with the low-temperature weathering of basalt that has been exposed to seawater for relatively long periods of time, either at the sea floor or within the upper part of the basement;

3 those involving the extrusion of hot lava directly on to the sea bed.

## 5.1 Hydrothermal activity: high-temperature basalt–seawater reactions

The hydrothermal processes can produce metal rich sediments (see also Chapters 15 and 16), and the presence of these deposits on the seabed has been known for more than 100 years. The processes involved are now believed to be responsible for some commercially exploited mineral deposits such as those found in the Troodos ore deposits in Cyprus (Richardson *et al.*, 1987) which are believed to be a relict mid ocean ridge system uplifted onto land. However, the most dramatic manifestation of seawater–rock interaction, the 'black smokers' and the extraordinary vent biological communities, have been identified only over the last few decades or so, usually by the use of manned and unmanned submersibles. Studies of these systems has shown that seawater convecting through newly generated oceanic crust at ridge-divergent plate boundaries during sea-floor spreading can play an important role in controlling the chemical mass balance of the oceans. Various lines of geological and geophysical evidence have now established that sea-floor spreading is the dominant process in the formation of the ocean basins. In this process, new oceanic crust (or lithosphere) is formed from molten rock (magma) at the spreading centres on the mid-ocean ridge system: see Fig. 5.1(a). The pre-existing, that is older, basalt sequences are porous, and faults and large transform fractures are found at both fast- and slow-spreading ridges. In

*Marine Geochemistry*, Third Edition. Roy Chester and Tim Jickells.
© 2012 by Roy Chester and Tim Jickells. Published 2012 by Blackwell Publishing Ltd.

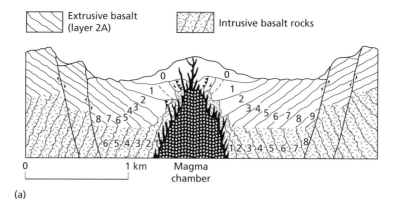

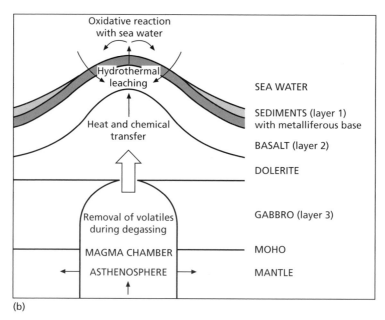

(b)

**Fig. 5.1** High-temperature hydrothermal activity at the spreading centres. (a) Preliminary structural model of the inner rift valley in the Famous area (Mid-Atlantic Ridge) showing the location of the magma chamber in relation to the generation of basalts (from Jones, 1978; after Moore et al., 1974). Relative ages of the basalts are indicated by the numbers 0–9, with 0 being the youngest. (b) The generation of hydrothermal solutions: the East Pacific Rise model (reprinted from Elderfield, 1976 with permission from Elsevier). Seawater penetrates the porous rock at the spreading ridge centre and circulates through the underlying crust. During this process the seawater undergoes both thermal and chemical transfer from the magma heat source, and also leaches elements from the rocks and sediments with which it comes into contact before re-entering the seawater reservoir via a series of venting systems as mineralized solutions.

these regions cold seawater penetrates this highly permeable pre-existing crust around the centres, sometimes to a depth of several kilometres, where it undergoes heat-driven circulation and comes into contact with the zones of active magma intrusion. During this process the seawater undergoes drastic changes in composition to form high-temperature hydrothermal solutions (up to ~350°C, but it should be pointed out that solutions do not necessarily boil due to the very high pressures at these depth in the ocean). The hot solutions emerge through the sea-floor venting system as hot springs that mix with the

**Table 5.1** Elemental composition of oceanic basalts*.

(a) Major element composition (wt% oxides).

| Element | Oceanic island basalts | | Mid-Atlantic Ridge basalts | |
| --- | --- | --- | --- | --- |
| | Average tholeiitic basalt | Average alkalic basalt | Tholeiitic basalt | High alumina basalt |
| $SiO_2$ | 49.36 | 46.46 | 50.47 | 48.13 |
| $TiO_2$ | 2.50 | 3.01 | 1.04 | 0.72 |
| $Al_2O_3$ | 13.94 | 14.64 | 15.93 | 17.07 |
| $Fe_2O_3$ | 3.03 | 3.37 | 0.95 | 1.17 |
| FeO | 8.53 | 9.11 | 7.88 | 8.65 |
| MnO | 0.16 | 0.14 | 0.13 | 0.13 |
| MgO | 8.44 | 8.19 | 8.75 | 10.29 |
| CaO | 10.30 | 10.33 | 11.38 | 11.26 |
| $Na_2O$ | 2.13 | 2.92 | 2.60 | 2.39 |
| $K_2O$ | 0.38 | 0.84 | 0.10 | 0.09 |
| $H_2O$ (total) | – | – | 0.59 | 0.29 |
| $P_2O_3$ | 0.26 | 0.37 | 0.11 | 0.10 |

(b) Minor element composition ($\mu g\,g^{-1}$).

| Element | Oceanic alkali basalts | Oceanic tholeiitic basalts | Basalts; Mid-Atlantic Ridge | Basalts; Mid-Indian Ocean Ridge | Basalt; Tonga Island Arc |
| --- | --- | --- | --- | --- | --- |
| Fe | 82 600 | 76 800 | 63 616 | 68 282 | 54 008 |
| Mn | 1 084 | 1 239 | – | – | – |
| Cu | 36 | 77 | 87 | 90 | 51 |
| Ni | 51 | 97 | 123 | 242 | 25 |
| Co | 25 | 32 | 41 | 73 | 30 |
| Ga | 22 | 17 | 18 | 20 | 13 |
| Cr | 67 | 297 | 292 | 347 | 75 |
| V | 252 | 292 | 289 | 340 | 230 |
| Ba | 498 | 14 | 12 | – | 14 |
| Sr | 815 | 130 | 123 | 131 | 115 |

* From original data sources listed by Riley and Chester (1971), and Chester and Aston (1976). The central ocean areas are dominated by tholeiitic pillow basalts of layer 2A, which are generated at the spreading ridges and which are moved away during sea-floor spreading so that their ages increase with increasing distance from the ridge crests.

overlying seawater which usually has a temperature of ~2°C. Confirmation of the existence of the hydrothermal solutions has been provided by photographic and visual (submersible dives) sightings of the venting of hot springs on to the sea floor at various spreading centres. These sightings have also shown the presence of specialized biological communities around the hot springs. These include dense beds of clams, mussels and giant vestimentiferan tubeworms, for which the primary food source is chemosynthetic bacteria which derive their energy needs from oxidizing reduced chemicals derived from the vents; this is an example of chemical, as opposed to photosynthetic, production, and may have implications for our understanding of the origin of life. For a detailed description of the biology of hydrothermal vent systems, and their status as environments for the origin of life, the reader is referred to the volumes edited by Childers (1988) and Holm (1992), respectively, and to Fisher et al. (2007).

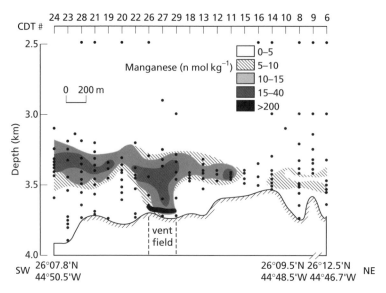

**Fig. 5.2** SW–NE transect for total reactive Mn at the TAG hydrothermal system on the Mid-Atlantic Ridge (from Klinkhammer *et al.*, 1986). Total reactive Mn is the Mn extracted from unacidified seawater samples by oxine after adjustment of the pH to 9, and represents mainly the dissolved fraction of the element. Reprinted with permission from Elsevier.

The dramatic effect that the venting of these high-temperature hydrothermal solutions can have on the chemistry of seawater is illustrated in Fig. 5.2, which shows the increase in dissolved Mn around a vent system at the TAG site on the Mid-Atlantic Ridge. Because of the importance of basalt–seawater interactions associated with venting systems such as this, oceanic rock is now recognized as a major reservoir in global geochemistry.

In addition to these remarkable features involving very hot waters, there are additional flows of hydrothermal waters that may be cooler (some <20°C) reflecting less interaction with hot basalts, although these interactions may still be geochemically significant and continue for tens of millions of years after the extrusion of the basalt (German and Von Damm, 2004).

It is now known that large-scale hydrothermal activity is always associated with the production of new oceanic crust at the centres of the mid-ocean ridges (Edmond *et al.*, 1982; German and Von Damm, 2004; Tivey, 2007), as was first predicted by Elder (1965). The process has far-reaching implications for Earth science, and such is its global extent that it has been estimated that the entire ocean circulates through the high-temperature ridge axial

zone, and so undergoes reaction with fresh basalt, to emerge via the hydrothermal venting system every 20–30 Ma. Such estimates are derived from heat flux measurements from sediment cores collected around mid ocean ridges and only 10% of the total heat flux at MORs may be associated with high temperature venting, with the majority of flow (approximately 1000 times the high temperature flows) associated with low temperature systems. The total flows through the high and low temperature hydrothermal systems are comparable to river inputs, with the flows through the low temperature systems dominant. The water exiting the vents often mixes with large amounts of ocean water and forms hydrothermal plumes as will be discussed later, and the ocean waters may circulate through these plumes on time scales of 4–8000 years (German and Von Damm, 2004). These calculations are inevitably uncertain, but they emphasize that, although hydrothermal processes are localized to MORs, they can have affects on the whole global ocean.

Studies of altered basalts dredged from the sea-floor have been used to infer processes within hydrothermal systems, and most recently some drilling of MORs has been used to understand these processes. Sampling of hydrothermal vent water is difficult and is usually done with a manned or unmanned submersible equipped with sampling gear designed for the water temperatures of >300°C. This sampling has led to an understanding of the broad processes involved in the venting systems and the summary following is based on the excellent review by German and Von Damm (2004). The chemistry of hydrothermal fluids varies from place to place and the very limited data available suggests it also varies with time within a particular vent system. Von Damm *et al.* (1985) identified a number of factors that contribute to variations in vent fluid composition.

1 Differences in rock type; for example, glasses react with seawater more rapidly than more crystalline forms.

2 Differences in temperature of the hydrothermal solutions at depth in the system.

3 Differences in the residence times, or flow rates, of the water in the fissure system: an increased residence time implies an increase in the time over which the water can react with the rock.

4 Differences in the depth at which the basalt–seawater reaction takes place, which affect the path length of the hydrothermal circulation.

5 Differences in the age of the hydrothermal systems; that is, in older systems the water is flowing through rocks that have already undergone considerable leaching.

Some hydrothermal fluids exit through an overlying sedimentary layer at the seabed and interactions with this layer can greatly modify the fluid composition, particularly if the sediments are calcareous and able to neutralize the acidic vent fluids (German and Von Damm, 2004) It was not possible to resolve the relative importance of these factors in a particular MOR system, but they serve to illustrate how the trace element composition of venting fluids can be influenced by local conditions. We can, however, describe a general series of reaction processes that dominate in these systems.

Seawater is drawn into a hydrothermal system and heated as it moves toward the hot basalt. A wide range of geochemical reactions ensue and the summary here represents those that are probably most important and ubiquitous. As temperatures exceed ~130°C, $CaSO_4$ (anhydrite) precipitates, removing calcium and a significant amount of the sulfate in seawater. Some sulfate is also reduced to sulfide by reactions with Fe(II) minerals. Magne-

sium is also removed from seawater by reactions with basalt to form clays such as smectite and chlorite which contain magnesium and hydroxide ions; a process that also releases $H^+$ ions into solution, thereby acidifying the waters once the carbonate buffer is consumed. Closer to the basalt heat source, if temperatures and pressures are high enough, the seawater separates into two fluid phases, one chloride-rich and one chloride-poor. After heating, these now hot acidic and reducing fluids move back toward the seabed interacting further with basalt and, if it is present, with overlying sediments.

The composition of venting fluids is summarized in Table 5.2, alongside that of seawater, and the wide range of compositions from the different systems should be noted. In comparison to seawater, vent

**Table 5.2** Chemical composition of known MOR vent fields and average seawater from German and von Damm (2004) based on slow to ultrafast spreading centres and excluding sediment covered systems, ultramafic and back-arc systems.

| Chemical Species | Units | Seawater | Vent fluids |
|---|---|---|---|
| T | °C | 2 | >2–405 |
| pH | 25°C, 1 atm | 7.8 | 2–>6.6 |
| Alkalinity | meq kg$^{-1}$ | 2.4 | 3.75–<2.3 |
| Cl | mmol kg$^{-1}$ | 540 | 30.5–1245 |
| SO$_4$ | mmol kg$^{-1}$ | 28 | <0–9 |
| H$_2$S | mmol kg$^{-1}$ | 0 | 0–110 |
| Si | mmol kg$^{-1}$ | 0.03–0.18 | 2.7–24 |
| Li | µmol kg$^{-1}$ | 26 | 4–2350 |
| Na | mmol kg$^{-1}$ | 464 | 11–983 |
| K | mmol kg$^{-1}$ | 10 | 0–59 |
| Rb | µmol kg$^{-1}$ | 1.3 | 0.16–31 |
| Cs | nmol kg$^{-1}$ | 2 | 2.3–364 |
| Mg | mmol kg$^{-1}$ | 52 | 0 |
| Ca | mmol kg$^{-1}$ | 10 | 1.3–109 |
| Sr | µmol kg$^{-1}$ | 87 | 29–133 |
| Ba | µmol kg$^{-1}$ | 0.14 | 1.6–>46 |
| Mn | µmol kg$^{-1}$ | <0.001 | 21–2750 |
| Fe | mmol kg$^{-1}$ | <0.001 | 0.007–15 |
| Cu | µmol kg$^{-1}$ | 0.007 | 0–150 |
| Zn | µmol kg$^{-1}$ | 0.012 | 0–600 |
| Co | µmol kg$^{-1}$ | 0.00003 | 0.02–0.42 |
| Pb | nmol kg$^{-1}$ | 0.01 | 183–376 |
| B | µmol kg$^{-1}$ | 415 | 356–1874 |
| Al | µmol kg$^{-1}$ | 0.02 | 0.1–18.7 |
| Br | µmol kg$^{-1}$ | 840 | 29–1910 |

fluids are acidic, reducing, silica-rich and deficient in $Mg^{2+}$, $SO_4^{2-}$ and alkalinity (unless it is added by reactions of vent fluids with carbonate-rich sediments before discharge), indeed it is likely that all $Mg^{2+}$, $SO_4^{2-}$ and alkalinity are removed from seawater within the vent system (see also Chapter 17). $Ca^{2+}$ is removed by the precipitation of anhydrite, but also released by reactions with basalt. The behaviour of K and Na is variable in terms of whether MORs are a source or sink and the behaviour of Cl is complicated by the phase separation of chloride discussed above. Many trace metals are massively enriched in vent fluids, particularly iron and manganese, due to leaching from the basalts by acidic, reducing solutions and complexation by chloride (see also Chapter 11). Any dilution with fresh seawater during transport through the basalt may lead to the precipitation of trace metals and highly non-conservative behaviour within the vent fluids. Dissolved Si is also released from the basalts as are trace gases such as methane, hydrogen, carbon monoxide and $^3He$. The discharges from hydrothermal systems also contain unique isotopic tracers of several elements, particularly Sr and Nd, and the geological record of the abundance of these isotopes can provide clues to the changes in hydrothermal activity over time (e.g. Elderfield and Shultz, 1996).

The hydrothermal systems described above are associated with spreading ridge crests where the sedimentation rates are low and where the venting fluids debouch, via chimneys, directly into seawater: sediment-starved systems. In contrast, under some conditions the spreading axis can be buried under a blanket of sediment: sediment-rich systems; one example of this is found in the Red Sea where the axis is covered by marine evaporites. Another example of this kind of hydrothermal system, which has been described by Von Damm *et al.* (1985), is located in the Guaymas Basin in the Gulf of California. North of the vent fields at 21°N, the East Pacific Rise (EPR) extends into the Gulf of California, and the spreading regime changes from a mature, open-ocean, sediment-starved type to an early opening, continental rifting type. An increase in sedimentation rate is associated with this change, and as a result the Guaymas Basin spreading axis is covered with a thick blanket of biogenous sediment that is rich in organic carbon derived from the highly productive overlying waters. The high-temperature (up to 315°C) Guaymas Basin venting solutions have a composition that is strikingly different from those characteristic of the sediment-starved, open-ocean systems, such as that found at 21°N on the East Pacific Rise. In particular, whereas the venting solutions at 21°N on the EPR are rich in the ore-forming elements (Fe, Mn, Cu, Zn, Pb, Co and Cd), have an acidic pH (3.3–3.8) and have low concentrations of ammonium, those from the Guaymas Basin are depleted in the ore-forming elements, have a more alkaline pH (5.9) and contain ammonium as a major ion. Von Damm *et al.* (1985) concluded that the Guaymas Basin venting solutions were formed as the result of a two-stage reaction process in which the hydrothermal solutions first reacted with the underlying basalt to produce a primary solution similar in composition to those found at the 21°N EPR sites, which then subsequently reacted with the biogenous sediments overlying the intrusion zone. During this second stage the pH of the primary hydrothermal solutions is raised by the dissolution of carbonate and by the thermocatalytic cracking of plankton carbon. As a result, sulfides are precipitated from the ascending solutions at depth in the sediment column, and the Guaymas Basin is thus a site of the active formation of a sediment-hosted massive sulfide deposit. The authors concluded that sediment-covered hydrothermal systems are probably not quantitatively important for the formation of oxidized metal-rich sediments.

High-temperature reactions can also take place between basalt and seawater when hot lava is extruded directly on to the sea bed. Bonatti (1965) has classified these submarine volcanic eruptions into two general types:

1 a quiet type, in which the reaction between seawater and lava is essentially prevented by the instantaneous formation of a thin crust of glass;

2 a more violent or explosive type, in which the lava is shattered on contact with seawater and the large surface area then available allows considerable hydration of the glass. In the explosive, hyaloclastic, eruption there is reaction between the basalt and seawater, with the result that new minerals such as palagonite, smectites and zeolites are formed and elements such as Ca, Na, K, Si, B, Mn, Zn and Cu are released into solution.

## 5.2 Hydrothermal activity: low-temperature basalt–seawater reactions

The upper 2–3 km of Layer 2 of the oceanic basement is a zone of chemical reaction between seawater and the crust as it moves away from the spreading centres. The extent to which an oceanic basalt undergoes low-temperature weathering reactions is a function of its age, that is, of the time it has spent in contact with seawater, which is related to its distance from the spreading centres. The low-temperature basalt–seawater reactions can take place both at the sea floor (as evidenced from dredged basalts) and at depth within Layer 2 of the oceanic crust (as evidenced from drilled basalts). Basalt–seawater reactions on the ocean floor take place in the presence of large volumes of water under oxidizing conditions. At depth within Layer 2, however, the rocks are in contact with much smaller volumes of water. In general, both field and laboratory experimental data indicate that the deeper basalts have undergone alteration, following reaction with seawater, in much the same way as surface basalts, although the reactions involved do not appear to have proceeded to the same extent.

Basalts dredged from the sea floor have invariably undergone some degree of reaction with seawater at the cold ambient bottom temperatures, and are often termed weathered basalts. During these low-temperature reactions, which have been reviewed by Honnerez (1981) and Thompson (1983), new mineral phases are formed and chemical transfer takes place between the rocks and seawater. Volcanic glass and bulk rock can be affected in different ways, but a number of general conclusions can be drawn regarding the low-temperature alteration of oceanic basalts (see e.g. Honnerez, 1981).

1 The oxidation and hydration of basalt is a ubiquitous phenomenon.

2 During basalt–seawater reactions there is often an uptake of K, Cs, Rb, B, Li and $^{18}O$ by the basalt (i.e. the rock acts as a *sink*, and the elements are usually incorporated into mineral phases formed during the alteration processes), and a loss of Ca, Mg and Si (i.e. the rock acts as a *source*).

3 Fe, Mn, Na, Cu, Ba and Sr exhibit no clear patterns, although the low-temperature weathering of basalt may provide a source for Fe and Mn to seawater (Elderfield, 1976).

4 Al, Ti, Y, Zr and the heavy rare earths show little or no change following basalt–seawater reactions.

5 Low-temperature alteration of the oceanic crust is a major sink for U, and according to Bloch (1980) may account for ~50% of the total amount of U supplied to the oceans at the present day.

*Hydrothermal Plumes.* The vent fluids from hydrothermal systems reach the ocean waters as acidic sulfide rich fluids at temperatures of up to 450°C and encounter cold 2°C alkaline oxic seawater. The resultant reactions can produce precipitation that form the chimneys, mounds and black and white 'smokers' so vividly revealed by cameras on submersibles. These chimneys form from the precipitation of calcium sulfate, silica and metal sulfides (German and Von Damm., 2004; Tivey, 2007). Fluids emanating from these mounds and chimneys are often enriched by many thousand-fold or more in iron and manganese compared to seawater (Table 5.2). As the fluids mix iron is rapidly precipitated on time scales of minutes to hours as sulfides initially, precipitates that can provide the black colour to black smokers, and then after oxidation as oxides. White smokers are the result of calcium sulfate precipitation. The precipitated iron can scavenge (Chapter 11) trace metals from the plume and hence as we shall see in Chapter 6, the enrichment of trace metals in vent fluids does not necessarily imply that these are a net source of metals to the ocean and in some situations, the MOR plumes represent net sinks for trace metals from the ocean. Manganese is also greatly enriched in the plume initially as relatively soluble Mn(II) which is much more slowly oxidized than Fe(II), because it involves a two electron transfer, with Mn(II) lifetimes estimated to be weeks to years. The high temperatures (and lower density) of vent fluids compared to ambient seawater cause them to rise until mixing equalizes density and thus forms a plume at heights of many hundreds of meters or more above the ridge crest. At slow spreading ridges such as the Mid Atlantic Ridge, the rift valley of the mid ocean ridge is relatively deep and the plume and resultant mineral deposits arising from the settlement of precipitated iron and associated trace metals may be contained within the rift valley. At faster spreading ridges, such as the East Pacific Rise, the MOR rift valley is much less deep and the plume will typically rise above the ridge and mix with the main ocean water body and disperse the deposits from the plum over a wider area (German and Von Damm, 2004).

Over time, hydrothermal flows may wax and wane; processes associated with episodes of basalt intrusion and changes in water flow paths within the hydrothermal system. Vent chimneys that lose their supply of hydrothermal fluids will cease to grow and eventually collapse, contributing to the metalliferous sediments around the MORs

## 5.3 The hydrothermal pathway: summary

**1** The oceanic crust is a major reservoir in global geochemistry and a number of different types of reactions occur between seawater and the basalts of Layer 2.

**2** These seawater reactions take place at a variety of temperatures and rock:water ratios, and can act as either source or sink terms in the marine budgets of some components.

**3** The most dramatic hydrothermal activity is found at the spreading ridge centres, where cold seawater circulates through hot, newly formed, basaltic crust. In this process, the composition of the seawater undergoes extensive changes before emerging at the sea bed in the form of white smoker or black smoker hot springs.

In the present chapter attention has been confined to the effects of hydrothermal activity on seawater, and the magnitude of the fluxes involved are discussed in Section 6.3. The effects that hydrothermal processes have on marine geochemistry, however, are much wider than this. In particular, the hydrothermal activity at spreading centres results in the formation of a series of mineral precipitates and in the generation of a unique type of deep-sea sediment. The chemical dynamics involved in the formation of the hydrothermal precipitates are discussed in Section 15.3.6 and 16.5.2. In this way, the full spectrum of hydrothermal activity in the oceans will be placed in a global marine context.

## References

Bloch, S. (1980) Some factors controlling the concentration of uranium in the World Ocean. *Geochim. Cosmochim. Acta*, **44**, 373–377.

Bonatti, E. (1965) Palagonite, hyaloclastites and alteration of volcanic glass in the ocean. *Bull. Volcanol.*, **28**, 257–269.

Childers, J.J. (1988) Hydrothermal vents. A case study of the biology and chemistry of a deep-sea hydrothermal vent of the Galapagos Rift. *Deep-Sea Res.*, **35**, 1677–1849.

Chester, R. and Aston, S.R. (1976) The geochemistry of deep-sea sediments, in *Chemical Oceanography*, J.P. Riley and R. Chester (eds), Vol. 6, pp.281–390. London: Academic Press.

Edmond, J.M., Von Damm, K.L., McDuff, R.E. and Measures, C.I. (1982) Chemistry of hot springs on the East Pacific Rise and their effluent dispersal. *Nature*, **297**, 187–191.

Elder, J.W. (1965) Physical processes in geothermal areas, in *Terrestrial Heat Flow*, W.H.K. Lee (ed.), pp.211–239. Washington, DC: American Geophysical Union. Monograph no. 8.

Elderfield, H. (1976) Hydrogenous material in marine sediments: excluding manganese nodules, in *Chemical Oceanography*, J.P. Riley and R. Chester (eds), Vol. 5, pp.137–215. London: Academic Press.

Elderfield, H. and Scultz, A. (1996) Mid ocean ridge hydrothermal fluxes and the chemical composition of the ocean. *Annual Rev. Earth and Planet. Sci.*, **24**, 191–224.

Fisher, C.R., Takai, K. and leBris, N. (2007) Hydrothermal vent ecosystems. *Oceanography*, **20**, 14–23.

German, C.R. and Von Damm, K.L. (2004) *Hydrothermal Processes in Treatise on Geochemistry*, H. Elderfield (ed.), Vol. 6, (series eds H.D. Holland and K.K. Turekian), pp.181–222. London: Elsevier.

Holm, N.G. (1992) *Marine Hydrothermal Systems and the Origin of Life*. Dordrecht: Kluwer Academic Publishers.

Honnerez, J. (1981) The aging of the oceanic crust at low temperature, in *The Sea*, C. Emillani (ed.), Vol. 7, pp.525–587. New York: John Wiley & Sons, Inc.

Jones, E.J.W. (1978) Sea floor spreading and the evolution of the ocean basins, in *Chemical Oceanography*, Vol. 4, J.P. Riley and R. Chester (eds), pp.1–74. London: Academic Press.

Klinkhammer, G., Elderfield, H., Grieves, M., Rona, P. and Nelson, T. (1986) Manganese geochemistry near high temperature vents in the Mid-Atlantic Ridge rift valley. *Earth Planet. Sci. Lett.*, **80**, 230–240.

Moore, J.G., Fleming, H.S. and Phillips, J.D. (1974) Preliminary model for extrusion and rifting at the axis of the Mid-Atlantic Ridge, 36°48 north. *Geology*, **2**, 437–440.

Richardson, C.J., Cann, J.R., Richardson, H.G. and Cowan, J.G. (1987) Metal-depleted root zones of the Troodos ore-forming hydrothermal systems, Cyprus. Earth planet. *Sci. Lett.*, **84**, 243–253.

Riley, J.P. and Chester, R. (1971) *Introduction to Marine Chemistry*. London: Academic Press.

Thompson, G. (1983) Hydrothermal fluxes in the ocean, in *Chemical Oceanography*, J.P. Riley and R. Chester (eds), Vol. 8, 270–337. London: Academic Press.

Tivey, M.K. (2007) Generatoin of seafloor hydrothermal vent fluids and associated mineral deposits. *Oceanography*, **20**, 50–65.

Von Damm, K.L., Edmond, J.M., Measures, C.I. and Grant, R. (1985) Chemistry of submarine hydrothermal solutions at Guaymas Basin, Gulf of California. *Geochim. Cosmochim. Acta*, **49**, 2221–2237.

# 6 The transport of material to the oceans: relative flux magnitudes

In the preceding chapters we have considered the three principal primary pathways by which material is transported to the World Ocean at the present day, that is, fluvial run-off, atmospheric deposition and hydrothermal exhalations. The magnitudes of the fluxes associated with each of these transport pathways are discussed below in order to consider the relative importance and impact of the different sources for different components.

## 6.1 River fluxes to the oceans

### 6.1.1 Introduction

The flux of material transported by a river reflects a complex interaction between hydrological and chemical factors in the catchment system, and it must be stressed that there are considerable uncertainties associated with all fluvial flux estimates. To make global flux estimations, average concentrations of constituents in individual river systems are often used, for example, in scaling-up procedures, but there are problems in assessing the extent to which the average values are meaningful. For example, concentrations of constituents in river water can be affected by a number of factors associated with variations in river flow. These include: long-term (e.g. annual) temporal variations at individual sampling stations, and spatial variations between individual stations; short-term fluctuations, for example, those in response to storm events; fluctuations as a result of rare events, such as those brought about by severe drought and catastrophic flooding; and seasonal variations in biological production. On the longer time scale river flows alter with climate (e.g. see Dai *et al.*, 2009). Changes to river management, particularly damming and land use changes, can alter fluvial sediment transport; overall the balance of these changes seems to have resulted in a net reduction of sediment transport of about 10% compared to fluxes in the absence of human activity (Syvitski *et al.*, 2005). As a result of factors like this, sampling campaigns rarely capture the full spectrum of flow and biogeochemical regimes within any particular river system.

The scaling-up of a single river data set to a global ocean scale, even if the average concentrations used are representative of that river system, can bias flux estimates because hydrological, climatic and lithological controls on the composition of river water vary widely from one river catchment to another. One way of overcoming this difficulty is to use data from major river systems. This approach, however, also has problems because the aggregate run-off from the 20 largest rivers in the world accounts for only ~30% of the global run-off. It therefore may be more reasonable to select rivers on the basis of particular catchment regimes rather than on the basis of size (see e.g. GESAMP, 1987). Despite these difficulties, however, it is still potentially rewarding to take a first look at the magnitude of fluvial fluxes to the oceans.

When the magnitude of the river-transported signal is considered it is necessary to distinguish between gross and net fluxes. In the present context, the following general definitions are adopted.

1 *Gross fluxes* are those transported by rivers to the marine boundary, which is taken here as the estuarine mixing zones.

*Marine Geochemistry*, Third Edition. Roy Chester and Tim Jickells.
© 2012 by Roy Chester and Tim Jickells. Published 2012 by Blackwell Publishing Ltd.

2 *Net fluxes* are those that are transported out of the estuarine mixing zone in an offshore direction and so are discharged into coastal seas.

The simplest way of estimating gross fluvial fluxes, and the one adopted here, involves a scaling-up procedure, which can be represented by equations such as

$$RF_g = X_d Q + X_p M_p \qquad (6.1)$$

where $X_d$ is the average dissolved content of an element $X$ in river water, $Q$ is the annual river-water discharge to the oceans ($38.5 \times 10^3 \, km^3 \, yr^{-1}$; Syvitski *et al.*, 2005) $X_p$ is the average content of river particulate material (RPM) and $M_p$ is the river particulate discharge to the oceans ($12.6 \times 10^9 \, tonnes \, yr^{-1}$; Syvitski *et al.*, 2005). For net fluxes, it is necessary to take account of the processes that occur in estuaries, and this is discussed below in relation to dissolved and particulate trace elements.

### 6.1.2 The gross and net fluvial fluxes of total suspended material

#### 6.1.2.1 Gross flux

The variation in suspended load between some major river systems is illustrated in Table 6.1(a). There have been various attempts to quantify the average global discharge of river suspended sediment to the land–ocean margins, and here we use the estimates of Syvitski *et al.*, (2005) (Table 6.1b). They estimate a current global flux of suspended matter of $12.6 \times 10^9$ tonnes $yr^{-1}$, and suggest this value is about 10% lower than that in pre-industrial times due to balance of the impacts of increased soil erosion and damming.

As noted earlier, there are substantial uncertainties in overall flux estimates and the compilations of Syvitski *et al.* (2005) have an overall uncertainty of about 15%.

Despite constraints imposed by uncertainties in the database, it is still possible to identify a number of general features in the geographical distributions of river suspended-sediment loads (i.e. excluding bedload, which probably makes up ~10% of the total sediment load) and their discharges to the land–ocean boundaries.

1 Most rivers have average suspended sediment loads in the range ~$10^2$–$10^3 \, mg \, l^{-1}$ (Milliman, 1981).
2 Mountain rivers have sediment yields that, on average, are around three times greater than those

**Table 6.1(a)** The river discharge of suspended sediment to the oceans: Suspended sediment discharge from some major rivers (Milliman and Meade, 1983).

| River | Annual suspended sediment discharge ($\times 10^6 \, t \, yr^{-1}$) |
|---|---|
| Hwang Ho (China) | 1080 |
| Ganges (India) | 1670 |
| Brahmaputra (Bangladesh) | |
| Yangtze (China) | 478 |
| Indus (Pakistan) | 100 |
| Amazon (Brazil) | 900 |
| Mississippi (USA) | 210 |
| Irrawaddy (Burma) | 265 |
| Mekong (Thailand) | 160 |
| Red (North Vietnam) | – |
| Nile (Africa) | 0 |
| Zaire (Africa) | 43 |
| Niger (Nigeria) | 40 |
| St Lawrence (Canada) | 4 |

**Table 6.1(b)** Suspended sediment discharge in the oceans by rivers from the continents (Syvitski *et al.*, 2005).

| Continental region | Drainage Area ($10^6 \, km^2$) | Discharge $km^3 \, yr^{-1}$ | Current Sediment discharge ($10^6 \, t \, yr^{-1}$) | Sediment yield ($t \, km^{-2} \, yr^{-1}$) |
|---|---|---|---|---|
| North America | 21 | 5820 | 1910 | 91 |
| South America | 17 | 1154 | 2450 | 144 |
| Europe | 10 | 2680 | 680 | 68 |
| Asia | 31 | 9810 | 4740 | 152 |
| Africa | 20 | 3800 | 800 | 40 |
| Australasia | 4 | 610 | 390 | 98 |

of plains rivers (Dedkov and Mozzherin, 1984) and now provide almost 90% of the sediment flux (Syvitski *et al.*, 2005).

3 The suspended loads transported to the ocean margins annually can vary over three orders of magnitude from one river to another (e.g., from the $1890 \times 10^{12}$ g yr$^{-1}$ for the Hwang Ho, to the $4 \times 10^{12}$ g yr$^{-1}$ for the St Lawrence: estimates from Holeman (1968)), and much of the suspended sediment delivered to the oceans is carried by a relatively small number of major rivers. Table 6.1(a) emphasizes the differences in sediment yield between rivers, with high sediment loads associated with high mountain regions and highly erodible sediments, such as in the Himalayas or Indonesia. Smaller rivers in such areas can produce a disproportionately large part of the total flux.

4 Overall, a large proportion of the total river suspended sediment discharged annually to the land–ocean boundary originates from the Asian continent (Table 6.1b). Most of the sediment flux to the world oceans goes into the Atlantic, Indian and Pacific Oceans, with rather little into the Arctic (Table 6.1c). This pattern is further illustrated in Fig. 6.1(a) based on an earlier but broadly comparable data set on particulate fluxes from major drainage basins.

To this particulate riverine flux can be added $2.9 \times 10^9$ tonnes yr$^{-1}$ from glacial sources, 90% from Antarctica and hence reaching the Southern Ocean (Raiswell *et al.*, 2006). Given the uncertainties in this number, and uncertainties in how much is trapped on land and in the coastal ocean (Raiswell *et al.*, 2006) we have excluded it from subsequent calculations which may mean that the overall sediment input is underestimated by ~20%. If iceberg transport carries material into the open ocean and bypasses

the coastal trapping, this underestimate may be more significant. Further studies of the potential importance of glacial inputs are certainly required.

### 6.1.2.2 Net flux

Processes operating in the estuarine environment severely modify the gross riverine particulate matter (RPM) flux delivered to the land–sea margins, and according to Judson (1968) it is probable that at the present time ~90% of the particulate material transported by rivers is trapped in estuaries (see Section 3.2.7). This average value will of course disguise significant global variability with trapping being particularly effective on large tectonically passive shelves. If this estimate of ~90% for the estuarine removal of suspended particulates is applied to the global estimate of ~$12.6 \times 10^9$ tonnes yr$^{-1}$ for the discharge of RPM (see Section 6.1.1) it would result in a *net* global river-transported flux of ~$1.26 \times 10^9$ tonnes yr$^{-1}$, very close to the estimate made independently by Bewers and Yeats (1977). Elemental fluxes can then be estimated by multiplying the gross or net fluxes of RPM by the average concentrations such as those listed in Chapter 3 (see Table 3.4). The resulting gross and net particulate fluxes of some major and trace elements are given in Table 6.2. Note this approach does assume that there is no fractionation of the particulate flux during estuarine removal, only sedimentation.

### 6.1.3 The gross and net fluvial fluxes of total dissolved material

#### 6.1.3.1 Gross fluxes

According to Meybeck (1979) the total dissolved load transported to the oceans by rivers is ~$3.7 \times 10^{15}$ g yr$^{-1}$, which is around 25% of the suspended sediment load, using the estimate of ~$12.6 \times 10^{15}$ g yr$^{-1}$ excluding bedload given above. Like the suspended sediment yields, there are patterns in the fluvial discharge of total solutes to the oceans. This was highlighted by Walling and Webb (1987), who produced a map showing the generalized global pattern of fluvial total solute yields: see Fig. 6.1(b). Global total solute discharge patterns reflect the combined influences of factors such as the magnitude of the river run-off, the catchment lithology and the

**Table 6.1(c)** Suspended sediment discharge by rivers to various ocean basins (Syvitski *et al.*, 2005).

| Ocean Basin | Current sediment discharge $10^6$ tonnes yr$^{-1}$ |
|---|---|
| Arctic | 420 |
| Atlantic | 3410 |
| Indian | 3290 |
| Pacific | 4870 |
| Mediterranean and Black Sea | 480 |
| Inland Seas | 140 |

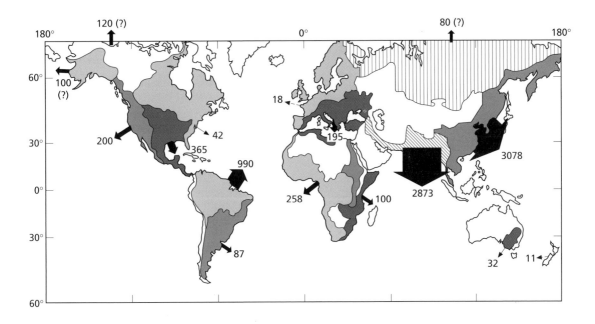

(a)

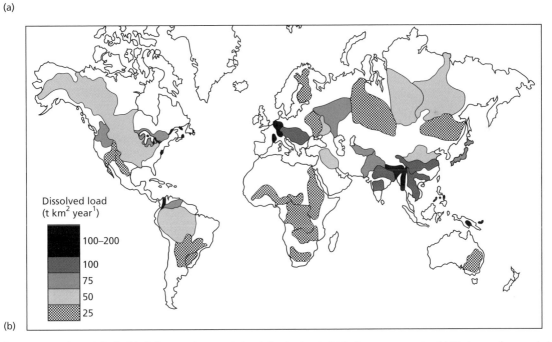

(b)

**Fig. 6.1** General trends in fluvial discharges of suspended and dissolved material to the ocean margins. (a) Discharge of suspended sediment (from Milliman, 1981); units in millions of metric tons (Mt) per year. (b) Discharge of total dissolved solutes (reprinted by permission from Walling and Webb, 1987). (c) Relative fluvial suspended sediment and water discharges from the continents (from Degens and Ittekkot, 1985). NA, North America; SA, South America; AS, Asia; AF, Africa; AR, Arctic USSR; OC, Oceania; EU, Europe.

climatic regime. Walling and Webb (1987) identified a number of general features in the global river discharge of total solutes that can be related to these run-off, geological and climatic controls.

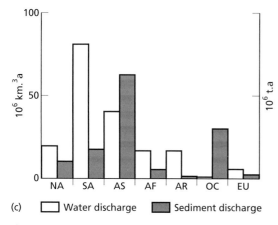

(c) ☐ Water discharge ■ Sediment discharge

**Fig. 6.1** *Continued*

**1** The large dissolved solute discharge values for Asian rivers result from their high run-off, which promotes enhanced transport.
**2** The relatively high loads for European rivers reflect the predominance of easily eroded sedimentary strata, including limestones, in their catchments.
**3** Generally low dissolved loads are found for the rivers of Africa and Australia, which are a consequence of the presence of ancient basement rocks, with a low weathering susceptibility, in their catchments.
**4** The extremely high dissolved solute loads for Burma (the Irrawaddy River) and Papua New Guinea (the Fly and Putari rivers) reflect a combination of high run-off, the availability of sedimentary rocks in the catchment regions and tropical temperatures, which enhance chemical weathering.

A comparison of fluvial water and sediment discharges from the continents is illustrated in Fig. 6.1(c).

**Table 6.2** Gross and net dissolved and particulate fluvial fluxes of elements to the world oceans in units of $10^9\,mol\,yr^{-1}$. The derivation of fluxes is described in the text. Part (a) of the table presents results for major ions and other elements with simple dissolved speciation, while part (b) presents the potentially more geochemically reactive trace metals.

| Element | Gross Particulate Flux | Net Particulate Flux | Gross Dissolved Flux | Net Dissolved Flux |
|---|---|---|---|---|
| (a) Major ions | | | | |
| Li | 46 | 4.6 | 10 | 10 |
| B | 82 | 8.2 | 35 | 35 |
| Na | 3900 | 390 | 9200 | 9200 |
| Mg | 6200 | 620 | 9400 | 9400 |
| S | | | 6700 | 6700 |
| Cl | | | 6400 | 6400 |
| K | 6400 | 640 | 1700 | 1700 |
| Ca | 6700 | 670 | 22900 | 22900 |
| Br | 0.8 | 0.08 | | |
| Rb | 15 | 1.5 | 0.73 | 0.73 |
| Sr | 21 | 2.1 | 27 | 27 |
| Cs | 0.6 | 0.06 | $3 \times 10^{-3}$ | $3 \times 10^{-3}$ |
| (b) Trace Metals | | | | |
| Al | 44000 | 4400 | 46 | 32 |
| V | 42 | 4.2 | 0.5 | 0.5 |
| Cr | 24 | 2.4 | 0.5 | 0.5 |
| Mn | 240 | 24 | 24 | 12 |
| Fe | 11000 | 1100 | 42 | 4.2 |
| Co | 4 | 0.4 | 0.1 | 0.05 |
| Ni | 19 | 1.9 | 0.5 | 0.5 |
| Cu | 21 | 2.1 | 0.9 | 0.9 |
| Zn | 68 | 6.8 | 0.35 | 0.35 |
| Cd | – | – | 0.027 | 0.135 |
| Pb | 9 | 0.9 | 0.015 | 0.015 |

Gross dissolved fluxes can be calculated using Equation 6.1 if the overall freshwater flow is known ($38.5 \times 10^3 \, km^3 \, yr^{-1}$, Syvitski *et al.*, 2005; see previous text) along with reliable average dissolved riverine concentrations (see Table 3.4) This is done for selected elements in Table 6.2.

As noted earlier, the sampling and analysis of natural water samples for trace metal analysis is a demanding task and much of the early data is erroneously high due to contamination. The results in Table 6.2 are based on the best recent compilation of riverine trace metal concentrations (see Chapter 3). These results reflect fluxes little affected by human impact. As noted in Chapter 3, some river systems are heavily contaminated by human actions, although these are often smaller flow river systems and so do not particularly perturb the large scale flux estimates. The results presented in Table 6.2 are generally similar (agreeing within a factor of ~3) with the compilation of Martin and Windom (1991) except for Mn (probably because the complex redox chemistry of this element makes estimates of fluxes difficult) and Cd and Pb where the differences may well reflect different assessments of the significance of anthropogenic contamination.

### 6.1.3.2 Net fluxes

The data given in Table 6.2 represent *gross* river-transported fluxes, that is, those that enter the estuarine mixing zone. Processes operating in estuaries can modify both the dissolved and particulate phases in the river-water inputs, and as a result the strengths of the river-transported signals entering an estuary can be very different from those that are exported to the ocean (see Section 3.2).

It was shown above that ~90% of the particulate material transported by rivers is trapped in estuaries at the present time. It is apparent, therefore, that the gross river particulate fluxes given in Table 6.2 must be strongly reduced in order to derive *net* fluxes for the export of particulate components from estuaries. In order to offer a first look at these net fluxes, the gross particulate fluxes have been reduced by a global factor of 90%; the data are listed in Table 6.2.

Some dissolved elements exhibit unambiguous conservative behaviour in the estuarine mixing zone. For these elements, which include those major components contributing to the salinity of seawater and

ions present as simple cations and anions in river water, the gross river fluxes will not be markedly changed during estuarine mixing, and those listed in Table 6.2 can be assumed to represent the net fluxes.

Quantification of the estuarine behaviour of particle reactive elements such as many trace metals is more complicated. Various attempts have been made over the past few years to estimate net river fluxes of dissolved trace elements from their gross fluxes by taking account of estuarine processes. For example, the zero-salinity end-member approach has been adopted in several investigations. This is an attempt to estimate the chemical composition of river water that has actually passed through the estuarine filter and reached coastal waters. Although there are a number of problems associated with this approach, it can still produce useful data. The zero-salinity end-member or, as it is sometimes called, the effective river end-member, can be identified from either estuarine or, under suitable conditions, coastal water data. A number of workers have used the estuarine mixing model outlined by Boyle *et al.* (1974) (see Section 3.2.3) to evaluate an effective river end-member by extrapolating linear element:salinity ratios back to a zero-salinity intercept. For example, Edmond *et al.* (1985) used this approach to evaluate the *net* river fluxes to the ocean from the Changjiang River and also quoted similar data for the Amazon. Bewers and Yeats (1977) used a similar approach on the St Lawrence, as did Kremling (1985) for rivers entering the North Sea and Bruland and Franks (1983) for the Eastern seaboard of the US.

The zero-salinity end-member approach yields estimates of the *net* flux of components from individual river-estuarine systems. These can then be extrapolated on to a global scale by some form of scaling-up procedure, although this is difficult because inevitably the behaviour of trace metals has been thoroughly studied in rather few estuaries. In Chapter 3 it was suggested that there is clear evidence for Fe and Al removal in most estuaries ranging from 70–95% and ~20–>50% removal respectively. For other trace metals the situation is more complex, but Martin and Windom (1991) do provide a useful estimate of the global significance of estuarine dissolved metal trapping. In this they distinguish between groups of metals that behave in the oceans similarly to the nutrients, and those they describe as under geochemical control (or scavenging control see

**Table 6.3** Estuarine removal of dissolved trace metals as a percentage of riverine dissolved inputs from Martin and Windom (1991). The negative sign indicates release rather than removal during estuarine mixing.

| Element | % removal |
| --- | --- |
| Al | 30 |
| Mn | 50 |
| Fe | 90 |
| Co | 50 |
| Ni | 0 |
| Cu | 0 |
| Zn | 0 |
| Cd | −500 |
| Pb | 0 |

Chapter 11). The nutrient-like metals of relevance here include Cd, Cu, Ni and Zn and those under geochemical control (and much more sensitive to estuarine removal) include Al, Co, Fe and Mn. For Pb, Martin and Windom (1991) suggest that while it is strongly controlled by geochemical removal, this operates principally in the freshwater regime and is followed by conservative behaviour within estuaries. Table 6.3 summarizes the removal estimates of Martin and Windom. In the case of Cd, uniquely, there is clear evidence of release from the particulate phase during estuarine mixing due to efficient chloride complexation of cadmium, and Martin and Windom argue this augments the dissolved flux by a factor of five.

In Table 6.2 these estuarine removal estimates along with the estimate of 90% particulate removal are combined to estimate net fluxes to the oceans for these selected trace metals.

## 6.1.4 The gross and net fluvial fluxes of organic carbon and the nutrients

### 6.1.4.1 Organic carbon

Major efforts over the last decade or so have greatly improved both the sampling and analysis of dissolved and particulate organic carbon (DOC and POC) in rivers. Much of this research has been reviewed by Perdue and Ritchie (2003). They suggest a global total organic carbon flux in rivers of 28–$31 \times 10^{12} \, mol \, C \, yr^{-1}$ with a DOC/POC ratio of 1.2,

resulting in DOC and POC fluxes of 13–14 and $15$–$17 \times 10^{12} \, mol \, C \, yr^{-1}$ respectively. These estimates are very similar to model based estimates of Seitzinger *et al.* (2005.)

As described in Chapter 3 this material includes a wide range of compounds so it is inevitably difficult to completely characterize its estuarine behaviour. However, we assume here again that 90% of the POC is trapped within the estuary, although about a third of this may be subsequently degraded and not buried as discussed in Chapter 3. In terms of the DOC, the evidence presented in Chapter 3 suggests most of this material is conservative within estuaries. Hence, the net flux to the oceans of POC and DOC are estimated to be 1.5–1.7 and $13$–$14 \times 10^{12} \, mol \, C \, yr^{-1}$ respectively, giving a total flux of $\sim 15 \times 10^{12} \, mol \, C \, yr^{-1}$.

### 6.1.4.2 The nutrients

The impacts of anthropogenic activity, particularly agricultural activity and waste water discharges, have greatly perturbed the riverine transport of nitrogen and phosphorus. Indeed it is now rather hard to estimate the natural transport fluxes and hence to actually estimate the scale of human perturbation. Seitzinger *et al.* (2005) estimate that fluxes of dissolved inorganic phosphorus (DIP) and nitrogen (DIN, principally as nitrate) have been increased by a factor of about 3, and other estimates all agree that the scales of human perturbation are very large. By contrast Seitzinger *et al.* (2005) suggest that the perturbations of the dissolved organic nitrogen and phosphorus discharges (DON and DOP) only represents a 25% increase over natural levels. In the case of dissolved silicon (DSi), human activity has little impact on the inputs to rivers which are dominated by natural weathering reactions. However, the increase in N and P concentrations serves to increase biological activity in rivers and the extensive damming of many rivers allows improved conditions for such biological activity. This activity includes the growth of diatoms – silica shelled phytoplankton – resulting in reduced DSi concentrations and fluxes to the oceans (e.g. Humborg *et al.*, 2000).

The scale of human impacts, and the resulting wide range of fluxes between river systems depending on the activity within catchments, also makes it difficult to estimate current fluxes from scaling up individual measurements or from models. The fluxes

**Table 6.4** Modern day gross fluvial fluxes of N, P and DSi $10^{12}$ mol yr$^{-1}$ based on Turner *et al.* (2003). Total N includes ammonium and dissolved and particulate organic N as well as nitrate, total P includes dissolved and organic P as well as DIP.

| Component | Flux |
|---|---|
| Nitrate | 1.2 |
| Total N | 1.5 |
| DIP | 0.08 |
| TP | 0.12–0.18 |
| DSi | 6.9 |
| Particulate Si | 126 |

shown in Table 6.4 are from an assessment by Turner *et al.* (2003). This paper compares these fluxes to those derived by others using rather different approaches which yield fluxes that differ from those estimated by Turner by up to a factor of two, and more in the case of total phosphorus. Such differences illustrate the incomplete nature of our estimates of global fluxes. As discussed in Chapter 3, nitrate and DSi generally appear to behave conservatively in estuaries, unless these are so large that significant denitrification takes place to remove nitrate. By contrast phosphorus is intensively cycled within estuaries between the particulate and dissolved phase with particulate P potentially trapped within the estuaries. Much of the particulate N and P material will be organic and although this may be retained with other particulate matter in estuaries it may, over longer time scales, be broken down by bacteria to its soluble forms. It seems reasonable therefore to assume that most of the fluvial nitrogen, phosphorus and dissolved silicon escapes estuarine trapping and reaches the coastal ocean. Hence the gross fluxes in Table 6.4 are equivalent to the net fluxes. Particulate Si is included for completeness in Table 6.4 and 90% of this is likely trapped within estuaries. Most particulate Si is, however, aluminosilicate material and essentially not bioavailable and so is not really relevant to nutrient fluxes.

## 6.2 Atmospheric fluxes to the oceans

### 6.2.1 Introduction

In order to estimate the *net* flux of a component from the atmosphere to the ocean it is necessary to know (i) its burden in the air; (ii) the rate at which it is deposited on to the sea surface; and (iii) the extent to which it is recycled back into the atmosphere.

#### 6.2.1.1 Atmospheric burden

To determine the atmospheric burden of a component in a slice of the atmosphere (e.g. a 1 m$^2$ column), data must be available on its concentration per unit of air (e.g. µg m$^{-3}$) and the height to which it is dispersed, that is, the scale height. The global atmospheric concentrations of many elements vary over one to three orders of magnitude, and are often geographically dependent. They may also vary with altitude; for instance the Sahara dust plume usually leaves the North African coast at a height of several kilometres. Because of these complications, the assessment of the atmospheric burdens of elements has often been restricted to a local scale. As more reliable data have become available, however, it has been possible to make first-approximation estimates of the global atmospheric burdens of some elements (see e.g. Walsh *et al.*, 1979). In making these estimates it is assumed that the atmosphere is in a steady state, that is, the rate of input of the component is equal to its rate of output, over the time scale of atmospheric aerosol residence times which are of the order of days. The global burden ($C_T$) of the component is then computed from an equation of the type

$$C_T = C_O A S \qquad (6.2)$$

in which $C_O$ is the surface concentration of an element (g or mol m$^{-3}$ of air), $A$ is the surface area of the atmosphere (m$^2$) and $S$ is the scale height, that is, the height to which the atmospheric component is dispersed (usually assumed to be between ~3000 and ~5000 m). While such an approach can be useful for some purposes, this kind of global averaging has many limitations given the large global gradients and improved understanding of processes is allowing improved global models of atmospheric aerosol transport and deposition to be developed (e.g. Fig. 4.5). Hence recent estimates of fluxes to the oceans tend to be based on models which include emissions, transport, atmospheric transformations and deposition.

#### 6.2.1.2 Rate of deposition

Air to sea fluxes result from the removal of material that is present in the atmosphere. The removal of gaseous components is described in Section 8.2, and attention at this stage will be confined to aerosols.

The deposition of aerosols from the atmosphere is controlled by a combination of 'dry' (gravitational settling and turbulent diffusion) and 'wet' (precipitation scavenging) processes. In both depositional modes the total (i.e. particulate + dissolved) concentrations reaching the sea surface from the atmosphere are dependent on the composition and atmospheric concentrations of the primary aerosol (Chapter 4). In addition to gradients in the deposition flux, the principal constraint on the manner in which atmospheric trace metals enter the oceanic biogeochemical cycles, however, is the degree to which they are soluble in seawater, and with respect to the particulate/dissolved metal speciation the processes associated with the two deposition modes are geochemically different. In the 'dry' mode, aerosols are delivered directly to the sea surface, and solubility is constrained by *particle ↔ sea-water* reactivity. In contrast, in the 'wet' deposition mode, trace metal solubility is constrained initially by *particle ↔ rainwater* reactivity, and as some rain waters can have a pH as low as <3, this can result in a number of trace metals being highly soluble prior to the aerosols reaching the sea surface.

*'Dry deposition'.* During their residence time in the atmosphere aerosols can pass through a number of cloud systems that do not generate rain, but rather evaporate. Within these cloud systems the aerosol particles may be subjected to several 'wetting' and 'drying' cycles, during which they can be coated by solution films having pH values as low as ~2.0 and reactive chemical species such as nitrogen compounds may be chemically altered by this cycling. The term 'dry' therefore is used in the present context to identify the aerosol deposition mode that does not involve an aqueous *deposition* phase. In the 'dry' deposition mode, therefore, although the aerosols may have passed through several 'wetting' and 'drying' cycles, they reach the sea surface by the direct air → sea surface route.

The 'dry' removal of particles from the atmosphere is a *continuous* process that is affected by a number

of factors, which include windspeed and particle size. Dry deposition rates increase markedly with particle size and 'dry' deposition therefore is especially important for the removal of large particles from the air. For example, McDonald *et al.* (1982) showed that the 'dry' deposition of sea-salts was dominated by large particles (MMD > 10 μm), which accounted for ~70% of the total salt deposition, although they made up only ~10% of the total mass. Dry deposition rates are usually estimated by multiplying an atmospheric concentration (mass or moles m$^{-3}$) by a deposition velocity usually in units of cm s$^{-1}$, to produce a flux in units of mass (or moles) m$^2$ time$^{-1}$. Seinfeld and Pandis (1998) present several representations of deposition velocity and all show a very strong and complex dependence on particle size, for instance being relatively constant for particles in the size range 0.1–1 μm and then with deposition velocities increasing by an order of magnitude with an increase in aerosol size from 1 to 10 μm.

*'Wet deposition'.* This involves the removal of both water-soluble gases and particulate material from the atmosphere by incorporation into precipitation to scavenge in cloud droplets (in-cloud) or falling rain (below cloud); processes that are random in time. In the most general sense, 'wet' deposition rates depend on the concentrations of a component in rain and the total amount of rain that falls on to a surface, and if both pieces of information are available then this is simplest means of calculating wet deposition. In this mode, therefore, aerosols reach the sea surface by the indirect air → rain → sea-surface route. To understand 'wet' deposition, it is therefore necessary to relate it to the elemental composition of rain water.

*Washout factors.* It is difficult to collect rain for compositional analysis at all over the oceans, because rainfall is such an episodic process, let alone to collect sufficient samples to derive a realistic average composition. The direct measurement of amounts of rainfall over the ocean is also difficult, although models can provide estimates of this and these can be well constrained by comparison to island data and ocean surface salinity measurements. Washout factors, or scavenging ratios, which have been measured at a few sites are often used to determine the degree to which a component is removed from the air by rain. The washout factor (WF) is calculated from an equation of the type

$$WF = C_r/C_a \qquad (6.3)$$

in which $C_r$ is the concentration of a component in the rain and $C_a$ is its concentration in low-level air; sometimes an air density term is included in the equation. Values of $WF$ for the elements studied most commonly, lie in the range $\sim 10^2 - \sim 10^3$. Measured aerosol concentrations over the ocean can then be multiplied by the washout factor to provide a flux estimate.

### 6.2.1.3 Extent of recycling

The recycling of particulate components across the sea surface can occur during the generation of sea-salts. Data are now available on the degree to which some elements are fractionated during the production of seasalt aerosol at the ocean surface with respect to bulk seawater (see Section 4.3), and for these the extent of recycling across the air–sea interface can be estimated with some degree of certainty. For example, Arimoto et al. (1985) used a combination of aerosol, rain and seawater data to estimate that at Enewetak (North Pacific) the percentage of wet deposition associated with recycled sea-salts can be substantial; values for individual elements included 15% (Zn), 30% (V) and 48% (Cu). For ions with major seawater sources, essentially all the gross flux to the oceans is probably derived from recycling of seaspray. Clearly, recycling must be taken into account when estimates are made of the *net* deposition fluxes of some elements to the sea surface from the atmosphere.

## 6.2.2 Atmospheric elemental fluxes

### 6.2.2.1 'Dry' deposition

Although the material reaching the sea surface by 'dry' deposition, that is, the direct air → sea-surface route, will undergo changes brought about by sea-water reactivity, it will initially have the same elemental composition as that of the material falling out from the parent aerosol: a particle size-dependent process, which changes with distance from the aerosol source.

### 6.2.2.2 'Wet' deposition: the elemental chemistry of marine rain water

In contrast to 'dry' deposition, aerosols involved in 'wet' deposition (precipitation scavenging), which reach the sea surface by the indirect air → rain → sea-surface route, can undergo considerable chemical changes before they reach the oceanic environment.

*Major ions in rain waters.* Data are now available on the major ion concentrations in rain waters from a number of marine locations (see Table 6.5). The principal major ions in world rain waters are dominated by the cations $H^+$, $NH_4^+$, $K^+$, $Ca^{2+}$ and $Mg^{2+}$, and the anions $SO_4^{2-}$, $NO_3^-$ and $Cl^-$. Scavenged sea-salt has a strong influence on the chemistry of precipitation over, and adjacent to, marine regions. Despite this, other sources impose fingerprints on the major ions in marine rain waters. For example, Church et al. (1982) have pointed out that even at remote marine locations alkali and alkaline earth cations in rain water can have a terrestrial dust source. The major anions in marine rain can also have contributions from non-sea-salt sources. For example, the sulfate in excess of that derived from sea-salt can arise from natural biogenic emissions,

**Table 6.5** Major ions in marine-influenced rain water (volume weighted mean) $\mu eql^{-1}**$.

|  | Amsterdam Island* | Bermuda† | Lewes, Delaware‡ |
|---|---|---|---|
| Percentage sea-salt (Na) | 97.7 | 80.3 | 54.1 |
| $\Sigma SO_4$ (% excess) | 29.2 (16.9) | 36.3 (51) | 62.5 (89) |
| $Na^+$ | 206.5 | 148 | 56.5 |
| $Mg^{2+}$ | 45.9 | 40 | 12.6 |
| $H^+$ (pH) | 8.8 (5.06) | 18.4 (4.74) | 53.4 (4.22) |
| $Ca^{2+}$ | 8.6 | 15.3 | 8.53 |
| $K^+$ | 4.4 | 4.03 | 1.91 |
| $NH_4^+$ | 1.8 | 4.54 | 18.9 |
| $Cl^-$ | 237.7 | 191 | 46.4 |
| $NO_3^-$ | 1.3 | 6.57 | 25.2 |

* Central Indian Ocean; data from Galloway and Gaudry (1984).
† North Atlantic; data from Church et al. (1982).
‡ Eastern coast USA; data from Church et al. (1982).
** the unit $\mu eql^{-1}$ Is widely used in the reporting of rain data because It is useful for estimating charge balance in the analysis. It Is the molar concentration x ion charge.

for example, from compounds such as dimethyl sulfide (see Section 4.1.4.3), and from terrestrial anthropogenic sources. In marine-dominated rain, the proportions of the major ions are therefore influenced by variations in the inputs of sea-salt, crustal dust and anthropogenic components, the proportions of which differ with distance from the continental sources (see e.g. Galloway *et al.*, 1982).

A number of trends found in the major ion chemistry of marine rain waters can be illustrated with respect to samples taken at three contrasting marine environments. These are: (i) Lewes, Delaware, on the mid-Atlantic USA coast; (ii) Bermuda, in the North Atlantic; and (iii) Amsterdam Island, a remote site in the southern Indian Ocean. Data for the major ions in rain waters from these three sites are given in Table 6.5. The data for USA coast and Bermuda sites were reported by Church *et al.* (1982), who assessed the marine influence on precipitation arising from storms that leave the North American continent and transit over the western Atlantic. The findings of the study showed that sea-salt contribution (by weight) to the major ions rose from 54% at the coastal site to 80% at Bermuda. In contrast, the sulfate decreased from the coastal site to Bermuda; but even at Bermuda ~50% of the total sulfate was in excess of the sea-salt sodium. At the remote Amsterdam Island site the excess sulfate made up ~15% of the total sulfate.

The acidity of rain water is an important environmental parameter, and the free acidity can result from a number of proton donors, such as strong acids (e.g. $H_2SO_4$, $HNO_3$), weak organic acids (e.g. acetic acid, formic acid) or metal oxides (e.g. Al, Fe): see for example Galloway *et al.* (1982). This acidity will be neutralized to some extent by bases such as ammonia, soil dust, calcium carbonate and seasalt alkalinity. The pH of natural rain water is generally acidic (~5.0–~5.5) as a result of the equilibration of atmospheric $CO_2$ with precipitation (see e.g. Galloway *et al.*, 1982; Pszenny *et al.*, 1982). Rainwater pH can also be strongly influenced by both water-soluble and particulate aerosol components scavenged from the air. For example, precipitation on Bermuda, downwind of North American industrial areas, is acidified relative to equilibration with $CO_2$ by a factor of about eight (Church *et al.*, 1982). Sea-salt in the Bermuda precipitation should not neutralize more than ~10% of the acidity, and Church *et al.*

(1982) concluded that most of the acidity in excess of the natural equilibration results from the long-range transport of sulfur and nitrogen precursors in the marine troposphere, with sulfuric acid dominating over nitric acid. According to Galloway and Gaudry (1984) precipitation on Amsterdam Island has two components: one from a seawater source and the other contributing to the acidity of the precipitation. The component giving rise to the acidity is substantially smaller than the seawater component, and the main proton donors are $H_2SO_4$, low molecular weight organic acids (HCOOH and $CH_2COOH$) and $HNO_3$, with the maximum contribution to acidity being 30%, 25% and 15%, respectively. Along with ammonium salts, seasalt alkalinity can also be a neutralizing agent. Galloway and Gaudry (1984) reported that there is an interaction between the alkaline seawater component and the acid component, which results in an average loss of ~10% of the original free acidity due to neutralization. The three principal sources for the acidic components in the rain water at Amsterdam Island were: (i) long-range transport from continental regions; (ii) long-range transport from marine regions; and (iii) local island emissions. The authors concluded that the $SO_4^{2-}$ source is derived from the oxidation of reduced marine sulfur components, but that continental sources possibly influenced the $NO_3^-$; it was not possible to assign sources to the organic acids.

Seawater components can neutralize part of the acidity in precipitation. A more dramatic effect on the acidity of precipitation, however, can result from the scavenging of crustal dusts. For example, Loye-Pilot *et al.* (1986) demonstrated that the pH of western Mediterranean rain water is strongly influenced by the type of material scavenged from the air. For example, rain waters associated with air masses that had crossed Western Europe, and had scavenged air influenced by European industrial emissions had pH values in the range 4.1–5.6. In contrast, the so-called 'red rains' associated with air masses that had crossed North African sources, and which had scavenged crust-dominated Saharan dust, had pH values as high as 6–7 as a result of the dissolution of calcium carbonate from the dusts.

*Major Ion Fluxes.* In Table 6.6, estimates of atmospheric inputs of major ions and some other chemically simple cations and anions are presented. The

**Table 6.6** Net Atmospheric dissolved and particulate fluxes of elements to the world oceans in units of $10^9\,mol\,yr^{-1}$. Part (a) of the table presents results for major ions and other elements with simple dissolved speciation, and part (b) presents the potentially more geochemically reactive trace metals.

| (a) Major Ions | Net Flux |
|---|---|
| Li | 1.5* |
| B | 7.1* |
| Na | 230* |
| Mg | 274* |
| S | 2530& |
| Cl | 420$ |
| K | 261* |
| Ca | 282* |
| Rb | 0.43* |
| Sr | 1.6* |
| Cs | 0.02* |

| (b) Trace Metals | |
|---|---|
| Al | 1326* |
| V | 0.84* |
| Cr | 0.80* |
| Mn | 6.2* |
| Fe | 309* |
| Co | 0.13* |
| Ni | 0.4–0.5** |
| Cu | 0.25–0.8** |
| Zn | 0.7–3.5** |
| Cd | 0.02–0.04** |
| Pb | 0.43** |

| Nutrients | |
|---|---|
| N | 4800@ |
| P | 18## |
| Si | 4900* |

*Fluxes calculated from model estimates of global dust flux to the oceans ($442 \times 10^{12}\,g\,yr^{-1}$; Jickells et al., 2005) and crustal abundances (Table 3.4). This assumes that the only source is crustal dust, if there is significant anthropogenic input these fluxes will be underestimated by reported enrichment factors for these elements in aerosols but Wiersma and Davidson (1986) suggest such anthropogenic enrichment is small.
& Brimblecombe et al. (1989).
$ Keene et al. (1999) total inorganic Cl emissions, minus seasalt and assuming 50% of emissions reaches the oceans.
@ Duce et al. (2008).
## Mahowald et al. (2008).
** Duce et al. (1991).

estimation of these particular fluxes is not straightforward. All these ions are relatively abundant in seawater and there are therefore relatively large emissions from seawater as seaspray that are then redeposited. As an example the net flux from land to sea of chloride is estimated in Table 6.6 as $420 \times 10^9$ moles $yr^{-1}$, but this represents only ~2% of total seasalt chloride production (Keene et al., 1999). The method of estimating the net fluxes in Table 6.6 is necessarily indirect but does avoid the difficulties created by this large scale recycling.

*Trace metals in rain waters.* Trace metals in rain waters are derived from material scavenged from the air. The total concentrations (i.e. dissolved + particulate) of trace metals at any individual site will reflect the type of aerosol scavenged. Data are now available on the trace-metal compositions of rain waters from a number of coastal and marine regions, and a selection of these is given in Table 6.7. There are problems, however, in directly comparing the trace-metal concentrations from different individual sites. One reason for this is that the concentrations of the metals, and major ions, in the rain can change during the course of a rain event (see e.g. Lim et al., 1991), and some trace metals can have higher concentrations in the early, relative to the later, precipitation. In addition, trace-metal concentrations can vary from one rain event to another at the same site. Because of this some authors quote their precipitation trace-metal data on a volume weighted mean (VWM) basis. This normalizes the precipitation concentration to the precipitation amount. Thus, according to Galloway and Gaudry (1984), it is as if all the precipitation at a single site collected over a given period was mixed in one container and the composition determined from that; most of the data sets given in Table 6.7 are expressed on a VWM basis.

The collection sites in Table 6.7 are ranked in terms of their remoteness from the major continental sources, and despite the problems inherent in comparing data sets from different sites, two overall trends in the distribution of trace metals in rain waters can be identified from the data in Table 6.7.
1 The concentrations of the non enriched elements (NEEs see Chapter 4), such as Al, are highest in the coastal rain waters where the atmosphere received inputs of crustal dust 'pulses', for example, the Mediterranean Sea.

**Table 6.7** Trace metal composition of marine-influenced rain waters (total trace metals; units, nmol l$^{-1}$).

| Trace metal | North Sea | | | Mediterranean Sea | | North Atlantic | | North Pacific: | South Pacific: |
|---|---|---|---|---|---|---|---|---|---|
| | Northeast coast, Scotland* (VWM) | North coast, Germany† (VWM) | Open-sea‡ (VWM) | South coast, France§ (VWM) | Sardinia¶ (VWM) | Bermuda$ (VWM) | Bantry Bay** (VWM) | Enewetak†† (VWM) | Samoa‡‡ |
| Al | – | – | 0.77 | 5.3 | 23 | – | 0.13 | 0.08 | 0.59 |
| Fe | 1.6 | 0.32 | 0.55 | – | 9.3 | 0.09 | 0.14 | 0.02 | 0.01 |
| Mn | 0.07 | 0.08 | 0.06 | – | 0.14 | 0.005 | $2 \times 10^{-3}$ | $0.2 \times 10^{-3}$ | $0.4 \times 10^{-3}$ |
| Cu | 0.04 | 0.03 | 0.015 | 0.04 | 0.05 | 0.01 | 0.01 | $0.2 \times 10^{-3}$ | $0.33 \times 10^{-3}$ |
| Zn | 0.2 | 0.38 | 0.11 | – | 0.24 | 0.02 | 0.12 | $0.8 \times 10^{-3}$ | 0.02 |
| Pb | 0.02 | 0.03 | 0.02 | 0.02 | 0.01 | $4 \times 10^{-3}$ | $2 \times 10^{-3}$ | $0.17 \times 10^{-3}$ | $0.07 \times 10^{-3}$ |
| Cd | $6 \times 10^{-3}$ | $4 \times 10^{-3}$ | $0.7 \times 10^{-3}$ | – | – | $0.5 \times 10^{-3}$ | $0.4 \times 10^{-3}$ | $0.02 \times 10^{-3}$ | – |
| Sb | – | 0.003 | 0.001 | – | – | – | – | – | – |
| Se | – | 0.006 | 0.004 | – | – | – | – | – | – |

* Balls (1989). † Stossel (1987). ‡ Chester et al. (1994). § Chester et al. (1997). ¶ Keyse (1996). $ Jickells et al. (1984). ** Lim et al. (1991). †† Arimoto et al. (1985). ‡‡ Arimoto et al. (1987). VWM, volume weighted mean.

2 The concentrations of the anomalously enriched elements (AEEs see Chapter 4), such as Pb and Zn, are highest in coastal regions, and decrease with the degree of remoteness of a site from the main continental regions.

These trends therefore reflect those found for the aerosols that are scavenged from the air by the precipitation (see Section 4.2.3).

### 6.2.2.3  Total atmospheric trace-element fluxes

Buat-Menard (1983) concluded that, in general, the net atmospheric fluxes of the AEEs (small particle size) to the sea surface primarily result from *wet* deposition over most marine areas, but that *dry* deposition is significant for sea-salt and mineral aerosols (large particle size). The data available in the literature tend to confirm these overall trends, although the situation is by no means absolutely clear. For example, at Enewetak, flux data (corrected for sea-surface recycling) showed that, although wet deposition exceeded dry deposition for Pb, V, Cd and Se, this was not the case for Cu and Zn. Further, at Enewetak, wet deposition was more important than dry deposition for Fe, although for Al dry deposition was an order of magnitude higher than wet removal.

Many of the early models used to estimate the atmospheric input of elements to the sea surface were inevitably somewhat crude, and total atmospheric deposition fluxes (*F*) were often calculated from an equation of the general type

$$F = CV \tag{6.4}$$

in which *C* is the mean atmospheric concentration of an element and *V* is the global deposition velocity. The simplest way of estimating the global deposition velocity of an element is by assuming that all deposition takes place by rain scavenging, which cleans the atmosphere around 40 times per year; see, for example, the model described by Bruland *et al.* (1974). In more complex models, attempts were made to take account of the actual deposition rates of elements to the sea surface, either by assuming a total deposition rate (see e.g. Buat-Menard and Chesselet, 1979) or by taking individual account of wet and dry deposition rates (see e.g. Duce *et al.*, 1976). Later, more advanced 'wet and dry' flux deposition models became available. For example,

Arimoto *et al.* (1985) determined the gross and net atmospheric fluxes of a series of crustal and enriched elements to the sea surface at Enewetak. The findings are of particular interest because the models take account of sea-surface recycling, and thus offer an estimate of the true *net* deposition of the elements to the ocean surface from the atmosphere. Advanced models of this type were applied subsequently to the Samoa (South Pacific) aerosol (Arimoto *et al.*, 1987). The data sets provided by Arimoto *et al.* (1985; 1987) therefore represent the best available estimates of the net deposition of trace metals to the sea surface.

A summary of some of the data given in the literature for the fluxes of trace elements to the sea surface is given in Table 6.8. With the exception of the data sets for Enewetak and Samoa, most of the calculations do not take account of sea-surface recycling, with the result that the fluxes, especially those for the anomalously enriched elements (AEEs), will tend to be overestimated. Further, the fluxes have been obtained by different methods. Despite constraints such as these, however, a strong geographical trend can be identified from the data in Table 6.8, indicating that the strengths of the air-to-sea fluxes decrease by orders of magnitude as the degree of remoteness of a site from the major aerosol sources increases. For example, this trend is well developed for Pb, for which the atmospheric input from the North Atlantic westerlies is over 50 times greater than that from the South Pacific westerlies. Note however, that trace metal fluxes dominated by anthropogenic emissions are sensitive to changes in atmospheric source strengths. This is particularly so for Pb, as discussed in Chapter 4, with the phasing out of the metal as a petrol additive. Witt *et al.* (2006) reported gradients in Pb concentrations between the northern and southern hemisphere to be much reduced after the year 2000 compared to the 1970s and 1980s. For other trace metals, there have been reductions in emissions from particular sources due to improved pollution control, although these have been offset to some extent by patterns of increasing population and industrialization. The most comprehensive analysis of trace metal fluxes is still that of Duce *et al.* (1991) and this is used for the fluxes in Table 6.13 later, although the fluxes of crustal bound metals are derived from more recent model calculations.

**Table 6.8** Atmospheric fluxes of trace metals to the sea surface; units, nmol cm$^{-2}$ yr$^{-1}$.

| | New York Bight* | North Sea† | Western Mediterranean‡ | South Atlantic Bight§ | Bermuda¶ | North Atlantic; Northeast Trades‖ | Tropical North Atlantic** | Tropical North Pacific: total net deposition†† | South Pacific: total net deposition‡‡ |
|---|---|---|---|---|---|---|---|---|---|
| Al | 220 | 1100 | 185 | 107 | 144 | 3590 | 185 | 44 | 5–67 |
| Sc | – | 0.1 | 0.02 | – | 0.01 | – | 0.02 | 0.004 | $1.3 \times 10^{-3}$ |
| V | – | 9.4 | – | – | 0.1 | – | 0.3 | 0.15 | – |
| Cr | – | 4 | 0.9 | – | 0.17 | 2 | 0.27 | – | – |
| Mn | – | 17 | – | 1.1 | 0.8 | 10 | 1.3 | 0.2 | 0.06 |
| Fe | 102 | 457 | 91 | 106 | 54 | 860 | 57 | 10 | 0.8–6 |
| Co | – | 0.7 | 0.06 | – | 0.02 | 0.2 | 0.05 | – | 0.004 |
| Ni | – | 4.4 | – | 6.6 | 0.05 | 1.1 | 0.3 | – | – |
| Cu | – | 20 | 1.5 | 3.4 | 0.5 | 0.8 | 0.4 | 0.2 | 0.07–0.12 |
| Zn | 21 | 137 | 16.5 | 11.5 | 1.1 | 2.3 | 2 | 1 | 0.04–0.1 |
| As | – | 3.7 | 0.7 | 0.6 | 0.04 | – | – | – | – |
| Se | – | 0.3 | 0.6 | – | 0.04 | – | 0.2 | 0.05 | 0.01 |
| Ag | – | – | 0.02 | – | – | – | 0.008 | – | – |
| Cd | 0.3 | 0.4 | 0.1 | 0.1 | 0.04 | – | 0.04 | 0.003 | – |
| Sb | – | 0.48 | 0.39 | – | 0.008 | – | 0.03 | – | – |
| Au | – | – | $0.25 \times 10^{-3}$ | – | – | – | $0.5 \times 10^{-3}$ | – | – |
| Hg | – | – | 0.025 | 0.12 | – | – | 0.01 | – | – |
| Pb | 19 | 13 | 15 | 3 | 0.5 | 0.15 | 1.5 | 0.03 | 0.006–0.012 |
| Th | – | 0.02 | 0.005 | – | – | – | 0.004 | 0.003 | $0.15 \times 10^{-3}$ |

*Duce et al. (1976). †Cambray et al. (1975). ‡Arnold et al. (1982). §Windom (1981). ¶Duce et al. (1976). ‖Chester et al. (1979). **Buat-Menard and Chesselet (1979). ††Arimoto et al. (1985). ‡‡Arimoto et al. (1987).
Data are all form the 1970s and 1980s when Pb emissions were close to their maximum.

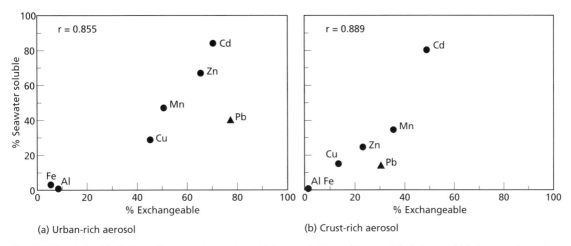

**Fig. 6.2** The relationship between the percentages of the total concentrations of trace metals that are soluble in sea water and the percentage in exchangeable associations in (a) anthropogenic-dominated and (b) crust-dominated aerosols (after Chester *et al.*, 1993). Reprinted with permission from Elsevier.

### 6.2.2.4 The fates of atmospherically transported trace elements in seawater

It is clear that the atmosphere provides an important pathway for the transport of trace metals to the oceans. The subsequent fate of the elements will depend on a number of factors. The initial constraint on the behaviour of atmospherically transported trace metals in the mixed layer will be imposed by the extent to which they are solubilized in seawater. This is important because the physical state (i.e. particulate or dissolved) of the metals affects both their subsequent involvement in the biogeochemical cycles and their residence times in seawater.

*'Dry' deposition.* Some of the early studies on the solubility of trace metals from aerosols indicated that the manner in which a metal is partitioned between crustal and non-crustal components exerts a major control on its fate in seawater (see e.g. Walsh and Duce, 1976; Hodge *et al.*, 1978). The general relationship between the solid state speciation of a trace metal in an aerosol (see Section 4.2.1.3) and its seawater solubility was confirmed by Chester *et al.* (1993), who showed that there is a well-developed relationship between the extent to which a trace metal is held in exchangeable associations in 'end-member' (crust-dominated and anthropogenic-dominated) aerosols and the extent to which it is

**Table 6.9** Seawater solubility of trace metals from atmospheric particulate aerosols; approximate percentage of total element soluble (after Chester *et al.*, 1996).

|  | Coastal regions: anthropogenic-dominated aerosol | Coastal regions: crustal-dominated aerosol | Open-oceans: 'mixed-source' aerosols |
|---|---|---|---|
| Al | <10 | <10 | <10 |
| Fe | <10 | <10 | <10–50 |
| Mn | 45 | 20 | 20–50 |
| Ni | 50 | <20 | 20–50 |
| Co | 25 | <20 | <20–25 |
| Cr | 12 | <10 | 10 |
| V | 30 | <20 | <20–85 |
| Cu | 35 | <10 | <10–85 |
| Pb | 50 | <10 | <10–90 |
| Zn | 70 | <10 | <10–75 |
| Cd | 85 | 80 | 85 |

soluble in seawater: see Fig. 6.2. On the basis of this 'percentage exchangeable versus percentage seawater soluble' relationship, Chester and Murphy (1990) divided the trace metals into a number of types; average percentage seawater solubility values for the crust-dominated and anthropogenic-dominated aerosols are given in Table 6.9.

*Type 1 (e.g., Al, Fe).* These metals are crust-controlled (i.e. they are NEEs and have $EF_{crust}$ values <10: see Section 4.2.1.1) in all marine aerosols. In both crust-dominated and anthropogenic-dominated aerosols, Al and Fe have less than ~10% of their total concentrations in exchangeable associations and also they are relatively insoluble in seawater (less than ~10% of their total concentrations, and possibly <1%, although as noted in Chapter 4, the absolute solubility of these elements may depend on the dust loading. However, because they are present in relatively high concentrations in marine aerosols, even a solubility of a few percent can release considerable quantities of the metals into seawater in a dissolved form. For example, the dissolution of Fe from aerosols can play an important role in primary productivity (see Section 9.1 and Chapter 11).

*Type 2 (e.g., Mn).* Manganese is usually crust-controlled (NEE) in marine aerosols, but unlike Al and Fe it is relatively soluble (~20%–~45%) from both crust-dominated and anthropogenic-dominated aerosols. The reason why Mn is more soluble than the type 1 trace metals is that it has a higher percentage of its total concentration (~20–~<50%) in exchangeable associations in all marine aerosols, reflecting different chemical speciation in the aerosol and soils – oxide rather than aluminosilicate phases.

*Type 3 (e.g., Cu, Zn, Cd and Pb).* These metals have higher $EF_{crust}$ values and a higher proportion of their total concentrations in exchangeable associations in anthropogenic-dominated aerosols, in which they behave as AEEs, than in crust-dominated aerosols, in which they behave as NEEs. Hence the metals are bound in rather simple and soluble chemical forms within the aerosol. The higher solubilities of the type 3 metals from the anthropogenic-dominated (~35% of the total ($\Sigma$) Cu, ~70% of the $\Sigma$ Zn, (~55–70% Cd and ~50% of the $\Sigma$ Pb) than from crust-dominated aerosols (less than ~10% of the $\Sigma$ Cu, $\Sigma$ Zn and $\Sigma$ Pb) is thus entirely consistent with their solid-state speciation signatures in the two aerosol 'end-members'.

'Wet' deposition. The trace-metal concentrations in rain waters reflect those in the aerosols that have been scavenged from the air so that the total precipitation fluxes, like the 'dry' fluxes, vary with the atmospheric concentrations of the metals. The dissolved/particulate speciation of trace metals following the 'dry' deposition of aerosols to the sea surface is a function of aerosol ↔ seawater reactivity, and essentially is constrained by the solid-state speciation of the metals in the aerosols (see previously). The dissolved/particulate speciation of the metals in 'wet' deposition, however, is a function of aerosol ↔ rain water reactivity, and so is constrained by both the solid-state speciation of the metals in the scavenged aerosols and the chemistry of the rain waters themselves. The aerosol ↔ rain water reactivity includes pH-dependent processes, such as adsorption/desorption and precipitation. As a result, although the overall dissolved–particulate speciation in rain waters depends on a number of complex interrelated factors, the solution pH exerts a fundamental control. This is illustrated in Fig. 6.3 for the solubility (dissolved/particulate concentrations × 100) of lead in rainwater collected on the Mediterranean coast. This region is on a boundary between high dust, low anthropogenic and low acidity air masses from the Sahara region and air masses impacted by emissions of acids and trace metals over Europe. Hence the more polluted air masses contain higher concentrations of some trace metals and these metals are more soluble due to the lower pH. Similar 'pH-solubility' relationships have been reported in other areas (e.g. Losno *et al.*, 1988; Lim *et al.*, 1994).

The average dissolved–particulate speciation of a series of trace metals in rain waters from several sites in the western Mediterranean is given in Table 6.10(a), and a number of general trends can be identified from these data (see e.g., Chester *et al.*, 1996), broadly similar to those seen for aerosols.

1 Aluminium and Fe, which are NEEs in marine aerosols, are relatively insoluble (less than ~20% of the $\Sigma$ Al and $\Sigma$ Fe) in rain waters from all the sites.

2 Manganese, which is an NEE in marine aerosols, is generally relatively soluble (~60–70%) in rain waters from all the sites.

3 The solubilities of some trace metals (e.g. Cu, Zn, Pb), which can switch character between NEE and AEE behaviour in marine aerosols as a function of their source (see Section 4.2), are variable but relatively high.

It is apparent, therefore, that aerosol trace-metal solubility trends in both seawater and rain water can be related to the solid-state speciation of the metals, with the exchangeable associations (i) being the most soluble; and (ii) making up a higher fraction of the total metal in anthropogenic-dominated than in

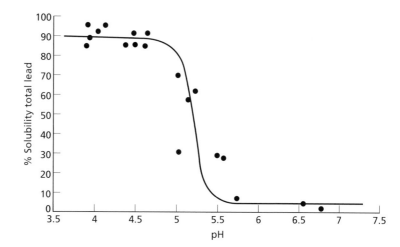

**Fig. 6.3** The pH–solubility relationship of Pb in western Mediterranean rain waters (after Chester *et al.*, 1990).

**Table 6.10** The dissolved–particulate speciation of trace metals in rain waters. (a) The average dissolved–particulate speciation of trace metals in rain waters from a number of Mediterranean Sea sites; data given as percentage of the total trace metal concentration in the dissolved phase (data from Guieu *et al.*, 1997).

|    | Cap Ferrat | Tour du Valat | Carpentras | Corsica |
|----|------------|---------------|------------|---------|
| Al | 18         | 19            | –          | 8       |
| Fe | –          | 11            | –          | 13      |
| Mn | 60         | 63            | –          | 67      |
| Ni | 54         | 58            | –          | –       |
| Co | 61         | 50            | –          | –       |
| Cu | 82         | 71            | 66         | 49      |
| Zn | –          | 68            | –          | 76      |
| Pb | 65         | 52            | 80         | 48      |
| Cd | 92         | 75            | –          | –       |

(b) Total trace-metal concentrations and $EF_{crust}$ values in rain waters that have scavenged European 'background' and Saharan dust aerosols; concentration units: $\mu g\,l^{-1}$, % = percentage total concentration in dissolved phase (data from Chester *et al.*, 1997).

| | Aerosol type scavenged | | | | | |
|----|---------------|--------------|----|---------------|--------------|------|
| | European 'background'; sample 1, pH 3.95 | | | Saharan dust; sample 12, pH 6.55 | | |
| | Concentration | $EF_{crust}$ | % | Concentration | $EF_{crust}$ | % |
| Al | 140    | 1.0  | 35 | 1462 | 1.0 | 1.15 |
| Co | 0.078  | 1.9  | 67 | 0.58 | 1.3 | 0.50 |
| Ni | 1.88   | 15   | 69 | 2.26 | 1.7 | 23   |
| Cu | 5.79   | 59   | 86 | 4.0  | 3.9 | 30   |
| Pb | 26.2   | 1248 | 90 | 3.46 | 16  | 52   |

crust-dominated aerosols (see Section 4.2.). The additional constraint that must be taken into account with respect to rain waters is the solution pH. This pH-driven solubility is also a function of the solid-state speciation of the metals in the scavenged aerosols. It must be remembered, however, that there is a coupling between the rainwater pH and the solid-state speciation in the scavenged aerosols because the type of aerosol scavenged has a strong influence on the pH. Thus, anthropogenic-dominated aerosols, which have relatively large fractions of the AEEs in exchangeable associations, give rise to rain waters with a low (more corrosive) pH and therefore a high AEE solubility. In contrast, crust-dominated aerosols, which have relatively small fractions of the AEEs in exchangeable associations, give rise to rain waters with a high (less corrosive) pH and therefore a lower solubility of any AEEs that are present. Although the solution pH and the trace metal solid-state speciation in the scavenged aerosols may be regarded as the primary controls, a number of other factors must be taken into account when assessing trace-metal solubility in rain waters. For example, Guieu et al. (1997) highlighted the importance of particle concentrations on the solubilities of trace metals from aerosols in rain waters.

It is unlikely that aerosols will retain their original identity as discrete particles in the upper water column following their deposition to the sea surface. Firstly the pH of seawater is about 8 and that of atmospheric aerosols and rainwater is much lower, and this change of pH will affect solubility Furthermore aerosol particles may be coated with organics or enter the gut system of filter-feeding organisms, where the pH can be very much lower than that in seawater and dissolved metals may also be complexed by organic matter in the seawater. Experimental studies therefore will reflect only the initial fates of atmospherically transported trace metals in the mixed layer. Therefore we will subsequently consider the total metal flux arriving in the oceans from the atmosphere, while recognizing that the impact of the metals depends on their subsequent behaviour in seawater which will be discussed in later chapters.

### 6.2.3 Atmospheric fluxes of organic matter

In terms of global cycles, the oceans provide a major sink for some atmospheric organic material, and although the data are still sparse several attempts have been made to estimate the global tropospheric burdens of organic matter. For example, Duce (1978) calculated that the non-methane global volatile organic carbon (VOC) burden is $\sim 50 \times 10^{12}$ g, and the total aerosol particulate organic carbon (POC) burden has been estimated to be $\sim 1 \times 10^{12}$–$5 \times 10^{12}$ g, of which $\sim 90\%$ is on particles with diameters $<1\,\mu$m (see e.g. Duce et al., 1983). In addition, the total amount of organic carbon cycling through the troposphere each year has been put at $\geq 800 \times 10^{12}$ g (see e.g. Duce, 1978; Robinson, 1978; Zimmerman et al., 1978). A number of first-order estimates have been made of the source–sink relationships of POM and VOM in the marine atmosphere. Thus, Williams (1975) calculated a wet deposition flux of carbon of $\sim 2.2 \times 10^{14}$ g yr$^{-1}$ to the ocean surface, which is somewhat smaller than that of $\sim 10 \times 10^{14}$ g yr$^{-1}$ estimated by Duce and Duursma (1977), although all are within the same order of magnitude but greater than the estimate of $0.3 \times 10^{12}$ g yr$^{-1}$ of Buat-Menard et al. 1979. The dry deposition flux of organic material (POM) to the oceans has been put at $\sim 6 \times 10^{12}$ g yr$^{-1}$ (Duce and Duursma, 1977). Assuming POC = POM $\times 0.7$, this is equivalent to a carbon flux of $\sim 4 \times 10^{12}$ g yr$^{-1}$. Thus, although it would appear that wet deposition is more important than dry fall-out in the fluxes of organic material to the sea surface, all the estimates should be treated with great caution. Further complications to the assessment of the fluxes of organic material to the ocean arise because the sea surface itself can act as a source, as well as a sink, for both POM and VOM through processes such as gas exchange and bubble bursting (see Sections 4.1.4.2 and 8.2). For example, Duce (1978) calculated that $\sim 14 \times 10^{12}$ g yr$^{-1}$ of organic carbon is produced by the ocean surface, with $\sim 90\%$ being found on particles $>1\,\mu$m in diameter. Even when recycling is taken into account, however, it would appear that on the basis of the estimate given by Williams (1975) for the oceanic wet deposition flux of carbon ($\sim 2.2 \times 10^{14}$ g yr$^{-1}$, $18 \times 10^3$ mol C yr$^{-1}$) the input of organic material to the ocean surface is greatly in excess of the output from the marine source. The overall result of this is that the oceans act as a major sink for atmospheric organic material (Duce et al., 1983) but this input is 'at most 1% of the biological productivity of the open ocean' (Buat-Menard et al., 1989) hence the rather limited attention that has been paid to this term in global biogeochemical budgets.

### 6.2.4 Atmospheric inputs of nutrients

The recognition of the importance of the atmospheric transport of nutrients to the oceans has led to considerable research particular for nitrogen deposition as discussed in Chapter 4. These inputs are approximately equally in the form of nitrate, ammonium and organic nitrogen, although the origin and bioavailability of the latter is particularly uncertain. All of these compounds are derived primarily from anthropogenic sources on continents, so their patterns of deposition (Fig. 6.4) are rather similar to those of other components with similar sources. The fluxes and significance of atmospheric inputs to the oceans has been reviewed by Duce *et al.* (2008) and these fluxes are presented in Table 6.6. Atmospheric nitrogen fluxes are greatly perturbed by human activity related to agricultural and combustion sources which emit $NH_3$ and $NO_x$ gases. For silicon and phosphorus there are no such gas phases and the perturbation of the atmospheric cycle is rather modest. There have been recent efforts to model the atmospheric P cycle and these provide the flux estimates in Table 6.6. The silicon cycle in the atmosphere is simply that of dust transport and deposition and is very small compared to the sources to the oceans from other inputs and so, although rather uncertain, is of limited geochemical significance.

## 6.3 Hydrothermal fluxes to the oceans

Although the most dramatic hydrothermal activity is associated with high-temperature processes at the spreading ridge centres, basalt–seawater reactions take place over a variety of temperatures and rock:water ratios (see Chapter 5), and all reaction types must be taken into account when attempts are made to derive hydrothermal fluxes to the oceans. Thompson (1983) distinguished four types of basalt–water reactions on the basis of their location, temperature and water:rock ratios.

1 Those involving surface basement rocks, which take place over long time periods away from the spreading centres at low temperature, high water:rock ratios.

2 Those involving deeper basement rocks, which take place within the basement over short time periods away from the spreading centres at low temperature, low water:rock ratios.

3 Those taking place at the mid-ocean ridge flanks (off-axis activity). These are associated with hydrothermal circulation, but occur at medium temperature, medium water:rock ratios during the so-called passive circulation phase where the temperature of the water decreases away from the spreading centres.

4 Those taking place at mid-ocean ridge axes (axial activity). These are associated with the dramatic venting of hydrothermal solutions at the spreading centres during the active phase of hydrothermal activity and take place at high temperature, low water:rock ratios.

Attention here will be focused on hydrothermal activity taking place at the crest and flank regions of the mid-ocean ridge system.

The first data available for the composition of hydrothermal venting solutions at the ridge crests was obtained from the low temperature, 'white smoker', Galapagos Spreading Centre (GSC), and Edmond *et al.* (1979) derived hydrothermal fluxes

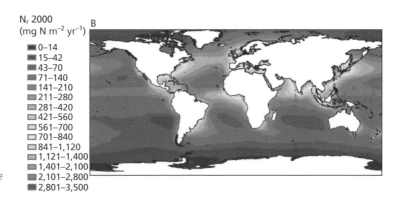

$N_r$ 2000
(mg N m$^{-2}$ yr$^{-1}$)

■ 0–14
■ 15–42
■ 43–70
■ 71–140
■ 141–210
■ 211–280
■ 281–420
□ 421–560
□ 561–700
□ 701–840
□ 841–1,120
■ 1,121–1,400
■ 1,401–2,100
■ 2,101–2,800
■ 2,801–3,500

**Fig. 6.4** Model estimates of total fixed nitrogen deposition to the oceans (Duce *et al.*, 2008). (See Colour Plate 2)

from these data. To do this the authors extrapolated the GSC venting solution composition to a 350°C high-temperature hydrothermal end-member and combined this with global heat-flux data estimated from hydrothermal $^3$He : transported-heat ratios, to estimate global ocean hydrothermal fluxes for those elements that behaved in a conservative manner during the mixing of hydrothermal solutions and seawater in the 'white smoker' GSC system (see Section 5.1). Subsequently, data were obtained for 'black smoker' systems, which vent high-temperature fluids directly to seawater (see Section 5.1): for example, Von Damm *et al.* (1985) made a detailed study of high-temperature venting fluids in the 21°N EPR region (see Section 5.1). The data obtained allowed estimates to be made of the hydrothermal fluxes of the ore-forming metals (e.g. Mn, Fe, Co, Cu, Zn, Cd and Pb); this was not possible for the low-temperature GSC venting fluids because a large proportion of these ore-forming metals were precipitated at depth within the crust in this 'white smoker' system.

The axial hydrothermal flux estimates from both the GSC and the EPR systems were obtained using $^3$He : heat data to assess the extent of the hydrothermal input to the World Ocean. Elderfield and Schultz (1996), however, drew attention to the fact that there is a conflict between the different methods used to estimate axial chemical fluxes. The authors reviewed the various geophysical and geochemical methods of estimating hydrothermal fluxes, and concluded that the best estimate of the axial high-temperature hydrothermal water flux is 3 ($\pm$1.5) $\times$ 10$^{13}$ kg yr$^{-1}$, which is better constrained than the $^3$He : heat data (1 $\times$ 10$^{13}$ to 16 $\times$ 10$^{13}$ kg yr$^{-1}$), but still very uncertain. This uncertainty arises principally because of uncertainties over the relative proportions of the total flux of heat that occurs as high temperature inputs and the much larger volume low temperature flow associated with the ridge flanks (German and Von Damm, 2003). The elemental flux estimates derived by Elderfield and Schultz (1996) based on Kadko *et al.* (1994) are presented in Table 6.11.

Elderfield and Schultz (1996) also pointed out that there is considerable uncertainty in the relationship between axial high-temperature effluent with a deeper source and other hydrothermal activity, such as off-axis low-temperature effluent with a shallower source. For example, the alkalis Li, K, Rb and Cs

**Table 6.11** Estimated high and low temperate hydrothermal fluxes to the oceans (Elderfield and Schulz, 1996) 10$^9$ mol yr$^{-1}$. Note positive numbers represent a flux into the ocean and negative numbers a sink.

| Element | High temperature hydrothermal flux | low temperature axial ridge fluxes |
|---|---|---|
| Li | 12–39 | –(2–11) |
| B | 1.1–4.5 | –(1.9–19) |
| Na | | |
| Mg | –1600 | –(700–1100) |
| S | +120–(–750)* | 800 |
| Cl | | |
| K | 230–690 | –(100–640) |
| Ca | 9–1300 | 200–550 |
| Br | | |
| Rb | 0.26–0.95 | –(0.19–0.28) |
| Sr | 0 | |
| Cs | $(2.9–6) \times 10^{-3}$ | $-(2–2.3) \times 10^{-3}$ |
| Si | 430–660 | 1300–1800 |
| Al | 0.12–0.6 | |
| V | 0 | |
| Cr | 0 | |
| Mn | 11–34 | |
| Fe | 23–190 | |
| Co | $(6.6–68) \times 10^{-4}$ | |
| Ni | | |
| Cu | 0.3–1.3 | |
| Zn | 1.2–3.2 | |
| Cd | | |
| Pb | $(2.7–110) \times 10^{-4}$ | |
| P | $-4.5 \times 10^{-2}$ | –3.2 |

* net sulfur change composed of sulfate sink of 840 $\times$ 10$^9$ mol yr$^{-1}$ and H$_2$S source of 85 to 960 $\times$ 10$^9$ mol yr$^{-1}$.

have large high-temperature fluxes, but these are balanced by low-temperature removal as the crust is altered as it moves away from the ridge crests. Elderfield and Schultz (1996) and German and Von Damm (2003) note that these flank fluxes should be regarded as no more than very approximate estimates, but it is apparent that for some elements the hydrothermal source–sink relationship appears to switch character between the axial and off-axis environments. For example, the axial source fluxes of B and Ba are approximately balanced by the same order of magnitude ridge-flank sinks and for U there appears to be extensive removal during hydrothermal circulation balanced by release later in the cycle. While these interactions lead to little or no net flux, the isotopic composition of an element can be modified during such hydrothermal cycling as appears to take place for Sr (German and Von Damm, 2003). In Table 6.11 estimated high temperature fluxes for selected elements are presented along with the available low temperature fluxes. The data available on the low temperature fluxes emphasizes the potential importance of these fluxes, and the many gaps within the table also emphasizes how much we do not yet know about these processes.

Three other factors must be taken into account when assessing the extent to which high-temperature axial hydrothermal fluxes affect the composition of seawater. Two of these factors are associated with the fact that hydrothermal exhalations form plumes above the venting sites as a result of the negative buoyancy of the hot hydrothermal fluids that rise above the sea floor. The third is related to the formation of sediments deposited around the venting centres.

### 6.3.1 Plume entrainment of seawater

As they rise, the high-temperature fluids associated with axial hydrothermal venting systems entrain ambient seawater, with a consequent continuous increase in plume volume until neutral buoyancy is achieved, following which the plume disperses laterally. According to Elderfield and Schultz (1996) the water flux associated with this plume entrainment is $1.8 \times 10^{17}$ to $3.4 \times 10^{17}$ kg yr$^{-1}$, which, when compared with the axial hydrothermal flux ($\sim 3 \times 10^{13}$ kg yr$^{-1}$), yields an entrainment ratio, that is, the ratio of entrained seawater to high-temperature fluid, of $\sim 10^4$. The global effect of this is that a very large volume of seawater passes through the plumes and is equivalent to an ocean water recycling time of $4 \times 10^3$ to $8 \times 10^3$ yr. There are, therefore, two seawater recycling stages associated with high-temperature axial hydrothermal activity. Stage 1, in which seawater recycles via circulation through the ridge system and changes composition as a result of high-temperature reactions involving the rocks it comes into contact with, to emerge at the sea bed as hydrothermal plumes. Stage 2, in which seawater circulates through the hydrothermal plumes in the water column, and changes composition by reacting with the plumes. These two stages operate on different time-scales. Thus, although ocean water

circulates through the mid-ocean ridge system in $\sim 10 \times 10^6$ yr, it is recycled through the hydrothermal plumes on a time-scale of thousands of years, which is only slightly longer than the physical ocean water mixing time. This plume recycling has very important geochemical implications for seawater composition. The plumes are rich in particulates, for example, from the oxidation and precipitation of $Fe^{2+}$ in the high-temperature venting solutions. These particles scavenge a wide range of dissolved trace metals, not only from the venting solutions themselves but also from the ambient seawater that is recycled through them. As a result of this, the hydrothermal plumes act as important chemical sinks in comparison to fluvial inputs for many of the dissolved trace metals in seawater, including oxyanions such as Mo, V and Cr for which hydrothermal activity is not a significant source: see Table 6.12. It may be concluded, therefore, that although hydrothermal activity at the spreading ridges acts as a source for many trace metals to the seawater (stage 1) when plume entrainment (stage 2) is included in the overall system, hydrothermal activity can act as a net water column (i.e. oceanic) sink, and not a net source, for some trace metals dissolved in seawater. However, as with all the hydrothermal fluxes estimated here, the considerable uncertainties of these estimates needs to be stressed. An additional issue is the dissolved and particulate distribution within the hydrothermal emissions, which will evolve over time in the plume. It must also be remembered that the dispersion and cycling of dissolved and particulate matter in the oceans may be different.

**Table 6.12** Hydrothermal plume removal fluxes from seawater; units, mol yr$^{-1}$ (data from Elderfield and Schultz, 1996).

| Element | Plume removal flux | Fluvial input flux |
| --- | --- | --- |
| Cr | $4.8 \times 10^7$ | $6.3 \times 10^8$ |
| V | $4.3 \times 10^8$ | $5.9 \times 10^8$ |
| As | $1.8 \times 10^8$ | $8.8 \times 10^8$ |
| P | $1.1 \times 10^{10}$ | $3.3 \times 10^{10}$ |
| Mo | $1.9 \times 10^6$ | $2.0 \times 10^8$ |
| Be | $1.7 \times 10^6$ | $3.7 \times 10^7$ |
| Ce | $1.0 \times 10^6$ | $1.9 \times 10^7$ |
| Nd | $6.3 \times 10^6$ | $9.2 \times 10^6$ |
| Lu | $0.6 \times 10^5$ | $1.9 \times 10^5$ |
| U | $4.3 \times 10^7$ | $3.6 \times 10^7$ |

### 6.3.2 Plume dispersion

A major problem in estimating the global ocean importance of hydrothermal exhalations lies in assessing the extent to which the plumes, which may change composition as a result of seawater entrainment, are dispersed away from the venting centres. Edmond *et al.* (1982) combined data on the distributions of hydrothermally derived sediments and ridge-produced $^3$He to demonstrate that the dispersal of hydrothermal solutions is controlled by the global oceanic circulation at mid-water depths. Other workers have used Mn as a hydrothermal tracer. For example, Klinkhammer (1980) used dissolved Mn profiles to show the existence of hydrothermal vents on the EPR. He also reported anomalously high dissolved Mn concentrations in the bottom waters of the Guatemala Basin ~1000 km from the crest of the EPR. Hydrographic evidence suggested that this Mn anomaly is a hydrothermal signal, thus providing evidence of the dispersal of hydrothermal solutions over large distances. The extent to which the plume dispersal takes place may be controlled by the topography of the ridge region where the venting solutions are formed initially. In this context, Klinkhammer *et al.* (1985) used dissolved Mn as a tracer for hydrothermal activity on the Mid-Atlantic Ridge between 11°N and 26°N, and found that the plumes formed were confined to the rift valley and did not spill over into the adjacent deep ocean basins. Clearly, therefore, the extent to which hydrothermal plumes are dispersed from their point of origin can vary considerably depending on local conditions. As a result of ridge geometry the solutions may be dispersed over much greater distances in the Pacific than in the Atlantic since in the Atlantic at slow spreading ridges, the rift valley of the mid ocean ridge is relatively deep and acts to contain the hydrothermal plumes (see Section 5.2).

### 6.3.3 The formation of hydrothermal sediments

In addition to precipitation and deposition within the hydrothermal venting systems (e.g. as sulfides) some elements are incorporated within hydrothermal sediments surrounding the active ridge systems, and so are prevented from large-scale oceanic dispersal. These sediments, which are considered in Section 16.5.2, are especially rich in Fe, and may also retain a large fraction of other metals released from hydrothermal vents. Some of these deposits, such as sulfides and sulfates may be buried, but others may redissolve and return to the water column. The metalliferous sediments by contrast are much less soluble.

## 6.4 Relative magnitudes of the primary fluxes to the oceans

### 6.4.1 Introduction

For many years it was thought that the oceans were a fluvially dominated system. This view has changed dramatically over the past few decades, however, as the importance of atmospheric deposition and hydrothermal activity at the spreading centres has become increasingly recognized. As a result quite different picture of the oceans has begun to emerge. In the three preceding sections an attempt has been made to estimate the fluxes of material delivered to the oceans from the major, global-scale, primary sources. These sources are fluvial, atmospheric and hydrothermal, although it must be remembered that hydrothermal activity also can act as a sink for some dissolved components as well as a source for others. The data sets for all these sources are still essentially crude, but it is now possible to make a number of general comparisons between them and the flux estimates for each of these components are presented in Table 6.13. Note as discussed the input of particulate matter form glaciers is not included in these estimates and may be significant.

### 6.4.2 Major elements and trace metals

It has been stressed already that the estimates in Table 6.13 are uncertain but several clear patterns are evident. The ranges of fluxes are very large and broadly reflect crustal abundances and solubilities. Hence, calcium which is crustally abundant, commonly found in rather readily weathered rocks such as limestone and readily soluble, has the highest flux of any of the listed elements and is dominated by fluvial inputs. Although Sr has many geochemical similarities to Ca, its low crustal abundance leads to a much lower flux. Elements such as Al are abundant but slow to weather and relatively insoluble, with the majority of their flux associated with particulate

**Table 6.13** Comparison of net fluxes from fluvial (Table 6.2; 6.4), atmospheric (Table 6.6) and hydrothermal (Table 6.11) combining high and low temperature data where available (otherwise only high temperature data) sources. $10^9 \, mol \, yr^{-1}$. Minus sign indicates a sink not a source.

| Element | Fluvial | Atmospheric | Hydrothermal |
|---|---|---|---|
| Li | 15 | 1.5 | +10 to −28 |
| B | 43 | 7.1 | −0.8 to −15 |
| Na | 9590 | 230 | |
| Mg | 10020 | 274 | −2300 to −3700 |
| S | 6700 | 2530 | 50−920 |
| Cl | 6400 | 840 | |
| K | 2340 | 261 | 50−130 |
| Ca | 23570 | 282 | 200−1850 |
| Br | | 3.4 | |
| Rb | 2.2 | 0.43 | 0.07−1.2 |
| Sr | 29 | 1.6 | 0 |
| Cs | 0.06 | 0.02 | $(0.9–3.7) \times 10^{-3}$ |
| Al | 4432 | 1326 | 0.12−0.6 |
| V | 4.7 | 3 | 0 |
| Cr | 2.9 | 3 | 0 |
| Mn | 36 | 24 | 11−34 |
| Fe | 1104 | 309 | 23−190 |
| Co | 0.45 | 0.5 | $(6.6–68) \times 10^{-4}$ |
| Ni | 2.4 | 0.4−0.5 | |
| Cu | 3 | 0.25−0.8 | 0.3−1.3 |
| Zn | 7.15 | 0.7−3.5 | 1.2−3.2 |
| Cd | 0.135 | 0.02−0.04 | |
| Pb | 1.05 | 0.43 | $(0.27–11) \times 10^{-3}$ |
| Organic C | $15 \times 10^3$ | $18 \times 10^3$ | |
| Total N | 1500 | 4800 | |
| $N_2$ fixation | 7140 | | |
| Total P | 120−180 | 18 | $-4.5 \times 10^{-2}$ |
| Total Si | 19500 | 4900 | 1700−2500 |

matter either in rivers or as atmospheric dust, much of which is geochemically rather inert.

Table 6.13 clearly illustrates that treating global cycles as fluvially dominated is an oversimplification. For soluble elements whose dominant global source is from rock weathering, fluvial fluxes are dominant, although atmospheric fluxes may represent ~10% of the total flux in some cases. Examples of such elements would be Li, B, Na, Mg, K, Ca, Rb, Sr, Si and Cs. This is probably also true for halogens like chloride and bromide although data for these elements is very limited and fluvial fluxes may include a significant component of seawater recycled through the atmosphere. For many of the transition metals fluvial, atmospheric and hydrothermal sources are

generally similar, for example those for Mn, Cu, Zn. However, for some of these metals removal in the hydrothermal plume may reduce the flux and the values here may be overestimates. Certainly for some trace metals such as Cr and V (Table 6.13) and some other metals present as oxyanions (Table 6.12) this is the case. The importance of atmospheric fluxes for many of these metals reflect increased emissions to the atmosphere as a result of human activity as reflected by their enrichment factors (Chapter 5).

Human activity has significantly modified the global S and N cycles and this is reflected in the modern day fluxes of both elements in Table 6.13. These anthropogenic emissions have particularly affected atmospheric fluxes, and result in modern day fluvial and atmospheric fluxes being of approximately comparable size. Uniquely for nitrogen there is an additional source via marine biological $N_2$ fixation which will be discussed further in Chapter 9, but an estimate of this is provided in Table 6.13 and shows it to be broadly comparable to anthropogenic and atmospheric sources. Duce *et al.* (2008) have noted the potential impact of the increased human inputs of nitrogen on the oceans which is likely to overall increase the productivity of the oceans, and this will be considered further in Chapter 9. Fluvial phosphorus fluxes have also been significantly increased by human activity as discussed in Chapter 3. Phosphorus has no significant volatile phase so all of the impacts of human activity are delivered via the fluvial input.

In assessing the impact of human activity on the oceans, the numbers in Table 6.13 are useful but need to be viewed in the context of the existing large inventories of metals and nutrients in the oceans. As an example, Duce *et al.* (2008) estimated that the present day flux of nitrogen to the oceans through the atmosphere was $4800 \times 10^9 \, mol \, yr^{-1}$ (Table 6.13). They also estimated the total nitrogen required to support ocean primary productivity as $628 \times 10^{12} \, mol \, yr^{-1}$, more than 100 times the atmospheric input. This productivity is sustained by the vast reservoirs of nitrogen in the deep waters of the open ocean which is slowly mixed into the surface where it is consumed by phytoplankton as we shall discuss in subsequent chapters. A similar situation applies to many of the trace metals and is best characterized by the residence times of these elements in the ocean (see Chapter 9 and 11).

The total organic carbon inputs to the oceans from both atmosphere and the ocean appear to be of similar magnitude, although of very different composition, the river inputs being largely associated with soil humic material while the atmospheric input of carbon is associated with some air borne soil material but also with the oxidation products of organic gases emitted by plants (see Chapter 4). As noted earlier, these organic carbon fluxes are very small compared to the internal cycles of organic carbon production and consumption.

The total organic loading in the atmosphere is of interest in terms of the global carbon cycle, but there is also the important issue of the delivery of specific organic compounds which may have particular, and in some cases harmful, impacts such as in the case of pesticides. Many organic compounds have significant vapour pressures at ambient temperatures and therefore have major gas-phase components. By using published concentration data, and taking account of particle/gas partitioning to assess deposition processes, Duce *et al.* (1991) estimated the fluxes of polychlorinated biphenyls (PCBs), hexachlorobenzene (HCB), and the pesticides hexachlorocyclohexanes (HCHs), dichlorodiphenyltrichloroethanes (DDTs), chlordane and dieldrin to the World Ocean in 1980s. The data are listed in Table 6.14. Table 6.14(a) gives the total deposition and mean fluxes of the organochlorines to different regions of the World Ocean; Table 6.14(b) provides estimates of the total deposition to the World Ocean in terms of the deposition mechanism ('wet' or 'dry'); and Table 6.14(c), compares the atmospheric and fluvial inputs of the organochlorines to the World Ocean. The major trends in the data are summarized in the following.

1 Because the major sources of the organochlorines are in the Northern Hemisphere, the dominant deposition of the compounds is to the North Atlantic and North Pacific.

2 Because of different sources and transport regimes, different organochlorines predominate in individual ocean basins. For example, HCH and DDT compounds have their highest deposition rates in the North Pacific, which probably arises from sources in Asia. In contrast, PCBs and dieldrin have higher deposition rates in the North Atlantic than in the North Pacific as a result of sources in North America

and Europe. Hexachlorobenzene and chlordane have deposition rates that are generally similar in the North Atlantic and North Pacific.

3 The mechanisms of air–sea exchange differ for some organochlorines. Although the magnitudes of the direct gas exchange are uncertain, such a mechanism can account for between ~25% and ~85% of the total air–sea exchange of the organochlorines. The direct 'dry' deposition of particle-bound organic material accounts for <5% of the total particle deposition, and the primary differences in deposition mechanisms between the compounds occur in 'wet' deposition. Overall, particle scavenging by 'wet' deposition is most significant for the PCB and DDT compounds, and gas scavenging is the predominant mechanism for the removal of HCH compounds in rain.

4 There is a general lack of data on the input of the organochlorines from large river systems, but on the basis of the limited data set given in Table 6.14(c), it is apparent that atmospheric input is the dominant transport route by which the organochlorines reach the World Ocean; however, it must be remembered that the estimates are still essentially very crude.

While all of the previous points are true, a feature of any consideration of specific organic compounds is that the patterns of use of individual compounds is constantly changing with changes in regulation, effectiveness and costs, so any particular distribution will change with time. However, it is reasonable to conclude that semi-volatile organic compounds emitted to the atmosphere via industrial or agricultural emissions are likely to undergo long range atmospheric transport, whereas emissions to soils are likely to be retained or only slowly moved through and degraded in soils. Hence, atmospheric transport of this group of compounds is particularly important.

### 6.4.3 Regional patterns

Table 6.13 represents a global synthesis of fluxes and is valuable in identifying the dominant pathways of inputs to the oceans and the long term balance between these inputs. However, it is self evident that there will be regional differences in the relative importance of these inputs.

**Table 6.14** The atmospheric input of organochlorine compounds to the World Ocean in the 1980s (data from Duce et al., 1991).

(a) Total deposition and mean fluxes*.

| Compound[†] | North Atlantic | | South Atlantic | | North Pacific | | South Pacific | | Indian | | Global | |
|---|---|---|---|---|---|---|---|---|---|---|---|---|
| | TD | MF | TD | MF | TD | MF | TD | MF | TD | MF | TD | MF |
| Σ HCH | 850 | 16 | 97 | 1.9 | 2600 | 30 | 470 | 4.3 | 700 | 10 | 4800 | 13 |
| HCB | 17 | 0.31 | 10 | 0.20 | 20 | 0.22 | 19 | 0.17 | 11 | 0.17 | 77 | 0.23 |
| Dieldrin | 17 | 0.30 | 2.0 | 0.04 | 8.9 | 0.10 | 9.5 | 0.09 | 6.0 | 0.09 | 43 | 0.11 |
| Σ DDT | 16 | 0.28 | 14 | 0.27 | 66 | 0.74 | 26 | 0.23 | 43 | 0.64 | 170 | 0.44 |
| Chlordane | 8.7 | 0.16 | 1.0 | 0.02 | 8.3 | 0.09 | 1.9 | 0.02 | 2.4 | 0.04 | 22 | 0.06 |
| Σ PCB | 100 | 1.8 | 14 | 0.27 | 36 | 0.40 | 29 | 0.26 | 52 | 0.77 | 240 | 0.64 |

*TD, total deposition (units, $10^6$g yr$^{-1}$); MF, mean flux (units, $\mu$g m$^{-2}$ yr$^{-1}$). † See text.

(b) Depositional processes to the World Ocean expressed as a percentage of total deposition.

| Compound* | Particle | | Gas | |
|---|---|---|---|---|
| | Dry | Wet | Dry | Wet |
| α-HCH | <0.1 | 0.1 | 38 | 62 |
| γ-HCH | <0.1 | <0.1 | 23 | 77 |
| HCB | 0.2 | 2.2 | 85 | 13 |
| Dieldrin | 0.3 | 13 | 54 | 33 |
| pp' DDT | 0.7 | 34 | 45 | 20 |
| Chlordane | 0.2 | 9.5 | 72 | 18 |
| Σ PCB | 0.6 | 23 | 65 | 11 |

*See text.

(c) Atmospheric and fluvial inputs to the World Ocean (units, $10^6$g yr$^{-1}$).

| Compound* | Atmospheric | Fluvial | Percentage atmospheric |
|---|---|---|---|
| Σ HCH | 4800 | 40–80 | 99 |
| HCB | 77 | 4 | 95 |
| Dieldrin | 43 | 4 | 91 |
| Σ DDT | 170 | 4 | 98 |
| Chlordane | 22 | 4 | 85 |
| Σ PCB | 240 | 40–80 | 80 |

*See text.

Hydrothermal inputs arising from mid ocean ridge systems and will have locally very large impacts around the systems. These impacts may have global scale impacts, such as the relatively fast cycling of all ocean water through the plumes from these systems. Alternatively the impacts may be relatively local such as the creation of highly metalliferous sediments around the mid ocean ridges. The scale of dispersal of impacts around hydrothermal impacts depends on the interplay between the rates of geochemical processes, such as iron and manganese precipitation, and ocean currents.

Atmospheric inputs also arise from geographically discrete areas. Figure 4.4 illustrates the global distribution of dust deposition which arises primarily from dust emissions from desert areas, particularly in North Africa and Asia. However, the emissions are distributed by atmospheric transport regimes that are much faster than ocean currents allowing emissions to be transported many thousands of kilometres through the atmosphere prior to deposition. Hence there are strong gradients in dust deposition fluxes away from sources, but also relatively efficient global transport. Hence, dust fluxes are higher in the Northern hemisphere than in the Southern Hemi-

sphere and particularly high in the tropical Atlantic. Figure. 6.5 illustrates atmospheric lead deposition to the oceans during the 1980s when lead was still extensively used as an automotive fuel additive. Emission sources were primarily in North America, Asia and Europe with strong gradients in fluxes away from these source regions, but again effective hemispheric and to some extent global transport. Thus deposition was highest in the temperate North Atlantic and North Pacific. As discussed in Chapter 4, these lead fluxes have decreased dramatically due to the elimination of lead additives from most fuels. However, these figures do illustrate the point that atmospheric transport acts to disperse emissions rather effectively reducing local impacts, but at the same time allow effective transport to remote regions, as is also evident for fixed nitrogen (Fig. 6.4).

Fluvial inputs all arrive in the ocean at the coast, generally through estuaries although for a few large rivers and some groundwater discharges the first interaction of ocean and freshwater takes place on the shelf. Hence the impacts of freshwater inputs are focused particularly on the coast and radiate from there, dispersed by ocean currents. This means the impacts of fluvial inputs are focused particularly on

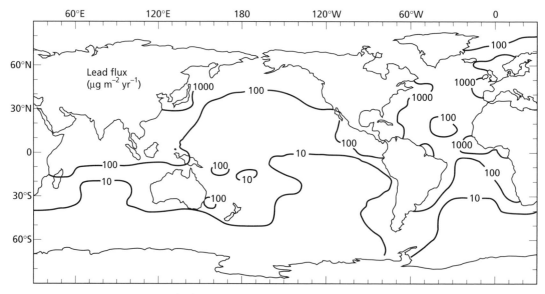

**Fig. 6.5** Global fluxes of lead (units, $\mu g\, m^{-2}\, yr^{-1}$) to the World Ocean (from Duce *et al.*, 1991).

the coastal system and less readily globally dispersed than atmospheric inputs. As discussed in the following, the coastal zone is particularly effective at trapping material, so the impacts of fluvial inputs of elements caught in this trap can be limited on the wider ocean scale.

## 6.5 The continental shelf

Beyond the estuaries lies a boundary zone to the open ocean formed by the continental shelf. This varies in width and depth depending on the geological environment and is bounded on the ocean side by a region known as the continental slope, where the seabed begins to drop relatively rapidly toward the abyssal depths. Using such a definition, the area of the shelf seas is about $27 \times 10^6 \, \text{km}^2$ (7.5% of the global ocean area ) with an average depth of about 130m, but deeper in polar regions due to the weight of ice on the adjacent land (Wollast, 2002; Constanza et al., 1997). The shelf seas can be only tens of km wide in some areas, for instance on tectonically active shelves such as the western coast of North and South America, and in other cases on more passive margins the shelf sea can be hundreds of km wide such as in the case of the North Sea (Jickells and Weston, 2010). During ice ages with sea level more than 100m lower than at present, the continental shelf seas would have been 75% smaller than at present (Martin and Windom, 1991).

The geochemical and biological processes in any particular shelf sea are unique, reflecting the interaction of ocean currents and land based inputs with the geomorphology of the particular shelf, and in that sense each shelf system is unique (see e.g. Jickells, 1998). However, with that caveat in mind, it is still possible to consider some general features of coastal systems and their role in ocean biogeochemistry.

The shelf seas have an importance to human society that is much greater than suggested by their area, providing food, mineral resources, recreation, transport and waste disposal services. Many of these activities can be in conflict leading to considerable stress on coastal systems worldwide, stresses that are likely to increase in the future with growing populations, industrialization and climate change (Constanza et al., 1997; Jickells, 1998; Crossland et al., 2005).

A particular feature of the coastal zone is its relatively high productivity (see Fig. 9.5) compared to open ocean waters $\sim 8.3 \times 10^{15} \, \text{g C yr}^{-1}$ or between 18 and 33% of global ocean productivity (Wollast, 2002).

This productivity reflects in part the efficient recycling of nutrients in coastal seas compared to open oceans by virtue of their shallow depth. Sinking organic matter is broken down either in the water column or at the seabed, and from there nutrients can be returned relatively easily to the surface waters in shelf seas compared to open ocean waters of greater depth. Another important contributor to the productivity is a relatively high supply of nutrients from outside the coastal zone to augment the internal recycling. This supply can come from rivers, groundwater, atmospheric deposition, direct discharges and in a few cases from hydrothermal sources. Another major source of nutrients is the input from mixing with open ocean waters. On some margins with active upwelling, for example, Benguela, California, North Africa coasts, this flux can be particularly large, but it is significant in almost all coastal systems. On a global basis, Wollast estimates a flux of nitrogen from the ocean to the shelf of $9000 \times 10^9 \, \text{mol yr}^{-1}$, which is six times the input from rivers (Table 6.13). Atmospheric inputs of nitrogen to shelf seas are usually smaller than river inputs (Spokes and Jickells, 2005). Hence as noted by Wollast (2002) 'the fertility of the coastal zone is largely due to nutrients provided by deep ocean waters and not continental inputs.' Of course this input from the ocean represents an internal cycle within the oceans rather than a 'new' input such as from rivers, but it is important to recognize the importance of this large natural input when considering the scale of human perturbation of the coastal system. Increasing nutrient inputs to coastal zones can lead to local areas of elevated primary productivity, harmful algal blooms and low oxygen concentrations in deep water (Jickells, 1998; Crossland et al., 2005; Diaz and Rosenberg, 2008; Middleburg and Levin, 2009). The latter are sometimes called 'dead zones', although there is in fact plenty of microbial life degrading organic matter and consuming oxygen, but rather little macroinvertebrate life, and appear to have increased in number and area since the 1960s probably due to increasing nutrient inputs (Diaz and Rosenberg, 2008).

The supply of organic matter to the sediments results in oxygen consumption and denitrification in sediment pore waters and sometimes the water column. Denitrification, which we will discuss further in Chapters 9 and 14, occurs under low oxygen conditions and converts nitrate into gaseous $N_2$ and $N_2O$. Seitzinger *et al.* (2006) estimate that 44% of all global denitrification takes place on the continental shelf, equivalent to the consumption of $18\,000 \times 10^9$ mole N $yr^{-1}$, which exceeds estimates of land based inputs to the coastal zones. This implies that not only are land based (fluvial + atmospheric) inputs of nitrogen trapped in the coastal zone, but that globally denitrification on the shelf is a major sink for oceanic nitrogen.

Denitrification means that the coastal zone is a uniquely effective sink for land based inputs of nitrogen to the oceans. The situation for other elements is less clear. The particle removal processes in estuaries discussed earlier will continue in shelf seas where energy levels are lower and residence times longer than in estuaries, providing time for the further flocculation and settlement of fluvial particles. Furthermore biological uptake of trace metals during photosynthesis offers an additional sink for dissolved metals, particularly those that are relatively inefficiently retained in estuaries, such as Cd, Cu, Ni and Zn. Some metals, such as Fe and Mn, may be remobilized from reducing sediments during sedimentary diagenesis (see Chapter 14). Overall Martin and Windom (1991) suggest '80–90% of the total trace element input to the oceans accumulates in ocean margins', although their definition of the margins includes the estuaries. While it is very difficult to develop comprehensive global budgets for continental shelf regions, it is clear that they represent effective traps for land derived sediments and that they are highly productive and the sedimentation of that productivity provides an additional mechanism to trap elements input to the coastal zone from land. Hence, for all the elements in Table 6.13 that are heavily involved in ocean biological cycles or particle reactive, the fluvial fluxes should be viewed as upper limits and the fluxes escaping the coastal seas to the open ocean are likely to be less than the figures presented.

As noted previously, during glacial periods the area of continental shelf would have been much less than at present and hence the efficiency of the shelf filter for fluvial inputs may have been reduced.

## 6.6 Relative magnitudes of the primary fluxes to the oceans: summary

1 The major primary sources of both particulate and dissolved components to the oceans on a global scale are river run-off, atmospheric deposition and hydrothermal exhalations, with glacial sources being locally predominant in some polar regions. Glacial inputs of particulate matter may be important but are very uncertain.

2 Of the major primary sources, river run-off and atmospheric deposition deliver their loads to the surface ocean, rivers to the ocean margins, and the atmosphere to the whole ocean surface. On a global basis, fluvial fluxes are generally greater than those resulting from atmospheric deposition, although there are some exceptions to this (e.g. atmospheric deposition dominates the input of Pb to almost the whole surface ocean). Retention in the estuarine and coastal zones, however, can mean that the net input of some fluvially transported components to the open ocean is less than that from atmospheric deposition. This atmospheric deposition retains its strongest surface water fingerprints for the 'scavenged-type' trace metals, such as Al, Mn and Pb.

3 Hydrothermal activity can act as a source for some components and a sink for others. When it acts as a source, high-temperature axial hydrothermal activity delivers components to mid-depth and bottom waters via the formation of vent plumes following the circulation of seawater through the crust. Although the extent to which the components in these plumes are globally dispersed is not yet known with certainty, it is now recognized that for some elements (e.g. Fe, Mn) hydrothermal inputs can match, and sometimes exceed, fluvial inputs. In addition to being a source, the hydrothermal plumes can act as an oceanic sink for some dissolved elements as seawater is cycled through them in the water column.

We have now identified the principal primary sources that supply material to the World Ocean, and have described the pathways along which the material travels in order to enter the seawater reservoir. In the following chapters, we will describe the physical and chemical nature of this reservoir, and discuss the mechanisms that keep it in motion.

# References

Arimoto, R., Duce, R.A., Ray, B.J. and Unni, C.K. (1985) Atmospheric trace elements at Enewetak Atoll: transport to the ocean by wet and dry deposition. *J. Geophys. Res.*, **90**, 2391–2408.

Arimoto, R., Duce, R.A., Ray, B.J., Hewitt, A.D. and Williams, J. (1987) Trace elements in the atmosphere of American Samoa; concentrations and deposition to the tropical South Pacific. *J. Geophys. Res.*, **92**, 8465–8479.

Arnold, M., Seghaier, A., Martin, D., Buat-Menard, P. and Chesselet, R. (1982) Geochimie de l'aerosol marin au-dessus la Mediteranee occidentale. *CIESMI J. Etud. Pollut. Mar. Mediterr.*, **VI**, 27–37.

Balls, P.W. (1989) Trace metal and major ion composition of precipitation at a North Sea coastal site. *Atmos. Environ.*, **23**, 2751–2759.

Bewers, J.M. and Yeats, P.A. (1977) Oceanic residence times of trace metals. *Nature*, **258**, 595–598.

Boyle, E.A., Collier, R., Dengler, A.T., Edmond, J.M., Ng, A.C. and Stallard, R.F. (1974) On the chemical mass-balance in estuaries. *Geochim. Cosmochim. Acta*, **38**, 1719–1728.

Brimblecombe, P., Hammer, C., Rodhe, H., Ryaboshapko, A. and Boutron, C. (1989) *Human Influence on the Sulphur Cycle in Evolution of the Global Biogeochemical Sulphur Cycle*. P. Brimblecombe and A. Yu Lein (eds), pp.77–121 SCOPE. Chichester: John Wiley & Sons, Ltd.

Bruland, K.W. and Franks, R.P. (1983) Mn, Ni, Zn and Cd in the western North Atlantic, in *Trace Metals in Sea Water*, C.S. Wong, E. Boyle, K.W. Bruland, J.D. Burton and E.D. Goldberg (eds), pp.395–414. New York: Plenum.

Bruland, K.W., Bertine, K., Koide, M. and Goldberg, E.D. (1974) History of metal pollution in southern California coastal zone. *Environ. Sci. Technol.*, **8**, 425–431.

Buat-Menard, P. (1983) Particle geochemistry in the atmosphere and oceans, in *Air–Sea Exchange of Gases and Particles*, P.S. Liss and W.G.N. Slinn (eds), pp.455–532. Dordrecht: Reidel.

Buat-Menard, P. and Chesselet, R. (1979) Variable influence of the atmospheric flux on the trace metal chemistry of oceanic suspended matter. *Earth Planet. Sci. Lett.*, **42**, 399–411.

Cambray, R.S., Jeffries, D.F. and Topping, G. (1975) *An Estimate of the Input of Atmospheric Trace Elements into the North Sea and the Clyde Sea (1972–73)*. Harwell: UK Atomic Energy Authority, Report AERE-R7733.

Chester, R. and Murphy, K.J.T. (1990) Metals in the marine atmosphere, in *Heavy Metals in the Marine Environment*, R. Furness and P. Rainbow (eds), pp.27–49. Boca Raton, FL: CRC Press.

Chester, R., Griffiths, A.G. and Hirst, J.M. (1979) The influence of soil-sized atmospheric particulates on the elemental chemistry of the deep-sea sediments of the North Eastern Atlantic. *Mar. Geol.*, **32**, 141–154.

Chester, R., Nimmo, M., Murphy, K.J.T. and Nicolas, E. (1990) Atmospheric trace metals transported to the Western Mediterranean: data from a station on Cap Ferrat. *Water Poll. Res. Rep.*, **20**, 579–612.

Chester, R., Murphy, K.J.T., Lin, F.J., Berry, A.S., Bradshaw, G.A. and Corcoran, P.A. (1993) Factors controlling the solubilities of trace metals from non-remote aerosols deposited to the sea surface by the 'dry' deposition mode. *Mar. Chem.*, **42**, 107–126.

Chester, R., Bradshaw, G.F., Ottley, C.J. *et al.* (1994) The atmospheric distributions of trace metals, trace organics and nitrogen species over the North Sea, in *Understanding the North Sea System*, H. Charnock, K.R. Dyer, J.M. Huthnance, P.S. Liss, J.H. Simpson and P.B. Tett (eds). *Philos. Trans. R. Soc. London, Ser. A*, **343**, 545–556.

Chester, R., Nimmo, M. and Keyse, S. (1996) The influence of Saharan and Middle Eastern desert-derived dust on the trace metal compositions of Mediterranean aerosols and rainwaters: an overview, in *The Impact of Desert Dust across the Mediterranean*, S. Guerzoni and R. Chester (eds), pp.253–273. Dordrecht: Kluwer Academic Publishers.

Chester, R., Nimmo, M., Keyse, S. and Corcoran, P.A. (1997) Rainwater–aerosol trace metal relationships at Cap Ferrat: a coastal site in the Western Mediterranean. *Mar. Chem.*, **58**, 293–312.

Church, T.M., Galloway, J.N., Jickells, T.D. and Knap, A.H. (1982) The chemistry of precipitation at the mid-Atlantic coast on Bermuda. *J. Geophys. Res.*, **87**, 11013–11018.

Constanza, R., D'Arge, R., De Groot, R., Farber, S., Grasso, M., Hannon, B., Limburg, K., Naeem, S., O'Neill, R.V., Pauelo, J., Raskin, R.G., Sutton, P. and van den Belt, M. (1997) The value of the world's ecosystem services and natural capital. *Nature*, **387**, 253–260.

Crossland, C.J., Kremer, H.J., Lindeboom, H.J., Marshall Crossland, J.I. and Le Tissier, M.D.A. (eds) (2005) *Coastal Fluxes in the Anthropocene*. Heidelberg: Springer.-Verlag.

Dai, A., Qian, T., Trenberth, K.E. and Milliman, J.D. (2009) Changes in continental discharge from 1948–2004. *J. Climate*, **22**, 2773–2792.

Dedkov, A.P. and Mozzherin, V.I. (1984) *Eroziya i stok Nanosov na Zemle*. Izdatatelstova Kazanskogo Universiteta.

Degens, E.T. and Ittekkot, V. (1985) Particulate organic carbon: an overview, in *Transport of Carbon and Minerals in Major World Rivers*, Part 3, E.T. Degens, S. Kempe and R. Herrera (eds), pp.7–27. Hamburg: Mitt. Geol.-Palont. Inst. Univ. Hamburg, SCOPE/UNEP, Sonderband 58.

Diaz, R.J. and Rosenberg, R. (2008) Spreading dead zones and consequences for marine ecosystems. *Science*, **321**, 926–929.

Duce, R.A. (1978) Speculations on the budget of particulate and vapour phase non-methane organic carbon in the global troposphere. *Pure Appl. Geophys.*, **116**, 244–273.

Duce, R.A. and Duursma, E.K. (1977) Inputs of organic matter to the ocean. *Mar. Chem.*, **5**, 319–339.

Duce, R.A., Hoffman, G.L., Ray, B.J. *et al.* (1976) Trace metals in the marine atmosphere: sources and fluxes, in *Marine Pollutant Transfer*, H. Windom and R.A. Duce (eds), 77–119. Lexington: Heath.

Duce, R.A., Mohnen, V.A., Zimmerman, P.R. *et al.* (1983) Organic material in the global troposphere. *Rev. Geophys. Space Phys.*, **21**, 921–952.

Duce, R.A., Liss, P.S., Merrill, J.T. *et al.* (1991) The atmospheric input of trace species to the World Ocean. *Global Biogeochem. Cycles*, **5**, 193–259.

Duce, R., LaRoche, J., Altieri, K. *et al.* (2008) Impacts of atmospheric nitrogen on the open ocean. *Science*, **320**, 893–897.

Edmond, J.M., Measures, C.I., McDuff, R.E. *et al.* (1979) Crest hydrothermal activity and the balance of the major and minor elements in the ocean; the Galapagos data. *Earth Planet. Sci. Lett.*, **46**, 1–18.

Edmond, J.M., Von Damm, K.L., McDuff, R.E. and Measures, C.I. (1982) Chemistry of hot springs on the East Pacific Rise and their effluent dispersal. *Nature*, **297**, 187–191.

Edmond, J.M., Spivack, A., Grant, B.C. *et al.* (1985) Chemical dynamics of the Changjiang estuary. *Cont. Shelf. Res.*, **4**, 17–30.

Elderfield, H. and Schultz, A. (1996) Mid-ocean ridge hydrothermal fluxes and the chemical composition of the ocean. *Annu. Rev. Earth Planet Sci.*, **24**, 191–224.

Galloway, J.N. and Gaudry, A. (1984) The composition of precipitation on Amsterdam Island, Indian ocean. *Atmos. Environ.*, **18**, 2649–2656.

Galloway, J.N., Likens, G.E., Keene, W.C. and Miller, J.M. (1982) The composition of precipitation in remote areas of the world. *J. Geophys. Res.*, **87**, 8771–8786.

German, C.R. and Von Damm, K.L. (2003) Hydrothermal Processes, in *Treatise on Geochemistry*, Vol. 6, H. Elderfield (ed.) pp.181–222. Amsterdam: Elsevier.

GESAMP (1987) *Land/Sea Boundary Flux of Contaminants from Rivers*. Paris: UNESCO.

Guieu, C., Chester, R., Nimmo, M. *et al.* (1997) Atmospheric input of dissolved and particulate metals to the North-Western Mediterranean Sea. *Deep Sea Res.*, **44**, 655–671.

Hodge, V., Johnson, S.R. and Goldberg, E.D. (1978) Influence of atmospherically transported aerosols on surface ocean water composition. *Geochem. J.*, **12**, 7–20.

Holeman, J.N. (1968) The sediment yield of major rivers of the world. *Water Resour. Res.*, **4**, 734–747.

Humborg, C., Conley, D.J., Rahm, L., Wulff, F., Cociasu, A. and Ittekkot, V. (2000) Silicon retention in river basins: Far-reaching effects on biogeochemistry and aquatic food webs in coastal marine environments. *Amnbio.*, **29**, 45–50.

Jickells, T., Knapp, A.H. and Church, T.M. (1984) Trace metals in Bermuda rainwater. *J. Geophys. Res.*, **89**, 1423–1428.

Jickells, T.D., An, Z.S., Anderson, K.K. *et al.* (2005) Global Iron Connections Between Desert Dust, Ocean Biogeochemistry and Climate. *Science*, **208**, 65–71.

Jickells, T.D. (1998) Nutrient biogeochemistry of the coastal zone. *Science*, **281**, 217–222.

Jickells, T.D. and Weston, K. (2010) Nitrogen cycle – external cycling, losses and gains, in *Treatise on Estuarine and Coastal Science*. Amsterdam: Elsevier.

Judson, S. (1968) Erosion of the land (What's happening to our continents?). *Am. Sci.*, **5**, 514–516.

Kadko, D., Baker, E., Alt, J. and Baross, J. (1994) *Global impact of submarine hydrothermal processes*. Final Report. Ridge/Vent Workshop.

Keene, W.C., Khalil, M.A.K., Erikson, D.J. III *et al.* (1999) Composite global emissions of reactive chlorine from anthropogenic and natural sources: reactive chlorine emissions inventory. *J. Geophys. Res.*, **104**, 8429–8440.

Keyse, S. (1996) *Factors controlling the solubility of trace metals in rainwater*. PhD Thesis, University of Liverpool.

Klinkhammer, G.P. (1980) Observations of the distribution of manganese over the East Pacific Rise. *Chem. Geol.*, **29**, 211–226.

Klinkhammer, G.P., Rona, P., Greaves, M. and Elderfield, H. (1985) Hydrothermal manganese plumes in the Mid-Atlantic Ridge rift valley. *Nature*, **314**, 727–731.

Kremling, K. (1985) The distribution of cadmium, copper, nickel, manganese and aluminium in surface waters of the open Atlantic and European shelf area. *Deep Sea Res.*, **32**, 531–555.

Lim, B., Jickells, T.D. and Davies, T.D. (1991) Sequential sampling of particles, major ions and total trace metals in wet deposition. *Atmos. Environ.*, **25**, 745–762.

Lim, B., Jickells, T.D., Colin, J.L. and Losno, R. (1994) Solubilities of Al, Pb, Cu and Zn in rain sampled in the marine environment over the North Atlantic Ocean and Mediterranean Sea. *Global Biogeochem. Cycles*, **8**, 349–362.

Losno, R., Bergametti, G. and Buat-Menard, P. (1988) Zinc partitioning in Mediterranean rainwater. *J. Geophys. Res. Lett.*, **15**, 1389–1392.

Loye-Pilot, M.D., Martin, J.-M. and Morelli, J. (1986) Influence of Saharan dust on the rain acidity and atmospheric input to the Mediterranean. *Nature*, **321**, 427–428.

Mahowald, N., Jickells, T.D., Baker, A.R. *et al.* (2008) The global distribution of atmospheric phosphorus deposition and anthropogenic impacts. *Global Biogeochem. Cycles*, **22**, GB4026, doi:10.1029/2008GB003240.

McDonald, R.L., Unni, C.K. and Duce, R.A. (1982) Estimation of atmospheric sea salt dry deposition; wind speed and particle size dependence. *J. Geophys. Res.*, **87**, 1246–1250.

Martin, J.-M. and Windom, H.L. (1991) Present and future roles of ocean margins in regulating biogeochemical cycles of trace elements, in *Ocean Margin Processes in Global Change*, R.F.C. Mantoura, J.-M. Martin and R. Wollast (eds), pp.45–67. Chichester: John Wiley & Sons, Ltd.

Meybeck, M. (1979) Concentrations des eaux fluviales en elements majeurs et apports en solution aux oceans. *Rev. Geol. Dyn. Geogr. Phys.*, **21**, 215–246.

Middleburg, J.J. and Levin, L.A. (2009) Coastal hypoxia and sediment biogeochemistry. *Biogeosciences*, **6**, 1273–1293.

Milliman, J.D. (1981) Transfer of river-borne particulate material to the oceans. In *River Inputs to Ocean Systems*, J.-M. Martin, J.D. Burton and D. Eisma (eds), pp.5–12. Paris: UNEP/UNESCO.

Milliman, J.D. and Meade, R.H. (1983) World-wide delivery of river sediment to the oceans. *J. Geol.*, **91**, 1–21.

Perdue, E.M. and Ritchie, J.D. (2003) Dissolved organic matter in freshwater in Surface and Ground Water, Weathering and Soils, (ed. J.J. Drever), Vol. 5 *Treatise on Geochemistry*, H.D. Holland and K.K. Turekian (eds), pp.273–318. Elsevier-Pergamon, Oxford.

Pszenny, A.A.P., MacIntyre, F. and Duce, R.A. (1982) Sea-salt and the acidity of marine rain on the windward coast of Samoa. *Geophys. Res. Lett.*, **9**, 751–754.

Raiswell, R., Tranter, M., Benning, L.G., Siegert, M., De'ath, R., Huybrechts, P. and Rayne, T. (2006) Contributions from glacially derived sediment to the global iron(oxyhyr)oxide cycle: implications for iron delivery to the oceans. *Geochim. Cosmochim. Acta*, **70**, 2765–2780.

Robinson, E. (1978) Hydrocarbons in the atmosphere. *Pure Appl. Geophys.*, **116**, 327–384.

Seinfeld, J.H. and Pandis, S.N. (1998) *Atmospheric Chemistry and Physics*. New York: John Wiley & Sons, Inc.

Seitzinger, S.P., Harrison, J.A., Dumont, E., Beusen, A.H.W. and Bouwman, A.F. (2005) Sources and delivery of carbon, nitrogen and phosphorus to the coastal zone: An overview of Global Nutrient Export from Watersheds (NEWS) models and their application. *Global Biogeochem. Cycl.*, **19**, doi:10.1029/2005GB002606.

Seitzinger, S., Harrison, J.A., Böhlke, J.K., Bouwman, A.F., Lowrance, R., Peterson, B., Tobias, C. and Van Drecht, G. (2006) Denitrification across landscapes and waterscapes: a synthesis. *Ecological Applications*, **16**, 2064–2090.

Settle, D.M. and Patterson, C.C. (1982) Magnitudes and sources of precipitation and dry deposition fluxes of industrial and natural lead to the North Pacific at Enewetak. *J. Geophys. Res.*, **87**, 8857–8869.

Spokes, L.J. and Jickells, T.D. (2005) Is the atmosphere really an important source of reactive nitrogen to coastal waters? *Continental Shelf Research*, **25**, 2022–2035.

Stossel, R.P. (1987) *Untersuchungen zur Nab und trokendposition von Schwermetallen auf der Insel der Pellworm.* Doctorgrades Dissertation, University of Hamburg.

Syvitski, J.P.M., Vörösmarty, C.J., Kettner, A. and Green, P. (2005) Impact of humans on the flux of terrestrial sediment to the global coastal ocean. *Science*, **308**, 376–380.

Thompson, G. (1983) Hydrothermal fluxes in the ocean, in *Chemical Oceanography*, Vol. 8, J.P. Riley and R. Chester (eds), pp.270–337. London: Academic Press.

Turner, R.E., Rabalais, N.N., Justic, D. and Dotch, Q. (2003) Global patterns of dissolved N,P and Si in large rivers. *Biogeochemistry*, **64**, 297–317.

Von Damm, K.L., Edmond, J.M., Grant, B., Measures, C.I., Walden, B. and Weiss, R.F. (1985) Chemistry of submarine hydrothermal solutions at 21°N, East Pacific Rise. *Geochim. Cosmochim. Acta*, **49**, 2197–2220.

Walling, D.E. and Webb, D.W. (1987) Material transport by the world's rivers: evolving perspectives, in *Water for the Future: Hydrology in Perspective*, pp.313–329. Wallingford: International Association of Hydrological Sciences, Publication 164.

Walsh, P.R. and Duce, R.A. (1976) The solubilization of anthropogenic atmospheric vanadium in sea water. *Geophys. Res. Lett.*, **3**, 375–378.

Walsh, P.R., Duce, R.A. and Fasching, J.L. (1979) Considerations of the enrichment, sources and fluxes of arsenic in the troposphere. *J. Geophys. Res.*, **84**, 1719–1726.

Wiersma, G.B. and Davidson, C.I. (1986) Trace metals in the atmosphere of rural and remote areas, in *Toxic Metals in the Atmosphere*, J.O. Nriagu and C.I. Davidson (eds), pp.201–266. New York: John Wiley & Sons, Inc.

Williams, P.J. (1975) Biological and chemical aspects of dissolved organic material in sea water, in *Chemical Oceanography*, Vol. 2, J.P. Riley and G. Skirrow (eds), pp.301–363. London: Academic Press.

Windom, H.L. (1981) Comparison of atmospheric and riverine transport of trace elements to the continental shelf environment, in *River Inputs to Ocean Systems*, J.-M. Martin, J.D. Burton and D. Eisma (eds), pp.360–369. Paris: UNEP/UNESCO.

Witt, M., Baker, A.R. and Jickells, T.D. (2006) Atmospheric trace metals over the Atlantic and South Indian Oceans: Investigations of metal concentrations and lead isotope ratios in coastal and remote marine aerosols. *Atmos. Environ.*, **40**, 5435–5451.

Wollast, R. (2002) Continental Margins – Review of Geochemical Settings, in G. *Ocean Margin Systems*, D. Wefer, D. Billet, I. Hebbe *et al.* (eds),pp.15–31. Heidelberg: Springer Verlag.

Zimmerman, P.R., Chatfield, R.B., Fishman, J., Crutzen, P.J. and Hanst, P.L. (1978) Estimates of the production of CO and $H_2$ from the oxidation of hydrocarbon emissions from vegetation. *Geophys. Res. Lett.*, **5**, 679–682.

# Part II
# The Global Journey: The Ocean Reservoir

# 7 Descriptive oceanography: water-column parameters

In the previous chapters, the various types of material that are brought to the oceans have been tracked along their transport pathways to the point where they cross the interfaces separating the sea from the other planetary reservoirs. Once they have entered the ocean reservoir these materials are subjected to a variety of physical, chemical and biological processes, which combine to control the composition of both the seawater and the sediment phases within it.

## 7.1 Introduction

In order to provide a framework within which to discuss the physical, chemical and biological oceanic processes, it is necessary first to describe the nature of the water itself, in terms of some of its fundamental oceanographic properties, and then to define the circulation mechanisms that keep it in motion. This circulation controls the physical transport of water masses, and their associated dissolved and particulate constituents, from one part of the oceanic system to another in the form of conservative signals. Superimposed on these are non-conservative signals, which arise from the involvement of the constituents in the major biogeochemical cycles that operate within the World Ocean. Components are eventually incorporated into bottom sediments mainly by the sinking of particulate material, and there is a continual movement of particulate components through the oceans. In order to understand how this dissolved and particulate throughput operates, a simple oceanic box model will be constructed. This will then be used in subsequent chapters, together with a variety of other approaches, as a framework within which to describe the oceanic distributions of a number of parameters that have proved especially rewarding for understanding the nature of the biogeochemical cycles in the sea; these parameters include dissolved oxygen and carbon dioxide, the nutrients, and both dissolved and particulate organic carbon. By means of this approach, the throughput of materials in the oceans will be described in relation to organic matter, suspended particulates and dissolved trace metals as they are transported to the interface that separates the water column from the sediment sink.

## 7.2 Some fundamental oceanographic properties of seawater

### 7.2.1 Introduction

The three fundamental properties of seawater that are of most interest to marine geochemists are salinity and temperature, which can be used to characterize water masses and which, together with pressure, fix the density of seawater, and density itself, which fixes the depth to which a water mass will settle and so drives thermohaline circulation which moves water and chemical components around the oceans. This circulation also moves heat around the oceans, and indeed the planet, and is also therefore a fundamental component of the climate system. Each of these properties will be described individually below. It must be remembered, however, that they are interlinked, and the equation of state of seawater is a mathematical expression of the relationship between the temperature, pressure, salinity and density of seawater; it is used, for example, for the calculation

*Marine Geochemistry*, Third Edition. Roy Chester and Tim Jickells.
© 2012 by Roy Chester and Tim Jickells. Published 2012 by Blackwell Publishing Ltd.

of the density of seawater from the other parameters. Recently, a new international framework has been agreed to define these relationships (McDougall *et al.*, 2009).

## 7.2.2 Salinity

Salinity is a measure of the degree to which the water in the oceans is salty, and is a function of the weight of total solids dissolved in a quantity of seawater. The major ions in seawater, that is, those that make a significant contribution to salinity, are listed in Table 7.1, and although the total salt content can vary, these constituents are always, or almost always, present in almost constant relative proportions. This reflects the very long residence times (see Chapter 11) of these major ions in seawater (hundreds of thousands of years or in some cases many millions of years) which therefore smoothes out any local effects of input, cycling and removal processes (the sources and sinks of the major ions in seawater are discussed in Section 17.2). This constancy of composition of seawater has extremely important implications for oceanographers.

Historically, there have been a number of definitions of salinity. According to an International Commission set up in 1899 salinity was defined as 'the weight of inorganic salts in one kilogram of seawater, when all bromides and iodides are replaced by an equivalent quantity of chlorides, and all carbonates are replaced by an equivalent quantity of oxides'. In theory, perhaps the simplest direct method for the measurement of salinity is to evaporate the water

**Table 7.1** The major ions of seawater (data from Wilson, 1975).

| Ion | Concentration $g\,kg^{-1}$ at $S = 35‰$ |
|---|---|
| $Cl^-$ | 19.354 |
| $SO_4^{2-}$ | 2.712 |
| $Br^-$ | 0.0673 |
| $F^-$ | 0.0013 |
| B | 0.0045 |
| $Na^+$ | 10.77 |
| $Mg^{2+}$ | 1.290 |
| $Ca^{2+}$ | 0.4121 |
| $K^+$ | 0.399 |
| $Sr^{2+}$ | 0.0079 |

to dryness and weigh the salt residue. In practice, however, accurate gravimetric methods are tedious and time-consuming, and in order to design a routine method for the determination of salinity the Commission made use of the concept of constancy of composition, which implies that it should be possible to use any of the major constituents as an index of salinity. It can be seen from Table 7.1 that chloride ions make up ~55% of the total dissolved salts in seawater, and reliable methods for the determination of chloride, using chemical titration with silver nitrate, were available at the time the Commission met. Chlorides, iodides and bromides have similar properties, and all of them react with silver nitrate to appear as chlorides in the titration.

An investigation was therefore made between *salinity* (S, in units of parts per thousand given the symbol ‰) and *chlorinity* (Cl, ‰), the latter being defined as 'the mass in grams of chlorine equivalent to the mass of halogens contained in one kilogram of seawater'. To examine the relationship, the salinities of nine seawaters were determined using an accurate gravimetric method. The chlorinities of the waters were then measured by titration, and the following relationship between salinity and chlorinity was established:

$$S(‰) = 1.805\,Cl(‰) + 0.030 \qquad (7.1)$$

This is termed *chlorinity salinity*, and was used for many years as the working definition of salinity.

The position began to change, however, with the introduction of the conductimetric salinometer, and a new investigation was carried out into the interrelations between the *measured* parameters (e.g. chlorinity, conductivity ratio, refractive index) and the *derived* parameters (e.g. salinity, specific gravity) of seawater. The data were assessed by a UNESCO Joint Panel (UNESCO, 1981), and on the basis of the relationship between chlorinity and the conductivity ratio a new equation for salinity was recommended:

$$S(‰) = 1.80655\,Cl(‰) \qquad (7.2)$$

This is termed *conductivity salinity*, and the equation can be used to determine salinity from the chlorinity obtained by the titration method.

However, the increasing use, and greater convenience, of high-precision electrical conductivity measurements has meant that chlorinity titration is now no longer the preferred method for the determination of salinity. In the light of this, the Joint Panel proposed a method for the determination of salinity from conductivity ratios, using a polynomial that relates salinity to the conductivity ratio $R$ determined at 15°C at one standard atmosphere pressure:

$$S(‰) = -0.08996 + 28.2970R_{15} + 12.80832R_{15}^2$$
$$- 10.67869R_{15}^3 + 5.98624R_{15}^4 - 1.32311R_{15}^5$$

$$(7.3)$$

where $R_{15}$ is the conductivity ratio at 15°C, and is defined as the ratio of the conductivity of a seawater sample to that of one having an $S(‰)$ of 35‰ at 15°C under a pressure of one standard atmosphere. Tables and computer programmes are available for the conversion of $R_{15}$ values into salinity, and a second polynomial has been provided to permit the conversion of conductivity ratios measured at other temperatures. Salinities determined from these polynomials are termed practical salinities. Practical salinity is based on conductivity ratios and hence has no dimension (units), and terms such as part per thousand (‰) have been abandoned to be replaced with $S = 2$ to $S = 42$; that is, the range over which practical salinities are valid. However, it is still useful for many purposes to use the old ‰ values (see e.g. Duinker, 1986).

The major elements that contribute to salinity are described as being conservative, that is, their concentration ratios remain constant, and their total concentrations can be changed only by physical processes. This is an operationally valid concept and it does appear that there are no significant variations in the ratios of sodium, potassium, sulfate, bromide and boron to chlorinity in seawater around the oceans at this time. Variations have been found, however, in the ratios of calcium, magnesium, strontium and fluoride to chlorinity (for a detailed review of the major elements in seawater: see Wilson, 1975). Recent very high precision and accuracy measurements of major ions in the oceans have demonstrated regional systematic very small variations in the ratios of major ions one to another. These variations in the ratios are generally very small and introduce corresponding small errors into calculations of density. There are also complications from the role of dissolved non-ionized species such as silicate which do not affect conductivity but do affect density. These small effects can be significant in terms of observed variations in ocean water density and hence are important to physical oceanographers and have contributed to the need for the recent redefinition of salinity (Millero *et al.*, 2008; McDougall *et al.*, 2009).

Corrections for these effects can now be made allowing a new SI unit of 'Absolute Salinity' to be derived that allows more straightforward and rigorous thermodynamic calculations of the physical properties of seawater such as heat capacity and the speed of sound (which is important for acoustic studies of the ocean and seabed). However absolute salinity cannot be measured in a straightforward way and is calculated indirectly based on measured practical salinity with units of $g\,kg^{-1}$ or $kg\,kg^{-1}$. The conversion of practical salinity, determined from conductivity measurements, to absolute salinity and the associated calculation of density from this and temperature, are routinely done using computer algorithms based on the fundamental equations presented in McDougall *et al.* (2009).

This revision of current practice for reporting salinity is too recent to have been adopted widely within the oceanographic community and the examples presented here continue to use the unit-less practical salinity scale. In practice the recent revisions are of considerable importance to the physical oceanographic community, but much less important for the marine chemistry community and McDougall *et al.* (2009) recommend the archiving of both practical and absolute salinity, since the former is the measured parameter.

Despite the fact that the major elements are more or less conservative in seawater, there are conditions under which their concentration ratios can vary considerably. Situations under which these atypical conditions can be found include those associated with estuaries and land-locked seas, anoxic basins, the freezing of seawater, the precipitation and dissolution of carbonate minerals, submarine volcanism, admixture with geological brines and evaporation in isolated basins.

Under *most* conditions, however, the major elements in seawater may be regarded as being present in almost constant proportions. However, the total salt content ($S$, ‰) can vary. This results from the operation of a number of processes. Those that *decrease* salinity include the influx of fresh water from precipitation (rainfall), land run-off and the melting of ice; and those that *increase* salinity include evaporation and the formation of sea ice (Bowden, 1975). A number of general trends can be identified in the distribution of salinity in the surface ocean.

**1** Salinity in surface ocean waters usually ranges between ~32 and ~37.

**2** Higher values are found in some semi-enclosed, mid-latitude seas and coastal waters where evaporation greatly exceeds precipitation and run-off. Examples of this are found in the Mediterranean Sea (range 37–39), and the Red Sea (range 40–41).

**3** In coastal waters run-off can result in a decrease in surface salinities.

The processes that cause the major variations in salinity occur at the surface of the ocean, and the most rapid variations usually are found within a few hundred metres or so of the air–sea interface. Regions in the water column over which rapid salinity changes take place are termed haloclines (see Fig. 7.1a). As there are no sources or sinks for salt in the deep layers of the ocean, the salinity derived at the surface can be changed only by the mixing of different water types with different initial salinities (see Section 7.3.4). In these deep-water layers, variations in salinity are smaller than those found at the surface. Different water masses, however, have individual 'salinity signatures', and these are extremely important in relation to deep-water circulation processes (see Section 7.3.3).

### 7.2.3 Temperature

Both horizontal and vertical temperature variations are found in the ocean. The main horizontal variations occur in the surface region, with temperatures ranging from ~28°C in the equatorial zones (and substantially higher in enclosed coastal waters) to as low as ~−1.9°C, the freezing point of seawater, in the polar seas. There is also a vertical temperature stratification in the water column, with in most cases the warm surface layer being separated from the main body of the colder deep ocean by the *thermocline*. There are thus three main temperature zones in the vertical ocean.

**1** *The surface ocean.* Here, the waters are heated by solar energy and the heat is mixed down to a depth of around 100–200 m, with the temperatures reflecting those of the latitude at which the water is found.

**2** *The thermocline.* Below the surface, or mixed, layer, the temperatures decrease rapidly with depth through a zone (the thermocline) that extends from the base of the surface layer down to as deep as ~1000 m in some places. A permanent main thermocline is found at low and mid-latitudes in all the major oceans, but it is absent at high latitudes. This main thermocline is usually less than 1000 m deep and its top is located at shallower depths in the equatorial than in the mid-latitude regions. Above the permanent thermocline, seasonal thermoclines are also found in parts of the ocean.

**3** *The deep ocean.* Under the base of the thermocline the deep ocean extends to the bottom of the water column. The waters here are cold, most being at <5°C, and although the temperature falls towards the bottom (to ~1°C) the rate of decrease is very much slower than that found in the thermocline. A generalized vertical temperature profile in the oceans, showing these various zones, is illustrated in Fig. 7.1(b). Figure 7.1 also shows salinity distributions, although these are somewhat less universally predictable than temperature because they depend on the balance of supply of freshwater from rivers, glaciers and rain and evaporation.

If a sample of seawater is brought to the surface from depth without exchanging heat with its surroundings, the temperature will fall as a result of adiabatic cooling and so will be lower than the in situ temperature. The temperature that it would have at the surface under atmospheric pressure is termed the potential temperature ($\theta$). As $\theta$ is not a function of depth it is a more useful parameter than in situ temperature for characterizing water masses and vertical motion in the oceans.

### 7.2.4 Density

The density ($\rho$) of seawater is a function of temperature, salinity and pressure, and the equation of state is a mathematical expression used to calculate the density from measurements of these parameters. The density of seawater exceeds that of pure water owing to the presence of dissolved salts, and the densities of most surface seawaters lie in the rather small range of 1024–1028 kg m$^{-3}$. As the values always start with

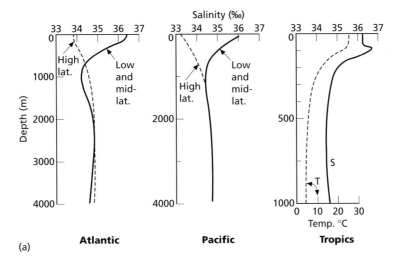

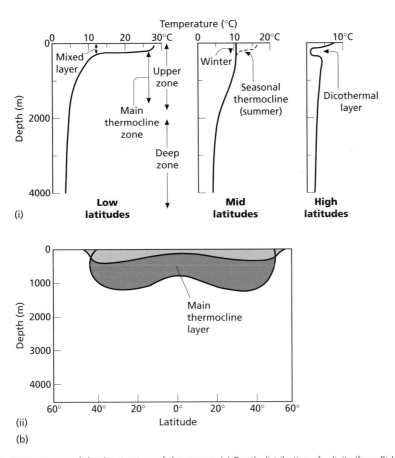

**Fig. 7.1** The salinity, temperature and density structure of the oceans. (a) Depth distribution of salinity (from Pickard and Emery, 1982). (b) (i) Depth distribution of temperature (from Pickard and Emery, 1982). Reprinted with permission from Elsevier. (ii) schematic representation of the thermocline (after Weihaupt, 1979). (c) (i) Depth distribution of density (from Pickard and Emery, 1982). Reprinted with permission from Elsevier. (ii) schematic representation of the pycnocline (after Gross, 1977).

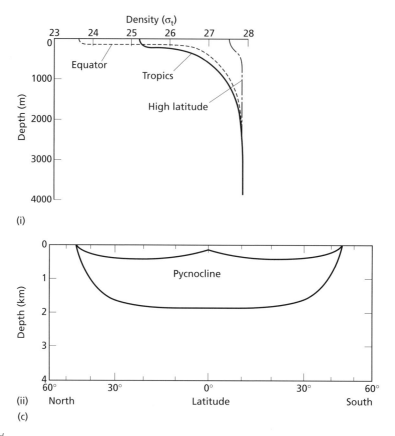

**Fig. 7.1** *Continued*

1000, however, it is common practice to shorten them by introducing the quantity:

$$\sigma_{s,t,p} = \rho_{s,t,p} - 1000 \qquad (7.4)$$

This is the in situ density, where subscripts $s,t,p$ indicate a function of salinity, temperature and pressure. Thus, the in situ density is the density of a seawater sample with the observed salinity and temperature, and the pressure found at the location of the sample. The effect of pressure on density, however, can often be ignored in descriptive oceanography, and for convenience atmospheric pressure is used, and the quantity

$$\sigma_t = \rho_{s,t,0} - 1000 \qquad (7.5)$$

is employed. This is termed 'sigma-tee', and seawater densities are most often quoted as $\sigma_t$ values. As these $\sigma_t$ values are a function of only temperature

and salinity, they therefore can be plotted on to temperature–salinity diagrams, which then can be used in the identification of water masses, since these are often characteristic of waters formed in certain regions (e.g. North Atlantic or Southern Ocean) and retain these characteristics as they travel through the oceans. Alternatively, potential density can be used on potential temperature–salinity diagrams, potential density ($\rho_e$) being the quantity

$$\sigma\theta = \rho_{s,\theta,0} - 1000 \qquad (7.6)$$

where $\theta$ is the potential temperature (see above). Lines of equal density in vertical or horizontal ocean sections are termed isopycnals, and isopycnal surfaces, or horizons, are often used in tracer circulation–distribution studies because they are the preferred surfaces along which lateral mixing takes place (see Section 7.4).

There is a vertical density stratification in the water column, with densities increasing with depth. The most dramatic change in density with depth is found in the upper water layer. In equatorial and mid-latitude regions there is a shallow layer of low-density water under the surface, below which the density increases rapidly. This zone of rapidly increasing density is termed the *pycnocline*, and it is an extremely important oceanographic feature because it acts as a barrier to the mixing of the low-density surface ocean and the high-density deep ocean. The pycnocline is formed in response to the combined rapid vertical changes in salinity (the halocline) and temperature (the thermocline). Over much of the open ocean outside of polar regions, temperature change is usually the main control on density change. It was pointed out above, however, that the thermocline is absent in high-latitude waters, and the pycnocline is not well developed here either: see Fig. 7.1(c). As a result, the water column at high latitudes is less stable than it is at lower latitudes, and in the absence of the mixing barrier the sinking of dense cold surface water can take place. The flow of warm less dense waters to these regions and the gravity-mediated sinking in them forms the basis of the thermohaline circulation in the oceans (see Section 7.3.3).

## 7.3 Oceanic circulation

### 7.3.1 Introduction

The oceanic water-column distributions of many dissolved constituents (e.g. the trace elements) are controlled by a combination of a transport process signal, associated with oceanic circulation, and a reactive process signal, associated with the major biogeochemical cycles. In order to evaluate the overall factors that control the distributions of the dissolved constituents in seawater it is necessary therefore to be able to distinguish between the effects of the two signals, that is, to extract reactive process information from the transport process, or *advective*, background. With this in mind, transport processes associated with oceanic circulation patterns will be described in the present section, the aim being simply to provide the basic information that is needed to interpret the reactive processes acting on dissolved constituents. This is especially relevant to the factors

that control the distributions of the nutrient and trace elements, topics considered in Chapters 9 and 11.

There is a wide spectrum of motions in the ocean and the processes that control them are complex. According to H. Leach (personal communication) the motions essentially can be subdivided into four types on the basis of the scales involved. These are:

1 the thermohaline circulation, which operates on a global ocean-scale of $>10^4$ km;
2 gyres, which operate on an ocean-basin scale of $10^3$ km;
3 mesoscale eddies, which operate on scales of 10 to $10^2$ km;
4 a variety of small-scale motions (e.g. surface waves, tidal currents, tidal mixing, Ekman layer transport), which operate on scales of <10 km to cm. Within the present context, however, it is only necessary to consider the most general features of oceanic circulation.

In the oceans, waters undergo both horizontal and vertical movements. The principal forces driving these movements are wind and gravity, and the resultant circulation patterns are modified by the effects of factors such as the rotation of the Earth, the topography of the sea bed and the positions of the land masses that form the ocean boundaries. Two basic types of circulation dominate the movement of water in the oceans. In the surface ocean the currents are wind-driven, and move in mainly horizontal flows at relatively high velocities. In the deep ocean circulation is driven largely by gravity (density changes) and the resultant thermohaline currents, which can have a vertical as well as a horizontal component, flow at relatively low velocities.

### 7.3.2 Surface water circulation

Circulation in the upper layer of the ocean is driven mainly by the response of the surface water to the movement of the wind in the major atmospheric circulation cells, and is constrained by the shape of the ocean basins. As the water is set in motion by the wind, it moves relative to a rotating earth and rotates fastest at the equator. This means that compared to a non-rotating sphere, a current moves at an angle relative to the wind – to the right (west) in the Northern Hemisphere and to the left (east) in the Southern Hemisphere, an effect known as the Coriolis Force. The currents that are generated across the air–sea interface decrease in strength with depth due to friction, falling to zero at about 100 m depth. This upper wind mixed layer of the ocean is often called the Ekman layer.

In terms of the major circulation patterns, the way in which the surface ocean responds to wind movement can be evaluated in terms of the two major wind systems, that is, the trades and the westerlies. In average terms, the trades move diagonally from the east towards the Equator in low latitudes, and the westerlies move diagonally from the west away from the Equator at mid-latitudes. The principal result of these wind movements, combined with the constraint imposed by the shape of the ocean basins, is that in each hemisphere a series of large anticyclonic water circulation cells, termed gyres, are set up in subtropical and high-(atmospheric) pressure regions. These gyres, which are the dominant feature of the surface water circulation system, are illustrated in Fig. 7.2; note that the water circulates clockwise in the Northern Hemisphere and anticlockwise in the Southern Hemisphere, with the

result that there is divergence around the Equator. The overall anticyclonic circulation at the ocean surface and the coriolis force act to produce a net surface water flow toward the centre of the ocean gyres, often called the Ekman transport, which results in sinking of waters within the centre of the gyres. These features give rise to the deepening of the thermocline within the gyres and shallowing at the Equator seen in Fig 7.1.

Boundary currents are developed on the landward sides of the gyres as the land masses are encountered. These currents are more intense on the western sides of the ocean basins due to the Earth's rotation, and the western boundary currents (e.g. the Gulf Stream and the Kuroshio in the Northern Hemisphere; and the Brazil and Agulhas Currents in the Southern Hemisphere) are narrow and fast-flowing, and form a sharp break between coastal and open-ocean waters. In contrast, the return flows on the eastern sides of the gyres are less intense and more diffuse in nature.

The main northern and southern gyre flows in the Atlantic and the Pacific are separated by the equato-

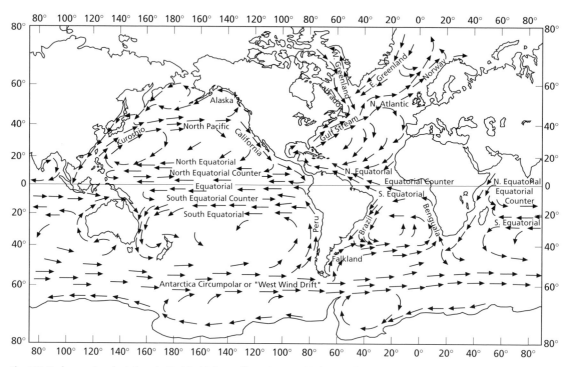

**Fig. 7.2** Surface water circulations in the World Ocean (from Stowe, 1979). Note the central gyre systems.

rial countercurrents. These are particularly well developed in the Pacific; for example, the strong subsurface Cromwell Current has been traced for over 12 000 km. In the Atlantic, the countercurrents are less intense on the eastern than on the western side of the ocean, and there is an important link between the North and South Atlantic surface waters on the western side where part of the South Equatorial Current crosses the Equator to join the North Equatorial Current: see Fig. 7.2. At their southern edges the South Atlantic, South Pacific and Indian Ocean gyres open out into the Antarctic and are bounded by the Antarctic Circumpolar Current, or West Wind Drift, Current: see Fig. 7.2. A circumpolar current of this type is not, however, found at high latitudes in the Northern Hemisphere because of the presence of the land masses there.

The Indian Ocean is a special case in that its northern land boundary is located at relatively low latitudes (~20°N). As a result, surface circulation in the northern Indian Ocean is strongly influenced by conditions on the surrounding land masses and is affected by the monsoon systems. During the winter months, the North Equatorial Current is developed, as it is in the Atlantic and the Pacific, but during the summer in the southwest monsoon the current is replaced by a flow in the opposite direction. Surface circulation patterns in the southern Indian Ocean are, however, similar to those in the South Atlantic and South Pacific.

In addition to the generation of surface currents, wind flow also can be responsible for vertical movement in the oceans in the form of upwelling and downwelling (sometimes termed 'Ekmann pumping'). Upwelling is the process by which deep water is transferred to the surface, and is extremely important in controlling primary productivity because it brings nutrients from depth to the surface waters (see Section 9.1). This type of upwelling can occur under various conditions, but it is classically associated with the eastern boundary currents (i.e. on the western margins of the continents) in subtropical coastal regions where the wind drives the water offshore, and in order to conserve volume it is replaced by subsurface water from depths of around 200–400 m. Coastal upwelling is found (i) on the western margins of the continents off West Africa, Peru and Western Australia; (ii) off Arabia and the east coast of Asia under the influence of the monsoons; and (iii)

around Antarctica. Equatorial upwelling occurs in non-coastal, or mid-ocean, areas where it can be initiated by, for example, diverging current systems, as happens in the equatorial Pacific (see Fig 7.2). In open-ocean areas where the thermocline acts as a mixing barrier, upwelling can take place owing to the erosion of the thermocline by turbulence as a result of ocean–atmospheric coupling mechanisms, such as storms. This turbulent vertical transport through the thermocline occurs in pulses, and can have a major effect on the output of deep-water nutrients to the euphotic zone in oligotrophic open-ocean areas (see Section 9.1.1). As noted above, there is net sinking (or downwelling) of water within the centres of the ocean gyres which acts to restrict the mixing of nutrients from deep water to the surface.

It may be concluded, therefore, that on a global scale wind-driven circulation, modified mainly by the Coriolis effect and constrained by the shape of the ocean basins, determines the surface current patterns in the oceans. In addition, tidal currents are present everywhere in the sea, but achieve a major importance only in coastal seas and estuaries.

### 7.3.3 Deep-water circulation

The movement of water in the deep ocean is mainly set up in response to gravity. The resultant thermohaline *circulation*, which arises from the density difference between waters, is much slower than the surface circulation and involves vertical as well as horizontal motion. In the simplest sense, dense water will sink until it finds its own density level in the water column where it will then spread out horizontally along a density surface, which provides a preferred water transport horizon as well as setting up a density stratification in the oceans. Thus, if new dense water is produced continuously at the surface, the displacement of subsurface water as the new water sinks will lead to the vertical and horizontal thermohaline circulation that keeps the deep ocean in motion.

Mixing between the surface and deep ocean must take place across the pycnocline, which acts as a mixing barrier. The pycnocline is well developed in tropical and temperate waters, but is less well developed, or is absent entirely, at high latitudes where the density of the cold surface waters is greater. It is here, in these high-latitude regions where the mixing

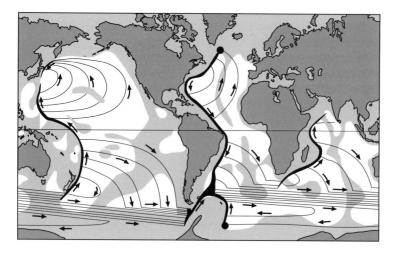

**Fig. 7.3** Deep-water circulation in the World Ocean (from Stommel, 1958). Note the strong western boundary currents. Reprinted with permission from Elsevier.

barrier does not operate efficiently, that the main sites are found for the sinking of the cold dense surface water, which is responsible for the ventilation of the deep sea.

A generalized deep-water circulation model was proposed by Stommel (1958), and this offers a convenient framework within which to outline the principal features in the transport of deep waters (Rahmstorf, 2002). Two primary sources of deep water are identified. Both are at high latitudes, one in the North Atlantic (the northern component) and one in the Antarctic (the southern component). The northern component deep-water source is in the Norwegian Sea, off the coast of southern Greenland and in the Labrador Sea. Here, the waters, which have a low surface temperature and a high salinity, sink to form the North Atlantic Deep Water (NADW). The southern component originates in the Weddell Sea, where cold, dense, sub-ice water sinks to form Antarctic Bottom Water (AABW). There is relatively weak deepwater formation in the North Pacific relative to Antarctica and the North Atlantic. This is because salinities in the North Pacific surface waters are relatively low (Fig 7.1) meaning that the cooling of surface waters at high latitude does not result in the formation of waters dense enough to sink to abyssal depths. By contrast in the North Atlantic the warm salty waters of the Gulf Stream move north and cool, forming dense waters that can sink to the depths of the ocean. The volumes of water involved in this circulation are enormous; about

$15 \times 10^6 \, \text{m}^3 \, \text{s}^{-1}$ in the North Atlantic and $21 \times 10^6 \, \text{m}^3 \, \text{s}^{-1}$ in the Southern Ocean, and the heat transport associated with these flows profoundly affects climate (Rahmstorf, 2002).

The principal sources of deep water are therefore located in the Atlantic and Antarctic, and it is possible to track the deep-water circulation path through the World Ocean from the Atlantic and Antarctic sources following the patterns suggested in Stommel's model, a characteristic feature of which is the existence of strong boundary currents on the western sides of the oceans due to the Coriolis Force (see Fig. 7.3). Figure 7.3 is a simplification of the global deep-water circulation but has proven a very useful framework within which to interpret patterns of global biogeochemistry. Taking the pathway proposed in the model, the NADW spills out from the northern basins and begins its global oceanic 'grand tour'. At the start of the tour the NADW flows down the North Atlantic in the strong boundary currents at the western edge of the land masses, crosses the Equator into the South Atlantic and becomes underlain by the AABW from the southern sinking source. Deep-water leaves the Atlantic–Antarctic by moving eastwards below the tip of southern Africa to enter the Indian Ocean via the Circumpolar Current. Some of the deep-water circulates within the Indian Ocean, moving northwards along a western boundary current. Transport out of the ocean into the Pacific (and indeed the Atlantic) is again via the Antarctic Circumpolar Current. In the Pacific the water is moved northwards along a western boundary current through the South Pacific into the North Pacific.

Using a variety of tracers (see Section 7.4) it is possible to estimate the time scales of this grand tour (Emerson and Hedges, 2008). The overall global oceanic deep-water 'grand tour' is therefore down the Atlantic to Antarctica (which takes ~200 yr), through the Antarctic into the Indian Ocean (a further approximately 200–300 yr to reach the top) and up through the South Pacific and then the North Pacific (taking about 500–660 yr to get from Antarctica to the Gulf of Alaska). Thus, the deep-water of the North Pacific is the oldest in the World Ocean in terms of time since it left the North Atlantic, and has an age of almost 1000 yr. This pathway along which the deep ocean is ventilated has an important impact on the distributions of constituents such as nutrients, trace metals and particulate matter in the ocean system, and this is discussed in subsequent sections.

The water that sinks from the surface to the deep ocean must be replaced and in Stommel's model it is assumed that there is a uniform upwelling throughout the oceans, a process that must be distinguished from the localized (e.g. coastal) wind-driven upwelling described above. On the basis of radiotracer data, Broecker and Peng (1982) estimated that the upwelling of subsurface water occurs at a rate of ~3 m yr$^{-1}$, although somewhat higher values of ~5 to ~7 m yr$^{-1}$ have been proposed by Bolin *et al.* (1983). This upwelling process is probably not uniform across the oceans but occurs in particular regions and may be associated with flows over complex topography (see Ledwell *et al.*, 2000; Garabato *et al.*, 2007). In order to reach the surface, deep-water has to cross the thermocline, which acts as a mixing barrier. Various studies, again using radiotracers, have demonstrated that the ventilation of the thermocline takes place, particularly in the equatorial regions of the oceans, from depths of at least 500 m (see e.g. Broecker and Peng, 1982). Large-scale upwelling of subsurface waters also occurs at high latitudes in the Antarctic, where the thermocline mixing barrier is absent.

Up to this point we have considered surface-water (wind driven) and deep-water (density driven) circulation as two separate systems. It is, however, artificial to divide the circulation into these components because nature can satisfy all the forcing mechanisms with a single integrated circulation (see e.g. Gordon, 1996). There have been a number of attempts to define this single circulation.

The principal feature within the surface ocean is the wind-driven gyre circulation, which retains water in the ocean basins. These gyres, however, are not closed and surface water is exchanged between the major oceans. This had led to the concept of a surface-water–deep-water circulation coupling, which acts as a global 'conveyor belt'. In this conveyor belt, the deep-water 'grand tour' transports colder water from the North Atlantic and Antarctic sinking centres to the North Pacific. The return flow of warm surface water from the Pacific occurs through the Indonesian archipelago and around the tip of South Africa.

### 7.3.4 Mid-depth water circulation: water types and water masses

So far, a brief outline has been given of the circulation patterns at the upper and lower levels of the oceanic water column. It is now necessary to consider what happens in the mid-depth, or intermediate, water column, and for this the concept of a water mass will be introduced. Because there are no appreciable sinks for heat and salt in the interior of the ocean, the temperature and salinity of a water body, both of which are conservative parameters, will have been fixed once it leaves the surface ocean and can be changed only by mixing with other waters having different properties. Such mixing is a very slow process relative to water mass transport rates and so water mass characteristics tend to be retained over distances of thousands of km and timescales of many hundreds of years. An exception to this is provided by heat transmitted through the oceanic basement, for instance at hydrothermal vents, which can raise the temperature of bottom waters.

Temperature and salinity therefore are very useful properties for characterizing seawaters of different origins, and a number of waters with different temperature–salinity signatures are found in vertical sections of the water column. The relationship between the temperature (*T*) and the salinity (*S*) of a water mass can be illustrated graphically on a *TS* diagram, or more usually a potential temperature–salinity diagram, which is one of the most widely-used diagnostic tools in physical oceanography.

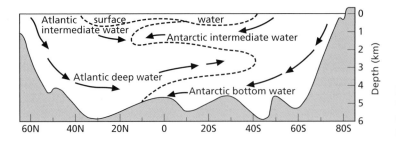

**Fig. 7.4** Cross-section of the Atlantic from Greenland to Antarctica, with schematic representation of water masses (from Stowe, 1979).

A water type is a body of unmixed water that is defined by a single temperature and a single salinity value, and therefore plots on a *TS* diagram as a single point. Thus, the temperature and salinity of a water type do not change. Changes do occur, however, on the mixing of water types and this leads to the identification of a water mass, which has a range of temperatures and salinities (and other properties) and is characterized on a *TS* diagram by a *TS* curve, instead of a single point. A well-defined water mass is one having a linear potential temperature–salinity relationship.

The structure of the water column can be elucidated in terms of water masses, and this can be illustrated with respect to the Atlantic: see Fig. 7.4. The North Atlantic Deep Water (NADW) is formed at the northern source and sinks to the bottom to be overlain by the Atlantic Intermediate Water (AIW). The NADW flows southwards across the Equator where it is underlain by the AABW. Surface and near-surface waters in the South Atlantic are underlain by the Antarctic Intermediate Water (AAIW), which is formed by the sinking of surface water at the Antarctic convergence (~50°S), and extends into the North Atlantic as far as ~10–20°N. Between ~20 and ~40°N there is an intrusion of high-salinity Mediterranean Water, which flows east to west.

## 7.4 Tracers

The water column is formed of a series of water types and water masses, which sink to a level dictated by their density, displacing other waters as they do so. As the formation of water types is a continuous process, the oceans are kept continually in motion as water circulates around the system. One of the principal thrusts in chemical oceanography over the past few years has been the use of various tracers to determine the rates at which the system operates. Tracers have also been used to elucidate the routes followed by components taking part in oceanic biogeochemical processes, and to study the rates at which the processes occur.

Conservative tracers have no oceanic sources, or sinks, and their distributions depend only on the transport of water. In contrast, non-conservative tracers undergo a variety of reactions within the system; these include biogeochemical cycling, radioactive decay, isotopic fractionation and exchange with the atmosphere. Tracers used to date have included temperature, salinity, dissolved oxygen, the nutrients and, more recently, trace metals, stable and radioactive isotopes and the chlorofluorocarbons or 'Freons'. For some oceanic experiments tracers are deliberately added to seawater; these include various dyes and substances such as $SF_6$ (sulfur hexafluoride).

The use of tracers such as these was considerably advanced by the pioneering work of the GEOSECS (Geochemical Ocean Sections Study) programme. This particularly was the case with respect to the penetration of anthropogenic tritium and $^{14}C$ from the atmosphere into the thermocline and deep water, thus allowing the surface waters of the World Ocean into which they initially entered to be labelled and traced during their oceanic passage. A detailed review of the findings that emerged from GEOSECS has been compiled by Campbell (1983). The study of tracers continued within the TTO (Transient Tracers in the Ocean), and continues now within the GEOTRACES programme (SCOR Working Group, 2007). Tracers also have been used extensively in the WOCE (World Ocean Circulation Experiment) programme.

Tracers are proving to be invaluable tools for oceanographers, and an excellent overview of their use in ocean studies has been provided by Broecker and Peng (1982); this is one of the classic texts in chemical oceanography, and will serve for many years as the 'tracer Bible'. For this reason, tracers will not be treated in detail in the present volume, and only a very general summary of their applications will be given in order to relate their use to the wider aspects of marine geochemistry.

The tracers used in oceanography can be divided into three broad groups: those used to identify water masses, those used to study water transport, and those used to study geochemical processes.

### 7.4.1 Tracing water-mass transport

Parameters that have been used to define water masses include temperature, salinity, dissolved oxygen, phosphate, nitrate and silica concentrations. However, dissolved oxygen, phosphate and nitrate are non-conservative tracers, that is, their concentrations change during the residence of the water in the deep sea. Water masses originate from the mixing of water types and to unscramble the mixtures it is necessary therefore to use conservative tracers, that is, those that are not changed by processes such as respiration or radioactive decay during their residence time in the deep sea. Tracers of this type include: (i) salinity (conductivity); (ii) temperature; (iii) $SiO_2$ and Ba (although these are influenced by mineral dissolution process in deepwater and so are not completely conservative; (iv) the isotopes $^2H$ and $^{18}O$; and (v) 'NO' and 'PO', which were introduced by Broecker (1974) and correct nitrate and phosphate for *in situ* biogeochemical changes within the deep sea. For a detailed treatment of the use of these various tracers in the identification of water masses, the reader is referred to Broecker (1981), Broecker and Peng (1982) and Campbell (1983). Other types of tracers also are used for water-mass studies. These include the chlorofluoromethanes (CFMs or Freons). The two CFMs ($CCl_3F$ (Freon-11) and $CCl_2F_2$ (Freon-12)) measured most commonly have no known natural sources and have been increasing in concentration in the atmosphere since their commercial production started in the 1930s. The CFMs enter the ocean across the air–sea interface (see Section 8.2), and their concentrations in surface seawater are functions of their concentrations in the overlying atmosphere and their water solubilities, which are dependent on the temperature and salinity of the seawater. Since these gases did not exist prior to the 1930s, their existence in particular water masses can provide information on when the water was last at the surface in exchange with the atmosphere. Furthermore since their initial production until their severe regulation and replacement with alternative compounds (due to their role in destroying stratospheric ozone), the global emission ratios of the various chlorofluorocarbons has varied in a well known way. Hence, the measurement of the concentration of several CFMs (and hence their ratios) in a water sample allows rather precise ageing of water masses over timescales of almost 100 years (e.g. Azetsu-Scott *et al.*, 2005).

Other tracers beside the CFMs offer ways to trace water movements, particularly radioisotopes, where the radioactive decay process provides a potential clock to measure rates of processes. The half-life of particular isotopes defines the time window over which these can be useful tracers. Radioisotope tracers can be classified into two types, depending on whether their distributions are influenced primarily by (i) transport in the water phase; or (ii) transport in the particulate phase. Water tracers include the following:
• the anthropogenically produced (from atmospheric nuclear bomb testing and radioactive reprocessing) transient tracers $^{90}Sr$ (half-life, 28.6 yr), $^{137}Cs$ (half-life, 30.2 yr), $^{85}Kr$ (half-life, 10.7 yr) and the Freons;
• the naturally produced steady-state tracers $^{39}Ar$ (half-life, 270 yr), $^{228}Ra$ (half-life, 5.8 yr) and $^{32}Si$ (half-life, 250 yr);
• and the mixed-origin tracers $^{14}C$ (half-life, 5730 yr) and $^3He$ (half-life, 12.4 yr).
*Particulate-phase tracers* include: $^{210}Pb$ (half-life, 22.3 yr), $^{228}Th$ (half-life, 1.9 yr), $^{210}Po$ (half-life, 0.38 yr) and $^{234}Th$ (half-life, 0.07 yr), $^{231}Pa$ (half-life $3.26 \times 10^4$ yr). The utility of these tracers is complicated by the role of particle water interactions and particle settling, but this of course also opens up the possibility of using these tracers to interrogate these biogeochemical processes, or indirectly to assess long-term changes in ocean circulation in the case of $^{231}Pa$ (Yu *et al.*, 1996).

Together, these two types of tracers yield information on ventilation and mixing in the subsurface ocean and on the origin, movement and fate of particulate matter in the system. In addition to these classic tracers, a number of individual chemical elements have been used to trace water-mass movements. For example, Measures and Edmonds (1988) used dissolved Al as a tracer for the outflow of Western Mediterranean Deep Water into the North Atlantic. Other tracers that have been used for the elucidation of circulation and mixing in the oceans include the rare-earth elements (REE) (see e.g. Elderfield, 1988) and Mn (see e.g. Burton and Statham, 1988).

### 7.4.2 Geochemical process tracers

Plotting down-column dissolved trace element concentrations against a conservative tracer, for example, salinity or potential temperature, and interpreting the shape of the curve can provide evidence on geochemical processes; for example, at steady state a concave shape with lower concentrations at depth indicates deep-water removal, or scavenging, of the element. However, the natural series radioisotopes are the most widely used geochemical process tracers. These isotopic clocks, which operate on time-scales ranging from a few days (e.g. $^{234}$Th, half-life, 24 days), through a few tens of years (e.g. $^{210}$Pb, half-life, 22 yr), to several thousand years (e.g. $^{230}$Th, half-life, $7.52 \times 10^4$ yr), have been used for a wide variety of biogeochemical purposes. These include tracing the origin of atmospherically transported components (e.g. $^{210}$Pb), gas exchange across the air–sea interface (e.g. $^{222}$Rn), trace element scavenging in the

water column (e.g. $^{230,228,234}$Th), trace element down-column fluxes (e.g. $^{234}$Th) and trace metal accumulation rates and bioturbation in sediments (e.g. $^{234}$Th, $^{210}$Pb). In addition, considerable use has been made in the past of artificial radionuclides, and this was revived following the Chernobyl incident to demonstrate the rapid transport of radionuclides from the accident to the deep sea (Fowler *et al.*, 1987).

Tracers have provided information on a wide variety of oceanographic topics, and from the point of view of their importance to our understanding of the processes involved in marine geochemistry it is worthwhile highlighting a number of these.

1 *Deep-sea residence time.* According to Broecker (1981) the great triumph of natural radiocarbon measurements in the oceans has been the establishment of a 1000 yr time-scale for the residence of water in the deep sea. This is illustrated in Fig. 7.5 although it should be noted that the water mass ages shown are complicated by slow equilibration times of $CO_2$ across the air sea interface and mixing of waters of different ages and so suggest somewhat longer transit times for deep water than described in Section 7.3.3 (See also Key *et al.*, 2004).

2 *Thermocline ventilation time.* Carbon-14 and $^3$He data have been used to establish the ventilation time of the main oceanic thermocline and with it the supply of nutrients from the deep interior of the ocean to the surface (Jenkins and Doney, 2003).

3 *The rates of vertical mixing.* In the more simple box models used by marine chemists, it is often the practice to separate a surface-ocean and a deep-

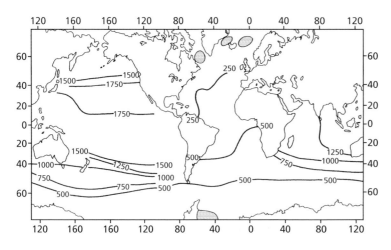

**Fig 7.5** Age of ~3 km deep ocean waters, that is, time since they were last at the ocean surface (based on Broecker, 1985).

ocean compartment. In order to apply such models to biogeochemical processes it is necessary to have at least an approximation of the rate at which the reservoirs mix. There are various problems involved in making such approximations, but Collier and Edmond (1984) used the $^{14}$C deep-water residence and a deep-water reservoir thickness of 3200 m to derive a mixing flux of 3.5 m yr$^{-1}$ for the mixing rate between the surface and deep ocean (see Section 7.5).

**4** *The dispersion of hydrothermal solutions.* Helium isotopes have been used to predict the mid-depth circulation of hydrothermal fluids vented at sea-floor spreading centres (see Section 6.3) and more recently the dispersal of these plumes within the oceans has been used to estimate mixing rates within the deep ocean (Garabato *et al.*, 2007).

**5** *Oceanic biogeochemical processes.* Important advances have been made in our understanding of geochemical processes through the use of the radionuclide 'time clock' tracers. Highly particle reactive thorium has revealed much about particle water interactions, including primary production and particle removal in surface waters using $^{234}$Th while the reversibility of deep sea particle water interactions has been elegantly interrogated using longer lived thorium isotopes. (Bruland and Lohan, 2003; also see Chapter 11).

**6** *Palaeooceanography.* Oxygen isotopes in carbonates have been used for the identification of past climatic conditions (see e.g. Broecker, 1974). Figure 7.6 illustrates the sort of records that can be derived from the analysis of foraminifera (small organisms with calcium carbonate skeletons) from deep ocean sediment cores for oxygen isotopes and other tracers. The variations in the oxygen isotope composition within the skeletons reflect changes in climate and ocean circulation (Boyle and Keigwin, 1986; Rahmstorf, 2002; Zachos *et al.*, 2001).

**7** *Atmospheric transport.* Lead-210 has been used as an atmospheric source indicator (see e.g. Schaule and Patterson, 1981).

**8** *Gas exchange.* Radon-222, and its disequilibrium with the parent $^{226}$Rn, has been used as a gas tracer in the oceans to model air–sea exchange processes (Nightingale and Liss, 2003).

**9** *Sedimentary processes.* Bacon and Rosholt (1984) and Thompson *et al.* (1984) have used thorium isotopes to elucidate trace element accumulation rates in deep-sea deposits. In addition, various isotopes

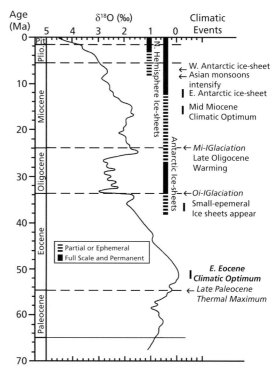

**Fig 7.6** Long-term changes in oceanic foraminifera carbonate δ$^{18}$O over 70 million years and their relationship to major changes within the Earth system based on Zachos *et al.* (2001).

have been used to investigate processes such as bioturbation, and to identify sources of individual components in sediments; the latter has involved elements such as the rare earths (see e.g. Bender *et al.*, 1971) and Sr isotopes (see e.g. Dasch *et al.*, 1971).

**10** *The oceanic cycles of elements.* The extent to which the isotopes of an element undergo fractionation is often characteristic of specific geochemical processes, and isotopic tracers have proved to be extremely useful in elucidating the marine cycles of a number of elements. An example of this is given in Worksheet 7.1, with respect to the marine cycle of Sr.

Even from this very brief survey it is apparent that the utilization of tracers has provided invaluable insights into the processes involved in the mixing and transport of waters in the oceans, and especially into the rates at which the system operates. It is this mixing, and the subsequent transport of the waters, that controls the water mass structure of the oceans.

**Worksheet 7.1:  The use of isotopes in the study of the marine cycles of elements**

An example of how isotopic ratios can be used as 'process tracers' in marine geochemistry has been described by Palmer and Edmond (1989). The oceanic composition $^{87}Sr$:$^{86}Sr$ ratio is recorded in marine carbonates and the ratio has varied with time over the past 75 Ma: see Fig. WS7.1.1 (Palmer and Elderfield, 1985). Strontium isotope rations can be measured to very high precision and accuracy so the small differences seen in Fig. WS7.1.1 are analytically real.

The outstanding feature of the curve is the increase in $^{87}Sr$:$^{86}Sr$ ratios over the past around 25 Ma. Strontium has a long residence time in the oceans (~5 million years: Broecker and Peng, 1982), and so the changes in the isotopic composition of seawater reflect long-term changes in global biogeochemistry. Palmer and Edmond (1989) argued that the defining processes for oceanic Sr isotopic abundance are a balance between riverine inputs and the modification of Sr isotopic composition by the cycling of seawater through hydrothermal systems. They then used this to estimate the present day hydrothermal flux of Sr and from a knowledge of the concentration of Sr in hydrothermal fluids they were able to calculate the water flux through the hydrothermal systems. As discussed in Chapter 5, this is a particularly uncertain and difficult parameter to calculate.

The relevant data is presented in Table WS7.1.1.

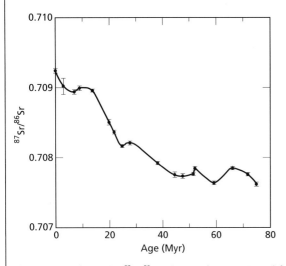

**Fig. WS7.1.1** The marine $^{87}Sr$:$^{86}Sr$ ratio over the past 75 Ma (after Palmer and Elderfield, 1985).

**Table WS7.1.1** Contemporary Global Sr Flux (Palmer and Edmond, 1989).

| Source | Sr concentration µmol kg⁻¹ | $^{87}Sr/^{86}Sr$ R | Sr Flux mol yr⁻¹ |
|---|---|---|---|
| River | 0.890 | 0.7119 | $33.3 \times 10^9$ |
| Hydrothermal | 126 | 0.7035 | |
| Seawater | 90 | 0.70916 | |

*continued*

At steady state, two equations (WS7.1.1 and WS7.1.2 following) can then be set up and solved to estimate the hydrothermal water flux.

$$\frac{FSr(riv) \times R(riv) + FSr(hydro) \times R(hydro)}{FSr(riv) + FSr(hydro)} = R(seawater) \qquad (WS7.1.1)$$

$$FSr(hydro) = [Sr] \times H \qquad (WS7.1.2)$$

where FSr is Sr flux through rivers(riv) or hydrothermal systems (hydro), R is strontium isotopic composition of rivers (riv), hydrothermal systems (hydro), seawater [Sr](hydro) is the concentration of Sr in hydrothermal fluids and H is the total water flow through hydrothermal systems.

Palmer and Edmond estimated H to be $1.3 \times 10^{14} \, kg \, yr^{-1}$. The authors then interpret the long-term changes in Sr isotope ratios in Fig. WS7.1.1 in terms of changes in the relative proportions of hydrothermal and weathering fluxes. Continental weathering is a major sink for atmospheric $CO_2$ and hence the long-term changes in Sr isotopes in the ocean have the potential to provide really important information about the way the Earth system has changed over million-year timescales.

While the calculations presented here are relatively simple, their interpretation is not necessarily so, since many components of the Earth system change on these time scales. Palmer and Elderfield (1985) suggested that the onset of widespread glaciations and/or the uplift of the Alps and Himalayas could explain the rise in Sr isotope ratios over the last 25 million years. However, Berner (2006) has noted that the weathering of volcanic and non-volcanic rocks yield Sr in river water with very different isotopic composition (0.703 and 0.713 respectively) and the relative importance of the weathering of these rock types over time could also significantly affect oceanic Sr composition. Vance *et al.* (2009) have suggested that the steady state geochemical models implied in the approach of Palmer and Edmond (1989) may be inappropriate given the long residence time of strontium if contemporary weathering rates are not representative because they reflect a system still responding to changing weathering rates following the end of the last glaciations. Derry (2009) has noted that cycling of strontium in low temperature hydrothermal systems (see Chapter 5) may also play a role in modifying the global strontium budget. Hence, records such as Fig. WS7.1.1 provide an exciting and tantalizing glimpse of long-term changes in the earth system, but their interpretation is challenging because of the number of factors that might have changed and the limitations of the geological record.

The transport of water during oceanic circulation also exerts a fundamental control on the distributions of both dissolved and particulate constituents in the system. This mixing is conservative, and compositional changes take place only as the result of the mixing of end-member waters that have different compositions acquired at their sources. Superimposed on this conservative circulation control is a second effect produced by the involvement of many constituents in the major biogeochemical processes that operate within the oceans, and which lead to non-conservative behaviour. It is possible therefore to make an important distinction between two genetically different types of signals in the oceans: (i) one resulting from the mixing and transport of water masses; and (ii) one resulting from internal biogeochemical oceanic reactivity. The information tracers provide therefore underpins much of the information in later chapters.

In order to gain an understanding of how these two types of signals interact to control the oceanic cycles of elements it is useful to set up ocean models designed to trace the pathways by which materials move through the system. In the following section a simple two-box model will therefore be outlined for use as a framework within which to describe the distributions of dissolved and particulate components in the oceans. Such models now form an essential component of much of marine biogeochemistry (e.g. Sarmiento and Gruber, 2006).

## 7.5 An ocean model

### 7.5.1 Introduction

Mixing in the water column involves a combination of advection and diffusion. Advection is a large-scale transport involving the net movement of water from one point in the ocean to another. This may occur either in a horizontal direction, for example, along a major current system, or in a vertical direction, for example, during upwelling or sinking. Superimposed on this net water transport are the effects of turbulent mixing, or turbulent exchange; this is a more or less random mixing, in which an exchange of properties takes place between waters by eddy, and to a much lesser extent molecular, diffusion without any overall net transport of the water itself.

There are, however, problems involved in understanding the real nature of vertical and horizontal mixing in the oceans. Some of these problems have been identified by Broecker and Peng (1982), who distinguished between isopycnal and diapycnal mixing. Isopycnal surfaces, along which potential density remains constant, are the preferred surfaces along which lateral mixing takes place. Diapycnal mixing results when water is carried in a perpendicular direction across isopycnal surfaces. In the interior of the ocean isopycnal surfaces are almost horizontal, but at high latitudes they rise towards the surface, where they can outcrop. The authors suggested therefore that the concept of isopycnal and diapycnal mixing should replace that of horizontal and vertical mixing.

Although a wide variety of models are available, many marine chemists prefer to adopt a discontinuous, or box model, approach to gain a first-order, or perhaps even a zero-order, understanding of biogeochemical processes within the ocean system. Box models of varying complexity have been used for the interpretation of the distributions of various components in the World Ocean, using tracer-derived data for the rates at which the system operates. The general principles involved in these models are described in the following.

In the *steady-state* ocean a component is removed from the system at the same rate at which it is added, that is, the input balances the output, and the concentrations at any point in the system do not change with time. To construct an oceanic model for a constituent it is necessary therefore to have data on (i) the magnitude and rates of the primary input mechanisms; (ii) the rates of exchange between the oceanic reservoirs involved; and (iii) the magnitudes and rates of the primary output mechanisms. The *primary* inputs to the oceans are mainly via river run-off, atmospheric deposition and hydrothermal exhalations (see Chapters 3–6). Exchange between oceanic reservoirs occurs through the advective transport of water during thermohaline circulation (downwelling and upwelling), by turbulent exchange and by the sinking of particulate matter from the surface to the deep ocean. Output from the system for most elements is via the burial of particulate material in sediments, although reactions of various kinds between seawater and the basement rocks also can act as sinks for some elements (see Chapter 5). In the steady-state ocean, the amount of a component entering a reservoir must balance the amount leaving. For example, if the concentration of an element in the surface ocean is to remain constant, the gain from primary inputs and upwelling must be matched by the outputs from downwelling and particle sinking (Broecker, 1974). Thus, in the system as a whole, the 'through flux' of a constituent entering via the primary input mechanisms must equal the total permanently lost to the output mechanisms.

### 7.5.2 A simple two-box oceanic model

It was shown in Section 7.3 that different processes control the circulation in the surface ocean (wind-driven) and the deep ocean (gravity-driven). The simplest two-box model therefore divides the ocean into these two reservoirs, that is, a warm surface water reservoir (~2% of the total volume of the ocean) and a cold deep-water reservoir (~80% of the total volume of the ocean), which are separated by the waters of the thermocline (~18% of the total volume of the ocean). Both reservoirs are assumed to be well mixed and interconnected, and they interact via vertical mixing, that is, the downwelling of surface water and the upwelling of deep water and the winter turnover of the cooled upper layer, and via the sinking of coarse particulate matter (CPM; see Section 10.4) from the surface ocean.

This type of two-box model is a useful first approximation of the ocean system because the main barrier to mixing in the water column is the thermocline, which separates the two main reservoirs, the waters of which have different densities. Two other factors also are important in distinguishing between the two reservoirs:

1 two of the principal primary input mechanisms, that is, river run-off and atmospheric deposition, supply material to the surface ocean;

2 the bulk of the phytoplankton and zooplankton biomass, which is involved in the oceanic cycles of many components, is found in the upper sunlit water layer of the surface reservoir (see Section 9.1).

A number of parameters are used to evaluate the rate of exchange of water between the surface and deep reservoirs. An average $^{14}C$ residence time of ~900 yr has been proposed for the deep ocean. This means that at the rate at which deep-water circulates in the World Ocean it takes ~900 yr for water to go through the cycle of sinking at the poles, circulating through the deep ocean and returning to the surface. By rounding up the estimate to 1000 yr, Broecker and Peng (1982) calculated that the yearly volume of water exchange between the deep and surface ocean is equal to a layer ~300 cm thick with an area equal to that of the ocean, that is, a mixing flux of 3 m yr$^{-1}$. Thus, the $^{14}C$ data indicate that upwelling brings an amount of water to the surface equal to an ocean-wide layer of ~300 cm. For comparative purposes, Broecker (1974) estimated that fluvial input to the ocean from continental run-off would yield a layer ~10 cm thick if it was spread over the entire ocean. The volumes of water exchanging between the surface and deep ocean are much larger than the inputs, and hence the internal cycling processes described in later chapters are very important in regulating the behaviour of chemical components in the ocean, in contrast for example to a small lake where the inputs, rather than internal cycling, may dominate the chemistry

In order to illustrate the principles involved in incorporating data of this kind into oceanic box models, a generalized two-box model that has been applied to the marine geochemical cycles of trace metals is illustrated in Fig. 7.7. Reducing the ocean to a two box system is of course a massive simplifica-

tion, but we will see in subsequent chapters that this can still provide a very useful description of oceanic geochemical processes. Furthermore, the principles embodied in Fig 7.7 can be extended to an ocean model consisting of multiple boxes (for instance to describe the three main ocean basins, the Atlantic, Pacific and Indian oceans), using the same equations, simply by adding terms to describe the water exchange between boxes.

## 7.6 Characterizing oceanic water-column sections

Various parameters that characterize the water in the oceans have now been described. From the viewpoint of marine geochemistry the question that must now be addressed is, 'Which of these parameters must be measured on a semi-routine basis during the occupation of water-column sections in order (i) to set the location in an oceanographic context, and (ii) to provide data that can be used to interpret biogeochemical processes?'

1 To set an individual seawater section into an oceanographic context, that is, to describe the nature of the water, it is usual to prepare vertical profiles of temperature, salinity, dissolved oxygen and the nutrients; other parameters also may be measured during individual investigations. These various parameters, either singly, or in combination, can be used for purposes such as defining the depth of the mixed layer, describing the structure and location of the thermocline, and identifying the different water masses (and therefore water sources) that are sampled by the section.

2 Data that are useful for the interpretation of biogeochemical processes include dissolved oxygen (to characterize redox-mediated reactions), the nutrients (to characterize the involvement of elements in biological cycles), total particulate matter (to characterize the involvement of elements in particle-scavenging reactions), chlorophyll (to characterize the standing crop of phytoplankton and infer the primary production), DOC and POC (to characterize the oceanic carbon cycle and global carbon flux), and a variety of the tracers listed in Section 7.4 (the choice depending on the nature of a specific investigation); however, it must be stressed that not all these parameters are measured on a purely routine basis.

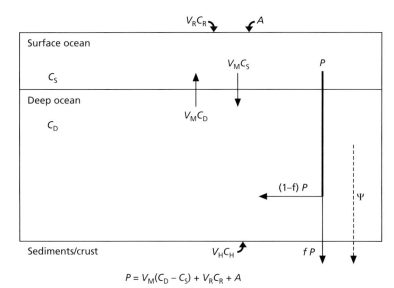

$$P = V_M(C_D - C_S) + V_R C_R + A$$

**Fig. 7.7** A generalized two-box ocean model applied to the marine geochemical cycles of trace elements (from Collier and Edmond, 1984): $P$, particulate matter flux out of the surface ocean; $V_R$, $C_R$, volume and concentration of an element in river water; $A$, atmospheric input; $C_S$, $C_D$, dissolved concentrations of an element in the surface and deep reservoirs; $f$, fraction of particulate matter preserved in sediments; $\psi$, additional particulate matter flux resulting from scavenging within the deep ocean; $V_H$, $C_H$, volume and concentration of an element in hydrothermal solutions. The particulate matter flux ($P$) is calculated by the mass balance of all other inputs and outputs to the surface ocean reservoir. Reprinted with permission from Elsevier.

Temperature, salinity, dissolved oxygen and water fluorescence (which can be calibrated to provide chlorophyll data) can now be continuously and automatically measured at sea. The other determinants mentioned can only be routinely analysed on individual water samples collected by sampling bottles that close and isolate a water sample from a particular depth which can then be and returned to the ship or home laboratory for analysis. The use of a variety of water-column parameters in a marine geochemical context is illustrated in Worksheet 7.2. Further examples of the manner in which oceanographic-related and process-related water-column parameters are used in marine geochemistry will be described later in the text, when the factors controlling the elemental compositions of the seawater and the sediment reservoirs are described.

## 7.7 Water-column parameters: summary

**1** The oceanic water column can be divided into two layers: a thin, warm, less dense surface layer that caps a thick, cold, more dense deep-water layer. The two layers are separated by the thermocline and the pycnocline, which act as barriers to water mixing.
**2** Currents in the surface layer of the ocean are wind-driven, and the major features of the surface circulation are a series of large anticyclonic circulation cells, termed gyres, which have boundary currents on their landward sides.
**3** Circulation in the deep ocean is gravity-driven. Cold, relatively dense, surface water sinks at high latitudes where the pycnocline is less well developed, or is absent entirely. The main features of the bottom circulation are strong boundary currents on the western sides of the oceans. The deep-water undergoes a global oceanic 'grand tour', which takes it down the Atlantic, through the Antarctic into the Indian Ocean, up through the South Pacific and into the North Pacific. As a result, the deep-water of the North Pacific is the oldest in the World Ocean, and acts as a sink for other deep-waters.
**4** For many geochemical purposes, the ocean can be represented by a two-box model that distinguishes between:
(a) a warm surface water reservoir (~2% of the total volume of the ocean) and
(b) a cold deep-water reservoir (~80% of the total volume of the ocean). The two layers are separated by the waters of the thermocline (~18% of the total volume of the ocean).
**5** The structure of the oceans is controlled by the mixing and transport of water masses, and information on the processes involved, and on the rates at which they operate, can be gained by using a variety of oceanic tracers.

**Worksheet 7.2: Characterizing the water column for geochemical studies**

The distribution of 'reactive' dissolved constituents in seawater is controlled by a combination of oceanic circulation patterns and the effects of internal reactive processes. A large proportion of the chemical signal for these constituents, however, can be determined by long-distance advection and the mixing of water masses of different end-member compositions, and it is necessary therefore to extract chemical information from the advective background. One way of doing this for trace elements is to relate the distribution of the element to an 'analogue' species that has a well-understood distribution. This 'advective–chemical' approach, which combines data for a variety of water-column parameters, was used by Chan *et al.* (1977) in their study of the distribution of dissolved barium in the Atlantic Ocean. This study can be used to illustrate how a variety of the seawater parameters described in the text can be used both to characterize the water column and to interpret the biogeochemical controls on the distribution of trace elements in the oceans. In this context, the use of a number of these water-column parameters is described below in relation to a variety of geochemical applications.

*'Species analogue' interpretation of trace-element data*
Physical circulation is the dominant control on the distribution of dissolved Ba, and many other dissolved constituents, in the basins of the Atlantic, and in order to extract chemical information from the advective background Chan *et al.* (1977) related the distribution of the element to those of other species that have well-understood distributions. The results showed that the water-column distributions of barium mimic those of the refractory nutrient silicate. This relationship is illustrated in Fig. WS7.2.1 for a station southeast of Iceland, and shows the close correspondence between silicate and Ba. The plot is based on concentrations of individual water samples collected at particular water depths at one sampling station.

   Chan *et al.* (1977) were able to use this Ba–silicate correlation to extract chemical information from the advective background by demonstrating that the marine geochemistry of barium is dominated by its involvement in the oceanic biogeochemical cycle in which it takes part, in a deep-water regeneration cycle associated with a slowly dissolving refractory, non-labile, phase (mimicked by silicate), rather than a more rapidly dissolving, labile, tissue phase (mimicked by nitrate and phosphate) that is regenerated at higher levels in the water column (see Chapter 9 and 11). The Ba–silica relationship is not an exact one, however, and this can be seen in property–property diagrams for GEOSECS station 37; this is located in the western Atlantic at a site where the major water-mass cores are well developed. The silica–Ba plot for this station is illustrated in Fig. WS7.2.2. It is apparent that this plot consists of straight-line segments connecting end-member waters of different compositions; for example, the Antarctic water masses (Antarctic Bottom Water AABW and Antarctic Intermediate Water AAIW) are strongly enriched in silica relative to barium compared with the North Atlantic deep-water (e.g. Lower North Atlantic Deep Water LNADW). Property–property plots such as this therefore can prove extremely useful for elucidating the complex interplay between chemical and physical features in the distributions of dissolved constituents in the ocean.

   In general, the oceanic distribution of Ba mimics that of silicate and further details on the physical features that affect Ba can be obtained by combining data for the element with that for silicate, which can be used to characterize water masses. Two such applications using this approach are described below.

*continued on p. 148*

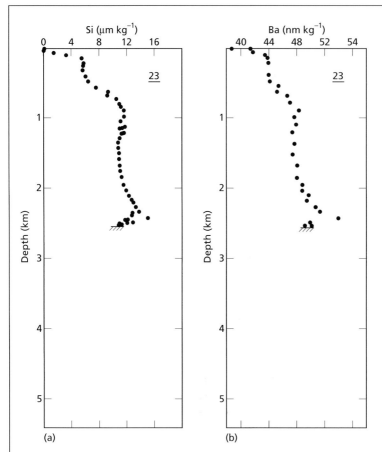

**Fig. WS7.2.1** Profiles of Si (a) and Ba (b) at station 23 (Reprinted from Chan *et al.*, 1977, with permission from Elsevier).

*The application of property–property plots to distinguish between water masses*
Silicate is the 'analogue' for dissolved Ba, and as the silicate concentrations of water masses can differ, silicate–property plots can be used to distinguish between individual water masses. This can be illustrated in terms of silicate–potential temperature plots for the waters of a series of high-latitude South Atlantic stations using results from analyses of samples from various water depths from several stations. In the region around the Weddell Sea the circulation is controlled by the interaction of the warmer, more silica-rich, Circumpolar Current (CPC) water with the colder, less silica-rich, Weddell Sea Water (WSW). The relationship can be displayed on a θ–Si plot and is illustrated in Fig. WS7.2.3 for deep waters from a number of stations lying between the South Sandwich Trench and the Atlantic Indian Ridge.

Information can be extracted from the type of plot shown in Fig. WS7.2.3 in terms of identifying both (1) overall water-mass trends and (2) variations at individual stations.
1 Overall water-mass trends show the make-up of the deep-water column. The waters at these high-altitude stations have a range of properties intermediate between those of the CPC and WSW end-members. The property–property plot shows a broad silicate maximum centred

*continued*

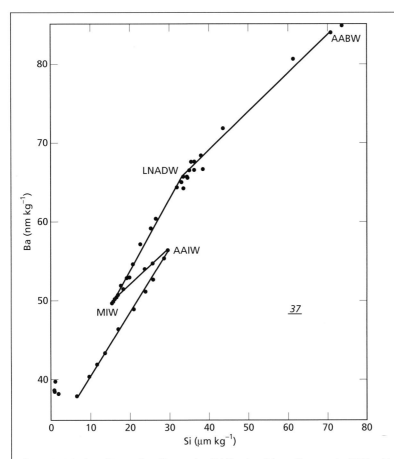

**Fig. WS7.2.2** Plot of Ba against Si at station 37 (Reprinted from Chan *et al.*, 1977, with permission from Elsevier).

around ~−0.2°C, with values that are close to those for the CPC (~129 μmol kg⁻¹) at stations 86 and 87. Below the silicate maximum there is clear evidence of the presence of colder, less silica-rich Weddell Sea Water (θ, −0.88; silicate, 111.5 μmol kg⁻¹), the most extreme values for the Weddell Sea Bottom Water being at station 89.

2 Variations at individual stations. On the basis of the data for the different stations, it would appear that the influence of the CPC proper on the deep-waters has its strongest effect at stations 86 and 87, but that it has a lesser effect on stations 84, 85 and 89.

*Characterizing the water column*

The hydrography of the Atlantic is dominated by the horizontal advection and vertical mixing of water mass cores formed at high latitudes. The formation processes in the Northern and Southern Hemispheres are different and therefore so are the physical and chemical properties of the waters produced in the two regions. In the North Atlantic the water masses are formed by the cooling and downwelling of surface waters that have tropical and subtropical origins, and the cores formed have *high salinities* and *low nutrient* concentrations. The waters are formed in a number of regions (see Section 7.3.3). The most extreme water types are produced

*continued on p. 150*

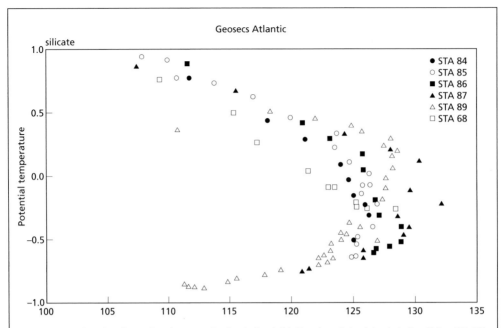

**Fig. WS7.2.3** Si–θ plots for stations between the South Sandwich Trench and the Atlantic Indian Ridge (84–90) and station 68 in the Argentine Basin (Reprinted from Chan *et al.*, 1977, with permission from Elsevier; not all original data points are plotted).

in late winter in the Norwegian and Greenland Seas via a modification of high-salinity water transported northwards through the passage around Iceland. The densest water formed is stored within deep basins, but there is a southwards return flow of waters of less extreme properties, which are determined by the source depths at which they originated and vertical mixing both in the overflow regions themselves and during descent to the deep-ocean floor. The main flows of these water masses are through the Denmark Strait, the Southwestern Faroe Channel and across the Wyville–Thompson Ridge between Faroe and Orkney. These water masses, together with one formed in the Labrador Sea, follow a number of complex trajectories and are then propagated southwards as a composite termed the North Atlantic Deep Water (NADW). Mediterranean Deep Water, which outflows through the straits of Gibraltar, forms an additional core that spreads across the North Atlantic and forms the upper boundary of the deep-water that moves into the South Atlantic.

The great complexity of the water column, which is characteristic of the formation areas of the NADW, can be illustrated by the chemical and hydrographic data obtained for the station southeast of Iceland (station 23). These data can be used to characterize the water column in terms of water masses, and to illustrate this plots of salinity and silicate against potential temperature are shown in Fig. WS7.2.4. Plotted in this way it is not possible to simply relate the data to sample depth and the results should be interpreted in conjunction with Fig. WS7.2.1.

On the basis of these plots the characterization of the water column at station 23 can be summarized as follows. The upper part of the column is occupied by high-salinity waters of a central Atlantic origin.

At ~800 m there is an inflection in potential temperature and salinity and an increase in silicate. This, together with other properties not illustrated here (e.g. a pronounced oxygen

*continued*

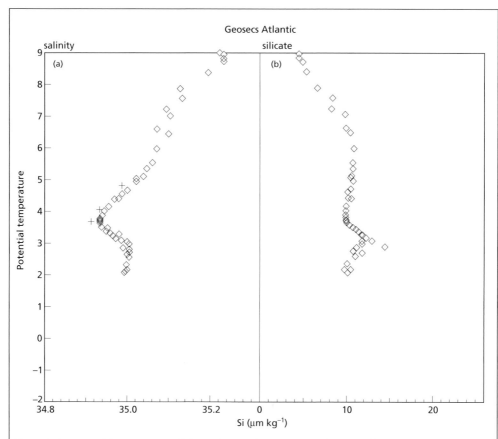

**Fig. WS7.2.4** Potential temperature–salinity and potential temperature–silicate plots for station 23 (Reprinted from Chan *et al.*, 1977, with permission from Elsevier).

minimum), indicates the presence of the core of the Mediterranean Intermediate Water (MIW). At depths below this there is a broad minimum in salinity and silicate (and an oxygen maximum), which shows the presence of the Labrador Sea Water. Below the salinity and silicate minima, that is, at depths >1600 m, the silicate values increase to reach a well-defined maximum (15.2 µm kg⁻¹) at ~2400 m, then fall off rapidly towards the bottom (11.0 µm kg⁻¹) at 2516 m. The values for potential temperature and salinity also decrease in a similar manner. These features were interpreted as representing the residual core of the Antarctic Bottom Water (AABW; high silicate, intermediate salinity) overriding the Arctic Bottom Water (ABW; low silicate, high salinity) from the Faroe Channel; that is, the AABW has penetrated to these high latitudes in the Northern Hemisphere. The θ–S properties of the ABW at station 23 indicate that the properties of the original Norwegian Sea Water, which has overflowed via the Faroe Channel, have been strongly modified by mixing with warmer, more saline Atlantic waters as it flowed down the slope.

# References

Azetsu-Scott, K., Jones, P.E. and Gershev, R.M. (2005) Distribution and ventilation of water masses in the Labrador Sea inferred from CFCs and carbon tetrachloride. *Mar. Chem.*, **94**, 55–66.

Bacon, M.P. and Rosholt, J.N. (1984) Accumulation rates of Th-230, Pa-231 and some transition metals on the Bermuda Rise. *Geochim. Cosmochim. Acta*, **48**, 651–666.

Bender, M., Broecker, W.S., Gornitz, V. *et al.* (1971) Geochemistry of three cores from the East Pacific Rise. *Earth Planet. Sci. Lett.*, **12**, 425–433.

Berner, R.A. (2006) Inclusion of the weathering of volcanic rocks in the GEOCARBSULF model. *Am. J. Sci.*, **306**, 295–302.

Bolin, B., Bjorkstrom, A. and Holmen, K. (1983) The simultaneous use of tracers for ocean circulation studies. *Tellus*, **35B**, 206–236.

Bowden, K.F. (1975) Oceanic and estuarine mixing processes, in *Chemical Oceanography*, J.P. Riley and G. Skirrow (eds), Vol. 1, pp.1–41. London: Academic Press.

Boyle, E.A. and Keigwin, L.D. (1986) Comparison of Atlantic and Pacific paleochemical records for the last 215,000 years: changes in deep ocean circulation and chemical inventories. *Earth Planet. Sci. Lett.*, **76**, 135–150.

Broecker, W.S. (1974) *Chemical Oceanography*. New York: Harcourt Brace Jovanovich.

Broecker, W.S. (1981) Geochemical tracers and oceanic circulation. In *Evolution of Physical Oceanography*, B.A. Warren and C. Wunsch (eds), *434–460*. Cambridge, MA: MIT Press.

Broecker, W.S. and Peng, T.-H. (1982) *Tracers in the Sea*. Palisades, NY: Lamont-Doherty Geological Observatory.

Broecker, W.S. (1985) *How to Build a Habitable Planet*. Palisades, NY: Eldigio Press, Lamont-Doherty Geological Observatory.

Bruland, K.W. and Lohan, M.C. (2003) Controls on trace metals in seawater, in *Treatise on Geochemistry*, Vol. 6, H. Elderfield (ed.) pp.23–47. Elsevier.

Burton, J.D. and Statham, P.J. (1988) Trace metals as tracers in the ocean. *Philos. Trans. R. Soc. London*, **325**, 127–145.

Campbell, J.A. (1983) The Geochemical Ocean Sections Study – GEOSECS, in *Chemical Oceanography*, J.P. Riley and R. Chester (eds), Vol. 8, pp.89–155. London: Academic Press.

Chan, L.H., Drummond, D., Edmond, J.M. and Grant, B. (1977) On the barium data from the GEOSECS Expedition. *Deep-Sea Res.*, **24**, 613–649.

Collier, R.W. and Edmond, J.M. (1984) The trace element geochemistry of marine biogenic particulate matter. *Prog. Oceanogr.*, **13**, 113–199.

Dasch, E.J., Dymond, J. and Heath, G.R. (1971) Isotopic analysis of metalliferous sediments from the East Pacific Rise. *Earth Planet. Sci. Lett.*, **13**, 175–180.

Derry, L.A. (2009) A glacial hangover, *Nature*, **458**, 417–418.

Duinker, J.C. (1986) Formation and transformation of element species in estuaries, in *The Importance of Chemical 'Speciation' in Environmental Processes*, M. Bernhard, F.E. Brinkman and P.J. Sadler (eds), pp.365–384. Berlin: Springer-Verlag.

Elderfield, H. (1988) The oceanic chemistry of the rare-earth elements. *Philos. Trans. R. Soc. London*, **35**, 105–126.

Emerson, S.R. and Hedges, J.I. (2008) *Chemical Oceanography and the Marine Carbon Cycle*. Cambridge University Press.

Fowler, S.W., Buat-Menard, P., Yokoyama, Y., Ballestra, S., Holm, E. and Nguyen, H.V. (1987) Rapid removal of Chernobyl fallout from Mediterranean surface waters by biological activity. *Nature*, **329**, 56–58.

Garabato, A.C.N., Stevens, D.P., Watson, A.J. and Roether, W. (2007) Short-circuiting of the overturning circulation in the Antarctic Circumpolar current *Nature*, **447**, 194–197.

Gordon, A.L. (1996) Communication between oceans. *Nature*, **382**, 399–400.

Gross, M.G. (1977) *Oceanography – A View of the Earth*. Englewood Cliffs, NJ: Prentice Hall.

Jenkins, W.J. and Doney, S.C. (2003) The subtropical nutrient spiral. *Global Biogeochemical Cycles*, **17** (4) doi:10.1029/2003GB002085.

Key, R.A., Kozyr, A., Sabine, C.L., Lee, K., Wanninkopf, R., Bulister, J.L., Feely, R.A., Millero, F.J., Mordy, C. and Peng, T-H. (2004) A global ocean carbon climatology: Results from Global Data Analysis Project (GLODAP). *Global Biogeochem. Cycl.*, **18**, GB4031, doi:10.1029/2004GB002247.

Ledwell, J.R., Montgomery, E.T., Polzin, K.L., St. Laurent, L.C., Schmidtt, R.W. and Toole, J.M. (2000) Evidence for enhanced mixing over rough topography in the abyssal ocean. *Nature*, **403**, 179–182.

Measures, C.I. and Edmonds, J.M. (1988) Aluminium as a tracer of the deep outflow from the Mediterranean. *J. Geophys. Res.*, **93**, 591–595.

McDougall, T.J., Feistel, R., Millero, F.J., Jacket, D.R., Wright, D.G., King, B.A., Marion, G.M., Chen, C.-T.A. and Spitzer, P. (2009) Calculation of the thermophysical properties of seawater, in *Global Ship-based Repeat Hydrography Manual*, IOCCP Report No. 14, ICPO Publication Series no. 134.

Millero, F.J., Feistel, R., Wright, D.G. and McDougall, T.J. (2008) The composition of standard seawater and the definition of the reference-composition salinity scale. *Deep-Sea Res. I*, **55**, 50–72.

Nightingale, P.D. and Liss, P.S. (2003) Gases in seawater, in *Treatise on Geochemistry*, Vol. 6 H. Elderfield (ed.) pp.50–88. Elsevier.

Palmer, M.R. and Elderfield, H. (1985) Sr isotope composition of seawater over the past 75 Myr. *Nature*, **314**, 526–528.

Palmer, M.R. and Edmond, J.M. (1989) The strontium budget of the modern ocean. *Earth Planet. Sci. Lett.*, **92**, 11–26.

Pickard, G.L. and Emery, W.J. (1982) *Descriptive Physical Oceanography*. Oxford: Pergamon Press.

Rahmstorf, S. (2002) Ocean circulation and climate during the past 120,000 years. *Nature*, **419**, 207–214.

Sarmiento, J.L. and Gruber, N. (2006) *Ocean Biogeochemical Dynamics*. Princeton University Press.

Schaule, B.K. and Patterson, C.C. (1981) Lead concentrations in the northeast Pacific: evidence for global anthropogenic perturbations. *Earth Planet. Sci. Lett.*, **54**, 97–116.

SCOR Working Group (2007) GEOTRACES – An international study of the global marine biogeochemical cycles of trace elements and their isotopes. *Chemie der Erde Geochemistry*. doi:10.1016/j.chemer.2007.02.001.

Stommel, H. (1958) The abyssal circulation. *Deep Sea Res.*, **5**, 80–82.

Stowe, K.S. (1979) *Ocean Science*. New York: John Wiley & Sons, Inc.

Thompson, J., Carpenter, M.S.N., Colley, S., Wilson, T.R.S., Elderfield, H. and Kennedy, H. (1984) Metal accumulation in northwest Atlantic pelagic sediments. *Geochim. Cosmochim. Acta*, **48**, 1935–1948.

UNESCO (1981) *UNESCO Technical Paper on Marine Science*, no. 38. Paris: UNESCO.

Vance, D., Teagle, D.A.H. and Foster, G.L. (2009) Variable Quaternary chemical weathering fluxes and imbalances in marine geochemical budgets. *Nature*, **458**, 493–496.

Weihaupt, J.G. (1979) *Exploration of the Oceans. An Introduction to Oceanography*. New York: Macmillan.

Wilson, T.R.S. (1975) Salinity and the major elements of sea water, in *Chemical Oceanography*, Vol. 1, J.P. Riley and G. Skirrow (eds), pp.365–413. London: Academic Press.

Yu, E.-F., Francois, R. and Bacon, M.P. (1996) Similar rates of modern and last-glacial ocean thermocline circulation inferred from radiochemical data. *Nature*, **379**, 689–692.

Zachos, J., Pagani, M., Sloan, L., Thomas, E. and Billups, K. (2001) Trends, rhythms and aberrations in global climate 65 Ma to present. *Science*, **292**, 685–693.

# 8     Air–sea gas exchange

The atmosphere is the major source of gases to sea-water. The atmosphere itself consists of a mixture of major, minor and trace gases, and the abundances of a number of these are given in Table 8.1. The ocean can act as either a source or a sink for atmospheric gases. The gases enter or leave the ocean via exchange across the air–sea interface, and are transported within the ocean reservoir by physical processes.

## 8.1 Introduction

Dissolved gases in seawater are important for a number of reasons. The air–sea exchange of several gases is very important in the climate system, most obviously for the air–sea exchange of $CO_2$. However, the exchange of other gases can also influence the climate system either directly via their absorption of radiation in the atmosphere, or indirectly via their role in atmospheric aerosol formation. The air–sea exchange is a key process in the cycle of oxygen in the oceans which is critical to many ocean biogeo-chemical processes and to sustaining much of the marine biosphere. Air–sea exchange is also impor-tant for many other biogeochemical cycles, such as the nitrogen cycle. In this chapter, the processes regu-lating air-sea exchange are considered and then the air–sea exchange of various key gases is described. Some gases behave in an essentially conservative way in the oceans, their concentrations modified only by mixing. Other gases are involved in major ocean biogeochemical cycles, of which $CO_2$ and $O_2$ are particularly important examples. Since the cycling of these latter two gases is intimately bound up with the overall cycle of carbon and nutrients, their internal cycles within the ocean are considered in Chapter 9.

## 8.2 The exchange of gases across the air–sea interface

The solubility of a gas in seawater is an important factor in controlling its uptake by the oceans; thus, the more soluble gases partition in favour of the water phase, whereas the less soluble gases partition in favour of the atmosphere. The solubilities of gases in seawater are a function of temperature, salinity and pressure. A considerable amount of very accu-rate data are now available on the solubilities of various gases (e.g. oxygen, nitrogen, argon) in sea-water, and equations have been derived to express the dependence of the solubility values on tempera-ture and salinity; an example of this is given in Work-sheet 8.1. All gases become more soluble, at least to some degree, in water as the temperature decreases, and this has important implications for the distribu-tions of gases in surface waters, which exhibit con-siderable temperature variations from the equatorial regions to the poles.

Kester (1975) drew attention to the importance of the concept of partial pressure as a useful means of representing the composition of a gaseous mixture, for example, the atmosphere. Thus, the total pressure exerted by a mixture of gases in a volume of the mixture is the sum of the partial pressures of the individual gases. This concept of partial pressure also can be applied to gas molecules dissolved in aqueous solution, and Henry's law describes the relationship between the partial pressure of a gas in solution ($P_G$) and its concentration ($C_G$); thus

$$P_G = K_G C_G \qquad (8.1)$$

where $K_G$ is the Henry's law constant. In a very sim-plistic manner, therefore, the rate of transfer of a gas

*Marine Geochemistry*, Third Edition. Roy Chester and Tim Jickells.
© 2012 by Roy Chester and Tim Jickells. Published 2012 by Blackwell Publishing Ltd.

**Table 8.1** The concentrations of some long lived gases in the atmosphere (after Richards, 1965).

Note that the concentrations of $CO_2$ $CH_4$ and $N_2O$ have varied significantly on glacial/interglacial time scales and are now being substantially enhanced by human activity.

| Gas | Concentration (%) | Gas | Concentration (p.p.m.) |
|-----|-------------------|-----|------------------------|
| $N_2$ | 78.084 | Ne | 18.18 |
| $O_2$ | 20.946 | He | 5.24 |
| $CO_2$ | 0.033 | Kr | 1.14 |
| Ar | 0.394 | Xe | 0.087 |
| | | $H_2$ | 0.5 |
| | | $CH_4$ | 2.0 |
| | | $N_2O$ | 0.5 |

from the atmosphere is proportional to its partial pressures in the two reservoirs, that is, the atmosphere and the sea, respectively. At equilibrium, when the partial pressure of a gas is the same in both the air and the water reservoirs, molecules enter and leave each phase at the same rate. When the partial pressure of the gas in one reservoir is higher than that in the other, however, there will be a *net* diffusive flow of gas into or out of the sea in response to the concentration gradient across the air–sea interface.

Following Liss (1983), Liss and Merlivat (1986) and Nightingale (2009), a net gas flux ($F$) across the air–sea interface therefore must be driven by a concentration difference ($\Delta C$) between the air and the surface water, with the magnitude and direction of the flux being proportional to the numerical value and sign of $\Delta C$; thus

$$F = K_{(T)w}\Delta C \qquad (8.2)$$

where $\Delta C$ is the concentration difference driving the flux ($F$) and the constant of proportionality $K_{(T)w}$, which links the flux and the concentration difference, has the dimensions of a velocity and may be termed the transfer coefficient, transfer velocity, or piston velocity.

The concentration difference ($\Delta C$) can be expressed as

$$\Delta C = C_a H^{-1} - C_w \qquad (8.3)$$

where $C_a$ and $C_w$ are the gas concentrations in the air and the water, respectively, and $H$ is the dimen-

sionless Henry's law constant (expressed as the ratio of the concentration of gas in air to its concentration in non-ionized form in the water, at equilibrium).

The total transfer velocity can be expressed as

$$1/K_{(T)w} = 1/\alpha k_w + 1/Hk_a \qquad (8.4)$$

where $k_a$ and $k_w$ are the individual transfer velocities for chemically unreactive gases in the air and water phases, respectively, $\alpha$ is a factor that quantifies any enhancement of gas transfer in the water as a result of chemical reactions ($\alpha = 1$ for an unreactive gas), and $H$ is as defined in Equation 8.3.

For most sparingly soluble gases (e.g. $O_2$, $CO_2$, $CH_4$, $SF_6$) the rate limiting step is transfer through the water phase as molecular diffusion is slower in water than in air and air-sea flux (F) simplifies to

$$F = k_w\Delta C = kw(CaH^{-1} - C_w) \qquad (8.5)$$

For highly reactive or highly soluble gases (e.g. $SO_2$ and $NH_3$), the atmospheric transfer term dominates.

$$F = k_a\Delta C \qquad (8.6)$$

There are some gases for which Equation 8.4 does not simplify to either 8.5 or 8.6 including dimethylsulfide – DMS (Nightingale, 2009).

The exchange or transfer of a gas across the air–sea interface is therefore dependent on:
1 the concentration difference across the air–sea interface, that is, any flux must be driven by a concentration gradient and the magnitude of the diffusive flux is proportional to this gradient (the coefficient of molecular diffusion);
2 the transfer coefficient (also known as the piston velocity or, transfer velocity).

In turn, these are dependent on physical factors such as wind velocity and temperature, and on the solubilities, diffusion rates and chemical reactivities (aqueous chemistry) of individual gases. Data are available on both the molecular diffusivities and solubilities of gases in seawater, and a summary of these is given in Table 8.2. The value of $k_w$ for different gases depends on the molecular diffusivity of gases usually paramerized as the Schmidt number 'Sc', which is the ratio of kinematic viscosity of seawater, and the molecular diffusivity of the gas in seawater, both well known physical constants. Thus if the rates of gas transfer are known for any one gas

**Worksheet 8.1:  The solubility of gases in seawater**

Weiss (1970) derived an equation to calculate the temperature and salinity dependence of gas solubilities in seawater from moist air. Thus:

$$\ln c^* = A_1 + A_2(100/T) + A_3 \ln(T/100) + A_4(T/100)$$
$$+ S‰[B_1 + B_2(T/100) + B_3(T/100)^2]$$

(WS18.1.1)

where $c^*$ is the solubility in seawater from water-saturated air at a total pressure of one atmosphere; the $A$s and $B$s are constants, the numerical values of which depend on the individual gas and the expression of the solubility; $T$ is the absolute temperature; and $S$ (‰) is the salinity. A data set derived from Equation (WS18.1.1) is given in Table WS8.1.1 showing how the solubility of oxygen in seawater varies with temperature and salinity. There are various ways of expressing gas solubilities, and in order to illustrate the temperature and salinity variations the values listed are given in $\mu mol\,kg^{-1}$.

**Table WS8.1.1** The solubility of oxygen in seawater; units, $\mu mol\,kg^{-1}$ (from Kester, 1975).

| | Salinity (‰) | | | | | | | | | | | | |
|---|---|---|---|---|---|---|---|---|---|---|---|---|---|
| $T$ (°C) | 0 | 4 | 8 | 12 | 16 | 20 | 24 | 28 | 31 | 33 | 35 | 37 | 39 |
| −1 | 469.7 | 455.5 | 441.7 | 428.3 | 415.4 | 402.8 | 390.6 | 378.8 | 370.2 | 364.6 | 359.0 | 353.5 | 348.2 |
| 0 | 456.4 | 442.7 | 429.4 | 416.5 | 404.0 | 391.9 | 380.1 | 368.7 | 360.4 | 354.9 | 349.5 | 344.2 | 339.0 |
| 1 | 443.8 | 430.6 | 417.7 | 405.3 | 393.2 | 381.5 | 370.1 | 359.0 | 351.0 | 345.7 | 340.5 | 335.4 | 330.3 |
| 2 | 431.7 | 418.9 | 406.5 | 394.5 | 382.8 | 371.5 | 360.5 | 349.8 | 342.0 | 336.9 | 331.8 | 326.9 | 322.0 |
| 3 | 420.2 | 407.9 | 395.9 | 384.2 | 372.9 | 361.9 | 351.3 | 340.9 | 333.4 | 328.5 | 323.6 | 318.8 | 314.1 |
| 4 | 409.3 | 397.3 | 385.7 | 374.4 | 363.5 | 352.9 | 342.6 | 332.5 | 325.2 | 320.4 | 315.7 | 311.1 | 306.5 |
| 5 | 398.8 | 387.2 | 375.9 | 365.0 | 354.4 | 344.1 | 334.1 | 324.4 | 317.4 | 312.7 | 308.1 | 303.6 | 299.2 |
| 6 | 388.7 | 377.5 | 366.6 | 356.0 | 345.8 | 335.8 | 326.1 | 316.7 | 309.8 | 305.3 | 300.9 | 296.5 | 292.2 |
| 7 | 379.1 | 368.2 | 357.7 | 347.4 | 337.5 | 327.8 | 318.4 | 309.3 | 302.6 | 298.3 | 294.0 | 289.7 | 285.5 |
| 8 | 369.9 | 359.4 | 349.1 | 339.2 | 329.6 | 320.2 | 311.1 | 302.2 | 295.8 | 291.5 | 287.3 | 283.2 | 279.2 |
| 9 | 361.1 | 350.9 | 341.0 | 331.3 | 322.0 | 312.9 | 304.0 | 295.4 | 289.2 | 285.0 | 281.0 | 277.0 | 273.0 |
| 10 | 352.6 | 342.7 | 333.1 | 323.7 | 314.6 | 305.8 | 297.2 | 288.9 | 282.8 | 278.8 | 274.8 | 271.0 | 267.1 |
| 11 | 344.5 | 334.9 | 325.5 | 316.5 | 307.6 | 299.1 | 290.7 | 282.6 | 276.7 | 272.8 | 269.0 | 265.2 | 261.5 |
| 12 | 336.7 | 327.3 | 318.3 | 309.5 | 300.9 | 292.6 | 284.5 | 276.6 | 270.8 | 267.0 | 263.3 | 259.6 | 256.0 |
| 13 | 329.2 | 320.1 | 311.3 | 302.8 | 294.4 | 286.3 | 278.5 | 270.8 | 265.2 | 261.5 | 257.9 | 254.3 | 250.8 |
| 14 | 322.0 | 313.2 | 304.7 | 296.3 | 288.2 | 280.4 | 272.7 | 265.3 | 259.8 | 256.2 | 252.7 | 249.2 | 245.8 |
| 15 | 315.1 | 306.5 | 298.2 | 290.1 | 282.3 | 274.6 | 267.1 | 259.9 | 254.6 | 251.1 | 247.7 | 244.3 | 240.9 |
| 16 | 308.5 | 300.1 | 292.0 | 284.2 | 276.5 | 269.1 | 261.8 | 254.7 | 249.6 | 246.2 | 242.8 | 239.5 | 236.3 |
| 18 | 295.9 | 288.0 | 280.3 | 272.9 | 265.6 | 258.5 | 251.7 | 245.0 | 240.1 | 236.9 | 233.7 | 230.6 | 227.5 |
| 20 | 284.2 | 276.7 | 269.5 | 262.4 | 255.5 | 248.8 | 242.3 | 235.9 | 231.2 | 228.2 | 225.2 | 222.2 | 219.3 |
| 22 | 273.4 | 266.3 | 259.4 | 252.7 | 246.1 | 239.7 | 233.5 | 227.5 | 223.0 | 220.1 | 217.3 | 214.4 | 211.6 |
| 24 | 263.3 | 256.5 | 250.0 | 243.6 | 237.3 | 231.3 | 225.4 | 219.6 | 215.4 | 212.6 | 209.9 | 207.2 | 204.5 |
| 26 | 253.8 | 247.4 | 241.2 | 235.1 | 229.2 | 223.4 | 217.7 | 212.2 | 208.2 | 205.6 | 202.9 | 200.4 | 197.8 |
| 28 | 245.0 | 238.9 | 232.9 | 227.1 | 221.5 | 215.9 | 210.6 | 205.3 | 201.5 | 198.9 | 196.4 | 194.0 | 191.5 |
| 30 | 236.7 | 230.9 | 225.2 | 219.7 | 214.2 | 209.0 | 203.8 | 198.8 | 195.1 | 192.7 | 190.3 | 188.0 | 185.7 |
| 32 | 228.9 | 223.4 | 217.9 | 212.6 | 207.5 | 202.4 | 197.5 | 192.7 | 189.2 | 186.9 | 184.6 | 182.3 | 180.1 |

**Table 8.2** The solubilities and molecular diffusivities of some gases in seawater (data from Broecker and Peng, 1982).

| Gas | 0°C | | 24°C | |
| --- | --- | --- | --- | --- |
| | Solubility ($cm^3 l^{-1}$) | Diffusion coefficient ($\times 10^{-5} cm^2 s^{-1}$) | Solubility ($cm^3 l^{-1}$) | Diffusion coefficient ($\times 10^{-5} cm^2 s^{-1}$) |
| He | 7.8 | 2.0 | 7.4 | 4.0 |
| Ne | 10.1 | 1.4 | 8.6 | 2.8 |
| $N_2$ | 18.3 | 1.1 | 11.8 | 2.1 |
| $O_2$ | 38.7 | 1.2 | 23.7 | 2.3 |
| Ar | 42.1 | 0.8 | 26.0 | 1.5 |
| Kr | 85.6 | 0.7 | 46.2 | 1.4 |
| Xe | 192 | 0.7 | 99 | 1.4 |
| Rn | 406 | 0.7 | 186 | 1.4 |
| $CO_2$ | 1437 | 1.0 | 666 | 1.9 |
| $N_2O$ | 1071 | 1.0 | 476 | 2.0 |

they can be calculated for others. The main resistance to gas transfer is concentrated in a thin layer near the air–sea interface, and a number of approaches have been used to investigate gas exchange rates across this interface. These include:

1 theoretical models of the gas transfer processes;
2 laboratory experiments designed to quantify the parameters required for the flux equations, for example, wind-tunnel experiments for investigating the dependence of transfer coefficients on windspeed;
3 field measurements of air–sea gas fluxes or transfer coefficients.

Each of these approaches is considered in the following sections.

*Theoretical models.* The ability to model the gas transfer process adequately is obviously important because it would permit transfer velocities to be determined from a knowledge of other parameters (e.g. windspeed), and also would allow measurements of the transfer velocity for one gas to be converted into equivalent values for another gas. Models of varying complexity have been used to describe the exchange of gases across the air–sea interface (Nightingale, 2009). These include the following.

1 The 'stagnant film model', in which the rate-controlling process is transport across the stagnant film layer by molecular diffusion.
2 The 'surface renewal model', which still retains the concept of a stagnant film, but where this film is periodically replaced by bulk seawater and the rate-

determining step for gas transfer is the rate at which the film is replaced.

There are more complex models that include the impact of factors such as bubbles, surfactants and breaking waves, but it is difficult to quantify the relevant terms in the equations in all but the simplest stagnant film models (Nightingale 2009). The classic stagnant film model (SFM) has been widely used in oceanography, and according to Broecker and Peng (1974), it does offer an adequate first-order approximation of the very complex processes that actually take place at the air–sea interface. For this reason the SFM is considered in more detail in Worksheet 8.2. In order to use the theoretical models it is necessary to have a knowledge of, among other parameters, air–sea gas transfer coefficients (or transfer velocities) ($k_w$). These can be measured under both laboratory and field conditions.

*Laboratory measurements.* Since it is difficult to fully describe air-sea exchange from first principles, laboratory experiments have been widely used to understand and quantify the processes involved. Most laboratory experiments involve the use of wind tunnels ranging from laboratory scale experiments to 50m tanks, although it is difficult in even the largest tanks to fully replicate the physical and biogeochemical complexity of the real ocean. Theses laboratory experiments document a major impact of wind speed on $k_w$. The relationship is not linear with marked increases in $k_w$ associated with transitions from smooth to rippled and to breaking wave water surfaces.

**Worksheet 8.2: The stagnant film model for gas exchange across the air–sea interface**

Following Broecker and Peng (1974; 1982), the concepts underlying the stagnant film model (SFM) can be summarized as follows. In the SFM it is assumed that the rate-limiting step to the transfer of a gas between the air and the water is molecular diffusion through a stagnant water film. The air above the film and the water below it are assumed to be well mixed, and the gas concentration at the top of the film is assumed to be in equilibrium with the air above it. If the partial pressure of the gas in the air yields a concentration of the gas at the top of the film that is different from that in the water below, then a concentration gradient is set up. Transfer of the gas through the film then results from molecular diffusion along the gradient from the high- to the low-concentration region.

The rate at which the diffusion occurs depends on a number of factors, which include the following.

1 The rate at which the gas diffuses through seawater. A list of the molecular diffusivities of dissolved gases in seawater is given in Table 8.2, from which it can be seen that the diffusivities increase with temperature.

2 The thickness of the film. As the thickness of the film increases, the time taken for a gas to diffuse through it also increases. The thickness of the film, and so the rate of gas exchange across it, is dependent on the degree of agitation of the sea surface by the wind; the stronger the wind, the thinner the film and the more rapid the exchange rate.

3 The difference in the concentration of a gas between the air and the sea (disequilibrium magnitude). The larger the difference, the greater the concentration gradient and the faster the molecular diffusive transfer.

The SFM is illustrated diagrammatically in Fig. WS8.2.1, with reference to radon exchange. The rate of transfer of the gas between the water and the air is controlled by the thickness of

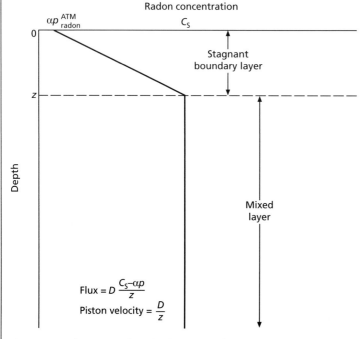

**Fig. WS8.2.1** The stagnant boundary layer model (from Broecker and Peng, 1974).

*continued*

the stagnant boundary layer, through which the gas is transferred only by molecular diffusion. The overlying air and the underlying water are assumed to be well mixed. In the flux equation $D$ is the molecular diffusivity of the gas, $z$ is the thickness of the film, $C_s$ is the concentration of the gas at the bottom of the film, $\alpha$ is the solubility of the gas, and $p$ is the partial pressure of the gas in the air.

There are difficulties in applying the SFM to 'reactive' gases, such as $CO_2$, for which chemical, as well as physical, transport across the stagnant layer is important. For example, $CO_2$ can react via hydration to give, on ionization, $HCO_3^-$ and $CO_3^{2-}$ (see Section 9.4). If this hydration were rapid enough for carbonate equilibria to be reached at all points as it crossed the film, then concentration gradients would be set up for all components in the system, that is, not simply for dissolved $CO_2$, and since the net rate of transport is the sum of all the individual carbon species along the individual gradients, the overall transfer rate would be enhanced. The effect of chemical enhancement on the exchange rate for $CO_2$ varies with the film thickness, although the average film thickness in the sea probably lies in the region where the chemical enhancement is negligible. Under some conditions, however, the air–sea exchange of $CO_2$ can be enhanced by the side effects of pH variations within the film, which are dependent on the equilibria in the carbonate system. The $\alpha$ term in Equation 8.4 allows for such effects.

*Field measurements.* A number of field techniques are available for the measurement of $k_w$, and these have been summarized by Nightingale (2009). An early example of how field data can be plugged into a theoretical model has been provided by Broecker and Peng (1974) who attempted to assess the applicability of the SFM to the exchange of gases across the ocean–atmosphere interface using radiocarbon and radon tracers to determine $k_w$ values. Their principal findings may be summarized as follows.

**1** The mean stagnant film thickness for the World Ocean was found to be between 40 ± 30 and 50 ± 30 μm, depending on the tracer technique used. The thickness appears to vary in inverse proportion to the square of the wind velocity, as predicted by Kanwisher (1963).

**2** From a knowledge of the wind velocity, mixed-layer thickness and partial pressure difference between the air and the sea surface, the net transfer rate of most gases between the atmosphere and the sea can be predicted to within an accuracy of ~±30%.

**3** Using the average film-layer thickness it is possible to compute the mean residence times of various gases in the atmosphere with respect to transfer to the mixed layer, and in the mixed layer with respect to transfer to the atmosphere. The average residence time for transfer from the atmosphere to the mixed sea-surface layer is ~300 yr for gases with solubilities in the 'normal' range (see Table 8.2); for $CO_2$ it is ~8 yr. The average residence time for transfer from the mixed layer of the sea to the atmosphere for most gases is ~1 month. For $CO_2$ the situation is more complex. The mixed layer of the ocean probably achieves chemical equilibrium with $CO_2$ in the atmosphere in 1.5 yr. This leads to the important conclusion that, because the replacement time of the mixed layer by mixing with underlying water is around a decade, the rate-limiting step in the removal of anthropogenic (fossil fuel) $CO_2$ from the atmosphere is vertical mixing within the sea, rather than gas transfer of $CO_2$ across the air–sea interface. Except for gases with very high solubilities (e.g. $SO_2$, $CO_2$) the sea is not an effective rapid sink for anthropogenic gases, as their entry into the system requires many hundreds of years.

Generally field studies of gas exchange have fallen into four groups. Budget calculations using [14]C and radon, direct measurements using biogeochemically inert gases, deliberate tracer experiments, such as using $SF_6$, and most recently for the first time direct measurements with new fast micrometeorological techniques (Nightingale, 2009). An example of the latter is shown in Fig. 8.1 and suggests encouraging agreement between the data (in this case for DMS) and various parameterizations of gas exchange rates, the latter generally derived from fitting to data rather than theoretical considerations. The most recent of

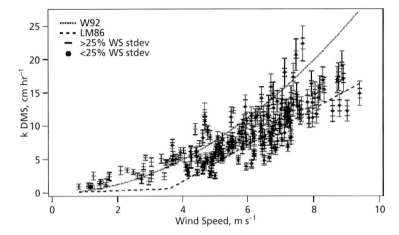

**Fig. 8.1** Direct measurements of DMS exchange rate versus windspeed alongside two model estimates (W92 – Wanninkof, 1992; LM86 – Liss and Merlivat, 1986) showing data approximates to model descriptions but uncertainties are large particularly at higher wind speeds (Huebert et al., 2004). Results are reported with different uncertainties by two slightly different approaches.

these by Nightingale et al. (2000) takes the form of a quadratic equation

$$k_w = 0.222U^2 + 0.333U \qquad (8.7)$$

where u is the wind speed measured 10 m above the surface. This strong non linear dependence on wind speed emphasizes the importance of high wind speed regimes in air–sea gas exchange.

Some of the constraints affecting the air–sea exchange of trace gases have been discussed above, but the overall process is extremely complex and for a detailed review of the subject the reader is referred to Nightingale (2009). This author considered a number of the constraints involved, including, in addition to windspeed, effects of waves, bubbles, humidity, rain and surfactants.

Once a gas is dissolved in seawater, that is, it has diffused across the interface into the liquid phase, the processes that determine its oceanic distribution depend on the nature of the gas itself. The distribution of those gases that generally are regarded as being *non-reactive* in seawater (e.g. nitrogen $N_2$ and the inert gases) is controlled by physical processes and by the effects of temperature and salinity on their solubility (Kester, 1975). It is usual to express variations in the non-reactive gases in terms of their percentage saturation. To do this, the observed concentration of the gas in the sea is given as the percentage of its solubility in pure water of the same temperature and salinity. Thus

$$\text{saturation} (\%) = 100 \times G/G' \qquad (8.8)$$

where G is the observed gas content in seawater, and G' is the solubility of the gas in pure water having the observed temperature and salinity (Richards, 1965; Riley and Chester, 1971).

Factors in addition to those described above affect the distributions of the *reactive* gases, such as oxygen and carbon dioxide, in the oceans. Although oxygen is in fact relatively chemically inert in seawater, it is involved in biological processes and these strongly affect its distribution in the water column. Carbon dioxide also takes part in biological processes, and the competing processes of photosynthesis (utilization of $CO_2$, liberation of $O_2$) and respiration (utilization of $O_2$, liberation of $CO_2$) are the cause of many of the in situ changes in the concentrations of the two gases in the sea. In addition, $CO_2$ is extremely reactive in seawater as discussed earlier. Oxygen and carbon dioxide therefore behave in a non-conservative manner in the oceans. These gases are considered, in the Chapter 9 and attention here will focus on other trace gases.

The oceans act as a source for some gases to the atmosphere and a sink for others. In the case of many of the gases for which the ocean is a source, the gas is produced by a wide variety of processes involving biological activity in the oceans (Nightingale and Liss, 2003). This may involve;

• direct biological production such as the production of $N_2O$ as a by-product of bacterial denitrification (see Chapter 9),

• indirectly such as the release of DMSP (Dimethylsulfoniopropionate, a compound which may have a

role in osmotic control in phytoplankton cells, UV absorption and deterring grazers) from cells which breaks down to form DMS, which breaks down in the atmosphere to form $SO_2$ (see also Chapter 4),
• via photochemical degradation of dissolved organic matter in the oceans, which itself ultimately has a biological origin (see Chapter 9), as in the case of COS emissions.

We do not understand the cycling of many gaseous compounds within the oceans or their air–sea exchange and research continues to identify new gases for which air–sea exchange is important. The estimation of the fluxes of even the better studied gases is uncertain but Table 8.3 illustrates the fluxes of some important gases, the relative significance of the fluxes compared to other sources, the production mechanism and the biogeochemical role of the gases.

The direct measurement of air–sea gas exchange is generally difficult. Improving flux measurements by improving the data base can be valuable (e.g. Watson et al., 2009), but the key challenge is really to understand the mechanism of production and from that derive a predictive capability based on parameters that can be readily mapped, such as for instance phytoplankton abundance based on satellite measurements from space. However, research to date has emphasized the complexity and diversity of processes that regulate the cycling of a wide variety of trace gases. The situation is much simpler for inert gases. In the case of noble gases such as nitrogen and argon, the ocean and atmosphere are approximately in equilibrium. Such equilibrium does not exist in the case of chlorofluorocarbons (CFM or CFCs). These are industrial chemicals produced since the 1930s which are almost inert except in the upper atmosphere where they contribute to ozone depletion, and hence their production is now severely restricted. Although these gases have very low solubilities, they do slowly dissolve in the oceans and hence there is a net atmosphere to ocean transfer. However, this oceanic sink is relatively weak with for instance only 1% of the emissions of CFC-11 (trichlorofluoromethane previously used mainly as a refrigerant) currently present in the oceans (Willey et al., 2004). The presence of these CFCs dissolved in seawater has been used as a water mass tracer (see Chapter 7). The solubility of these gases is highest at lowest seawater temperatures and these are also areas of deep water mass formation. Hence, these chlorofluorcarbons are most effective as tracers of the formation and subsequent movement of deep water on timescales of decades.

The ocean represents a net sink for $CO_2$ absorbing about 30% of current anthropogenic emissions by

**Table 8.3** Sea to air fluxes of some biogeochemically important gases, the strength of this source compared to other sources, the marine production mechanism and the biogeochemical role (adapted from Nightingale and Liss, 2003; plus Carpenter, 2003). Where a role is uncertain it is indicated by ?. A negative number indicates an air-to-sea flux. Fluxes are as mass of gas unless specified as that of a particular component of the gas, for example, TgS.

| Gas | Net annual Sea-to-air Flux | % of total | Production Mechanism | Atmospheric Role |
|---|---|---|---|---|
| DMS | 15–33 TgS | 80* | phytoplankton | cloud formation, acidity |
| COS | 0.3 TgS | 40 | photochemistry | cloud formation, acidity |
| $CH_3Cl$ | 0.2–0.4 Tg | 7–14 | phytoplankton? | ozone cycling |
| $CH_3I$ | 0.1–0.4 Tg | 50–80 | algae | ozone cycling, aerosol formation |
| $N_2O$ | 11–17 TgN | 60–90# | denitrification & nitrification | greenhouse gas, ozone cycling |
| $CH_4$ | 15–24Tg | 3–5# | bacterial | greenhouse gas, atmospheric oxidizing capacity |
| CO | 10–650 TgC | 3–20 | photochemical | atmospheric oxidizing capacity |
| Hg | $10^{-3}$Tg | 20 | biological | pollution |
| $CO_2$ | $-1.7 \pm 0.5 \times 10^3$TgC | −30 | respiration | greenhouse gas |

* Terrestrial sources of DMS are small, however the DMS is oxidized to $SO_2$ and the sulfur cycle has been heavily modified by human activity. Natural annual terrestrial emissions are 10 TgS and these have been increased massively by human activity with annual anthropogenic emissions of 93 TgS in the 1980s although these have declined somewhat since (Brimblecombe and Lein, 1989).
# Anthropogenic activity has increased global emissions of $N_2O$ and $CH_4$ by 1.6 and 3 fold (IPCC, 2007) approximately, and the figures in this table are as percentage of current emissions.

biological uptake and net physical air-to-sea transfer. The combustion of fossil fuels has resulted in a small decline in atmospheric $O_2$ (0.003% over a decade, i.e. a decline from the atmosphere being 20.95% $O_2$ to 20.947%; Manning and Keeling, 2006). However, overall the flux of oxygen between the atmosphere and ocean is approximately in balance although this does involve very large fluxes mediated by biological and physicochemical factors (Gruber *et al.*, 2001). The cycling of oxygen and carbon is considered in the next chapter.

## References

Brimblecombe, P. and Lein, A.Y. (1989) *Evolution of the Global Sulphur Cycle, SCOPE 39*. New York: John Wiley & Sons, Inc.

Broecker, W.S. and Peng, T.-H. (1974) Gas exchange rates between air and sea. *Tellus*, **26**, 21–35.

Broecker, W.S. and Peng, T.-H. (1982) *Tracers in the Sea*. Palisades: Lamont-Doherty Geological Observatory.

Carpenter, L.J. (2003) Iodine in the marine boundary layer. *Chem. Rev.*, **103**, 4953–4962.

Gruber, N., Gloor, M., Fan, S.-M. and Sarmiento, J.L. (2001) Air-sea flux of oxygen estimated from bulk data: Implications for the marine and atmospheric oxygen cycles. *Global Biogeochem. Cycl.*, **15**, 783–803.

Huebert, B.J., Blomqvist, B.W., Hare, J.E., Fairall, C.W., Johnson, J.E. and Bates, T.S. (2004) Measurement of the sea-air DMS flux and transfer velocity using eddy correlation. *Geophys. Res. Lett.*, **31**, doi:10.1029/2004GL021567.

IPCC (2007) Solomon S., Qin, D, Manning, M., Chen, Z., Marqis, M., Averyt, K.B., Tignor, M. and Miller, H.L., *Climate Change 2007: The Physical Science Basis. Contribution of working group I to the fourth assessment report of the intergovernmental panel on climate change*. Cambridge: Cambridge University Press.

Kanwisher, J. (1963) On the exchange of gases between the atmosphere and the sea. *Deep Sea Res.*, **10**, 195–207.

Kester, D.R. (1975) Dissolved gases other than $CO_2$. In *Chemical Oceanography*, J.P. Riley and G. Skirrow (eds), Vol. 1, pp.497–589. London: Academic Press.

Liss, P.S. (1983) Gas transfer: experiments and geochemical implications, In *Air–Sea Exchange of Gases and Particles*, P.S. Liss and W.G. Slinn (eds), pp.241–298. Dordrecht: Reidel.

Liss, P.S. and Merlivat, L. (1986) Air–sea exchange rates: introduction and synthesis, in *The Role of Air–Sea Exchange in Geochemical Cycling*, P. Buat-Menard (ed.), pp.113–227. Dordrecht: Reidel.

Manning, A.C. and Keeling, R.F. (2006) Global oceanic and land biotic carbon sinks from the Scipps atmospheric oxygen flask sampling network. *Tellus*, **58B**, 95–116.

Nightingale, P.D., Malin, G., Law, C.S., Watson, A.J., Liss, P.S., Liddicoat, M.I., Boutin, J. and Upstill-Goddard, R.C. (2000) In situ evaluation of air-sea gas exchange parameterizations using novel conservative and volatile tracers. *Global Biogeochem. Cycl.*, **14**, 373–387.

Nightingale, P.D. and Liss, P.S. (2003) Gases in Seawater in Nightingale, Gases in seawater, in *The Oceans and Marine Geochemistry, Vol. 6, Treatise on Geochemistry*, Holland, H.D. and Turekian, K.K. (eds), pp.49–81. Oxford: Elsevier-Pergamon.

Nightingale, P.D. (2009) Air-Sea Gas Exchange, in Surface Ocean-Lower Atmosphere Processes. Geophysical Research Series. *Am. Geophys.*, **187**, 69–97.

Richards, F.A. (1965) Dissolved gases other than carbon dioxide, in *Chemical Oceanography*, 1st edn, J.P. Riley and G. Skirrow (eds), pp.197–225. London: Academic Press.

Riley, J.P. and Chester, R. (1971) *Introduction to Marine Chemistry*. London: Academic Press.

Watson, A.J., Schuster, U., Bakker, D.C.E. *et al.* (2009) Tracking the Variable North Atlantic Sink for Atmospheric $CO_2$. *Science*, **328**, 1391–1393.

Wanninkhof, R. (1992) Relationship between wind speed and gas exchange over the ocean, *J. Geophys. Res.*, **97**, 7373–7382.

Weiss, R.F. (1970) The solubility of nitrogen, oxygen and argon in water and sea water. *Deep Sea Res.*, **17**, 721–735.

Willey, D.A., Fine, R.A., Sonnerup, R.E., Bullister, J.L., Smethie, W.M. Jr and Warner, M.J. (2004) Global oceanic chlorofluorocarbon inventory, *Geophys. Res. Lett.*, **31**, doi:10.1029/2003GL018816.

# 9 Nutrients, oxygen, organic carbon and the carbon cycle in seawater

The key biological process in the oceans is the production and consumption of organic matter. The production is dominated by photosynthesis by phytoplankton, free floating microscopic (generally <10 μm) algae; except in shallow coastal areas where benthic primary production by micro and macro algae can be important. The organic matter fuels the growth of higher organisms from zooplankton to whales. The breakdown of organic matter by respiration is used as an energy source by organisms from bacteria to whales, but it is the former that dominate the biogeochemical recycling of organic matter. Photosynthesis or primary production requires light and so is constrained to the sunlit parts of the oceans (the euphotic zone), usually to a depth where ~1% of the surface light persists. This depth varies throughout the oceans reflecting the amount of particulate matter in the water column, but in the open ocean is of the order of 100 m. Respiration does not require light and so can go on throughout the water column. This difference illustrates that the basis of most life in the oceans is ultimately supported by photosynthesis which only occurs in a rather shallow top layer of the oceans. The reversible reactions involved in photosynthesis and respiration can be rather simplistically described by the following conceptual equation;

$$CO_2 + H_2O \underset{\text{respiration}}{\overset{\text{photosynthesis}}{\rightleftharpoons}} CH_2O + O_2 \quad (9.1)$$

This equation demonstrates the linkages of the carbon and oxygen cycles. Photosynthesis also requires numerous nutrient elements as will be discussed in the following, and hence the carbon and oxygen cycles are linked to those of nutrient elements such as nitrogen and phosphorus. In this chapter the cycling of carbon, oxygen and nutrients will be considered and the linkages between them considered. The cycling of nutrients is a major control on the overall patterns of primary production in the oceans which will also be discussed.

## 9.1 The nutrients and primary production in seawater

### 9.1.1 Introduction

Organic matter is much more complex than is implied by Equation 9.1 and contains many more elements than just carbon, hydrogen and oxygen. Thus for photosynthesis, algae require a range of nutrient elements. Some of these are rather abundant in the oceans relative to the biological requirement (e.g. potassium and sulfur) while others are present at low concentrations relative to the biological demand and hence their supply can limit primary production rates. These potentially limiting nutrients are traditionally considered to be nitrogen, phosphorus and silicon, although research in recent years has emphasized the important role of some other elements particularly iron and possibly some other trace metals. There is a marked difference in concentrations between the traditional nutrients (N, P and Si present in $\mu mol\,l^{-1}$ concentrations) sometimes now referred to as macronutrients and trace metals such as iron present in $nmol\,l^{-1}$ concentrations and sometime now referred to as nanonutrients. The trace metals are discussed in detail in Chapter 10 but the

*Marine Geochemistry*, Third Edition. Roy Chester and Tim Jickells.
© 2012 by Roy Chester and Tim Jickells. Published 2012 by Blackwell Publishing Ltd.

role of iron will also be discussed in this chapter. However, the main focus here will be on N, P and Si along with carbon and oxygen.

## 9.1.2 Nutrient cycling

### 9.1.2.1 Nutrient chemistry

*Nitrogen nutrients.* Nitrogen is present in a wide range of oxidation states in seawater principally as: (i) molecular nitrogen (oxidation state zero); (ii) fixed inorganic salts, such as nitrate nitrogen ($NO_3^-$ oxidation state +5), nitrite nitrogen ($NO_2^-$ oxidation state +3) and ammonia ($NH_3/NH_4^+$ oxidation state −3; the proportions of the two ammonia forms depends on pH but ammonium is the dominant form); (iii) a range of organic nitrogen compounds associated with organisms, for example, amino acids and urea; plus the less well characterized degradation products of organic matter usually referred to as dissolved organic nitrogen (DON) and (iv) particulate nitrogen. Some other trace gases, particularly $N_2O$ which is a powerful greenhouse gas, are also important in the global N cycle.

Nitrogen is brought to the oceans from fluvial and atmospheric sources, by diffusion from sediments and by in situ nitrogen fixation. Nitrogen fixation is the process by which organisms can assimilate, or fix, molecular nitrogen. The strong binding between the N atoms in $N_2$ gas means that breaking these bonds in order to biologically fix this nitrogen demands a lot of energy and specialized enzymes. Nitrogen fixation in the oceans can be carried out by only a few phytoplankton species, mainly the cyanobacteria, and most oceanic phytoplankton are not capable of nitrogen fixation. Nitrogen fixation was thought to be a minor process in the oceans, but recent research has demonstrated that this is not the case (Mahaffey *et al.*, 2005; Arrigo, 2005) and the fluxes of N from fluvial and atmospheric inputs and nitrogen fixation are all important sources of fixed nitrogen for the oceans (see Section 6.4).

Nitrogen fixation is restricted to a relatively small number of phytoplankton and bacteria and because of this the majority of phytoplankton utilize 'pre-fixed' nitrogen in the form of dissolved species, with a preference for nitrate, nitrite and ammonium. Since the cellular nitrogen is usually in a low oxygen state mostly associated with proteins, taking up reduced forms of nitrogen such as ammonium minimizes a cell's energy use. Hence ammonium is usually the preferred nutrient source for phytoplankton, but it is often present at very low concentrations relative to nitrate, and hence nitrate uptake may dominate in such circumstances. Some phytoplankton can also utilize at least some forms of DON, such as urea and amino acids, to satisfy their nitrogen demands.

The input of fixed nitrogen to the oceans by in situ fixation and by fluvial and atmospheric fluxes is an important control on marine productivity on long time-scales, and hence on ocean–atmosphere $CO_2$ exchange, which affects global climate. However, most nitrogen for photosynthesis is supplied by internal cycling within the oceans. The utilization of the fixed nitrogen by phytoplankton takes place in the euphotic zone and some of the nitrogenous nutrients are released in a soluble form within this zone. The remainder of this particulate organic matter is transported out of the euphotic zone by sinking particulate matter, and a large fraction of this nitrogen is then released back into the solution at depth. The dissolved and particulate organic nitrogen (DON and PON) released in the water column at depth, or in the euphotic zone, is remineralized via bacterial mediation unless it is taken up directly by phytoplankton within the euphotic zone. The final inorganic end-product of remineralization is nitrate, the thermodynamically stable form of nitrogen within the well oxygenated ocean waters. The oxidation of ammonium to nitrite and then nitrate is termed nitrification, and is mediated by nitrifying bacteria. The reverse process, which is termed denitrification, is mediated by denitrifying bacteria, mainly in anoxic sediments and ocean regions of low water column dissolved oxygen (see later), the dominant mechanism for the removal of fixed nitrogen from the biosphere. This loss was until recently assumed to occur by a single process but it is now clear that there are several processes by which nitrogen gas can be formed; classical denitrification (Equation 9.2) and a process called annamox (Equation 9.3) as well as possibly some other reactions (Arrigo, 2005; Brandes *et al.*; 2007, Jickells and Weston, 2011). Classical denitrification is actually a multistep process involving first reduction of nitrate to nitrite ($NO_2^-$) then to

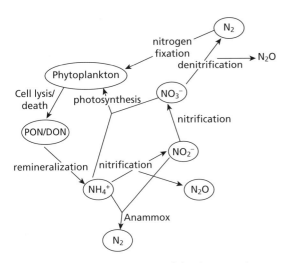

**Fig. 9.1** A simplified representation of the nitrogen cycle.

$N_2O$ and finally to nitrogen gas. Leakage from the last two steps results in the release of the greenhouse gas $N_2O$ as does oxidation of ammonium to nitrite. Equation 9.2 therefore represents a simplified description of the overall denitrification process.

The nitrogen cycle is summarized in Fig. 9.1.

$$4NO_3^- + 5CH_2O \rightarrow 2N_2 + 3H_2O + CO_2 + 4HCO_3^-$$
Classical Denitrification

(9.2)

$$NH_4^+ + NO_2^- \rightarrow N_2 + H_2O$$ 

(9.3)

Anammox

*Phosphorus nutrients.* The overall chemistry of phosphorus is simpler than that of nitrogen with only one oxidation state available. There are a variety of forms of phosphorus in seawater. These include dissolved inorganic phosphorus (predominantly orthophosphate ions, $HPO_4^{2-}$ but also other acid dissociation products of phosphoric acid with the proportions dependent on pH), dissolved organic phosphorus (DOP) and particulate phosphorus; however, phytoplankton normally satisfy their phosphorus requirements by direct assimilation of orthophosphate, although there is evidence that they can also utilize some of the DOP (Paytan and McLaughlin, 2007). Orthophosphate is often called simply phosphate and all analytical techniques measure all the acid dissocia-

tion products of phosphoric acid. Phosphate, like nitrate, is also released back into the water column during the oxidative destruction of organic tissues. Most of the regeneration of phosphorus probably takes place via bacterial decomposition, which leads to the formation of orthophosphates.

*Silicon nutrients.* Some phytoplankton and zooplankton such as diatoms and radiolarians require silicate for opal shell formation. Hence the supply of silicon influences the growth of some groups of organisms and not others. Silicon is therefore important in regulating the species composition, but not necessarily the overall productivity of the oceans. It is supplied to the oceans in both dissolved and particulate forms via river run-off, atmospheric deposition and glacial weathering (especially from the Antarctic), and dissolved silicon in seawater is probably present as orthosilicic acid, $Si(OH)_4$, which is often referred to as silicate. The particulate forms of the element include a wide variety of silicon and aluminosilicate minerals, together with diatom and radiolarian shells, which contain silica in the form of opal (see also Section 15.2). Opaline silica from the skeletal or hard parts of the organisms is released back into the water column by simple physical dissolution without any direct bacterial involvement since the waters of the oceans are undersaturated with respect to opal. However, the coating of inorganic opal by biological material such as residual cell membranes is known to affect its dissolution, so even this essentially chemical dissolution is bacterially mediated. Aluminosilcate material delivered to the oceans tends to have a very low solubility and is not available for phytoplankton growth, so the cycle is dominated by silicate supply and opal cycling. The marine Si cycle has been comprehensively reviewed by Ragueneau *et al.* (2000).

When considering the macronutrients in terms of their oceanic chemistries, it is useful to distinguish silicate from nitrate and phosphate. The reason for this is that both the latter two nutrients are involved in nutrition and are incorporated into soft tissues, whereas silicate is used only in the building of hard skeletal parts. None the less, all three nutrients are initially removed from solution by organisms, mainly phytoplankton, during primary production in the euphotic zone.

### 9.1.2.2 New production

An important concept to emerge in the nutrient field has been the distinction between new and regenerated production (see e.g. Dugdale *and* Goering, 1967; Eppley *and* Peterson, 1979). This concept is related to the way in which nutrients are supplied to the euphotic zone, and the processes involved can be illustrated for nitrogen. The supply of nitrogenous nutrients required during primary production can be related to two different types of sources:

1 A 'new' supply to the euphotic layer from river run-off, atmospheric deposition, upwelling of deep-water , mainly in the form of nitrate, and nitrogen fixation.

2 A regenerated supply from the short-term recycling processes within the euphotic layer itself. This supply is normally assumed to be mainly in the form of ammonium, with lesser amounts of urea and amino acids, arising from the excretory activities of animals and the metabolism of heterotrophic microorganisms, that is, this supply is derived from phytoplankton via the food web. This assumption reflects a view that phytoplankton preferentially take up reduced forms of N (e.g. ammonium), and that uptake rates are much faster than are nitrification rates to nitrate. This assumption has allowed the measurement of the relative rates of 'new' and 'regenerated' production based on uptake rates of isotopically labelled nitrate and ammonium by water column communities during incubations of water samples, with nitrate uptake equated to 'new' production. However, recent estimates of rather rapid nitrification rates (Yool *et al.*, 2007) in the surface waters of the ocean now challenge the validity of estimates of new and regenerated production made in this way, as will the supply of fixed nitrogen from outside the euphotic zone that is not present as nitrate, for example, atmospheric inputs of ammonium and organic nitrogen.

The concept of new and regenerated production, ignoring nitrification, is illustrated diagrammatically in Fig. 9.2. Despite potential complications associated with the newly recognized importance of nitrification, this concept has proven useful, particularly in the linkage of the carbon and nitrogen cycle. Eppley *and* Peterson (1979) pointed out that in an ideal closed system, with steady-state standing stocks and fluxes, the cycling of nutrients through an enclosed food web could continue indefinitely. In the real ocean system, however, there is a loss of organic material from the euphotic zone to deep water, for example, by the sinking of faecal pellets and 'marine snow' (see Section 10.4), and this is compensated by the input of new nutrients into the euphotic zone. The assumption is that the system is in steady state (i.e. inputs balance losses) over some time scale, often assumed to be annual. If such balance does not exist,

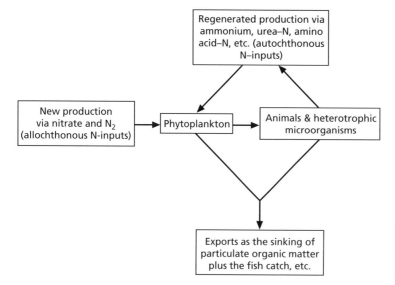

**Fig. 9.2** The production system of the surface ocean, illustrating the concepts of new and regenerated production (from Eppley and Peterson, 1979).

primary production, and or, fixed nitrogen concentrations will change over time, that is, the system is not at steady state. We will consider the long term steady state assumption later, but here it should be noted that steady state on an annual timescale does not necessarily imply a steady state balance on shorter time scales. Primary production associated with the newly available nitrogen (e.g. nitrate) is termed new production, and that resulting from the nitrogen recycled in the euphotic zone (e.g. ammonium) is referred to as regenerated production. Thus, the organic matter sinking out of the euphotic zone (termed export production) is equivalent to the new production (Martin *et al.*, 1987; Falkowski *et al.*, 1998); that is, the new production is that part of the primary production which is available for export, and it is this that drives the downward flux of organic matter (the global carbon flux) to deep waters (see Sections 9.4 and 10.4), an important issue since this represents a source of carbon to deep water and a sink for atmospheric $CO_2$. New production, as a percentage of the total primary production, ranges from less than ~10% in the oligotrophic waters of the subtropical gyres to greater than ~20% in coastal upwelling regions (see Section 9.1.6), that is, most of the primary production in surface waters is driven by internal recycling (Bruland, 1980; Falkowski *et al.*, 1998). However, during the spring bloom (see later) the primary source of nitrate is often that supplied from outside the euphotic zone during deep winter water column mixing and hence, is characterized by a relatively high proportion of new production. Regenerated production arises from nutrients that undergo short-term recycling in the euphotic zone. In contrast, new production requires the input of new nutrients into the euphotic zone (Falkowski *et al.*, 1998); for example, from (i) atmospheric deposition, (ii) a deep-water flux via eddy diffusion and vertical advection and, for nitrogen, and (iii) fixation (see e.g. Jenkins *and* Goldman, 1985; Duce, 1986).

### 9.1.3 Primary production and the distribution of nutrients in the upper oceans

As discussed in Chapter 7, the water circulation patterns in the ocean are different above and below the permanent thermocline. Physical transport processes have a profound impact on nutrient distributions, so

it is useful to separate surface and deep water nutrient distributions, although the underlying biogeochemical processes operating are similar throughout the oceans.

The measurement of macronutrient concentrations has been undertaken for many years by many nations on research cruises. The provision of agreed standards and other intercomparison programmes has allowed the data to be pooled within international data bases to allow the large scale three-dimensional distributions of the macronutrients to be described.

In the upper layers of the ocean the role of primary production is central to the cycling of nutrients. Nutrients are taken up in the euphotic zone and released back into seawater following the remineralization of sinking detritus at depth. The nutrient concentrations thus build up in deeper water from where they can be brought to the surface again via physical processes, such as upwelling and the seasonal changes in depth of the wind-mixed upper layer of the ocean. This process of nutrient and carbon transport from the surface to deep ocean is sometimes termed the 'biological pump'. The overall effect of these processes is to generate vertical oceanic nutrient distributions that exhibit a characteristic 'surface-water depletion, deep-water enrichment' profile (Fig. 9.3). The extent of the surface depletion depends on the balance of supply and demand and a particularly important issue is the extent to which deep waters rich in regenerated nutrients are returned to surface waters. In the highly stratified tropical ocean gyre regions (Chapter 7), such resupply is very limited and inorganic nutrient concentrations fall to concentrations that are barely detectable even with modern highly sophisticated analysis, although the concentrations of dissolved organic N and P may still be detectable (see later). Under such conditions the rates of supply of inorganic nutrients comes to limit phytoplankton growth.

Vertical water-column profiles are of particular interest in marine geochemistry because they offer clues to the main processes controlling the cycling of a particular constituent. Components showing surface water depletion and deep water enrichment are said to show nutrient-like behaviour reflecting a dominant role for biological cycling. Furthermore, nitrate and phosphate can be used as analogues for other elements that are taken into organic tissue phases, and silicate can be used as an analogue for

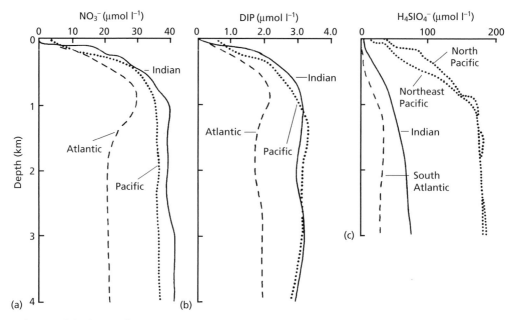

**Fig. 9.3** The vertical distribution of nutrients in the ocean basins (based on Andrews *et al.*, 2004).

elements incorporated into the skeletal parts of organisms. In this way it is possible to distinguish between organic tissue and skeletal trace-metal-particulate carrier phases; this type of approach is discussed in detail in Section 11.3.

### 9.1.4 Primary production: plankton

Plankton are microscopic free-floating, or weakly swimming, organisms that are passively distributed by ocean currents. Eukaryotes are unicellular, or multicellular, organisms having nuclear membranes and DNA that is arranged in the form of chromosomes. Prokaryotes (including bacteria and archea) are unicellular organisms that do not possess nuclear membranes and do not have their DNA confined within a cell nucleus. Plankton are subdivided into phytoplankton (the 'grass' of the sea) and zooplankton (the 'grazers' of the sea). Phytoplankton are autotrophs; that is, they manufacture their organic constituents from inorganic compounds and do not rely on other organisms as an energy source. Important members of the phytoplankton include diatoms (with opal skeletons), cocolithophores (with calcium

carbonate skeletons), silicoflagellates (opal skeletons) and dinoflagellates (which have organic membranes but no mineral skeleton), all of which are eukaryotes (see Fig. 9.4).

The only major group of autotrophic prokaryotes are the cyanobacteria. These cyanobacteria are unique in the oceans because they can fix dissolved gaseous nitrogen ($N_2$), whereas other phytoplankton can only utilize 'pre-fixed' forms of nitrogen (see Section 9.1). Zooplankton are heterotrophs, that is, they obtain energy supplies and organic substrates by feeding directly, or indirectly, on autotrophs. The zooplankton include copepods, pterpods, radiolarians and foraminifera; the latter two are protists (i.e. unicellular eukaryotes).

Marine autotrophs (e.g. the algal biomass) are primary producers. The photo-autotrophic biomass (phytoplankton) is the most important primary source of organic carbon in the oceans, and primary production is the initial stage in the marine food chain, which subsequently involves a number of trophic levels. For example, herbivorous zooplankton (the 'grazers') consume the phytoplankton (the 'grass') during secondary production. These

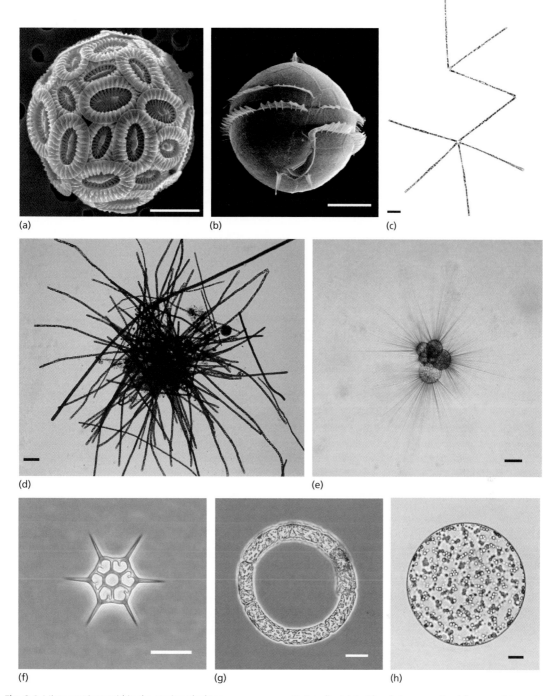

**Fig. 9.4** Micro-orgaisms within the marine plankton.
a) *Syracosphaera molischii*, a coccolithophorid;
b) *Protoperidinium cerasus*, a heterotrophic dinoflagellate;
c) a chain of cells of the pennate diatom *Thalassionema frauenfeldii*; d) a filamentous colony of filaments of the cyanobacterium *Oscillatoria*. The filaments form a low oxygen microclimate in the centre of the colony which allows the fixation of nitrogen; e) the foraminiferan *Globigerina bulloides*; f) half of the siliceous skeleton of the silicoflagellate *Dictyocha speculum*; a living cell is sandwiched in between two of these plates; g) part of a coiled chain of cells of the diatom

*Guinardia striata*. The chains can extend for several tens of cells in a long coil; h) the spherical non-motile stage of the prasinophyte *Halosphaera minor*. Scale bars: a) 1 μm; b) 10 μm; c), f), g), h) 20 μm; d), e) 200 μm. a), b) Scanning electron micrographs; c), d), e), h) Light microscope, bright field; f), g) Light microscope, phase contrast. a), b) from samples preserved in formalin; c), d), f), g) From samples preserved in Lugol's Iodine; e), h) Live specimens. Picture credits: a) Pauhla McGrane; b)–h) Robin Raine. (See Colour Plate 3) herbivorous zooplankton are, in turn, fed upon by carnivorous zooplankton and fish species (tertiary production). There are also specialized chemosynthetic organisms associated with mid-ocean ridges (Chapter 5), but these are not globally significant in the marine carbon cycle.

A number of size-based classifications have been proposed for the planktonic community. The size categories identified by Lalli *and* Parsons (1993) include the following: femtoplankton (0.02–0.2 μm), picoplankton (0.2–2.0 μm), nanoplankton (2.0–20 μm) and microplankton (20–200 μm). Viruses are part of the femtoplankton, and bacteria are placed in the picoplankton. Both bacteria and viruses are involved in the degradation of primary production and respiration. It was the identification of the nanoplankton, however, which changed our view of the oceanic plankton community. Because of their relatively small size, the nanoplankton were often missed by the early collection techniques, and it is now thought that small phytoplankton and zooplankton are responsible for a large fraction of oceanic primary production and its consumption in secondary production

The process of photosynthesis, in which organic compounds are synthesized from inorganic constituents present in seawater during the growth of marine plants (phytoplankton), is usually termed primary production, although this is not a biochemically rigorous definition. In the process, which includes the absorption of solar energy and the assimilation of carbon dioxide, the chlorophyll-containing plants synthesize complex organic molecules from inorganic starting materials, the principal products being oxygen and food substances such as carbohydrates (see Equation 9.1). It is now the convention to distinguish between new primary production, which arises from nutrients transported into the euphotic zone (e.g. from upwelling and atmospheric deposition), and regenerated production, which utilizes nutrients that are recycled within the upper layers of the ocean (see Section 9.1.2).

Photosynthesis requires light in the wavelength range 370–720 nm, and even in clear tropical waters only ~1% of the visible light energy penetrates to a depth of ~100 m. Photosynthesis is therefore generally restricted to the upper 100 m or so of the open-ocean water column, although it may, in fact, be inhibited at the surface layer itself because of the effects of high light intensity, and as a result the zone of *maximum* production is often found a few metres below the surface. Gross primary production decreases with depth, but the loss of carbon through respiration remains constant with depth because phytoplankton respire during both the hours of sunlight and darkness. The depth at which gross primary production and respiration balance is termed the compensation depth, and at this position net primary production is zero. Net primary production is positive in the water column above the compensation depth, and this is the region referred to as the euphotic zone. In the aphotic zone, which lies below the compensation depth, respiration exceeds gross production and there is a net loss of organic material.

### 9.1.5 Primary production: seasonality and controls

In its simplest form a limiting factor for a specific activity may be defined as the critical minimum required to sustain that activity. In addition to light availability, the factors that limit primary production include nutrient availability and zooplankton grazing. Primary production in the oceans is thus controlled by a complex combination of physical (e.g. light, temperature), biological (e.g. growth rate) and chemical (e.g. availability of nutrients) variables.

The availability of nutrients can limit phytoplankton growth and the limiting nutrients were traditionally considered to be the macronutrients nitrate and phosphate. Other chemicals are required for growth but these are usually present in relatively high concentrations compared to the amounts needed by phytoplankton, so are unlikely to limit primary production. Silicate can limit diatom growth, and hence species abundance and possibly carbon export rates, but not overall productivity. Since some phytoplankton species are capable of nitrogen fixation and hence can relieve nitrogen limitation, it has been argued that the long term limitation on ocean productivity over geological timescales is phosphorus supply. However, on shorter timescales direct experiments often suggest phytoplankton production is limited by nitrogen supply (Falkowski *et al.*, 1998; Mills *et al.*, 2004) In addition, there is a growing realization of the importance of the micronutrients, especially iron, in nutrient limitation (see later). It must be stressed, however, that the concept of a limiting nutrient strictly refers only to new production, because productivity can be maintained in the presence of low nutrient concentrations by recycling. Most phytoplankton have an absolute requirement for one or more nutrients, but the concept of a 'limiting nutrient' is complex. Nutrient availability can regulate rate processes, such as photosynthesis, or the final yield of a plant crop. In practice, the distinction between the two concepts have often been blurred with the result that nutrient limitation is sometimes applied to growth rates of phytoplankton and sometimes to the limitation of the standing crop, although the limitation of the standing crop is not necessarily accompanied by a severe limitation of growth rates (see e.g. Cullen *et al.*, 1992). Nutrient depletion can lead to low concentrations of several nutrients and co-limitation of production by the availability of several nutrients and light may be the norm (Arrigo, 2005).

Inputs of nutrients from outside the oceans were discussed in earlier chapters (e.g. rivers and the atmosphere) and these help set the long term availability of nutrients in the system over very long timescales of tens of thousands of years. However, on shorter timescales it is the supply of nutrients to nutrient-depleted surface waters from the deep ocean, where nutrient concentrations are relatively high, that dominates the nutrient supply. The balance of nutrient demand by phytoplankton and supply via physical mixing sets the stage for the broad patterns of spatial and seasonal variations in primary production in the oceans. These are described below, building up an argument from first principles that can then be compared to the actual patterns observed. Throughout this discussion we will consider the role of inorganic nutrients only. In inorganic nutrient depleted surface waters, there are often significant amounts of dissolved organic nitrogen and phosphorus (DON and DOP). However, the bulk of this DON and DOP appears to be unavailable to phytoplankton of short time scales, although specific compounds within the DON and DOP may be used. The DON and DOP are discussed further in Section 9.4.

As noted earlier, in highly stratified tropical gyres, nutrient supply from below is low and both total primary production and new production rates are relatively low. Seasonality is not well developed in such environments, even the passage of major storms such as hurricanes rarely produce mixing of surface waters with sub-euphotic zone waters. As a result tropical ocean gyre waters are characterized by relatively low rates of primary production throughout the year. These waters are also characterized by intense euphotic zone nutrient recycling (low 'new production').

Phytoplankton take up nutrients across their cell walls. If we simplify phytoplankton geometry to spheres, then it is clear that smaller spheres (phytoplankton) have relatively more surface area to their volume than larger spheres. This larger surface area : volume ratio should make it easier to take up nutrients at low concentrations across the cell wall. Although the shapes of phytoplankton are more complex than simple spheres, in low nutrient waters small phytoplankton appear to be favoured probably because they are better able to take up nutrients at low concentrations into the cell by diffusion. Hence tropical ocean gyre waters can be characterized as having low primary production rates (sometimes described as oligotrophic) that persist throughout the year and a community dominated by small phytoplankton with intense grazing and recycling leading to little export of carbon.

In high latitude waters conditions are very different. In winter cooling of the water column eliminates water column density stratification allowing wind driven mixing to supply deepwater nutrients to the surface waters. However, light levels are low (or even zero at very high latitudes) and phytoplankton are also mixed deep into the water column by the same mixing processes that affect the waters themselves, severely light limiting photosynthesis. In spring, the water column begins to warm and stabilize allowing phytoplankton in the euphotic zone access to improved light levels in a nutrient rich environment. Under such conditions phytoplankton can grow very fast, often doubling their biomass on a daily basis. Initially at this time grazer numbers are low, restricted by a lack of food supply over the winter and inherently slower growth rates than the algae. This situation allows a very rapid increase in phytoplankton abundance known as the *spring bloom*, and this annual greening of the temperate ocean can be clearly seen in satellite imagery (Fig. 9.5). This spring bloom soon begins to exhaust the nutrient supply and grazer numbers begin to increase and the combination of

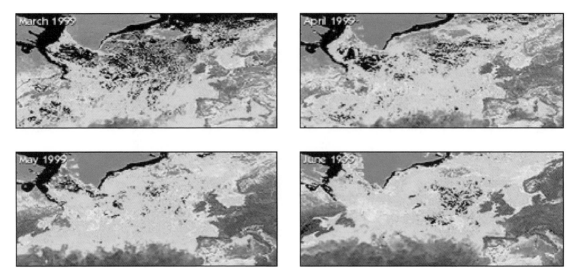

**Fig. 9.5** Satellite images of the seasonality of North Atlantic phytoplankton stocks in surface waters based on satellite imagery. (See Colour Plate 4)

both processes mean that the spring bloom ends. Primary production continues through the summer season albeit at a lower rate fuelled mainly by recycling; rather like the situation in the tropical gyres. In some locations, water column cooling and increased winds in autumn allow some mixing of nutrients while light levels still allow rapid growth, and in such circumstances an autumn bloom can occur, before the winter conditions return. The seasonality of temperate waters is therefore very different to that of tropical waters. There are also other differences driven by the same balance of physical processes. Under spring bloom conditions, phytoplankton with fast growth rates are favoured. Small size is not an advantage because nutrients are not limiting and large phytoplankton capable of fast nutrient uptake and a relative resistance to grazing are favoured. As a result, the spring bloom in many areas is dominated by large diatoms. Such phytoplankton have high sinking rates and hence are readily lost from the euphotic zone when they die, leading to export of carbon and nutrients, a loss that is balanced on an annual basis by the resupply from deep winter mixing. Thus the temperate oceans are characterized by strong seasonality in production, species composition and recycling efficiency. At intermediate latitudes, the seasonality is intermediate, and in some regions there is a small autumn bloom driven

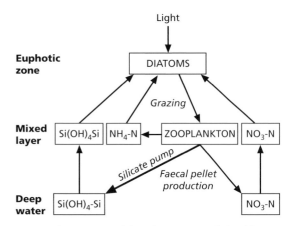

**Fig. 9.6** The operation of the 'silicate pump'. Idealized flow diagram illustrating the flow of nitrogen and silicon ('silicate pump') in a diatom-dominated grazed system (after Dugdale *et al.*, 1995). In this type of system the 'silicate pump' for the export of silicate to depth occurs via zooplankton faecal pellets following diatom phytoplankton grazing. Reprinted with permission from Elsevier.

by mixing. Very simplified versions of these contrasting seasonal cycles are illustrated in Fig. 9.7 later.

Dugdale *et al.* (1995) have described the operation of a 'silicate pump' (Fig. 9.6) which operates in diatom dominated communities to compare the relative efficiency with which organic material produced by photosynthesis in such phytoplankton communi-

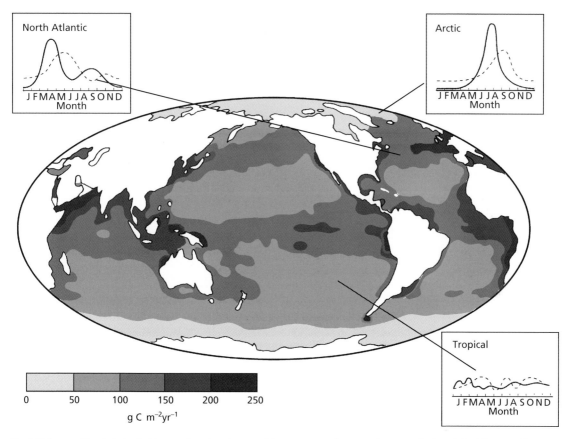

**Fig. 9.7** Seasonality of phytoplankton at three locations overlain on a map of ocean primary production derived from satellite imagery from Antoine *et al.* (1996). Solid lines represent phytoplankton seasonality and dashed lines zooplankton.

ties is exported to deep water, with the oligtrophic ocean gyres where recycling of nutrients within the euphotic zone dominates. The efficiency of the export of organic matter reflects in part the relatively large size of diatoms and their rapid sinking, since the efficiency of export represents a balance between rates of sinking and rates of organic matter degradation. However, a more general role for mineral 'ballast' of any sort (calcium carbonate or opal biogenic material or even mineral dust) has been suggested. In this proposal it is suggested that the ballast encourages the more rapid sinking of material to the deep ocean and hence the more efficient export of organic carbon, although field evidence to support the importance of ballast is equivocal (e.g. Thomalla *et al.*, 2008).

This seasonality analysis describes the theoretical two end members of an ocean that in reality is a gradient from polar waters where complete darkness eliminates all winter production to the centre of the ocean gyres. Between these two extremes lies a gradient which is seen in the relative magnitude and timing of the spring bloom, the significance of the export of primary production to deep water and the species composition (see Fig. 9.5 and Fig. 9.7).

In regions of regular deepwater upwelling (see Chapter 7) such as off the coast of California, Peru, Namibia and North Africa, nutrient supply and light levels are suitable for sustaining primary productivity throughout the year leading to continued high production conditions similar to the spring bloom in all seasons. These areas are some of the

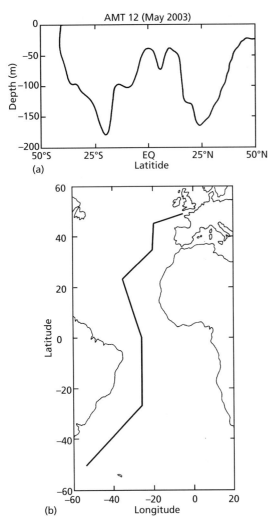

(a)

(b)

**Fig. 9.8** Vertical distribution of nitrate in the upper ocean (a) along a north south transect in the Atlantic Ocean. Line represents boundary of upper very low nitrate water with deeper nitrate rich water drawn at 1 micromole l⁻¹ (b) based on Robinson et al. (2006). Reprinted with permission from Elsevier.

most productive of deep ocean environments and sustain some of the world's major oceanic fisheries. Along the Equator there is also an upwelling (see Fig. 9.8) and this leads to enhanced production. In coastal waters, enhanced nutrient supply from on land and from upwelling, coupled to limited loss to deep water of sinking organic matter because the shelf waters are shallower, also allows rather high primary production. In coastal waters the seasonal forcing by light and mixing still creates a gradient from highly

seasonal systems at high latitude with a strong spring bloom, to low seasonality in tropical systems.

### 9.1.6 High nutrient low chlorophyll regions and iron limitation

To date, we have identified two major ocean biogeochemical regions;
**1** the ocean gyres with low standing stocks of algae and production that continues throughout the year, and
**2** the high latitude seasonally variable systems; although the reality is a gradient between these extremes.

For many years oceanographers puzzled over other areas of the ocean such as the Southern Ocean where despite seasonally abundant light, nutrient levels remain high and phytoplankton production low suggesting production limited by something beside the macronutrients and light. Such areas have become known as 'high nutrient low chlorophyll' or HNLC regions. These regions also include the temperate North Pacific and the tropical central east Pacific, but by far the biggest and most important HNLC region is the Southern Ocean which is almost entirely HNLC in character.

There have been various ideas advanced to explain the HNLC regions, including a role for light and grazer limitation. While these factors probably play a role, research over the last 20 years has clearly demonstrated the key role of iron limitation. Iron, a constituent of many oxidizing metalloenzymes, respiratory pigments and proteins, is essential for life. Phytoplankton require iron for the synthesis of chlorophyll and nitrate reductase, and the idea that iron can act as a limiting nutrient on primary production is not a new one. For example, in his review article on the topic, Martin (1992) quotes references to the importance of iron in limiting phytoplankton production as far back as the 1930s. However, there are a number of problems with respect to evaluating the importance of iron as a limiting nutrient. One of the most important of these is the low concentrations of iron in open-ocean waters and the associated difficulties of making accurate measurements of its concentration, and avoiding contamination during manipulation of samples. Improvements in analytical techniques and contamination control have revolutionized our understanding of metal chemistry in the

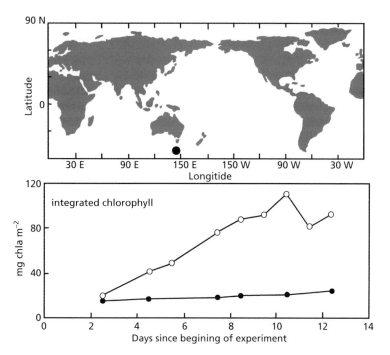

**Fig. 9.9** Results from the SOIREE experiment in which iron is added to a patch of seawater within the Southern Ocean HNLC iron limited waters at a site indicated by dot on the map. The graphs show the chlorophyll concentrations (a measure of phytoplankton biomass) within (open circles) and outside (filled circles) the area to which Fe was added (Boyd *et al.*, 2000).

oceans as will be discussed in Chapter 10. These improvements have allowed the reliable measurements of iron concentrations. Research into the role of iron has also catalysed the development of many new techniques in ocean biogeochemistry. Perhaps the most important of these is the development of techniques to conduct in situ experiments in open ocean waters, such as the addition of iron, and the tracking of the water (and the biological response within that water to the iron addition) as it subsequently moves around under the influence of wind and currents (Fig. 9.9). Such experiments get around problems with adding iron, or other substances, to relatively small sample bottles where the plankton community is already perturbed by confinement. Open ocean in situ iron addition experiments have unequivocally demonstrated that iron availability is the limiting factor for primary production in HNLC regions (Falkowski *et al.*, 1998; Jickells *et al.*, 2005; Boyd *et al.*, 2007). Furthermore it is clear from iron addition experiments that as well as increasing productivity overall, iron addition selectively favours larger faster growing diatoms and hence also increases carbon export to deep water. In many parts

of the HNLC regions, iron may co-limit primary production with other nutrients or light, but it is clear that iron supply is critical, as illustrated in Fig. 9.9.

As discussed in Chapter 4, the main source of iron to the oceans is associated with crustal dust and this is derived from desert regions and carried by the prevailing wind patterns: see Fig. 4.4. Most deserts are in tropical Northern Hemisphere regions, and hence the largest HNLC regions associated with very low dust supply which leads to iron limitation are in the Southern Ocean remote from such sources. Iron limitation can be relieved locally by iron release from the sediments of islands or continental margins, leading to enhanced production downstream of islands (Boyd *et al.*, 2007). Elsewhere the production is dependent on an iron supply from deepwater and this is usually inadequate in the HNLC regions to allow phytoplankton to grow fast enough to remove all the macronutrients; hence the HNLC status.

We know from the paleo record that dust levels were higher during the peak of the ice ages due to higher winds and drier climates, and it has been argued that the resulting increasing productivity could have been responsible for the drawdown of atmospheric $CO_2$ and hence created the ice age. More recent studies suggest that the dust supply changes could not in themselves have triggered the ice age but the increased dust associated with the ice age climate did contribute significantly to the lower $CO_2$ and the colder climate (Jickells *et al.*, 2005).

Iron also plays another important role in ocean biogeochemistry. The enzymes required for nitrogen fixation have a very high iron requirement. Hence iron supply can affect nitrogen fixation rates. Nitrogen fixation is confined to warm highly stratified waters and so the atmospheric dust supply of iron is particularly important, although P supply can also be limiting. The surface water phosphate concentrations in the tropical North Atlantic are lower than in other ocean gyres and it has been suggested that this may reflect increased Fe supply promoting nitrogen fixation and an increased uptake of phosphorus (Mahaffey *et al.*, 2005).

### 9.1.7 Primary production: global patterns

The geographical distribution of primary production in the oceans is illustrated in Fig. 9.7. This figure is derived from modelling and satellite imagery. Earlier estimates were based on field measurements which are time consuming and expensive and hence rather limited. Thus the satellite and model derived estimates are greatly preferable but are indirect and a particular limitation is that the satellites only see the surface layer of the oceans. If productivity is higher at depth then it may not be completely represented. This situation can arise because of photo-inhibition of productivity in the surface layer or because light penetrates to a depth where nutrient supply is rather higher and hence production is increased despite declining light levels.

Annual ocean primary production was estimated by Antoine *et al.* (1996) on the basis of the data in Fig. 9.7 to be $33\,Gt\,C\,yr^{-1}$ (Table 9.1), although other estimates are closer to $50\,Gt\,C\,yr^{-1}$ (Falkowski *et al.*, 1998; Tables 9.2 *and* 9.5). Any such estimate is necessarily uncertain and primary productivity can vary from year to year, but in subsequent discussions we will assume a productivity of $50\,Gt\,C\,yr^{-1}$. Antoine *et al.* (1996) also estimated productivity by 'ocean province' (Table 9.1) with their oligotrophic, mesotrophic and eutrophic classification approximating to gyre + HNLC, high latitude seasonal systems and coastal + upwelling respectively. This compilation emphasizes that the oligotrophic gyres are the largest ocean biogeochemical province, but that the higher latitude regions of seasonally varying production are more productive. The relatively small coastal and upwelling regions are the most productive per unit area.

The *f*-ratio concept (Section 9.1.2) was introduced by Eppley *and* Peterson (1979) to express the proportion of new to total production; the *f*-ratio is highest in eutrophic upwelling zones (values as high as ~0.8) and lowest in oligotrophic zones (values as low as ~0.1). Hence the proportion of the export of organic matter and the 'new' production (f ratio) increase with increasing primary production (Dunne *et al.*, 2007) or looked at another way, the low production systems are characterized by very efficient internal recycling. This pattern increases the importance of mesotrophic and eutrophic regions for the export of carbon and nutrients to deep waters (see also Section 9.4.5).

**Table 9.1** Oceanic primary production based on the division of the oceans into provinces according to their annual mean levels of chlorophyll concentration in the upper layer (Antoine *et al.*, 1996).

| Province | Chlorophyll $\mu g\,l^{-1}$ | Area % | % | Production $Gt\,C\,yr^{-1}$ | $g\,C\,m^{-2}\,yr^{-1}$ |
|---|---|---|---|---|---|
| Oligotrophic | <0.1 | 56 | 44 | 14.5 | 91 |
| Mesotrophic | 0.1–1 | 42 | 48 | 15.7 | 131.5 |
| Eutrophic | >1 | 2.4 | 8.5 | 2.8 | 422 |
| Total | | 100 | 100 | 33 | |

## 9.2 The distribution of nutrients in the deep oceans

### 9.2.1 Redfield ratios

In a classic study, Redfield (1934; 1958) showed that the concentrations of the major nutrients such as nitrate and phosphate, are present in a constant ratio one to another in seawater despite large variations in concentration with both depth and location. This ratio is also similar to the average N:P ratio (stoichiometry) in phytoplankton a relationship that has become known as the *Redfield ratio*. This observation has been interpreted as demonstrating the remarkable ability of phytoplankton, which are just a few μm across and only live for a day, to ultimately control the chemistry of the oceans. The Redfield ratio has been extended to include other elements including C, O, Si and even Fe and the constancy first described by Redfield more than 75 years ago has been largely confirmed by subsequent study (see Fig. 9.10). However, there are two general exceptions to this.

1 In anoxic regions in which nitrate is used in the destruction of organic matter (see Chapter 14), phosphate can increase with a corresponding decrease in nitrate.

2 Nitrate concentrations can approach, and sometimes reach, zero concentrations in nutrient-starved regions that have low concentrations of phosphate (see Section 9.1.5).

The overall constancy of the Redfield ratio has proven very useful for modelling and calculating relationships between nutrients and carbon, but the fundamental reasons why it should be the value that it is (N:P, 16:1) is not clear, particularly since there are many process that can alter the ratio, for example nitrogen fixation. It is now known that the ratio of N:P in cells is variable between species and within species, depending on the conditions under which they are growing (Arrigo, 2005). The actual value of the Redfield ratio therefore really represents the average of the whole phytoplankton and bacteria population and the global constancy reflects the long residence times of nutrients and carbon within the oceans (~$10^4$ years, see below) relative to the internal ocean water mixing times (~$10^3$ years) meaning that mixing will average out most local variability (Falkowski and Davis, 2004). This also implies that the Redfield ratio can change from a particular value if the balance of cycling processes for individual components change.

*Nitrate, phosphate and silicate.* Vertical water-column profiles of nutrients in the major oceans are illustrated in Fig. 9.3.

There are considerable general similarities between the profiles of the nutrients which can be identified as follows:

1 A surface layer in which nitrate and phosphate are heavily depleted by biological uptake.

2 A layer in which the concentrations increase rapidly with depth as a result of their regeneration from the sinking biomass. Sometimes a layer of maximum concentration is found for both nutrients between ~500 and ~1500 m.

3 A layer in which the concentrations are relatively high and increase only slowly or not at all with depth.

Although silicate is incorporated into the shells of organisms, and not their soft tissues, it still has a nutrient-type distribution profile in the water

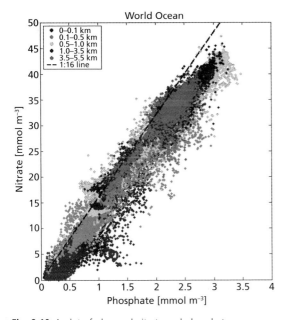

**Fig. 9.10** A plot of observed nitrate and phosphate concentrations in the world oceans producing the so called Redfield ocean N:P ratio, courtesy of Dr T. Tyrrell and based on Tyrrell (1999). (See Colour Plate 5)

column. The silicate profile, however, does differ from those of nitrate and phosphate in that the return of silicate to the water column by shell dissolution can occur at different depths to that of the soft tissue regeneration because it is driven by geochemical dissolution rather than bacterial regeneration. Since silicate is generally undersaturated everywhere in the oceans, silicon shell material dissolves at all depths, although this is a relatively slow process compared to bacterial degradation. Hence biogenic silica is regenerated more slowly than nitrate or phosphate, and silicate concentrations increase throughout the water column.

Based on data such as that in Fig. 9.3 we can estimate average nitrate, phosphate and silicate concentrations in the global ocean which are 30, 2.3 and $100\,\mu mol\,l^{-1}$, respectively (Broecker and Peng, 1982). This information can be used to estimate the residence time of these components in the oceans. This is the average time each nutrient spends in the oceans before removal. This calculation assumes steady state, that is, the inputs equal the loss, a common biogeochemical assumption but one that is rather difficult to test. The recent discovery of large amounts of marine nitrogen fixation and of new methods of nitrogen loss from the oceans have led to several re-evaluations of the marine nitrogen cycle and to debates about whether it is in steady state (Brandes et al., 2007).

The residence time is calculated by dividing the nutrient inventory by the input rate. Formalizing this mathematically;

Residence Time = Inventory/Input.

At the steady state by definition, input = output, so either could be employed but in practice it is usually easier to use input.

To illustrate the calculation, consider silicate with an average concentration in the oceans of $100\,\mu mol\,l^{-1}$, an ocean volume of $1.4 \times 10^{21}\,l$ and an input rate of $25.5 \times 10^{12}\,mol\,yr^{-1}$ (Chapter 6, summary table).

Si Residence Time = $5.5 \times 10^{3}$ years.

The residence times of P and N are then $14 \times 10^{3}$ and $2.2 \times 10^{3}$ years respectively. There are many uncertainties and approximations in such calculations, reflecting choices with regard to inputs, particularly such as whether to include all forms and inputs of nutrients, or only riverine dissolved nitrate, phosphate and silicate, in both inputs and inventories. However, all calculations demonstrate that residence times are long compared to ocean mixing times and that the residence time of N is significantly shorter than P and Si.

## 9.2.2 The horizontal distribution and circulation of nutrients in the oceans

In Chapter 7, attention was drawn to the fact that the chemical signals of non-conservative constituents in the oceans are controlled by circulation patterns that transport the constituents from one part of the ocean to another and mix water masses. Superimposed on this physical water transport are the effects of internal oceanic reactions arising from the involvement of the constituents in biogeochemical cycles. The nutrients provide a classic example of how these physical, geochemical and biological processes interact. The shapes of the vertical oceanic profiles of nitrate, phosphate and silicate described above were interpreted in terms of the involvement of the nutrients with biota. During this, they are removed from solution into particulate phases of the biomass, including both tissue and skeletal parts of organisms, in the upper ocean and are subsequently regenerated back into solution at depth, that is, a non-conservative signal. The horizontal distributions of the nutrients illustrate the control by circulation patterns. As demonstrated in Fig. 9.3 nutrient concentrations are higher in the deep waters of the Pacific than the Atlantic. This type of nutrient distributions can be related to deep-water circulation patterns (see Chapter 7), and Broecker (1974) has suggested that the following sequence governs the nutrient build-up in the deep waters of the Pacific. The surface water become relatively nutrient-depleted by phytoplankton growth and is cooled at high latitudes. It then sinks, moving away from the water mass formation zones toward the oceans deep interior (see Chapter 7). As this water flows through the deep ocean along the global conveyor belt (Chapter 7), its nutrient concentrations are increased by the productions of the decomposition of sinking particulate organic matter.

Thus the 'oldest' waters (since they were last at the surface) in the Pacific have the highest nutrient concentrations. The distributions of the nutrients in the oceans therefore are controlled by an interaction, or balance, between oceanic circulation and biological activity, over time scales of many hundreds of years.

Eventually mass balance requires the return of the deep water to the surface either via slow upwelling across the oceans, or in specific upwelling areas.

From Fig. 9.3 it is clear that the effect of this process is that the nitrate and phosphate concentrations in the deep North Pacific are approximately twice those in the deep Atlantic. In reality there is a gradient in deep water nutrient concentrations throughout the oceans as shown in Fig. 9.11. The increases in silicate concentrations along the global 'grand tour' are substantially larger than for nitrate and phosphate. This has been interpreted as reflecting the slower dissolution of silica with depth, meaning that more particulate silicon reaches the deep ocean relative to nitrogen and phosphorus which are bound within the more rapidly degraded organic matter and are degraded at shallower depths. While this difference in dissolution contributes to the differences in the behaviour of silicate and nitrate, it may not be the only reason for the greater enhancement of silicate relative to nitrate and phosphate between the Atlantic and Pacific. It has been suggested that the Southern Ocean plays a key role in the differentiation of the two elements. The Southern Ocean is an area of major opal sedimentation. This reflects both the importance of opal export in this region and the characteristic high Si/N ratio of this material due to iron limitation which encourages diatom growth with particularly large Si-rich frustules. This major recycling zone for silicate within the global deepwater cycling path may play an important role in creating the characteristic deepwater silicate distribution (Sarmiento *et al.*, 2007).

## 9.3 Dissolved oxygen in seawater

The vertical and horizontal distributions of oxygen in the oceans reflect a balance between: (i) input across the air–sea interface from the atmosphere; (ii) involvement in biological processes; and (iii) physical transport. The various factors that control the distribution of dissolved oxygen in the sea lead to a number of pronounced features in its vertical profiles. These are illustrated in Fig. 9.12, and can be summarized as follows.

**1** Oxygen in the surface, or mixed, layer is derived from exchange with the atmosphere (see Chapter 8) so that its concentration is determined largely by its solubility in seawater. Because of this, on a global

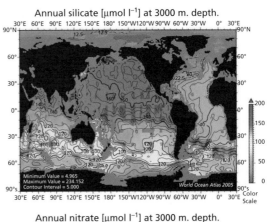

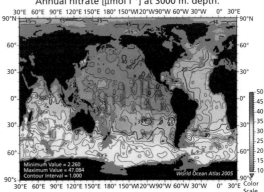

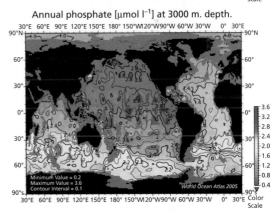

**Fig. 9.11** Map of nutrient concentrations (in the world oceans $\mu$mol l$^{-1}$) at 3000 m depth from *NOAA World Ocean Atlas*. (See Colour Plate 6)

basis the concentrations of dissolved oxygen in seawater are greater in cold high-latitude waters than in those from the warmer subtropical regions, because oxygen is more soluble at lower temperatures (Chapter 8). This atmospheric supply is

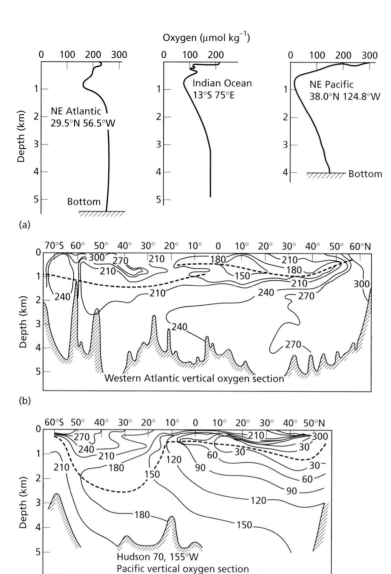

**Fig. 9.12** The vertical distribution of dissolved oxygen in the oceans (modified form Kester, 1975; which lists the original data sources). (a) Vertical profiles for the Atlantic, Indian and Pacific Oceans. (b) Vertical section-western Atlantic. (c) Vertical section – central Pacific. The broken lines in (b) and (c) indicate the depth of the oxygen minimum.

supplemented by oxygen released during photosynthesis, a process that may be represented simplistically by Equation 9.1, the key point being that the production of organic matter in photosynthesis produces oxygen. This process requires light so is restricted to the upper water column. The exact depth to which sufficient light penetrates is variable but usually the limit of photosynthesis is the upper 100 to 150 m. The products of photosynthesis are mostly recycled within the euphotic zone but some escapes and sinks to depth either as particles or as dissolved organic matter. The breakdown of this organic matter by bacteria goes on throughout the water column regenerating nutrients (see Section 9.2) and consuming oxygen.

The rate of oxygen exchange with the atmosphere is usually much faster than the rate at which the internal surface layer biogeochemical and physical processes take place. As a result, photosynthesis usually does not lead to an excess of oxygen in the surface layer; nonetheless, the surface ocean usually does have a slight oxygen supersaturation (~5%), which may result from the trapping of air bubbles (Broecker, 1974). Under some conditions, however, the exchange of photosynthetically produced oxygen with the atmosphere can be blocked (e.g. by seasonal stratification, a density cap formed by summer warming of the surface layer), with the result that photosynthesis can produce an oxygen supersaturation. For example, a summer subsurface shallow oxygen maximum (SOM) has been reported in a number of nutrient-poor oligotrophic marine regions (see e.g. Shulenberger and Reid, 1981). The development of this subsurface oxygen maximum can be used to help provide estimates of photosynthesis integrated over the time of seasonal stratification (Jenkins and Goldman, 1985). Such integration is particularly valuable because conventional methods provide direct estimates of photosynthesis only over timescales of hours, and extrapolation from such short term spot measurements to regional budgets is necessarily very uncertain.

2 Below the zone in which photosynthesis takes place, there is a decrease in dissolved oxygen owing to its consumption as a result of respiration and the decay of organic matter. The change in concentration at these depths may be either gentle or sharp, depending on the physical structure of the water column. The development of a shallow oxygen minimum immediately below the euphotic zone has again been used to provide estimates of rates of photosynthesis, respiration and nutrient cycling (Johnson *et al.*, 2010).

3 *Oxygen minima* are a characteristic feature in most ocean areas. As discussed in Chapter 7, water masses deep within the ocean are formed at high latitudes mostly, and then sink and advect away from the region of formation carrying with them surface water oxygen saturations. These water masses are then isolated from contact with the atmosphere and will lose oxygen through the bacterial degradation of sinking organic matter as they 'age'. The waters formed at high latitudes sink to the bottom of the ocean and are only slowly and inefficiently mixed with waters above them. Hence the waters at the bottom of the ocean are colder and have been at the surface of the ocean more recently than the interme-

diate waters of the ocean. The intermediate waters at around 1000 m have therefore been subject to the slow removal of oxygen for longer than the deep water and hence have the lowest oxygen concentrations (Fig. 9.12).

As discussed in Chapter 7, the deep waters of the world ocean 'age' (in the sense of time since they were last at the surface) from the North Atlantic to the North Pacific. During the 'grand tour' therefore dissolved oxygen becomes depleted, with the overall result that concentrations decrease from the waters of the deep Atlantic to those of the deep Pacific (see Fig. 9.13). This is the exact mirror image of the increase in nutrient concentrations along the 'grand tour' discussed above and is driven by exactly the same process of the deepwater breakdown of organic matter (see Section 9.2). Recalling from Chapter 7 that the grand tour takes many hundreds of years, it is clear that the rates of oxygen consumption are relatively low and apart from some particular oxygen minimum zones in intermediate waters discussed below, the deep waters of the world ocean do not become oxygen depleted in the contemporary ocean.

In some particular regions of the ocean, the oxygen minimum zone is particularly evident and leads to the development of hypoxic regions where oxygen concentrations fall to very low levels (Fig. 9.14). The boundaries of these zones depend on the oxygen concentration considered, and in Fig. 9.14 this boundary is set at $<20\,\mu M$ $O_2$; although in some parts of these hypoxic zones oxygen falls to essentially zero. Paulmier and Ruiz-Pino (2009) estimate that the ocean volumes with oxygen concentration less than 4.5, 45 and $90\,\mu M$ $O_2$ are 0.46, 18.6 and $38.3 \times 10^6$ km$^3$, respectively. Once the oxygen falls to such low levels, biogeochemical cycles change markedly and reduced species such as Mn (II) are favoured over oxidized forms such as Mn(IV) (see Chapter 11 and 14) and probably more importantly the nitrogen cycle is changed and denitrification (and annamox) processes replace oxygen based breakdown of organic matter (see Section 9.1). Indeed these extreme hypoxic zones probably represent the dominant global sink for nitrogen in the oceans. The development of these hypoxic zones is controlled by the balance of the lateral oxygen supply and the vertical supply of degradable organic matter. All the hypoxic zones in Fig. 9.15 are associated with limited

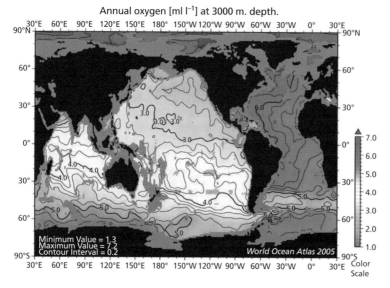

Fig. 9.13 Map of oxygen concentrations (ml l⁻¹) in the world oceans at 3000 m depth from. Garcia *et al.*, (2006) *World Ocean Atlas 2005, Volume 3: Dissolved Oxygen, Apparent Oxygen Utilization, and Oxygen Saturation.* S. Levitus (ed.) NOAA Atlas NESDIS 63, U.S. Government Printing Office, Washington, D.C. (See Colour Plate 7)

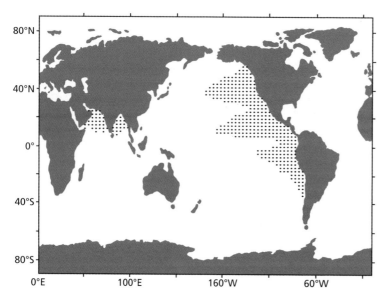

Fig. 9.14 Map of the distribution of intense oxygen minimum zones (OMZ) in the world ocean, stippled areas, defined here as $O_2$ concentrations $<20\,\mu mol\,l^{-1}$ based on Paulmier and Ruiz-Pino (2009). Reprinted with permission from Elsevier.

vertical mixing and slow water flows and some are associated with enhanced organic matter fluxes characteristic of upwelling. However, it seems that the limited lateral (or vertical supply) of well oxygenated waters is the key cause of the development of low oxygen regions (Helly and Levin, 2004; Paulmier and Ruiz-Pino, 2009). The 'younger' waters in the Atlantic have generally higher oxygen concentrations than the other ocean basins and in low oxygen regions, concentrations do not fall to such extreme low levels. However, regions of relatively low oxygen are evident off the west African and Benguela coasts with oxygen concentrations falling to $<100\,\mu M$. Furthermore there is evidence that the extent of these zone is increasing, at least in the Atlantic, in accord with model predictions where the decrease in oxygen is driven by increased global warming; resulting in increased stratification, reduced mixing, and also

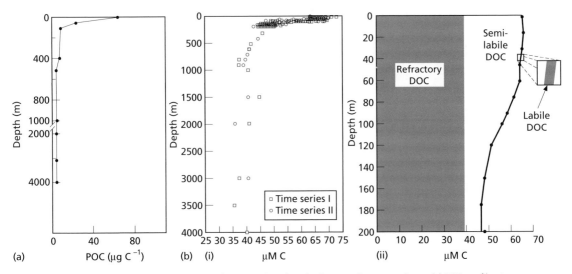

**Fig. 9.15** Vertical profiles of dissolved and particulate organic carbon in the oceanic water column. (a) POC profile at 36°20′N : 67°50′W in the Atlantic Ocean (after Williams, 1971). (b) DOC profiles at the JGOFS Equatorial Pacific Time Series site at 0°, 140°W in the central Pacific (from Carlson and Ducklow, 1995). (i) DOC profile from 0 to 4000 m. (ii) Partitioning of the bulk DOC pool in the upper 200 m into refractory, semilabile and labile pools.

a decrease in the oxygen concentrations in surface waters before they sink (Stramma *et al.*, 2008). There is also evidence from the paleo record that large parts of the ocean have been completely devoid of oxygen in the past when ocean circulation patterns were different (Stramma *et al.*, 2008).

There are some ocean areas where water exchange is physically restricted, and so the supply of dissolved oxygen, is also restricted. Here, the oxidative decomposition of organic matter utilizes, and can sometimes exhaust, the dissolved oxygen so that anoxic (zero oxygen) conditions are set up. Such anoxic conditions actually prevail in only a very small proportion of the World Ocean, examples being found in fjords, certain basins (e.g. the Gotland Basin in the Baltic, the Black Sea) and in deep-sea trenches (e.g. the Cariaco Trench). Seasonal stratification and increasing rates of organic matter supply from nutrient enrichment can also lead to localized regions of seasonal oxygen depletion in coastal waters often described as 'dead zones' reflecting their major impact on fish and macro-invertebrates. While such areas have existed in areas such as the Baltic before human impacts, there is evidence that the nature and scale of these 'dead zones' is growing as a result of human impact on the environment in areas such as

the Gulf of Mexico (Diaz and Rosenberg, 2008; Middleburg and Levin, 2009).

In order to evaluate the amount of oxygen that is biologically consumed in the deep ocean the concept of oxygen utilization is often used. This is a measure of the oxygen that has been utilized, rather than the amount that remains in the waters, and is based on the premise that in surface water the oxygen is present at equilibrium, or saturation, values with the overlying atmosphere. In contrast to surface waters, deep waters are highly undersaturated with respect to dissolved oxygen due to bacterial breakdown of organic matter, and a measure of the amount of oxygen that has been utilized can therefore be obtained from the difference between the saturation (based on water temperature and salinity) and the observed oxygen contents. It is usual, however, to use the term apparent oxygen utilization (AOU) because surface water can be up to ~5% supersaturated with dissolved oxygen (see above). The AOU is thus a measure of the change in dissolved oxygen that has taken place after the waters have left the surface. Values of AOU are lowest in the Atlantic (less utilization) and highest in the Pacific (more utilization), as the deep water 'grand tour' progresses along its path and utilization and water-mass mixing take place. It must be stressed, however, that most oxygen utilization occurs in surface waters, where air-sea exchange can replace consumed oxygen.

In addition to the broad ocean wide patterns of deep water oxygen concentration discussed above, oxygen concentrations also reflect water mass structures within ocean basins. According to Kester (1975), ocean-scale sections of the vertical and horizontal distributions of dissolved oxygen reflect a number of major features in water circulation patterns. This is evident in the oxygen distribution diagrams given in Fig. 9.12(b) and (c) and these can be used to illustrate in a general the relationship between dissolved oxygen and water masses. In this context, Kester (1975) identified a number of features in these diagrams.

1 Antarctic Intermediate Water can be seen as an intrusion of oxygen-rich water in both the Atlantic and Pacific Oceans, extending from the surface at ~50°S to ~800 m at ~20°S.

2 North Atlantic Deep Water is evident as oxygen-rich deep water extending from ~0 to ~2000 m at ~60°N to an oxygen maximum at ~3000 m in the south equatorial Atlantic.

3 The northward flow of North Atlantic Deep Water and Antarctic Bottom Water into the deep Pacific gives rise to a progressive decrease in the oxygen content from south to north.

4 Waters at intermediate depths are more deficient in oxygen in the North Pacific than in the North Atlantic.

### 9.3.1 The nutrients and oxygen in the oceans: summary

1 Nitrate and phosphate are involved in nutritional processes and are incorporated into the soft-tissue phases of plankton during photosynthesis, whereas silicate is involved only in building the hard skeletal parts of organisms.

2 New nutrients are supplied to the euphotic zone by processes such as upwelling and erosion of the thermocline from below, and from external sources such as atmospheric deposition; these nutrients take part in new production and are transported out of surface waters via sinking particulates. Regenerated nutrients are recycled within the euphotic zone and take part in regenerated production.

3 The vertical water-column profiles of the nutrients exhibit a characteristic surface-depletion–depth-enrichment effect as a result of their uptake in surface waters during primary production and their release at depth following the remineralization of sinking detritus. The remineralization of organic matter consumes oxygen leading to a decrease in oxygen associated with the increase in nutrients. In surface waters oxygen is maintained close to saturation values by air-sea exchange

4 The horizontal distribution of the nutrients and oxygen is controlled by water circulation patterns, upon which are superimposed the effects of internal biogeochemical reactivity. As a result, there is a steady push of nutrients towards the deep Pacific, which lies at the end of the deep-water global transport path, leading to their build-up in these waters and an associated low oxygen level.

## 9.4 Ocean carbon cycle

### 9.4.1 Organic matter in the sea

#### 9.4.1.1 Introduction

There are two general approaches to studying organic matter in the oceans. The first is based on estimating the bulk concentration of organic carbon and the other is based on directly measuring particular compounds in ocean waters. Most of the organic carbon is chemically uncharacterized, despite considerable research effort. This reflects the difficult of characterizing a very large and complex mixture of organic compounds formed mostly from the bacterial and photochemical degradation of organic matter produced within the oceans. The bulk concentration of particulate and dissolved organic carbon is usually measured using methods that involve conversion of the organic matter to $CO_2$ and this $CO_2$ is then quantified. The results are therefore presented as concentrations of organic carbon, rather than as concentrations of organic matter.

Here we will consider the overall cycling of bulk dissolved and particulate organic carbon before briefly considering the characterization of this material.

For analytical convenience, the organic matter in aquatic systems is usually divided into two principal fractions.

1 The fraction passing through a 0.45 µm membrane filter includes material in true solution, together with colloidal components, and is termed dissolved organic matter (DOM); its carbon content is classed as dissolved organic carbon (DOC) and there is an associated dissolved organic nitrogen (DON) and phosphorus (DOP) pool. It has been estimated that as much as 50% of the DOM in seawater may be present as high molecular weight (MW > 1000) colloids (Hansell and Carlson, 2002; Benner, 2002).

2 The filter-retained material is referred to as particulate organic matter (POM), and its carbon content is termed particulate organic carbon (POC).

Some authors also distinguish a third fraction, termed volatile organic carbon (VOC).

Estimates of the amount of organic carbon in the POC and DOC reservoirs have varied over the years, and a summary of some of the most recent findings are given in Table 9.2. From the data in this table it is apparent that the DOC pool is the principal reservoir for organic carbon in the oceans. This seawater DOC provides a major dynamic pool in the global carbon cycle. The POC pool is dominated by the detrital (non-living) fraction. Little organic carbon is stored in the living plankton biomass, and on a whole ocean basis this reservoir is trivial compared with the seawater DOC and the non-living POC (Olsen *et al.*, 1985); however, in the euphotic zone, where most primary production occurs, plankton and bacteria become more important.

## 9.4.2 The sources of oceanic organic matter

The organic matter in seawater can be divided into two genetic classes. That having an external source is termed *allochthonous*, and that originating in the ocean itself, that is, from internal sources, is referred to as *autochthonous*. It is clear form Table 9.2 that autochthonous sources dominate.

## 9.4.3 The sources of organic matter

Allochthonous organic matter is brought to the oceans chiefly by river run-off, with smaller inputs from atmospheric transport; these inputs have been discussed in Chapter 3 and 4 respectively.

### 9.4.3.1 The sources of autochthonous organic matter in the oceans

The principal mechanism by which organic carbon is initially produced in the oceans is in situ photosynthesis by phytoplankton, which may be represented crudely by an equation of the general type Equation 9.1, although of course in reality the organic matter formed is vastly more complex. The resulting biotic carbon may form part of the POC (particulate organic carbon) pool, or it may enter the DOC (dissolved organic carbon) pool through

**Table 9.2** Sources, reservoirs and sinks for organic carbon in the oceans (approximate values).

|  | Data |  |  |
| --- | --- | --- | --- |
| Sources ($10^{12}$ mol C yr$^{-1}$) |  |  |  |
|  | Atmospheric input | 18 | Chapter 6 |
|  | Fluvial input: | 15 | Chapter 6 |
|  | Net primary productivity | 4230 | Chapter 9 |
| Reservoirs ($10^{12}$ mol C) | POC (non-living) | 0.25 | E and R (2003) |
|  | POC (living biomass) | 0.8–1.7 | ditto |
|  | DOC | 57 | ditto |
| Sinks ($10^{12}$ mol C yr$^{-1}$) | Nearshore sediments | 9 | ditto |
|  | Pelagic sediments | 1 | ditto |
|  | All marine sediments | 10 | ditto |

Eglinton and Repeta (2003).

processes such as the passive exudation of organics by phytoplankton, release during cell rupture due to grazing or bacterial/viral lysis of cells, excretion by zooplankton and organic matter decay reactions. The DOC and POC is subsequently modified by cycling many times through a variety of bacterial and photochemical processes (Hansell and Carlson, 2002). The $CO_2$ involved in photosynthesis may be present as free dissolved $CO_2$, or as dissolved bicarbonate or carbonate ions (see Section 9.5), and because the concentration of total $CO_2$ is high relative to phytoplankton requirements, it does not limit photosynthesis.

*Autochthonous POC: the living POC fraction.* This is composed largely of plankton, and these biota have been described in Section 9.1

*Autochthonous POC: the non-living fraction (detritus).* This consists mainly of dead organisms, faecal material, organic aggregates of various types, and other complex organic particles. In open-ocean waters much of this non-living POC is derived from phytoplankton. Cawet (1981) has identified four major processes that are involved in the formation of non-living POC: (i) the direct formation of detritus (e.g. organic fragments, faecal pellets); (ii) the agglomeration of bacteria; (iii) the aggregation of organic molecules by bubbling in the surface layers; and (iv) the flocculation or adsorption of DOC on to mineral particles. Fellows *et al.* (1981) have also suggested that bacteria, and other micro-organisms, can repackage soluble nutrients into POC. This DOC → bacteria → POC route can result in increases in POC at depth in the water column and so enhance its supply to the deep-water pool.

*Autochthonous DOC.* Dissolved organic carbon is the principal reservoir for organic carbon in the oceans. It appears to be produced predominantly from autochthonous sources based on both its bulk and isotopic composition. The bulk of DOC from phytoplankton is rapidly recycled with bacteria estimated to consume >50% of primary production, but a small fraction of the DOC is either released as, or converted into, forms that escape rapid bacterial degradation and come to dominate the DOC, as discussed further below (Eglinton and Repeta, 2003; Hansell and Carlson, 2002).

### 9.4.4 The distribution and composition of oceanic organic matter

#### 9.4.4.1 The distribution and composition of POM in the oceans

The POM in the oceans consists of living organisms and dead material (detritus), and can originate from both marine and terrestrial sources.

With respect to autochthonous POC, some aspects of the major element composition of organic tissue are relatively well understood. One reason for this is that the C:N:P ratios of both plant and animal tissue are fairly constant. These are often termed the Redfield ratios (see Section 9.2.1), after Redfield (1934; 1958). These ratios indicate that the major plant nutrients, phosphate and nitrate, change concentrations in seawater in a fixed stoichiometry that is the same as the N and P stoichiometry of plankton. Thus, the nutrients are removed from seawater during photosynthesis in the proportions required by the biomass. The elemental composition of phytoplankton and zooplankton is very similar, and C:N:P proportions of 106:16:1 are commonly used for their organic tissue material. These are the Redfield ratios and represent the preformed composition of the organic material that undergoes oxidative destruction in the euphotic zone according to the relationship $C:N:P:O_2 = 106:16:1:138$, which corresponds to the consumption of 138 moles of oxygen to produce 106 atoms of carbon, 16 atoms of nitrogen and 1 atom of phosphorus. The exact value of the Redfield ratio is subject to periodic revision based on improving data and understanding of the processes involved in setting the Redfield ratios (e.g. Paulmier *et al.*, 2009) but these changes to the ratios are minor. Recently, as discussed in Section 9.2, it has been argued that the actual numerical values of the Redfield ratio reflect a balance between production and consumption processes for organic matter and have no particular biological significance. Hence, the constancy of the ratio within the oceans reflects efficient rates of water mixing relative to rates of biogeochemical processes (Arrigo, 2005; Falkowski and Davis, 2004). Despite these developments, the Redfield ratios have remained a valuable and enduring paradigm of marine biogeochemistry.

The soft, that is, non-skeletal, parts of organisms are composed mainly of proteins, carbohydrates, lipids and, in higher plants, lignins. The oceanic living biomass is dominated by plankton, and the structures of some of the major metabolites (i.e. substances that take part in the processes of metabolism) found in these organisms are illustrated in Worksheet 9.1. Whitehead (2008) has suggested an approximate biochemical composition of various organisms, which emphasizes the differences between marine bacteria, phytoplankton and zooplankton, and the much greater difference between these short lived small marine organisms and larger longer lived terrestrial plants (Table 9.3).

Terrestrial, that is, allochthonous, sources also provide a variety of particulate, and dissolved, organic matter to the oceans. This organic matter covers a wide variety of compounds; including hydrocarbons, fatty acids, carbohydrates, fatty alcohols and sterols, natural polymers such as lignin, cutin and chitin, and possibly amino acids. Some of these compounds are exclusive to terrestrial biota. For example, lignins are complex phenolic polymers that are unique to the tissues of vascular land plants (see Table 9.3). Other compounds can be derived from both terrestrial and marine sources, but sometimes can be distinguished from each other on the basis of their detailed chemistries. This 'biomarker' approach can be illustrated with reference to a number of compounds. For example: (i) terrestrial *n*-alkanes derived from plant waxes have their principal homologues (i.e. similar carbon structures but differing carbon chain lengths) in the series $C_{23}$–$C_{35}$, whereas those derived from marine plankton are dominated by homologues in the series $C_{15}$–$C_{21}$; (ii) terrestrial fatty acids derived from plant waxes have homologues in the series $C_{14}$–$C_{36}$, whereas those from marine plankton sources are generally in the range $C_{12}$–$C_{24}$ (see Peltzer and Gagosian, 1989).

Only a small fraction of deep-water POM, probably only a few per cent, is composed of living material, the remainder being *refractory* in character (Volkman and Tanoue, 2002). The refractory deep-water POM consists of two fractions: (i) a small-sized fraction, which makes up the bulk of the POM, and (ii) a larger sized fraction, consisting mainly of faecal pellets and organic debris (marine snow), which is utilized in the foodchain. The refractory POM is considered in more detail in Section 9.5.5, where the oceanic organic carbon cycle is described.

The POC (living + non-living components) constitutes an average of less than ~5% of the TOC in seawater. In most surface waters (0 to ~300 m) the living fraction is the most important source of POC. In these surface waters POC concentrations vary both geographically and seasonally following the patterns of primary production, and appear to lie in the range ~1.8 to ~25 μM C l$^{-1}$ (~0.02–0.3 mg C l$^{-1}$) – see Cawet (1981). In deeper waters the POC concentrations fall, and range from ~0.33 to ~2.5 μM (~0.004 to ~0.03 mg C l$^{-1}$): see Cawet (1981), and have an average of ~0.8 μM. Thus, there is an overall decrease in the concentrations of POC with depth in the water column: see Fig. 9.15. Further, in deepwaters the POC, much of which is non-living and refractory, is considerably less variable in concentration than it is in the surface layer. Nonetheless, variation in the concentrations of POC with depth in the water column can occur when the profile section includes more than one water mass.

### 9.4.5 The distribution and composition of DOM in the oceans

Dissolved organic carbon makes up, on average, ~95% of the TOC in the oceans. The analysis of seawater DOC, which usually consists of oxidizing all of the organic C to $CO_2$ and then measuring its concentration, has proven challenging but over the last decade or so a body of very reliable data has become available and a clear pattern of behaviour is evident. DOC concentrations are highest in surface waters and particularly so in ocean gyres with surface water concentrations in the range 60–90 μmol l$^{-1}$ and deep water concentrations 30–50 μmol l$^{-1}$ (Fig. 9.15). This pattern is consistent with the formation of DOC primarily by processes associated with primary production in surface waters. There is a consistent gradient of concentrations of DOC in the deep ocean; with concentrations of about 50 μmol l$^{-1}$ in the North Atlantic, 40 μmol l$^{-1}$ and 35 μmol l$^{-1}$ in the North Pacific, consistent with the slow breakdown of DOC along the deep water circulation paths described earlier for nutrients. It is possible to use the $^{14}C$ content of DOC to estimate an average age of this material since its formation and this is about 5000 years; suggesting that most of the DOC in the deepwater is resistant to degradation on timescales of thousands of years. Thus as noted earlier most of the

**Worksheet 9.1: Some organic compounds found in plankton.**

A variety of metabolites, that is, substances that take part in the processes of metabolism, are associated with marine organisms. The most important of these are described in the Worksheet, and a number of organic structures are illustrated.

*Proteins*
Proteins, a major structural component of living tissue, are complex nitrogenous organic compounds, which are polymers of α-amino acids assembled in chains. Globular proteins, in the form of enzymes, catalyse biochemical reactions. In addition to acting as enzymes, proteins serve as antibodies, transporters, receptors and metabolic regulators. The most abundant amino acids found in the proteins of plankton are glycine, alanine, glutamic acid and aspartic acid.

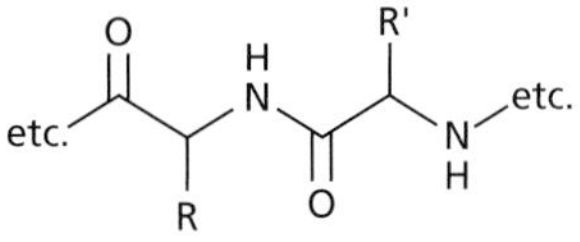

Biopolymers of amino acids e.g.

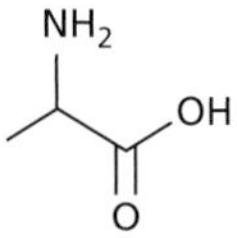

**Alanine** 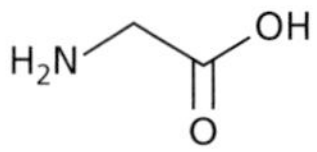                    **Glycine**

*Carbohydrates*
Carbohydrates include sugars, starches and cellulose. Carbohydrates form the supporting tissues of phytoplankton and also act as energy sources; for example, catabolism of carbohydrates can provide most of the energy requirement for the cell. Carbohydrates can be classified into monosaccharides, which are subdivided into pentoses (e.g. ribose) and hexoses (e.g. glucose), disaccharides (e.g. sucrose), which are sugars, and polysaccharides, which include starches, cellulose and the mucopolysaccharide chitin.

**Glucose**

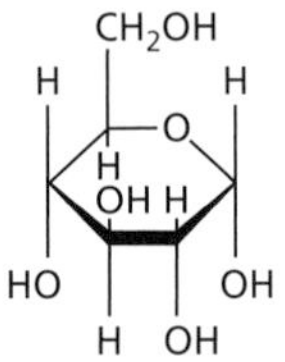

monosaccharide

*continued*

**Cellulose**

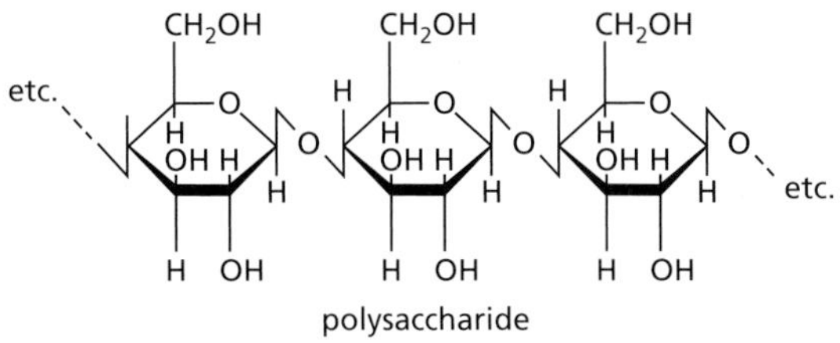

CH$_2$OH        CH$_2$OH        CH$_2$OH

etc.

polysaccharide

**Chitin**

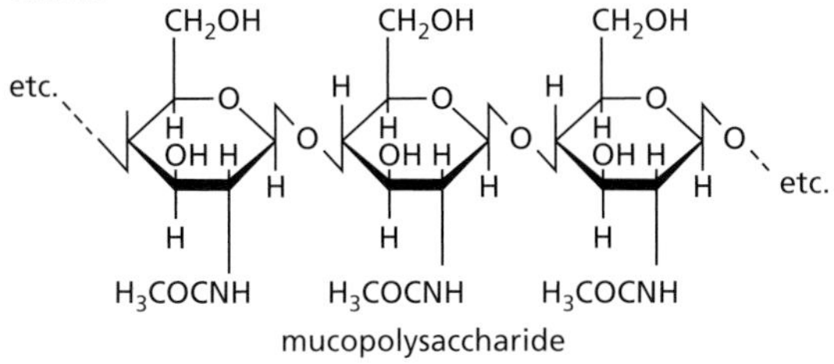

CH$_2$OH        CH$_2$OH        CH$_2$OH

etc.

H$_3$COCNH      H$_3$COCNH      H$_3$COCNH

mucopolysaccharide

*Lipids*

Lipid is a general term that applies to all substances produced by organisms that are insoluble in water but can be extracted by some kind of organic solvent. In this broad definition the term lipid covers a wide variety of compound classes, including the photosynthetic pigments. More usually, lipids refer to fats and waxes, phospholipids and glycolipids, sterols and some hydrocarbons (e.g. *n*-alkanes).

*Fats*, which are utilized mainly as energy stores in the energy budget of organisms, are triglycerides formed by the combination of glycerol and fatty acids.

*Wax esters*, which have a function as protective coatings, are esters of fatty acids and long-chain alcohols. They also can act as food reserves in many species of marine organisms and are abundant in zooplankton.

*Phospholipids* are glycerides (fatty acid esters of glycerol) with both phosphoric acid and fatty acid units, and are the most abundant lipids in phytoplankton. They can serve as membrane constituents, buoyancy controls and thermal or mechanical insulators.

*Glycolipids* are combinations of lipids and sugars and are found in cell membranes and walls.

*Sterols* are the hormonal regulators of growth, respiration and reproduction in most marine organisms, and also may act as membrane rigidifiers. In addition, sterols can be present as lipoproteins (e.g. cholesterol).

*Pigments*

Phytosynthetic coloured pigments such as chlorophyll (green), and the carotenoids (e.g. carotene and fucoxantin; orange/red), are used by plants to absorb and transfer light energy during photosynthesis. Most phytoplankton have chlorophyll as their primary photosynthetic pigment, and four major types of chlorophylls (Chl) have been identified (Chl *a*, Chl *b*, Chl *c* and Chl *d*), although for chlorophylls *c* and *d* there are a number of compounds with distinct structures. Chl *a* is the primary pigment in algae, higher plants and cyanobacteria, with other forms acting as accessory pigments to pass on light energy captured by the *a* form. Photosynthetic bacteria

*continued on p. 190*

**Wax ester**

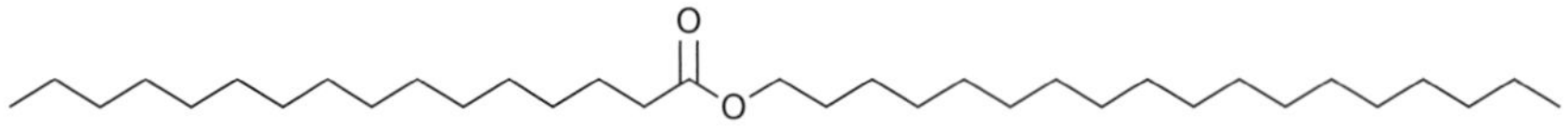

**Phospholipid**

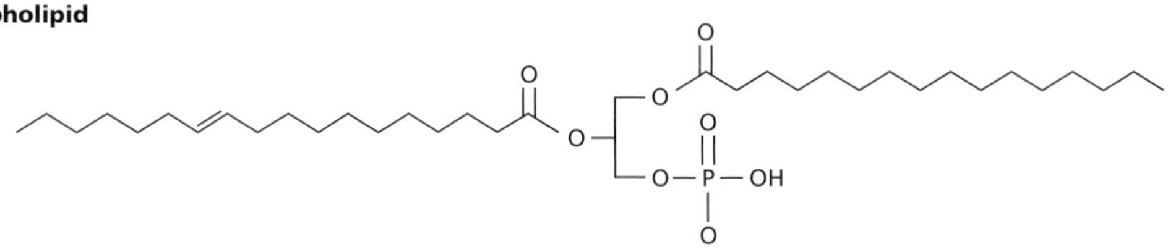

1-palmitoyl-2-palmitoleoyl-3-phosphatidic acid

**Sterols**

e.g. Cholesterol

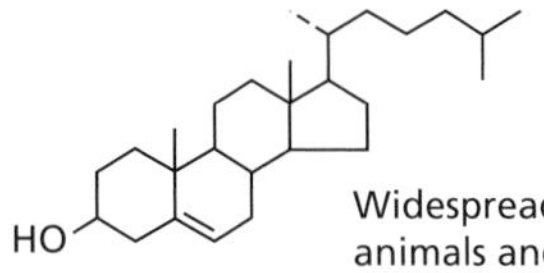

Widespread distribution in
animals and plants

e.g. Dinosterol

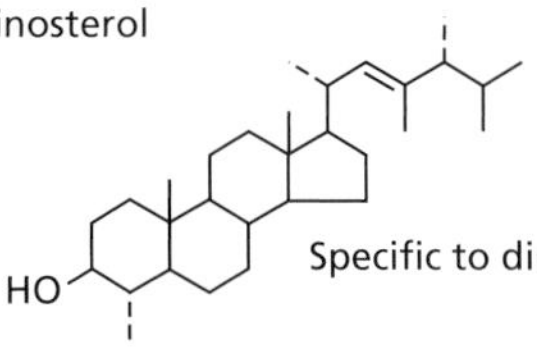

Specific to dinoflagellate algae

utilize a distinct set of pigments known as *bacteriochlorophylls* (*a* to *e*), with the *a* form being the primary pigment. The distribution of chlorophyll is not homogeneous throughout the oceanic water column, but is characterized by a global-scale maximum often found near the bottom of the euphotic zone in the vicinity of the thermocline. The concentrations of chlorophylls in surface waters are generally higher in euphotic than in oligotrophic regions, and chlorophyll *a* has been used as an indicator of primary production.

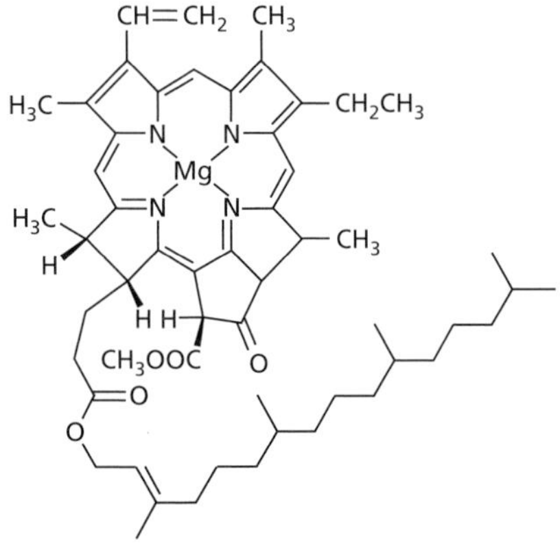

Chlorophyll *a*

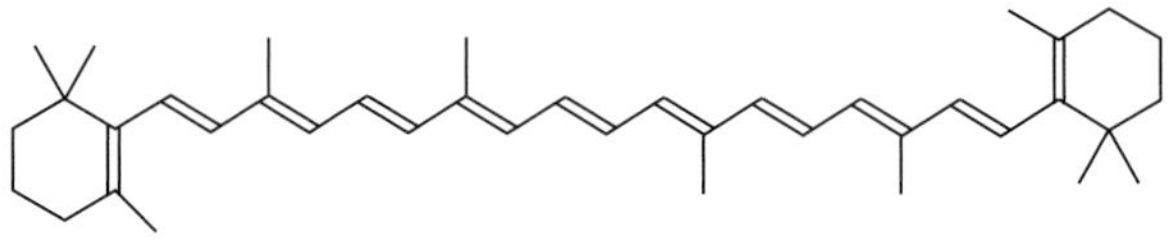

β-Carotene

*continued*

*Nucleic acids*

Ribonucleic acid (RNA) and deoxyribonucleic acid (DNA) are involved in the storage and transmission of genetic information and are made up from nucleotide units composed of a phosphate, a pentose sugar and a nitrogen-containing organic base. The DNA encodes the amino acid sequences of all the proteins synthesized by a cell. The RNA molecules are vital intermediates in the production of protein from this genetic code.

*Other metabolites*

Various other organic compounds are also found in plankton. These include vitamins, co-enzymes and hydrocarbons. According to Gagosian *and* Lee (1981) the principal hydrocarbons in algae are alkenes, together with lesser amounts of *n*-alkanes, branched alkanes and cyclic alkanes. The *n*-alkanes can be used to distinguish marine from terrestrial organic matter inputs (see Section 9.4.4).

**Table 9.3** Major biochemical composition by % C for various organisms (Whitehead, 2008).

| Material | Protein | Polysaccharide | Lipid | Pigment | Nucleic Acid | Lignin | Tannin |
|---|---|---|---|---|---|---|---|
| Bacteria | 55–70 | 3–10 | 5–20 | 2–5 | 20 | 0 | 0 |
| Phytoplankton | 25–50 | 3–50 | 5–10 | 3–20 | 20 | 0 | 0 |
| Zooplankton | 45–70 | 3–5 | 5–20 | 1–5 | 20 | 0 | 0 |
| Vasular Plants | 2–5 | 37–55 | <3 | 5–20 | <1 | 15–40 | <20 |
| Wood | <1 | 40–80 | <3 | 0 | <1 | 20–35 | <45 |

DOC formed is rather rapidly recycled, but a small pool of highly resistant material is formed and then degraded only very slowly (Fig. 9.15). Most of this DOC is also of relatively low molecular weight (<1000) as characterized by ultrafiltration techniques. (see reviews by Benner, 2002; Hansell and Carlson, 2002).

Only a small fraction, perhaps <20%, of the DOM in seawater has been chemically well characterized (see also Section 14.2.3). This is the labile fraction and consists mainly of compounds such as lipids, carbohydrates, amino acids, urea and pigments, all of which typically are associated with the biochemistry of living organisms (see Worksheet 9.1), and the biomass itself is probably the principal source of these compounds (Benner, 2002). In general, there are three pathways by which the autochthonous POC produced in oceans by biota can contribute to the oceanic DOC pool: (i) exudation by phytoplankton; (ii) excretion by zooplankton; and (iii) post-death organism decay processes. During their lifetime phytoplankton release some of their phytosyntheti-cally fixed carbon to the surrounding waters by metabolic processes. Following the death of both phytoplankton and zooplankton, decomposition occurs via the action of autolytic enzymes present in the tissues and by bacteria that have colonized the material. During these processes the DOC released into the water can include both biologically labile and refractory fractions (Carlson, 2002). Estimates of the amount of photosynthetically fixed carbon released as DOC vary, but Hansell (2002) suggests about 20% of net community production is released as DOC with possibly higher percentages during the spring bloom, leading this author to an estimate of overall DOC production of $1.2 \times 10^{15} \, \mathrm{g \, C \, yr^{-1}}$.

Information on the concentrations of some individual compounds in the oceans are also available; for instance for particular pollutants such as DDT, or an important metabolite such as urea, but these invariably represent only a tiny fraction of the bulk DOC. The majority of the DOC is uncharacterized. Table 9.4 presents a recent summary of the information available on the composition of DOC. Note that

**Table 9.4** Average chemical characteristics of dissolved organic matter isolated from surface and deep ocean waters (based on Benner, 2002). HMW – high molecular weight.

| Chemical Characteristic | Surface Water | Deep Ocean |
| --- | --- | --- |
| Humic substances (% of DOC) | 5–25 | 15–25 |
| HMW DOC (% of DOC) | 25–40 | 20–25 |
| Total neutral sugars (% DOC) | 2–6 | 0.5–2 |
| Total amino acids (% DOC) | 1–3 | 0.8–1.8 |
| Total amino sugars (% HMW DOC) | 0.4–0.6 | 0.04–0.07 |
| Solvent extractable lipids | 0.3–0.9 | – |

the composition is defined in part by the methods of extraction. Some DOC can be extracted and concentrated onto absorbents and then eluted under certain conditions. Such materials are often called *humic* substances a term originally used for a rather better defined freshwater DOC component, but more generally now used for complex and highly degraded organic material. Once extracted, this fraction can then be at least partially chemically characterized. An alternative extraction technique involves ultrafiltration which can separate and concentrate high molecular weight fractions (>1000 molecular weight) which can again be partially chemically characterized. Note that these two fractions may overlap. Table 9.4 illustrates how poorly characterized the DOC is, but also that surface waters where most DOC is produced have higher and more variable concentrations of biologically labile components such as sugars and amino acids.

The analysis of specific compounds in ocean waters has provided limited information on the composition of DOC and some examples are presented in the following.

*Lipids.* These include *n*-alkanes ($C_{16}$–$C_{32}$), pristane, fatty acid esters, aliphatic hydrocarbons, triglyceride, sterols and free fatty acids, principally palmatic, oleic, myristic and stearic acids (see e.g. Ehrardt *et al.*, 1980; Kennicutt *and* Jeffrey, 1981; Kattner *et al.*, 1983; Parrish *and* Wangersky, 1988). Hydrocarbons are present in all marine organisms, but usually account for only ~1% of the total lipids. The nonvolatile natural hydrocarbons (>$C_{14}$) in the marine environment include saturated aliphatic hydrocarbons (e.g. *n*-alkanes, regular branched isoprenoids, branched alkanes), unsaturated aliphatic hydrocar-

bons (e.g. *n*-alkenes), saturated alicyclic hydrocarbons (e.g. cyclanes), unsaturated alicyclic hydrocarbons (e.g. cyclenes) and aromatic hydrocarbons (e.g. retene, perylene). The pollutants include *n*-alkanes, isoalkanes and aromatic compounds from oil pollution, polycyclic aromatic hydrocarbons (PAHs), polychlorinated biphenyls (PCBs), and biocides resistant to bacterial decay such as DDT and pentachlorophenol (PCP). The *n*-alkanes are the dominant constituents of natural hydrocarbons in the marine environment (Saliot, 1981). The *n*-alkanes in seawater originate from: (i) natural internal sources, that is, the oceanic biomass; (ii) natural external terrestrial sources (mainly associated with higher plant metabolism); and (iii) anthropogenic sources (e.g. oil pollution). These sources can sometimes be distinguished by characterizing the *n*-alkanes on the basis of their carbon numbers and their carbon preference index (CPI). In this context

$$CPI = \frac{\text{sum of odd carbon } n\text{-alkane concentrations}}{\text{sum of even carbon } n\text{-alkane concentrations}}$$

Thus, a CPI of 1 indicates no carbon preference (see e.g. Kennicutt *and* Jeffrey, 1981). *n*-Alkanes in the range $C_{17}$–$C_{22}$ are indicative of a phytoplankton source, whereas the heavier *n*-alkanes in the range $C_{28}$–$C_{32}$ originate from terrestrial plants (see e.g. Eglinton and Hamilton, 1963). The *n*-alkanes of plants and organisms generally show an odd carbon preference. In contrast, the *n*-alkanes of oils usually have a CPI of around 1.

*Carbohydrates.* The dissolved carbohydrates reported to be present in seawater during an algal bloom include laminaribiose, laminaritriose, glycosylglycerols, sucrose and raffinose (Sugugawa *et al.*, 1985). These low molecular weight carbohydrates also are found in dinoflagellate cells, and are thought to have been derived from phytoplankton.

*Amino acids.* These include dissolved free amino acids (DFAA), dissolved combined amino acids (DCAA) and dissolved total amino acids (DTAA). Lee and Bada (1977) reported that in the equatorial Pacific and the Sargasso Sea the concentrations of DCAA were much higher than those of DFAA, and showed a maximum in the euphotic zone where the concentrations of phytoplankton and zooplankton were higher. Further, the concentrations of the DCAA were higher at all water depths in the productive equatorial Pacific than in the oligotrophic Sargasso Sea. Lee and Bada (1977) also found that DFAA concentrations were small and relatively invariant in the water column. Liebeziet *et al.* (1980), however, reported that there was a significant variation in the concentrations of DFAA with depth in the Sargasso Sea. These authors provided data on the distributions of individual DFAA (threonine, serine, glutamic acid, aspartic acid, glycine, alanine) and found that concentrations were enriched at the upper boundaries of the seasonal pycnocline. These enhancements were attributed to increased auto- and heterotrophic activity, leading to the production of dissolved organic compounds in the sharp density layer, where there is a concentration of bacteria and zooplankton that take advantage of the energy-rich compounds concentrated at the discontinuities in the water column.

*Halogenated organics.* A number of organisms are known to produce halogen-containing organic compounds, with bromine rather than chlorine being the dominant halogen present. This topic has been reviewed by Fenical (1981). Some of these compounds are important in the transfer of halogens from the ocean to the atmosphere as discussed in Chapter 8.

*Dissolved organosulfur compounds.* These compounds include:
**1** those having natural sources (e.g. dimethyl sulfide (DMS: see Section 4.1.4.3), dimethyl disulfide (DMDS),

carbon disulfide, methylmercaptan (MeSH), dibenzothiophene (DBT), sulfur-containing amino acids;
**2** those derived from anthropogenic sources (e.g. DBT from oil spills and diphenylsulfone).

*Dissolved organic nitrogen and phosphorus.* The dissolved organic matter also contains organically bound nitrogen and phosphorus (DON and DOP) and even in surface waters of the ocean, where inorganic forms of N and P are undetectable, significant amounts of DON and DOP may be present. The concentrations of DON and DOP are usually determined by the measurement of total nitrogen and phosphorus and then subtraction of inorganic N and P, a process that can lead to relatively large uncertainties in DON and DOP concentrations when inorganic concentrations are high relative to organic concentrations.

The ratios of DOC:DON range from 9–18 in surface and deep waters and DOC:DOP ratios range from 180–570 in surface waters to 300–600 in deep waters (Benner, 2002) . These ratios indicate that the oceanic dissolved organic matter is depleted in N and P relative to Redfield ratios (C:N:P of surface DOC 300:22:1 and deep-water 444:25:1 compared to Redfield ratio 106:16:1, (Benner, 2002; see also Section 9.2) suggesting that these nutrient elements are preferentially recycled compared to C, as might be expected in a nutrient depleted environment.

Bronk (2002) suggests average concentrations of DON in surface waters are 5.8 μM and in deep water 3.9 μM, suggesting there is a deep water recalcitrant DON pool augmented in surface waters by about 1–2 μM. In tropical gyre waters inorganic N concentrations fall to much less than 1 μM and there is clear evidence of nitrogen limitation, so the bulk of this surface water DON is probably still not readily and rapidly available for phytoplankton uptake. Although the bulk of the DON (as with DOC) is poorly characterized a small component of DON in surface waters (5–10%) is known to be composed of highly biologically labile compounds such as urea and amino acids, so the DON in surface waters is probably best characterized as a relatively inert background material augmented by biologically labile and semi-labile material derived from phytoplankton and bacteria which may be subsequently available to other bacteria and phytoplankton. DOP data is more limited but the concentrations of DOP in deep water are generally rather low and only a small fraction (<20%) of total dissolved phosphorus. In surface waters the DOP percentage rises to >60%, and in many tropical gyre waters DOP concentrations are considerably higher than the almost undetectable phosphate concentrations. In coastal temperate waters at least, there is evidence for seasonal production of DOP (Karl and Björkman, 2002). It has been proposed that the DOP pool can be characterized by labile, semilabile (lifetime six months) and refractory (lifetime 6–12 years) pools, and the semilabile fraction is particularly significant for sustaining productivity in tropical gyres (Roussenov et al., 2006).

The description of dissolved organic compounds given above has been confined mainly to the bulk water column. It must be stressed, however, that a considerable concentration of both DOC and POC is found in the sea-surface microlayer (see Section 4.3). For example, the total DOC is enhanced in the microlayer, relative to bulk seawater, by a factor of about 1.5 to about 3, and the individual compounds contributing to this enrichment include surfactant lipids (including hydrocarbons), carbohydrates and amino acids. The composition of the dissolved and particulate material in the microlayer is discussed in Section 4.3.

*To summarize.* Although the jury are still out with respect to the origin, absolute concentrations, and therefore the overall distribution of DOC in the oceans, a number of general conclusions can be drawn on the basis of our present knowledge.

1 The depth profile of DOC, DON and DOP in the water column generally shows higher values in the surface layer, indicating net production in the euphotic zone. Concentrations decrease to lower concentrations in deep water with small systematic oceanic gradients. There is evidence of a seasonal production of DOC, DON and DOP.

2 As DOM is cycled through the heterotrophic bacteria, the residual material becomes more refractive and eventually biologically resistant and, in a very broad sense, the DOM in seawater can be divided into at least three pools.

(a) A highly labile DOM pool, containing biologically available organic matter that turns over on time-scales of hours to days;

(b) a semilabile pool, containing organic matter that turns over on a seasonal time-scale;

(c) a refractory pool, which is 'old' and turns over on a relatively long time-scale of several thousands of years.

The refractory DOM is representative of the vast majority of deep-water DOM.

3 The percentage of refractory material increases in concentration from surface to deep waters. In surface waters the accumulation of labile DOC depends on the coupling of production and consumption (bacterial utilization) processes. When they are tightly coupled it is unlikely that highly labile DOC will accumulate in the euphotic zone, and most of the labile DOC will in fact be in the semi-labile class.

### 9.4.6 The marine organic carbon cycle

The concentrations of both POC and DOC are higher, and more variable, in surface than in deep waters. Deep and surface waters therefore may be considered to be two distinct organic carbon reservoirs, and the main features in the marine organic carbon cycle can be interpreted in terms of a surface water → deep water → sediment global ocean journey, involving (i) fluvial, atmospheric and primary production carbon sources, (ii) transport down the water column and (iii) incorporation into the sedimentary sink; changes that occur in the sediment sink itself are considered in Chapter 14. The data for organic carbon in the oceans are continually being refined, but a recent estimate of the fluxes into and out of the ocean, and the size of the various organic carbon reservoirs within the ocean, is presented in Table 9.2. The movements of organic carbon between, and within, the reservoirs are discussed below and summarized in Table 9.5. It is clear from Table 9.5 that the continental margin is both more productive (see Section 6.5) and more efficient

**Table 9.5** Marine organic carbon cycle $10^{12}$ mol C yr$^{-1}$ (from Sarmiento and Gruber, 2006). Continental Margin is area <1000 m deep, it represents 8% of the total ocean area of $3.5 \times 10^{14}$ km$^2$.

| Process | Deep Ocean | Continental Margin | Total |
|---|---|---|---|
| Primary Production | 3750 | 712 | 4462 |
| Export below euphotic zone | 583 | 233 | 816 |
| Flux to sea floor | 28 | 183 | 211 |
| Burial | 1.5 | 14.5 | 16 |

at burying organic carbon (see the following) than the open ocean.

It is important to recognize the large uncertainties in this, and indeed most ocean, budgets. For instance, while it is possible to accurately measure the rate of growth of algae over a few hours in a collected water sample, scaling this measurement up to longer time periods and over large areas of the ocean is difficult. Measuring sinking fluxes within the water column is possible using sediment traps, large cones that collect falling material (see Chapter 12), but the material in the oceans is sinking relatively slowly compared to horizontal currents, making its collection difficult and inefficient: an analogy is trying to collect snow-flakes in a vertical rain gauge in a strong wind.

It can be seen from the data in Table 9.5 that the photosynthetic fixation of $CO_2$ is by far the most important input of organic carbon to the oceans. At first sight, this may seem surprising because the living biomass itself makes up only a small portion of the total organic carbon in the oceans. However, most of the photosynthetic organic carbon is rapidly recy-cled in the upper water column (the surface reser-voir). The cycle is not fully closed, and some of the organic carbon escapes (overall ~18% but more in the continental margin) to be added to both the DOC and POC oceanic pools in the water below the euphotic zone. This loss is equivalent to the 'new' and 'export' production discussed in Section 9.1. This export occurs via sinking organic particles and mixing of higher DOC waters into the deep ocean. Hansell (2002) suggests DOC export contributes about 20 ± 10% of the total export, with particle export therefore dominant. The particle export occurs mainly associated with relatively large fast sinking particles including faecal *pellets* and various aggregates of smaller particles often bound together by organic material and sometime referred to as marine snow. These particles sink through the ocean on timescales of tens of days, and the rates vary in time and space, driven in part by surface water pro-cesses. In addition there is a population of small particles within the deep ocean which are so small and light that they have very low sinking velocities and contribute little to the overall sinking rate (see also Chapter 12).

Sinking particles are subject to continued bacterial degradation as they sink through the deep ocean, releasing nutrients and consuming oxygen as we have already seen. Fluxes of organic carbon decrease exponentially with depth as the more readily degraded fractions are lost most quickly. In the end <1% of primary production in the open ocean reaches the sediments, and most of this is recycled on the surface of the seabed rather than ultimately being buried. The degradation processes within the water column selectively degrade the more labile organic matter such as amino acids (Volkman and Tanoue, 2002), and hence not only affect the total flux, but also influence the composition of the organic matter changes during sinking. The proportion reaching the seafloor is higher on the continental margins, in part simply because they are shallower, but also because the more productive shelf seas tend to favour larger faster sinking phytoplankton. The higher flux to the seafloor is also associated with more efficient burial processes, hence the dominance of the continental margin in ocean carbon burial (see Section 14.2).

*To summarize.* The principal feature in the oceanic organic carbon cycle is the phytosynthetic produc-tion of carbon in the surface waters and its recycling at relatively shallow depths. This biologically medi-ated loop in the recycling of organic carbon in the upper water column is related intimately to the fate of nutrient elements and involves the internal oceanic formation, and subsequent destruction, of POC. However, the biological loop is not 100% efficient, and a small fraction of the POC formed in surface layers escapes to deep water as large-sized faecal material and 'marine snow'. The large-sized organic aggregates are the principal driving force behind the global carbon flux. They also act as carrier phases for trace elements that have been removed from solu-tion, and their down-column transport provides a mechanism for the delivery of particulate-associated constituents to the surface of the sediment reservoir. This mechanism operates on an ocean-wide scale, and it is now recognized that the global carbon flux is one of the major processes controlling both the throughput of material in the ocean system and the trace-element composition of seawater. Table 9.2 suggests the inputs to the ocean of organic carbon slightly exceed the removal by burial, implying the oceans are a small net sink for oxygen on a global scale. However, such a conclusion needs to be tem-pered by the realization that the terms in all the budgets are uncertain and the residual terms such as burial are part of a much larger cycle and are particularly uncertain.

## 9.5 Dissolved carbon dioxide in seawater: dissolved inorganic carbon and the dissolved CO$_2$ cycle

### 9.5.1 Introduction

Carbon dioxide is transferred between ocean water, the oceanic biosphere and the atmosphere via both the photosynthesis of marine plants and the formation of carbonate shells by plants and animals, and therefore is involved in the formation of both soft organic tissues and hard skeletal material. As carbon dioxide is removed from the waters by photosynthesis it is being continuously added from the atmosphere, and at greater depths in the water column it is regenerated by the oxidative destruction of organic matter, increasing in concentrations below the surface. This leads to depletion of total carbon dioxide concentrations in surface waters and higher concentrations at depth, a distribution somewhat like that of nutrients discussed earlier. The biological formation of calcium carbonate from bicarbonate produces CO$_2$. The exchange of CO$_2$ between the atmosphere and ocean was discussed in Chapter 7, where it was noted that CO$_2$ exchange differed from that of most gases because of reactions with carbonate and bicarbonate ions in seawater and here we will consider the interactions of these ions and hence the regulation of the ocean inorganic carbon system. The inorganic and organic carbon systems are closely connected within the ocean and are in turn connected to the global carbon system and atmospheric CO$_2$. The realization of the threat of global climate change posed by increasing atmospheric CO$_2$ has made studies of the ocean carbon cycle particularly important. The chemistry of the ocean inorganic carbon system will now be discussed and this will then lead onto a broader discussion of the global C cycle and climate change.

### 9.5.2 Parameters in the seawater CO$_2$ system

The chemical cycle of CO$_2$ in the oceans is governed by a series of equilibria, which can be expressed as follows (see Millero, 2007).

**1** The CO$_2$ in the atmosphere equilibrates with seawater via exchange across the air–sea interface; thus

$$CO_{2\,(g)} \leftrightarrow CO_{2\,(aq)} \tag{9.4}$$

**2** The aqueous CO$_2$ equilibrates with water via carbonic acid, which is present at very low concentrations compared to CO$_{2\,(aq)}$ and bicarbonate and carbonate ions,

$$CO_{2\,(aq)} + H_2O \leftrightarrow H^+ + HCO_3^- \text{ (bicarbonate ion) } K_1$$
$$\tag{9.5}$$

and

$$HCO_3^- \leftrightarrow H^+ + CO_3^{2-} \text{ (carbonate ion) } K_2 \tag{9.6}$$

**3** The other key equation involves the solubility product of calcium carbonate (CaCO$_3$)

$$Ca^{2+} + CO_3^{2-} \leftrightarrow CaCO_3 \text{ (solid) } K_{sp} \tag{9.7}$$

This series of interactions is often described as the carbonate system. The approximate ratios of the ions are CO$_2$:HCO$_3^-$:CO$_3^{2-}$ 1:100:10 (Denman *et al.*, 2007). A system such as this with a mixture of weak acid (carbonic acid) and associated anions (bicarbonate and carbonate) acts as pH buffer system keeping ocean pH rather stable at about 8. This compares to freshwater systems where pH can be much more variable in cases where there is no carbonate buffer system. However, the addition of acids, including CO$_2$ and carbonic acid to seawater does slowly acidify the oceans as will be discussed later.

The equilibrium constants for reactions 9.4–9.7 have now been determined to great accuracy. These equilibria are significantly influenced by temperature, ionic strength (salinity) and pressure, and all these factors need to be included to accurately characterize the distribution and behaviour of the carbonate system (for a detailed discussion see Millero, 2007). The individual concentrations of bicarbonate and carbonate cannot be directly measured. However, knowing these equilibrium constants and with adjustments for the effects of temperature, salinity and ionic strength, it is possible to fully characterize the carbonate system by measurements of two additional parameters in a seawater sample out of four that can be measured; pH, partial pressure of CO$_2$, total alkalinity (TA) and total inorganic carbon (TCO$_2$) (Millero, 2007), TA and TCO$_2$ are defined in the following.

Traditionally, pH has been defined operationally as

$$pH = \log_{10} \alpha_H \qquad (9.8)$$

where $\alpha_H$ is the activity of the hydrogen ion. Thus, pH is the negative log of the hydrogen ion concentration, and the pH of a solution is a measure of its acidity in terms of some operational scale. In terms of modern electrochemical theory, however, the situation is more complex than this because of the interactions of the carbonate system with other major ions in seawater. These days, pH can now be measured with great accuracy and precision using electrode and spectroscopic methods (Millero, 2007).

The partial pressure of $CO_2$, pCO2 is usually measured by equilibrating a seawater sample with an atmosphere above and then measuring the $CO_2$ concentration in the atmosphere.

Total inorganic carbon (TCO$_2$), is given by the sum of the concentrations (C) of all the species:

$$\sum CO_2 = c_{CO_2} + c_{H_2CO_3} + c_{HCO_3} + c_{CO_3} \qquad (9.9)$$

TCO$_2$ can be measured by acidifying seawater to convert all the ions to $CO_2$ and then measuring this component in one of several ways.

### Alkalinity

From the pH range given above it can be seen that seawater is slightly alkaline in character, a property that arises from the dissolution of basic minerals in seawater. The total alkalinity (TA) is the buffering capacity of natural waters and is equal to the charges of all the weak ions in solution (Stumm *and* Morgan, 1981). Total alkalinity is an important physicochemical property of seawater and, as the major reservoir of inorganic carbon, plays a critical role in several chemical and biological processes, including photosynthesis, respiration and precipitation/dissolution of calcium carbonate.

The alkalinity of seawater can be expressed in terms of the amount of strong acid necessary to bring its reaction to some standard specified end-point in a given volume of solution. Alkalinity is therefore determined using titrimetric techniques, and historically total alkalinity (TA) has been defined as the number of equivalents of strong acid required to neutralize a known amount of seawater to an end-point corresponding to the formation of carbonic acid from carbonate. This definition emphasizes that all bases in seawater contribute to the alkalinity, although in practice only bicarbonate, carbonate and boric acid are significant.

For the determination of alkalinity by titration with a strong acid (e.g. HCl) an end-point was therefore selected at which the bicarbonate, carbonate and borate ions are completely combined with protons to form $H_2CO_3$ and $H_3BO_3$, and the expression for the total alkalinity of seawater (in units of equivalents per litre, or eq. $1^{-1}$) was therefore commonly given as follows (see e.g. Dickson, 1981):

$$TA = [HCO_3^-] + 2[CO_3^{2-}] + [B(OH)_4^-] + [OH^-] - [H^+] \qquad (9.10)$$

which defines the equivalence point for an alkalinity determination at which

$$[H^+] = [HCO_3^-] + 2[CO_3^{2-}] + [B(OH)_4^-] + [OH^-] \qquad (9.11)$$

### 9.5.3 Oceanic distributions of the carbonate system

Considerable effort has been expended over the last 20 years to generate a high quality large scale series of measurements of the ocean carbonate system and these now yield a picture entirely consistent with our broader understanding of ocean biogeochemistry. The overall patterns of the distribution of the carbonate system reflect surface water uptake of total $CO_2$ associated with photosynthesis and release at depth during organic matter breakdown (Fig. 9.16). The pCO$_2$ concentrations are generally maintained close to equilibrium with the atmosphere in surface waters and increase with depth due to organic matter breakdown, which also lowers the pH of deep waters with values as low as 7.5 in the highest pCO$_2$ waters at intermediate depths in the Pacific Ocean (Millero, 2007). Figure 9.16 also illustrates the effect of the global deepwater circulation 'grand tour' with total alkalinity and TCO$_2$ concentration being higher in the older waters of the deep North Pacific compared to the North Atlantic as is the case for macronutrients (see Section 9.2; Key *et al.*, 2004). In surface waters algal blooms result in direct and quite large scale drawdown of pCO$_2$. For example Bakker *et al.*, (2007) showed that when passing from low

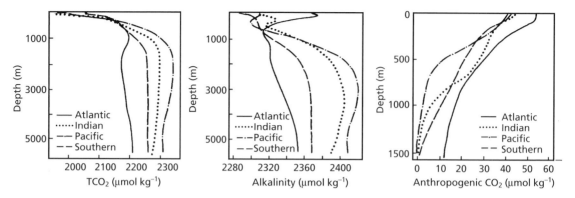

**Fig. 9.16** Vertical profiles of average $TCO_2$, Alkalinity and calculated Anthropogenic $CO_2$ in ocean waters from Atlantic, Indian, Pacific and Southern Oceans based on Key *et al.*, (2004).

productivity Southern Ocean waters into highly productive waters subject to natural iron fertilization from the nearby Crozet island, chlorophyll concentrations increased from <0.5 to $2\,\mu g$ chlorophyll $l^{-1}$ and $pCO_2$ concentrations decreased by 50 to $100\,\mu atm$, a decrease of 10–20% compared to saturation values.

The solubility of calcium carbonate increases with increasing pressure, decreasing temperature and increasing $pCO_2$ and all these factors combine to move ocean surface waters, which are supersaturated with respect to calcium carbonate, to a state of undersaturation at depth as will be discussed further in Chapter 15.

The present distribution of the carbonate system reflects both the large scale natural system and the recent perturbation due to the burning of fossil fuel. It is now possible to directly measure the increasing alkalinity and $pCO_2$ concentrations in the surface ocean associated with rising atmospheric $CO_2$ concentrations. Bates (2007) showed that in various time series spanning periods of 5–22 years from 1988 onwards from the Atlantic and Pacific, total alkalinity and $pCO_2$ concentrations increased at rates ranging from 0.4–$2.22\,\mu mol\ kg^{-1}yr^{-1}$ and 0.7–$25\,\mu atm\ yr^{-1}$, respectively. Thus the surface of the ocean is already directly responding to the increasing atmospheric $CO_2$ concentrations, although recent studies have emphasized that the uptake is variable in space and time, probably reflecting physical and biological interannual variability in the oceans (Watson *et al.*, 2009). Using various tracers and

models it is also possible to separate the natural and fossil fuel components of the carbonate system (Fig. 9.16). Although such a derivation is necessarily rather uncertain compared to the directly measured parameters, the figure does illustrate that most anthropogenic $CO_2$ is still stored in the surface ocean, reflecting the relatively slow ocean mixing times compared to the rapid atmospheric $CO_2$ release during recent decades. The deeper penetration of anthropogenic $CO_2$ in the Atlantic reflects the fact that this water was more recently at the surface and hence subject to the direct uptake of anthropogenic $CO_2$.

### 9.5.4 The uptake of $CO_2$ by the oceans

The chemical driving force for ocean–atmosphere $CO_2$ exchange is the difference between the partial pressure of $CO_2$ ($pCO_2$) in the surface ocean waters and the overlying air. This difference is termed $\Delta pCO_2$ (see also Section 8.2) and is a measure of whether the seawater is undersaturated, or supersaturated, with respect to $CO_2$. When the $pCO_2$ in seawater exceeds that in the overlying atmosphere, the net flow of $CO_2$ should be into the air (fugicity), and when the $pCO_2$ in seawater is less than that in the atmosphere, net $CO_2$ drawdown should occur. Even at equilibrium, there will continue to be exchange of $CO_2$ but the net flux will approach zero. The $\Delta pCO_2$ values are rarely in practice zero, and it is evident that the atmosphere is usually out of equilibrium with the surface ocean. One reason for the

$pCO_2$ differences between the two reservoirs arises because the gas transfer rates of $CO_2$ to and from the atmosphere are much slower than the rates of temperature changes and biological processes, both of which are linked intimately with the fate of $CO_2$ in seawater.

The oceans exchange $CO_2$ with the atmosphere is regulated by three mechanisms (Denman *et al.*, 2007);

**1** Physical exchange regulated by changes in the solubility of $CO_2$, which is principally a function of temperature. Hence warming of waters induces $CO_2$ fluxes from the ocean to atmosphere and cooling has the reverse effect. This is sometimes called the solubility pump.

**2** Biological cycling of $CO_2$ via photosynthesis and respiration as discussed in Section 9.1 and 9.2. This process is often called the organic carbon pump.

**3** Calcium carbonate precipitation, where organisms such as coccolithophores, foraminifera and corals precipitate calcium carbonate according to the reaction

$$Ca^{2+} + 2HCO_3^- \leftrightarrow CaCO_3(s) + H_2O + CO_2 \qquad (9.12)$$

Note this reaction in moving from left to right removes one unit of alkalinity as carbonate solid while producing one unit of carbon dioxide. Hence the growth of carbonate shelled organisms is a source of $CO_2$ to the atmosphere and a sink for alkalinity. This process is often called the calcium carbonate pump.

The organic carbon and calcium carbonate pumps will alter the surface seawater $pCO_2$ which will promote air sea exchange by Equation 9.4, but this exchange can be regulated by purely physical processes such as temperature change as well.

A very important issue in terms of the contemporary increase in atmospheric $CO_2$ is the transfer of carbon into the deep ocean where, as we have seen, water residence times are of the order of hundreds of years. Hence it is not simply ocean uptake of $CO_2$, but its subsequent transfer to deep ocean waters that matters. $TCO_2$ concentrations in the deep ocean are ultimately regulated by mixing back to the surface and the interaction with calcium carbonate sediments, but the timescales of these processes are very long compared to the current very rapid changes in atmospheric $CO_2$ due to human activity.

The dissolution of $CO_2$ in the oceans is greatly increased by the reactions of $CO_2$ with bicarbonate and carbonate ions (Equations 9.4–9.6). This enhanced solubility is often characterized by the 'Revelle factor' (R), named after a pioneer in work on the ocean carbonate system. The Revelle factor relates the fractional change in seawater $pCO_2$ to the fractional change in $TCO_2$ (essentially bicarbonate + carbonate) at equilibrium after a perturbation of the $CO_2$ system (Solomon *et al.*, 2007).

$$R = \frac{(dpCO_2 / pCO_2)TA, T, S}{(dT_{CO_2} / T_{CO_2})TA, T, S} \qquad (9.13)$$

where $T_{CO2}$ is the total concentration of carbon dioxide in all its forms, $pCO_2$ is the partial pressure of carbon dioxide gas, $TA$ is the total alkalinity, $T$ is the temperature and $S$ the salinity. This homogeneous buffering reaction determines how much additional $CO_2$ can be dissolved in surface seawater in response to a $pCO_2$ increase in the atmosphere. The buffer factor varies with temperature and has a value of ~10. Thus, a change of ~10% in atmospheric $pCO_2$ produces a change of only ~1% in dissolved $CO_2$. However, R changes as the $CO_2$ in the oceans rises. For example, as the $CO_2$ content of the atmosphere increases, the concentration of dissolved $CO_2$ will also increase to re-establish equilibrium. Because $CO_2$ is in equilibrium with bicarbonate, carbonate and hydrogen ions, the solution of the gas will be accompanied by a decrease in pH, and the value of R increases; that is, the resistance of the aqueous system to change increases, and the ocean absorbs proportionally less $CO_2$ and the atmospheric fraction increases. Thus the oceans have a very large capacity to absorb atmospheric $CO_2$, but this capacity decreases as $CO_2$ inputs increase.

The increasing concentrations of $CO_2$ in surface waters also lead to slow acidification of the seawater system, despite the effects of the carbonate buffer. Since the start of the industrial revolution it is estimated that ocean surface water pHs have decreased by about 0.1 unit, corresponding to a 30% increase in $H^+$ concentrations. This acidification reduces the carbonate ion concentration and will, if $CO_2$ emissions from human activities continue at a high rate, over time move the ocean surface waters from being slightly supersaturated with respect to calcium carbonate to undersaturated (Equation 9.7),

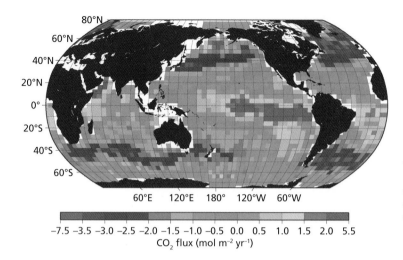

**Fig. 9.17** Mean annual net air-sea flux for $CO_2$ (mol $CO_2$ m$^{-2}$ yr$^{-1}$) for 1995. Red-yellow areas indicate where the ocean is a source of atmospheric $CO_2$ and blue-purple areas where it is a sink (Takahashi *et al.*, 2002). Available from IPCC. Reprinted with permission from Elsevier. (See Colour Plate 8)

particularly with respect to the more soluble polymorph of calcium carbonate, aragonite (see Chapter 15). The implications of ocean acidification for marine organisms, particularly those which produce calcium carbonate skeletons, are uncertain but one that is now generating very widespread concern (Raven *et al.*, 2005).

The ocean $pCO_2$ sink depends on wind speed, temperature and biological activity and hence varies widely across the oceans. Based on almost one million measurements of surface water $pCO_2$, Takahashi *et al.* (2002) have mapped the estimated the global air-sea $CO_2$ fluxes (Fig. 9.17). This map is an annual composite and there is substantial seasonal variability. The main feature of the map is the strong source region for ocean to atmosphere net $CO_2$ exchange associated with upwelling of deep $CO_2$ rich water particularly in the tropical Pacific and Indian Oceans and the strong sinks (i.e. areas of atmosphere to ocean net exchange) at higher latitudes which Takahashi *et al.* attribute to both a strong biological sink and to cooling of waters moving polewards. The overall fluxes calculated from this vast data set are entirely consistent with estimates from global carbon models such as that presented in Fig. 9.18.

### 9.5.5 CO$_2$ and world climate

The context for much of the modern interest in atmospheric and ocean $CO_2$ is the increase in atmospheric $CO_2$ concentrations over recent decades due to fossil fuel combustion and deforestation. This increase has been recorded in the now iconic record created by Keeling and collaborators with the longest record available from the Maunoa Loa observatory on Hawaii (Fig. 9.19).

The ocean contains a very large reservoir of readily exchangeable carbon (which excludes the major rock reservoirs that can only exchange on timescales of millions of years) and hence plays a key role in the global carbon cycle, as illustrated in Fig. 9.18. This figure illustrates the huge scale of the natural global carbon cycle and the very significant perturbation of this cycle by human activity. The global emissions of $CO_2$ from industrial activity are rather well known as is the increase in atmospheric $CO_2$. The fluxes of $CO_2$ from land use changes are more difficult to estimate, but this can be done, albeit with a significant uncertainty. It is then clear that a significant component of the $CO_2$ emissions have been removed from the atmosphere and estimates suggest that both the land biota and ocean uptake make important contributions to this uptake. A best recent estimate from the Intergovernmental Panel on Climate Change (IPCC) of the global carbon cycle is reproduced in Table 9.6. As noted previously, there are good reasons to believe that the ocean uptake may decrease into the future as the alkalinity of surface waters falls, and there may also be a reduction in the efficiency of terrestrial $CO_2$ uptake. This concern makes the

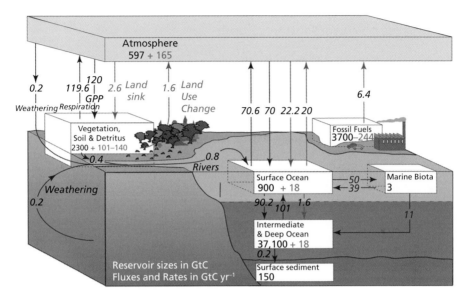

**Fig. 9.18** Global carbon cycle (Denman *et al.*, 2007). Available from IPCC. (See Colour Plate 9)

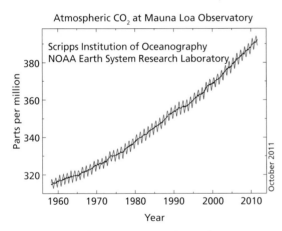

**Fig. 9.19** Atmospheric $CO_2$ concentrations at the Mauna Loa Observatory (Available from www.esrl.noaa.gov/gmd/ccgg/trends/, last accessed 1 October, 2011).

**Table 9.6** The 1990s global carbon budget Gt ($10^{15}$g) C yr$^{-1}$ (based on Denman *et al.*, 2007). Negative sign indicates fluxes in the reverse direction.

| Source | Flux |
| --- | --- |
| Atmospheric increase | 3.2 ± 0.1 |
| Emissions (Fossil fuel + cement) | 6.4 ± 0.4 |
| *Net ocean-to-atmosphere flux* | −2.2 ± 0.4 |
| Net land-to-atmosphere flux | −1.0 ± 0.6 |

The net land to atmosphere flux comprises an emission of 1.6 Gt C yr$^{-1}$ from land use changes such as deforestation and a sink from terrestrial biological uptake of 2.6 Gt C yr$^{-1}$ to yield a net flux of 1 Gt C yr$^{-1}$. Note, however, the terrestrial and ocean sinks are subject to significant interannual variability which may reflect large scale changes in ocean productivity related to climate phenomena such as *El Niño* or even large scale volcanic eruptions which can produce sufficient atmospheric dust to reduce solar radiation for photosynthesis at the surface of the earth. *El Niño* (or the El Nino-Southern Oscillation, ENSO) is a global scale climate phenomena involving a major reorganization of atmospheric circulation over the tropical South Pacific leading to reduced

issue of reducing $CO_2$ emissions from human activity an even more pressing one for society. The budget in Table 9.6 is for the 1990s for which complete data is available, but the IPCC report also indicates that the flux rates of industrial emissions have continued to increase as have the $CO_2$ concentrations into the period 2000–2005 while the land and ocean fluxes have remained essentially constant.

upwelling on the Peruvian coast over timescales of several years (Solomon *et al.*, 2007).

The biogeochemical cycles of C, N, P, S and O play significant roles in controlling the global environment, and the oceans, which cover three-quarters of the Earth's surface, are important in these cycles. This is especially true for carbon, for which the oceans are a major reservoir, and one of the most important recent thrusts in oceanography has been to elucidate the role played by the oceanic carbon flux in the global carbon cycle. This role has received increasing attention because of the effects that anthropogenic intervention can have on climate cycles by influencing the atmospheric levels of carbon dioxide and other 'greenhouse' gases.

Climate describes the long-term behaviour of the weather. The Earth's climate has changed significantly over time including the swings between glacial and interglacial climate through the Holocene, as evidenced in ice cores and deep-sea sediments. The driving force for both weather and climate is solar energy. Climate is dependent upon the radiative balance of the atmosphere, and is governed by interactions between the atmosphere and a number of terrestrial reservoirs, which include the oceans, the ice-caps, the land surfaces and various ecosystems. Heat is redistributed around the planet by the global wind system and oceanic currents. Because of their capacity for heat storage, the oceans provide the main natural control on the heat-retaining properties of the atmosphere. As a result, the oceans play a dominant role in determining climate, and water movements (both horizontal and vertical) in the oceans act like a large flywheel that drives global climate. In this way, long-term changes in the interactions between the atmosphere and the oceans may account for much of the climatic variation over the past few thousand years as the 'flywheel' has speeded up or slowed down.

The oceans operate on different time-scales to the atmosphere. For example, the ocean eddies correspond to cyclonic features in the atmosphere, but whereas the latter exist on scales of days/weeks, ocean eddies can last for many months. Other oceanic features that are involved in the transport of heat persist for considerably longer. Thus, ocean gyres take around a decade to circumnavigate an ocean basin. Further, the global journey associated with the deep circulation conveyor belt operates over a time-scale of millennia (see Chapter 7), which means that the deep ocean can sequester heat absorbed at the surface for periods of 1000 yr or more.

Climate also can be affected by the additional warming of the ocean caused when the activities of humankind enhance the natural 'greenhouse' effect. The mean temperature of the Earth is constrained by the balance between energy from the sun coming in, visible radiation (sunlight) which warms it, and infrared radiation going out, which cools it down. This is the radiative forcing on climate. There are a number of factors which can change the Earth's radiation balance, and these are termed the climate forcing agents. Apart from changes in solar radiation itself, the most important climate forcing agents are associated with the 'greenhouse effect'.

The atmosphere is relatively transparent to the shortwave solar radiation involved in warming the Earth, but many atmospheric trace gases absorb some of the longwave (infrared) radiation, emitted from the surface, which cools the Earth. As a result, the atmosphere acts like a blanket, preventing much of the infrared radiation from escaping into space and so makes the Earth warmer. This is analogous to a greenhouse in which the glass allows sunlight in, but prevents some of the infra-red radiation leaving. The 'greenhouse effect' is a natural phenomenon that prevents the Earth from freezing as the trace gases in the atmosphere absorb emitted heat, and the mean temperature of the Earth is at present ~32°C warmer than it would be if the natural 'greenhouse' gases were not present. However, as the result of anthropogenic intervention the concentrations of a number of the 'greenhouse' gases have increased markedly over the past 100 years, and are expected to produce significant global warming in the next century. In fact, there is evidence that greenhouse warming is already taking place; for example the global mean air temperature has increased by ~0.3–0.6°C over the twentieth century.

The principal natural 'greenhouse' gas is water vapour, but this is not influenced significantly by emissions related to human activity, and it is the gases which have strong anthropogenic sources that give rise to concern. Of these, carbon dioxide ($CO_2$), methane ($CH_4$), nitrous oxide ($N_2O$) and the halocarbons, especially the chlorofluorocarbons (CFCs), and in the lower atmosphere ozone ($O_3$), are the most important. The effectiveness of a 'greenhouse' gas in influencing the Earth's radiative budget depends on its concentration in the atmosphere and its ability to absorb outgoing terrestrial radiation (Watson *et al.*, 1990).

Carbon dioxide is the least effective 'greenhouse' gas per kilogram emitted to the atmosphere and the global warming potential (GWP) index defines the time-integrated warming effect of a unit mass of a 'greenhouse' gas relative to that of $CO_2$ (GWP = 1). The GWPs of the 'greenhouse' gases, extrapolated over a 100-year period, increase in the order (GWP): carbon dioxide (1) ) < methane (21) < nitrous oxide (310) < CFCs (3500–7300) (Solomon *et al.*, 2007). However, the *contribution* of a 'greenhouse' gas to global warming is a function of the product of the GWP and the amount of a gas emitted, and 1990 emissions for the 'greenhouse' gases have been estimated to be: carbon dioxide (26 000 Tg) >> methane (300 Tg) >> nitrous oxide (6 Tg) >> CFCs (7 Tg); $1 Tg = 10^6$ metric tons $= 10^{12} g$. Carbon dioxide therefore has the greatest potential contribution to global warming. Thus, in 2005 the relative contribution of the key greenhouse gases in the radiative forcing of the climate are estimated to be: carbon dioxide (63%), methane (18%), chlorofluorocarbons CFCs (12%), and nitrous oxide (6%) (Solomon *et al.*, 2007). At present, the increase in the concentration of carbon dioxide is therefore the most important single agent in the radiative forcing of the climate.

The importance of greenhouse gas concentrations is elegantly demonstrated by the long ice core records from Antarctica which show that the regular cycles of glacial interglacial change (as recorded by the abundance of the heavy isotope of hydrogen deuterium in the ice – δD and the volume of ice as recorded in ocean benthic foraminifera oxygen isotope change – $δ^{18}O$ ) is exactly in phase with the concentration changes of the three main greenhouse gases $CO_2$, $CH_4$ and $N_2O$ as measured in ice cores (Fig. 9.20; IPCC, TS1). The scale of the recent increases in the greenhouse gases due to anthropogenic emissions is also evident in this figure.

### 9.5.6 The uptake of $CO_2$ by the oceans: synthesis

The three major reservoirs involved in carbon dioxide exchanges are the atmosphere, the terrestrial biosphere and the oceans. The oceans are the largest of the rapidly exchanging carbon reservoirs and contain about 60 times as much carbon as the atmosphere, most of which is dissolved $CO_2$ in the form of bicarbonate. Not all of the excess, that is, anthropogenic $CO_2$ injected into the atmosphere enters the oceans, but because they do act as a sink for the gas much research has been directed towards predicting the environmental impact that excess $CO_2$ will have both on the ocean system itself and on the global climate in general.

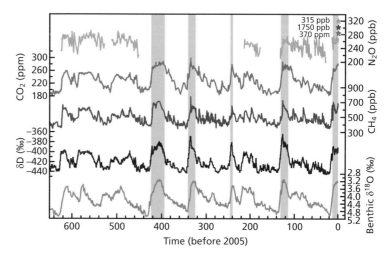

**Fig. 9.20** Changes in concentrations of the gases $N_2O$, $CH_4$ and $CO_2$ in bubbles in Antarctic ice over the last 600 000 years, the temperature of the ice as formed which is recorded as changes in the hydrogen isotope (δD) composition of the ice and the estimate of global ice volume based on the oxygen isotope composition of benthic foraminifera $δ^{18}O$. Available from the IPCC, modern day concentrations of $N_2O$ (315 ppb), $CH_4$ (1750 ppb) and $CO_2$ (370 ppm) are indicated by star symbols for comparison. (See Colour Plate 10)

There are three stages in the uptake of excess $CO_2$ by the oceans.

**1** The thin, well mixed, surface layer will establish equilibrium with $CO_2$ in the atmosphere relatively quickly (a few years), so that on time-scales of decades the concentration of $CO_2$ in the atmosphere is controlled mainly by exchange with the oceans. The capacity of the surface ocean to take up excess $CO_2$ is enhanced by the buffer mechanism that converts the gas to bicarbonate.

**2** The capacity of the surface ocean to take up additional $CO_2$ is increased by the transport of $CO_2$ to deep water, which effectively reduces surface water $pCO_2$. This occurs by two main processes.

(a) By the sinking of organic debris via the 'biological pump'. The mechanism operates on short timescales at all latitudes, and it has been estimated that it removes $8–15\,Gt\,C\,yr^{-1}$ from the surface ocean (Emerson and Hedges, 2008).

(b) By downward water mixing associated with the oceanic thermohaline circulation: the 'physical pump'. This process operates at high latitudes on long time-scales associated with deep-water transit, and in most oceanic regions only the top thousand or so metres of the ocean have yet acquired significant amounts of excess $CO_2$.

Because most anthropogenic $CO_2$ is emitted at low and mid-latitudes in the Northern Hemisphere, and because there is limited north to south atmospheric transport, it might be supposed that on short time-scales much of this $CO_2$ leaves the surface seawater layer via the 'biological pump'. Although it is extremely important in the natural carbon cycle, Watson *et al.* (1990) questioned the efficiency of the 'biological pump' to sequester excess $CO_2$ because marine biota do not respond directly to increases in $CO_2$. Hence, while there is a large natural biological $CO_2$ exchange between the ocean and the atmosphere, this will not offset anthropogenic $CO_2$ emissions unless the biological pump is increased alongside the increase in atmospheric $CO_2$ concentrations.

**3** Both the 'physical pump' and the 'biological pump' mechanisms trap $CO_2$ in the deep waters, and this can be returned to the surface ocean only by the long-timescale thermohaline circulation. As the $CO_2$ content of the deep-water rises, however, water masses now supersaturated with respect to carbonate minerals will become undersaturated and carbonates that are in contact with them will begin to dissolve. Thus, the capacity of the ocean to absorb $CO_2$ will be further enhanced on a very long timescale by the utilization of $CO_2$ during the dissolution of sedimentary carbonate, with the result that when the waters are returned to the surface they can absorb further $CO_2$; for a detailed discussion of the effect of deep-sea calcite and organic carbon preservation on atmospheric $CO_2$ concentrations, see Archer and Maier-Reimer (1994).

With respect to the question 'how will the oceans cope with excess $CO_2$?': the conclusions drawn by Broecker and Peng (1982) provide a useful summary of the possible consequences of an oceanic invasion of $CO_2$: a summary that can be related to the various stages involved in the uptake of excess $CO_2$.

**1** The times required for the ocean system to equilibrate with the carbon dioxide in the atmosphere are about a few years for the surface ocean, several tens of years for the main thermocline, and hundreds of years for the deep sea. Thus, the thermocline acts as a barrier for the oceanic equilibrium with atmospheric $CO_2$, and equilibrium between the atmosphere and the total ocean will take many centuries.

**2** The dissolution of calcite from deep-sea sediments, which raises the alkalinity, affects the bottom waters and therefore only equilibrates with the atmosphere when the waters are brought to the surface in a time period that requires several thousands of years. Further, the excess calcium released from the carbonates in a dissolved form will only be removed from seawater in many tens of thousands of years.

It is apparent, therefore, that the various reservoirs, or boxes, in the oceans equilibrate with the atmosphere on very different timescales, ranging up to tens of thousands of years. Broecker and Peng (1982) concluded, therefore, that as a consequence of this, carbon dioxide added to the atmosphere by human activity over the next two centuries will have an effect on ocean chemistry even on geological timescales.

# References

Andrews, J.E., Brimblecombe, P., Jickells, T.D., Liss, P.S. and Reid, B. (2004) *An Introduction to Environmental Chemistry*. Blackwell Publishing Ltd.

Antoine, D., André, J.-M. and Morel, A. (1996) Oceanic primary production 2. Estimation at global scale from satellite (coastal zone color scanner) chlorophyll. *Global Biogeochem. Cycl.*, **10**, 57–69.

Archer, D. and Maier-Reimer, E. (1994) Effect of deep-sea sedimentary calcite preservation on atmospheric $CO_2$ concentration. *Nature*, **367**, 260–263.

Arrigo, K.R. (2005) Marine microorganisms and global nutrient cycling. *Nature*, **437**, 349–354.

Bakker, D.C.E., Nielsdóttir, M.C., Morris, P.J., Venebales, H.J. and Watson, A.J. (2007) The island mass effect and biological carbon uptake for the sub-Antarctic Crozet Archipelago. *Deep-Sea Res. II*, **54**, 2174–2190.

Bates, N.R. (2007) Interannual variability of the ocean $CO_2$ sink in the subtropical gyre of the North Atlantic over the last two decades. *J. Geophys. Res.*, **112**, C09013, doi:10.1029/2006JC003759.

Benner, R. (2002) Chemical composition and reactivity, in *Biogeochemistry of Marine Dissolved Organic Matter*, D.A. Hansell and C.A. Carlson (eds) pp.59–90. Academic Press.

Boyd, P.W. *et al.* (2000) Phytoplankton bloom upon mesoscale iron fertilization of polar Southern Ocean waters. *Nature*, **407**, 695–702.

Boyd, P.W. *et al.* (2007) Mesoscale Iron Enrichment Experiments 1993–2005: Synthesis and future directions. *Science*, **315**, 612–617.

Brandes, J.A., Devol, A.H. and Deutsch, C. (2007) New developments in the marine nitrogen cycle. *Chem. Rev.*, **107**, 577–589.

Broecker, W.S. (1974) *Chemical Oceanography*. New York: Harcourt Brace Jovanovich.

Broecker, W.S. and Peng, T.H. (1982) *Tracers in the Sea*. Palisades, NY: Lamont-Doherty Geological Observatory.

Bronk, D.A. (2002) Dynamics of DON in *Biogeochemistry of Marine Dissolved Organic Matter*. D.A. Hansell and C.A. Carlson (eds) pp.153–247 Academic Press.

Bruland, K.W. (1980) Oceanographic distributions of cadmium, zinc, nickel and copper in the North Pacific. *Earth Planet. Sci. Lett.*, **47**, 176–198.

Carlson, C.A. (2002) Production and removal processes in *Biogeochemistry of Marine Dissolved Organic Matter*. D.A. Hansell and C.A. Carlson (eds), pp.91–151. Academic Press.

Carlson, C.A. and Ducklow, H.W. (1995) Dissolved organic carbon in the upper ocean of the central Equatorial Pacific, 1992: Daily and finescale vertical variations. *Deep Sea Research II*, **42**, 639–656.

Cawet, G. (1981) Non-living particulate matter, in *Marine Organic Chemistry*, E.K. Duursma and R. Dawson (eds), pp.71–89. Amsterdam: Elsevier.

Cullen, J.J., Lewis, M.R., Davies, C.O. and Barber, R.T. (1992) Photosynthetic characteristics and estimated growth rates indicate grazing is the proximate control of primary production in the equatorial Pacific. *J. Geophys. Res.*, **971**, 639–654.

Denman, K.L. *et al.* (2007) Coupling between changes in the climate system and biogeochemistry, in *Climate Change 2007: The Physical Science Base*. Contributions from Working Group 1 to the fourth assessment report of the Intergovernmental Panel on Climate Change. S. Solomon *et al.* (eds). Cambridge University Press.

Diaz, R.J. and Rosenberg, R. (2008) Spreading dead zones and the consequences for marine ecosystems. *Science*, **321**, 926–929.

Dickson, A.G. (1981) An exact definition of total alkalinity and a procedure for the estimation of alkalinity and total inorganic carbon from titration data. *Deep Sea Res.*, **28**, 609–623.

Duce, R.A. (1986) The impact of atmospheric nitrogen, phosphorous, and iron species on marine biological productivity, in *The Role of Air–Sea Exchange in Geochemical Cycling*, P. Buat-Menard (ed.), pp.479–529. Dordrecht: Reidel.

Dugdale, R.C. and Goering, J.J. (1967) Uptake of new and regenerated forms of nitrogen in primary productivity. *Limnol. Oceanogr.*, **12**, 196–206.

Dugdale, R.C., Wilkerson, F.P. and Minas, H.J. (1995) The role of a silicate pump in driving new production. *Deep Sea Res. I*, **42**, 697–619.

Dunne, J.P., Sarmiento, J.L. and Gnanadesikan A. (2007) A synthesis of global particle export from the surface ocean and cycling through the ocean interior and on the seafloor. *Global Biogeochem. Cycl.*, **21**, GB4006, doi:10.1029/2006GB002907.

Eglinton, G. and Hamilton, R.J. (1963) The distribution of alkanes, in *Chemical Plant Taxonomy*, T. Swain (ed.), pp.187–218. New York: Academic Press.

Eglinton, T.I. and Repeta, D.J. (2003) Organic matter in the contemporary ocean, in *Treatise on Geochemistry*, Vol. 6, H. Elderfield (ed.) pp.145–180 H.D. Holland and K.K. Turekian (eds). Oxford: Elsevier-Pergamon.

Emerson, S.R. and Hedges, J.I. (2008) *Chemical Oceanography and the Marine Carbon Cycle*. Cambridge University Press.

Eppley, R.W. and Peterson, B.J. (1979) Particulate organic matter flux and planktonic new production in the deep ocean. *Nature*, **282**, 677–680.

Ehrardt, M., Osterroht, C. and Petrick, G. (1980) Fatty-acid methyl esters dissolved in seawater and associated with suspended particulate material. *Mar. Chem.*, **10**, 67–76.

Falkowski, P.G., Barber, R.T. and Smetacek, V. (1998) Biogeochemical controls and feedbacks on ocean primary production. *Science*, **281**, 200–206.

Falkowski, P.G. and Davis, C.S. (2004) Natural proportions. *Nature*, **431**, 131.

Fellows, D.A., Karl, D.M. & Knauer, G.A. (1981) Large particle fluxes and the vertical transport of living carbon in the upper 1500m of the northeast Pacific Ocean. *Deep Sea Res.*, **28**, 921–936.

Fenical, W. (1981) Natural halogenated organics, in *Marine Organic Chemistry*, E.K. Duursma and R. Dawson (eds), pp.375–393. Amsterdam: Elsevier.

Gagosian, R.B. and Lee, C. 1981 Processes controlling the distribution of biogenic organic compounds in seawater. In Marine Organic Chemistry, E.K. Duursma and R. Dawson (eds) pp.91–123 Elsevier Amsterdam.

Hansell, D.A. (2002) DOC in the Global Ocean Carbon Cycle in *Biogeochemistry of Marine Dissolved Organic Matter*. D.A. Hansell and C.A. Carlson (eds) pp.685–715. Academic Press.

Hansell, D.A. and Carlson, C.A. (2002) *Biogeochemistry of Marine Dissolved Organic Matter*. Academic Press.

Helly, J.J. and Levin, L.A. (2004) Global distribution of naturally occurring marine hypoxia on continental margins. *Deep-Sea Res. I*, 51, 1159–1168.

Jenkins, W.J. and Goldman, J.C. (1985) Seasonal oxygen cycling and primary production in the Sargasso Sea. *J. Mar. Res.*, 43, 465–491.

Jickells, T.D. *et al.* (2005) Global iron connections between desert dust, ocean biogeochemistry and climate. *Science*, 308, 67–71.

Jickells, T. and Weston, K. (2011) Nitrogen cycle – external cycling losses and gains, in *Treatise on Estuarine and Coastal Science*, D. McLusky and E. Wolanski (eds).

Johnson, K.S., Riser, S.C. and Karl, D.M. (2010) Nitrate supply from deep to near-surface waters of the North Pacific subtropical gyre. *Nature*, 465, 1062–1065.

Karl, D.M. and Björkman, K.M. (2002) Dynamics of DOP in *Biogeochemistry of Marine Dissolved Organic Matter*. D.A. Hansell and C.A. Carlson (eds) pp.249–366. Academic Press.

Kattner, G.G., Gercken, G. and Hammer, K.D. (1983) Development of lipids during a spring plankton bloom in the northern North Sea. *Mar. Chem.*, 14, 163–173.

Kennicutt, M.C. and Jeffrey, L.M. (1981) Chemical and GC-MS characterisation of marine dissolved lipids. *Mar. Chem.*, 19, 367–387.

Kester, D.R. (1975) Dissolved gases other than $CO_2$, in *Chemical Oceanography*, Vol. 1, J.P. Riley and G. Skirrow (eds), pp.497–589. London: Academic Press.

Key, R.M. *et al.* (2004) A global ocean carbon climatology: Results from Global Data Analysis Project (GLODAP). *Global Biogeochem. Cycl.*, 18, doi:10.1029/2004GB002247.

Lalli, C. and Parsons, T.R. (1993) *Biological Oceanography*. Oxford: Pergamon Press.

Lee, C. and Bada, J.L. (1977) Dissolved amino acids in the equatorial Pacific, Sargasso Sea and Biscayne Bay. *Limnol. Oceanogr.*, 22, 502–510.

Liebeziet, G., Bolter, M., Brown, I.F. and Dawson, R. (1980) Dissolved free amino acids and carbohydrates at pycnocline boundaries in the Sargasso Sea and related microbial activity. *Oceanol. Acta*, 3, 357–362.

Mahaffey, C., Michaels, A.F. and Capone, D.G. (2005) The conundrum of marine $N_2$ fixation. *Am. J. Sci.*, 305, 546–595.

Martin, J.H. (1992) Iron as a limiting factor in oceanic productivity, in *Primary Productivity and Biogeochemi-cal Cycles in the Sea*, P.G. Falkowski and A.D. Woodhead (eds), pp.123–137. New York: Plenum Press.

Martin, J.H., Knauer, G.A., Karl, D.M. and Broenkow, W.W. (1987) VERTEX: carbon cycling in the northeast Pacific. *Deep Sea Res. I*, 34, 267–285.

Middleburg, J.J. and Levin, L.A. (2009) Coastal hypoxia and sediment biogeochemistry. *Biogeosciences*, 6, 1273–1293.

Millero, F.J. (2007) The marine inorganic carbon cycle. *Chem. Rev.*, 107, 308–341.

Mills, M.M., Ridame, C., Davey, M., La Roche, J. and Geider, R.J. (2004) Iron and phosphorus co-limit nitrogen fixation in the eastern tropical North Atlantic. *Nature*, 429, 29204.

Olsen, J.S., Garrels, R.M., Berner, R.A., Armentano, T.V., Dyer, M.I. and Taalon, D.H. (1985) The natural carbon cycle. In *Atmospheric Carbon Dioxide and the Global Carbon Cycle*, J.R. Trabalka (ed.), pp.175–213. Washington, DC: U.S. Department of Energy.

Parrish, C.C. and Wangersky, P.J. (1988) Iatroscan-measured profiles of dissolved and particulate marine lipids classes over the Scotian Slope in the Bedford Basin. *Mar. Chem.*, 23, 1–15.

Paulmier, A., Kriest, I. and Oschlies, A. (2009) Stoichiometries of remineralisation and denitrification in global biogeochemical models. *Biogeosciences*, 6, 923–935.

Paulmier, A. and Ruiz-Pino, D. (2009) Oxygen minimum zones (OMZs) in the modern ocean. *Progr. Oceanogr.*, 80, 113–128.

Paytan, A. and McLaughlin, K. (2007) The oceanic phosphorus cycle. *Chem. Rev.*, 107, 563–576.

Peltzer, E.T. and Gagosian, R.B. (1989) Oceanic geochemistry of aerosols over the Pacific Ocean, in *Chemical Oceanography*, Vol. 10, J.P. Riley and R. Chester (eds), pp.282–338. London, Academic Press.

Ragueneau, O. *et al.* (2000) A review of the Si cycle in the modern ocean: recent progress and missing gaps in the application of biogenic opal as a paleoproductivity proxy. *Global and Planetary Change*, 26, 317–365

Raven, J. *et al.* (2005) Ocean acidification due to increasing carbon dioxide The Royal Society.

Redfield, A.C. (1934) On the proportion of organic derivatives in sea water and their relation to the composition of plankton, in *James Johnstone Memorial Volume*, pp.177–192. Liverpool: Liverpool University Press.

Redfield, A.C. (1958) The biological control of chemical factors in the environment. *Am. J. Sci.*, 46, 205–221.

Robinson, C. *et al.*, (2006) The Atlantic Meridional Transect (AMT) Programme: A contextual view 1995–2005. *Deep-Sea Res. II*, 53, 1486–1515.

Roussenov, V., Williams, R.G., Mahaffey, C. and Wolff, G.A. (2006) Does the transport of dissolved organic nutrients affect export production in eth Atlantic Ocean? *Global Biogoechem. Cycl.*, 20, GB3002, doi:10.1029/2005GB002510.

Saliot, A. (1981) Natural hydrocarbons in sea water, in *Marine Organic Chemistry*, E.K. Duursma and R. Dawson (eds), pp.327–374. Amsterdam: Elsevier.

Sarmiento, J.L. and Gruber, N. (2006) *Ocean Biogeochemical Dynamics*. Cambridge University Press.

Sarmiento, J.L., Simeon, J., Gnanadesikan, A., Gruber, N., Key, R.M. and Sclitzer, R. (2007) Deep ocean biogeochemistry of silicic acid and nitrate. *Global Biogeochem. Cycl.*, **21**, doi: 10.1029/2006GB002720.

Shulenberger, E. and Reid, J.L. (1981) The Pacific shallow oxygen maximum, deep chlorophyll maximum, and primary productivity, reconsidered. *Deep Sea Res.*, **28**, 901–919.

Solomon, S., Qin, D., Manning, M., Chen, Z., Marqis, M., Averyt, K.B., Tignor, M. and Miller, H.L. (2007) IPCC 2007, *Climate Change: The physical science basis*. Contribution of Working Group I to the fourth assessment report of the intergovernmental panel on climate change. Cambridge: Cambridge University Press.

Stramma, L., Johnson, G.C., Sprintall, J. and Mohrholz, V. (2008) Expanding oxygen-minimum zones in the tropical oceans. *Science*, **320**, 655–658.

Stumm, W. and Morgan, J.J. (1981) *Aquatic Chemistry*. New York: John Wiley & Sons, Inc.

Sugugawa, H., Handa, N. and Ohta, K. (1985) Isolation and characterisation of low molecular weight carbohydrates in seawater. *Mar. Chem.*, **17**, 341–362.

Takahashi, T. et al. (2002) Global sea-sir CO2 flux based climatological surface ocean pCO2, and seasonal biological and temperature effects. *Deep-Sea Res. II*, **49**, 1601–1622.

Thomalla, S.J., Poulton, A.J., Sanders, R., Turnewitsch, R., Holligan, P.M. and Lucas, M.I. (2008) Variable export fluxes and efficiencies for calcite, opal, and organic carbon in the Atlantic Ocean: A ballast effect in action? *Global Biogeochem. Cycl.*, **22**, GB1010, doi:10.1029/2007GB002982.

Tyrrell, T. (1999) The relative influences of nitrogen and phosphorus on oceanic primary production, *Nature*, **400**, 525–530.

Volkman, J.K. and Tanoue, E. (2002) Chemical and biological studies of particulate organic matter in the ocean. *J. Oceanogr.*, **58**, 265–279.

Watson, R.T., Rodhe, H., Oeschger, H. and Siegenthaler, U. (1990) Greenhouse gases and aerosols, in *Climate Change: the IPCC Scientific Assessment*, J.T. Houghton, G.J. Jenking and J.J. Ephraums (eds), pp.1–40. Cambridge: Cambridge University Press.

Watson, A.J. Schuster, U., Bakker, D.C.E., *et al.* (2009) Tracking the Variable North Atlantic Sink for Atmospheric $CO_2$. *Science*, **328**, 1391–1393.

Whitehead, K. (2008) Marine organic geochemistry, in *Chemical Oceanography and the Marine Carbon Cycle*, S.R. Emerson and J.I. Hedges(eds) pp.264–302. Cambridge University Press.

Williams, P.M. (1971) The distribution and cycling of organic matter in the ocean. In *Organic Compounds in Aquatic Environments*, S.D. Faust & J.V. Hunter (eds) pp.145–63. New York: Marcel Dekker.

Yool, A., Martin, A.P., Fernández, C. and Clark, D.P. (2007) The significance of nitrification for oceanic new production. *Nature*, **447**, 999–1002.

# 10        Particulate material in the oceans

Lal (1977) estimated that the total mass of suspended material in the oceans is ~$10^{16}$ g, which is equivalent to an average seawater concentration of only ~10–20 $\mu g \, l^{-1}$. This suspended material moves through the ocean system, but the journey it undertakes is a dynamic one and its concentration and composition are subject to continuous change as a result of processes such as aggregation, disaggregation, zooplankton scavenging, decomposition and dissolution (Gardner *et al.*, 1985). As it undertakes this journey, the particle microcosm plays a vital role in regulating the chemical composition of seawater via the removal of dissolved constituents (e.g. trace elements and nutrients) from solution and their down-column and lateral transport to the bottom-sediment sink as well as the supply of nutrients and carbon to deep ocean waters via the degradation of sinking organic matter (Chapter 9). Indeed, such is the extent to which the behaviour of dissolved trace metals is dominated by interactions with the suspended solids that Turekian (1977) referred to the phenomenon as the *great particle conspiracy*. Here then, perhaps lies the key to Forchhammer's (1865) 'facility with which the elements in seawater are made insoluble'. In the present chapter attention will be paid to the sources, distribution and composition of the *total suspended material* (TSM) in the sea, and in following this route an attempt will be made to decipher the role that the TSM plays in the major oceanic biogeochemical cycles.

## 10.1 The measurement and collection of oceanic total suspended matter

A number of techniques have been used to measure the concentrations of TSM in seawater. Some of these are indirect, that is, the concentrations of TSM are measured in situ in the water column, but no actual samples are collected; these techniques include those based on optical phenomena, such as light absorption (transmissometry) and light scattering (nephelometry). Other techniques involve the direct collection, and subsequent analysis, of samples of TSM, for example, by filtration or centrifugation. In addition, one of the most important recent advances in the direct collection of TSM has come through the introduction of the sediment trap, a device that collects material as it sinks down the water column. An important advantage of this type of device is that it can retain the large-sized, relatively rare particles that dominate the vertical TSM flux (see Section 12.1).

## 10.2 The distribution of total suspended matter in the oceans

Jerlov (1953) used light-scattering measurements to make one of the first major surveys of oceanic TSM. From the data obtained he was able to identify a number of overall trends in the distribution of the TSM, including the recognition of what were subsequently termed *nepheloid layers*, that is, layers of relatively turbid water that can extend hundreds of metres above the sea bed. This work was followed by a series of investigations made by Russian scientists in the 1950s, and the results of these have been summarized by Lisitzin (1972). The next principal development came in the 1960s from work carried out by American groups, mainly at the Lamont-Doherty Geological Observatory, who used optical methods to study the oceanic distribution of TSM. One of the major findings to emerge from this work was the confirmation of the presence of the nepheloid layers in many regions of the World Ocean

*Marine Geochemistry*, Third Edition. Roy Chester and Tim Jickells.
© 2012 by Roy Chester and Tim Jickells. Published 2012 by Blackwell Publishing Ltd.

(see e.g. Ewing and Thorndike, 1965; Connary and Ewing, 1972; Eittreim et al., 1976). More recent studies have often focused on particle fluxes rather than water column concentrations (see Lampitt and Antia, 1997; and Chapter 12) or attempts to model the many complex and often biologically mediated processes that operate to control the aggregation, disaggregation and sinking of particulate matter in the oceans (e.g. Stemmann et al., 2004).

Much of the indirect data on the distribution of oceanic TSM were obtained in the form of optical parameters. Subsequently, however, Biscaye and Eittreim (1977) converted nephelometer data from the Atlantic Ocean water into units of absolute TSM concentrations. In this way, they were able to identify a number of distinctive features in the vertical and horizontal distributions of TSM in the water column, and these are illustrated in Fig. 10.1 in the form of a generalized vertical profile. From this figure it can be seen that the distribution of oceanic TSM can be described in terms of a *three-layer model* in which the main features are: (i) a surface-water layer; (ii) a clear-water minimum layer; and (iii) a deep-water layer. This three-layer model offers a convenient framework within which to describe the distribution of TSM in the World Ocean.

### 10.2.1 The surface–water layer

In surface waters, TSM concentrations are higher, and more variable, in coastal and estuarine regions than they are in the open ocean. This results largely from the combined effects of; (1) an input of externally produced particulates via river run-off and atmospheric transport, and (2) the internal generation of particulates from primary production and (3) tidal resuspension, all of which have their strongest signals in coastal waters.

1 Although on a global scale ~90% of river suspended material (RSM) is retained in estuaries (see Section 3.2.7.1), the material that does escape contributes to the higher concentration in coastal waters. Further, some estuaries can transport *plumes* of suspended material for considerable distances out to sea in the surface–water layer. As an example for a particularly large river, at high discharge suspended particulate matter concentrations $> 100\,mg\,l^{-1}$ can occur up to 100 km offshore in the Amazon plume (Nittrouer et al., 1986) and these plumes can now be clearly seen in satellite images.

2 Particulate organic material, including both living and dead fractions, makes up a major proportion of TSM in many surface waters of the oceans away from local sources such as river plumes, and most of it results from the photosynthetic fixation of $CO_2$ during primary production. Hence its abundance relates to that of chlorophyll (see Chapter 9). However, the actual concentrations of TSM exhibit large temporal and spatial variations. For example, Chester and Stoner (1972) reported a range of $< 100$– $> 3000\,\mu g\,l^{-1}$ of TSM in a variety of coastal waters from the World Ocean. Away from the coastal regions, the lowest TSM concentrations are found in open-ocean areas of low productivity (oligotrophic zones), especially at the centres of the central gyre systems, where they can fall to values as low as $< 10\,\mu g\,l^{-1}$. There are also inter-oceanic variations in the concentrations of TSM in the mixed layer. For

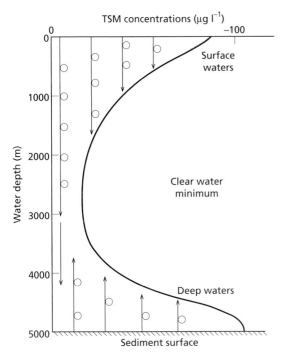

**Fig. 10.1** Typical oceanic TSM profile for a water column with a well-developed nepheloid layer (from Chester, 1982; after Biscaye and Eittreim, 1977). Although the arrows representing settling are drawn as straight lines, vertical particle settling (represented by circles) can be affected by horizontal advective processes. Reprinted with permission from Elsevier.

example, Chester and Stoner (1972) reported an average TSM concentration of ~75 μg l⁻¹ for open-ocean surface waters of the Atlantic and Indian Oceans, and according to Lal (1977) concentrations in the Pacific are about two or three times lower than this.

## 10.2.2 The clear-water minimum

This was the term used by Biscaye and Eittreim (1977) to describe the subsurface region in which there is a decrease in the concentrations of TSM. The decrease is caused by the destruction of particulate material as it sinks from surface waters, or is moved out from the ocean margins by advective processes, and the depth at which the minimum is reached varies from one area of the ocean to another. The decrease in particulate concentrations with depth is largely the result of processes that affect the organic matter (oxidative destruction) and shell fractions (dissolution) of the TSM. The overall result of these processes is a decrease in the concentrations of TSM with depth below the surface. Apart from the fact that the concentrations are lower, however, the distribution of TSM around the clear-water minimum parallels that in surface waters. Thus, TSM concentrations at the clear-water minimum represent the particle contribution resulting from the downward transport of material from surface waters and therefore reflect a balance between surface primary production, variable rates of particle setting and down-column particle destruction–dissolution processes. The distribution of TSM at the clear-water minimum in the Atlantic Ocean is illustrated in Fig. 10.2(a).

## 10.2.3 The deep-water layer

If the sinking of particulate matter from the surface reservoir were the only source of TSM to deep waters, then the concentrations of TSM would be expected to continue to fall all the way down the water column as the destructive processes continued, and the clear-water minimum would not be developed. In fact, it is now known that in many oceanic areas there is an increase in TSM concentrations below the minimum, and this therefore requires an additional (i.e. non-surface water) source of particulate matter. Much of this additional source comes

from the resuspension of bottom sediments into turbid 'nepheloid layers'. In these layers the concentration of particles decreases upwards away from the sediment source, to fall away completely at the clear-water minimum. However, even in deep waters there is a contribution to the TSM from particles sinking through the minimum. In order therefore, to assess the spatial distribution of material in the nepheloid layers, Biscaye and Eittreim (1977) identified two particle populations below the clear-water minimum.

*The gross particulate standing crop.* This was identified as the total TSM below the clear-water minimum. There are two main features in the distribution of the gross particulate standing crop.

1 Relatively low concentrations are found in the central portions of both the North and South Atlantic, reflecting the low surface-water TSM concentrations in the main gyres.

2 The concentrations in the western Atlantic basins are higher than those in the eastern basins, sometimes by as much as an order of magnitude. In the western basins, the maximum TSM concentrations are coincident with the axes of the deep-water boundary currents (see Section 7.3.3).

The distribution of the gross particulate standing crop therefore reflects the effects of two different kinds of processes: those that control the surface-water distribution of TSM and those that regulate the additional supply of particulates necessary to sustain the three-layer water-column model.

*The net particulate standing crop.* This was defined as the amount of TSM below the clear-water minimum that is in excess of the clear-water concentration itself. Thus, the authors were able to obtain a much clearer assessment of the deep-water, or abyssal, particulate signal by excluding 'noise' from the surface-derived source. The distribution of the net particulate standing crop in Atlantic deep-waters is illustrated in Fig. 10.2(b). On the basis of these data Biscaye and Eittreim (1977) concluded that the abyssal signal results from processes that raise particles into near-bottom waters and maintain them in suspension. These processes include supply from the continental shelves by turbidity currents and advective transport, and the direct resuspension of sediment by bottom currents. It was shown in Section 7.3.3 that a characteristic feature of deep-water

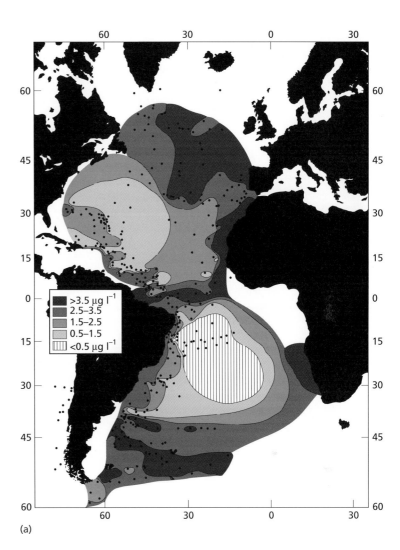

**Fig. 10.2** Distribution of TSM in the Atlantic Ocean (from Biscaye and Eittreim, 1977). (a) Concentration of TSM at the clear-water minimum. (b) Net particulate standing crop in the deep water. Reprinted with permission from Elsevier.

(a)

Legend in figure:
- $>3.5\ \mu g\ l^{-1}$
- $2.5–3.5$
- $1.5–2.5$
- $0.5–1.5$
- $<0.5\ \mu g\ l^{-1}$

circulation is the existence of strong boundary currents on the western sides of the oceans, and it is these erosive boundary currents that are the most effective agents in sediment resuspension, and thus in the formation of nepheloid layers; these layers have particle concentrations that are usually in the range ~200–500 μg l⁻¹. One problem associated with the development of nepheloid layers, however, has been that currents with enough energy to erode the bottom, and to maintain the particles in suspension, have not been found concurrently with high concentrations of particles in the layers themselves. It has been known for some time that bottom-current

velocities can vary, and can include local episodes of high-speed currents. Further, data obtained from the High Energy Benthic Boundary Layer Experiment (HEBBLE) have shown that there is a high temporal and spatial variability in nepheloid-layer particle concentrations, leading to the inference that strong episodic sediment-modifying events, or 'storms', have occurred on the sea bed. Such deep-sea sediment transport storms have, in fact, been identified by Gross *et al.* (1988) in the Nova Scotia Rise area of the North Atlantic. Four storms of high kinetic energy and near-bed flow were observed over a one-year period. These storms were associated with

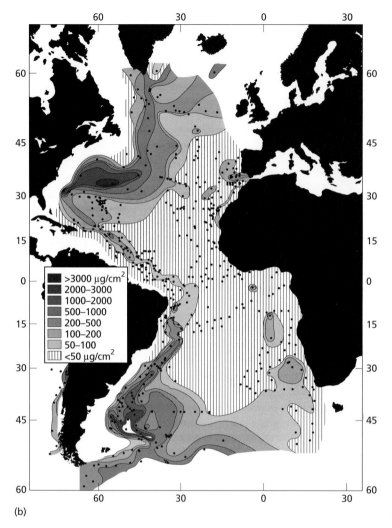

(b)

**Fig. 10.2** *Continued*

particle concentrations >2000 μg l⁻¹, and the authors suggested that the occurrence of a few of these large episodic events per year could account for most of the suspended load in the deep-sea nepheloid layer.

It may be concluded, therefore, that the spatial and temporal distributions of TSM in the oceans are controlled by the classic oceanographic parameters such as the transport of material from the continents, primary production and the major water circulation patterns. The vertical distribution of TSM in the water column can be described in terms of a three-layer model in which the main particulate sources are the surface ocean, from which particles sink downwards, and the bottom sediments, from which particles are lifted by erosive currents and moved upwards. An intermediate, or clear-water minimum, layer is developed at that part of the water column where these two principal supply mechanisms have their smallest effects, the surface source trailing out downwards and the bottom source falling off upwards. In addition to these mainly vertical supply processes, the lateral advection of particles at mid-depths, for example, from shelf sediments giving local concentration increases within the clear water region of the order of 10 μg l⁻¹, can modify the simple three-layer down-column TSM profile (see Section

10.2). Such margin processes, which are relatively common but intermittent, may extend for more than 100 km offshore and can affect not only the water column particulate loads, but also the dissolved chemistry (Sherrell *et al.*, 1998; McCave *et al.*, 2001).

## 10.3 The composition of oceanic total suspended matter

Oceanic TSM consists of a mixture of components, some of which have external sources and some of which are produced internally. In the present section attention is confined largely to the surface-water TSM, and the composition of the down-column flux is considered in Chapter 12.

### 10.3.1 Externally produced components of TSM

These are formed on the continents, in estuaries, or in the air, and are brought to the oceans mainly via river run-off, atmospheric deposition and locally via ice transport.

Crustal weathering products are a ubiquitous, although often minor, component of surface-water TSM, the principal crust-derived minerals being aluminosilicates (e.g. clay minerals and feldspars) and quartz. Chester and Aston (1976) gave data on the distributions of aluminosilicates in surface waters from a number of contrasting oceanic regions. These data showed that in the China Sea, which receives a relatively large fluvial input, aluminosilicate material made up ~60% of the TSM. Off the coast of West Africa, where there is a relatively large aeolian input arising from dust pulses originating in the Sahara desert (see Chapter 4), aluminosilicates accounted for ~10% of the total suspended solids. Krishnaswami and Sarin (1976) reported relatively high concentrations of aluminosilicates in high-latitude North and South Atlantic waters, where they probably resulted from glacial weathering. In open-ocean waters, however, aluminosilicates generally make up less than ~5% of the TSM. The overall pattern to emerge from studies such as these is that the highest concentrations of aluminosilicates are found in regions where the terrestrial source signals are strongest, and the lowest concentrations are found in the open-ocean central gyres remote from the land masses.

In addition to crustal weathering products externally produced material in oceanic TSM includes organic suspensions, flocculated metal-organic colloids and precipitated iron and manganese oxides formed in fluvial and estuarine environments.

### 10.3.2 Internally produced oceanic TSM components

These include biological material (tissues and shells) and inorganic precipitates, together with various resuspended sediment components.

Oceanic TSM components that are produced in the biosphere include both organic matter and shell material. The factors controlling the distribution of organic matter in the oceans have been described in Section 9.2, and the highest concentrations are usually associated with regions of high primary production around the major ocean basins. Particulate organic matter (POM), which is composed of living organisms (phytoplankton, zooplankton, etc.) and detritus (dead organisms, faecal debris, etc.) makes up a large fraction of the TSM found in many surface waters of the World Ocean, usually greater than ~50% of the total solids (see e.g. Chester and Stoner, 1972; Krishnaswami and Sarin, 1976; Masuzawa *et al.*, 1989). The shell material secreted by marine organisms consists largely of either calcium carbonate or opal (see Chapter 9 and 15). Together, these shell components make up a major fraction of surface-water TSM, but there are geographical trends in their distributions; for example, opaline shells can dominate in some high-latitude regions where diatoms are the predominant phytoplankton, whereas carbonate shells can dominate, particularly in the central gyres, at lower latitudes (see also Chapter 15).

Some components of TSM are also produced within the ocean itself from components dissolved in seawater; these include material such as barite (barium sulfate), a number of carbonates, and iron and manganese oxyhydroxides. The precipitation of some of these compounds may be influenced by biological processes, although the minerals are not directly produced. For instance, barite appears to precipitate within degrading organic matter. In the lower and mid-water column, hydrothermal precipitates consisting of minerals such as chalcopyrite, sphalerite and hydrous iron and manganese oxides are also added to the TSM population (see Chapter

15). There are also some anthropogenic TSM components including tar and plastics.

Data for the elemental compositions of a number of the principal components that contribute to the oceanic particulate population are given in Table 10.1. These are:

1 marine organisms, which are representative of the internally produced TSM biomass population;

2 faecal pellets, which offer an estimate of the composition of large-sized aggregates that leave surface waters (see Section 10.4);

3 river and atmospheric particulate material, which represent the composition of continentally derived, that is, external, mainly natural, material transported to the ocean system.

It must be stressed, however, that these are average values, and that wide variations can be found in the elemental compositions of the individual components themselves.

The overall chemical composition of bulk oceanic TSM is controlled by the proportions in which the various components described above are present Thus, the concentrations are high in surface waters from the China Sea (large river input), intermediate in the eastern margins of the North Atlantic (large aeolian input, intense primary production), and low in the open-ocean regions of the Atlantic and Indian Oceans (low external inputs, low primary production). Concentrations of metals that form a major part of aluminosilicates, such as Al, have surface water particulate concentrations in the open ocean of about $100\,ng\,l^{-1}$, while other trace metals such as Cu are present at $<10\,ng\,l^{-1}$.

Using data such as in Table 10.1 it is possible to identify components of the TSM that can be used to identify sources. Thus for example Al is abundant in soil dust and aerosols but not in biological tissue, so is a good tracer of external soil inputs. The significance of the inputs of other elements via this route can be estimated from the average ratio of those elements to Al in soil dust. The Si and Ca (if corrected for soil components) can be used as a tracer of biological shell inputs and the organic carbon content (or a surrogate element that is exclusively associated with the organic component of cells such as iodine) for phytoplankton organic matter. As an example of this approach Jickells *et al.* (1990) estimated that on average in the Sargasso sea in the tropical North Atlantic the surface water TSM was ~60% organic matter, 14% calcium carbonate and <5% clays. While this source apportionment does not total exactly 100% due to uncertainties particularly in the estimation of the organic carbon content, the calculation clearly emphasizes that even in a region with a relatively high aeolian dust loading, biogenic particles dominate the TSM. This study was also able to show that relative importance of the sources varied seasonally with a biogenic component maximum associated with the spring bloom and a clay maximum in summer when biological production was low and dust transport at a maximum.

## 10.4 Total-suspended-matter fluxes in the oceans

In terms of the three-layer distribution model described in Section 10.2, it is apparent that TSM sinks through the water column from the surface layer to provide a *downward* particulate signal. This signal decreases in strength with depth, but below the clear-water minimum it encounters the outriders of an *upward* particulate signal from the sea bed. In addition to these vertical signals, the water-column TSM profiles can be modified by *laterally* advected particulate signals. These various signals interact, both to govern the throughput of particulate material in the ocean system and to control the net output of the material to the sediment sink. This *particulate throughput* is the key mechanism controlling the rates at which many dissolved trace metals are removed from the ocean reservoir, but before attempting to understand the way in which this 'great particle conspiracy' operates it is necessary to understand the nature of the particulate fluxes involved.

According to McCave (1984) the size distribution of suspended particulates is a function of a number of variables, which include the source and nature of the particles, the physical and biological processes that cause their aggregation, and the age of the suspension. In practice, most of the particles suspended in seawater have diameters $<2\,\mu m$ (McCave, 1975), and those with diameters $>20\,\mu m$ are rare (Honjo, 1980). Particles are transported out of the mixed or surface layer down the water column by gravitational and Stokes settling. Stokes' Law describes particle settling in terms of particle density and size and the viscosity of the medium. Settling rates increase

**Table 10.1** The elemental composition of oceanic total suspended material (TSM); units, $\mu g\,g^{-1}$. Some principal components of oceanic TSM.

| Element | Microplankton: North Pacific* | Phytoplankton: Monteray Bay, California* | Zooplankton: Monteray Bay, California* | Zooplankton: North Pacific* | Zooplankton: North Atlantic† | Bulk plankton, average‡ | Bulk plankton: marine organisms, average§ | Faecal pellets | Faecal pellets¶ | River Particulate matter** | Soil-sized aerosols†† |
|---|---|---|---|---|---|---|---|---|---|---|---|
| Al | 72–108 | 7–2850 | <8–313 | 9–31 | – | 202 | 159 | 20800 | – | 94000 | – |
| Fe | 1030–4000 | 49–3120 | 54–1070 | 90–1720 | 567–1467 | 306 | 862 | 21600 | 24000 | 48000 | 52000 |
| Mn | 3.4–32 | 2.1–30 | 2.2–12 | 2.9–7.1 | 0–23 | 7.7 | 9.3 | 2110 | 243 | 1050 | 1312 |
| Cu | 40–104 | 1.3–45 | 4.4–23 | 6.2–58 | 10–90 | 14 | 27 | 650 | 226 | 100 | 157 |
| Ni | 11–12 | <0.5–13 | <0.5–13 | 5–13 | 15–77 | 12 | 17 | – | 20 | 90 | 91 |
| Co | <1 | <1 | <1 | <1 | 8–20 | – | <1 | 15 | 3.5 | 20 | 9 |
| Cr | <4 | <21 | <1 | <1 | – | – | <1 | – | 38 | 100 | 85 |
| V | <3 | <3 | <3 | <3 | – | – | <3 | 76 | – | 170 | 145 |
| Ba | 51–70 | 5–500 | 4–257 | 51–70 | – | 55 | 60 | 192 | – | 600 | 487 |
| Sr | 6800–9650 | 53–3934 | 83–810 | 380–3000 | 57–520 | – | 862 | 1430 | 78 | 150 | 101 |
| Pb | 17–39 | <1–47 | <1–12 | 22–14 | 0–123 | – | 20 | – | 34 | 100 | 465 |
| Zn | 285–4190 | 3–703 | 53–279 | 60–750 | 120–400 | 131 | 257 | <20 | 950 | 250 | 683 |
| Cd | 1.0–2.2 | 0.4–6 | 0.8–10 | 1.9–3.5 | 2–9 | 22 | 4.6 | – | 19.6 | 1 | – |
| Hg | 0.11–0.53 | 0.10–0.59 | 0.07–0.16 | 0.04–0.45 | – | – | 0.16 | – | – | – | – |

* Martin and Knauer (1973). † Martin (1970). ‡ Collier and Edmond (1983). § Chester and Aston (1976). ¶ Spencer et al. (1978). Fowler (1977). ** Martin and Whitfield (1983). †† Chester and Stoner (1974).

with the square of the radius so increase rapidly with size. Calculations suggest that particles less than $10\,\mu m$ in diameter will sink at $<0.5\,m\,d^{-1}$ and hence take many years to reach the seabed. Particles with a diameter $>100\,\mu m$ by contrast will sink at $>10\,m\,d^{-1}$. Clays and phytoplankton cells are $<100\,\mu m$, and generally $<10\,\mu m$ in size. However, relatively rapid transport of particles from the surface layers is observed which is in excess of that predicted from Stokes' law for such small particles. Evidence for the accelerated sinking of particles has come from a number of sources. These include the measured, or inferred, down-column transit times of minerals, biogenic components, stable trace metals and radionuclides, all of which indicate that relatively large-diameter particles are required to sediment the phases out of seawater. A striking example was provided by the sinking of radionuclide fallout from the Chernobyl nuclear accident at rates greater than $100\,m\,d^{-1}$ (Fowler *et al.*, 1987).

As most TSM is composed of particles with diameters $<2\,\mu m$, it must be assumed that this removal is preceded by the aggregation of the particles into larger units. This requirement for a 'fast' particle settling mechanism has been considered by various authors. In a benchmark paper, McCave (1975) made a theoretical study of the transport of TSM in the oceans and concluded that the principal feature in the vertical down-column flux of particles is that relatively rare, rapidly sinking, large particles contribute most of the *flux*, whereas more common smaller particles contribute most of the *concentration*. Brun-Cottan (1976) also concluded that small particles ($<5\,\mu m$ diameter) might behave conservatively in the water column and undergo lateral movement in a given water mass, but that it is mainly the larger ($>50\,\mu m$ diameter) particles, or aggregates, formed in surface waters that fall directly to the sea bed. The large particles are formed by aggregation in the surface layers, via the formation of faecal pellets during zooplankton grazing or the formation of marine snow (see the following). It must be stressed, however, that there is a continual break-up of the larger particles, and Cho and Azam (1988) have suggested that it is the action of free-living bacteria that gives rise to the large-scale production of the small particles at the expense of the large aggregates. Radionuclide scavenging studies (see Chapter 11) also demonstrate this process of deep-water aggregation and disaggregation (Anderson, 2003). The continual cycle of aggregation and disaggregation leads to the production of an oceanic particle population that has no specific size-class populations within it. Nonetheless, in the context of both biogeochemical reactivity (e.g. the removal of dissolved elements by active biological uptake and passive scavenging) and particle flux transport (which is dominated by large-sized material), it is convenient to envisage the microcosm of particles as being divided into two general populations.

1 A common, small-sized population (probably $<5\,\mu m$ diameter), termed *fine particulate matter* (FPM) by Lal (1977), which undergoes large-scale horizontal transport; that is, a suspended population (concentration $\sim20$–$50\,\mu g\,l^{-1}$, sinking rate $<1\,m\,day^{-1}$), which dominates the standing stock of TSM.

2 A rare, large-sized population, referred to as *coarse particulate matter* (CPM) by Lal (1977), which consists mainly of particle aggregates $>50\,\mu m$ in diameter that undergo vertical settling; that is, a sinking population. This is the prime carrier in down-column transport from the surface water to the ocean depths, sinking at rates of tens to hundreds of metres per day. Water sampling is thought to sample this rapidly sinking particle population rather poorly.

It is now widely recognized that *faecal material* (pellets and debris) makes up a significant fraction of the rain of particles from the surface layer to deep water. Faecal material, however, is not the only type of biogenic particulate material to fall from the surface. For example, *marine snow* is a term that has been applied to relatively large amorphous aggregates of biological origin. To some extent this may be analogous to faecal material, because faecal pellets are known to be an important component of this marine snow (see e.g. Shanks and Trent, 1980). However, marine snow also includes other components, and Honjo *et al.* (1984) prefer the term *large amorphous aggregates* (LAA) to describe the large-sized biogenic particles. In this terminology, therefore, LAA consist of the debris of phytoplankton and zooplankton, including faecal material, which is hydrated into a matrix of aggregates to which are attached other components, such as microorganisms and clay particles. More recently some researchers have used the term Transparent Exopolymer Particles (TEP) for these particles and have shown that they are formed in part by aggregation of

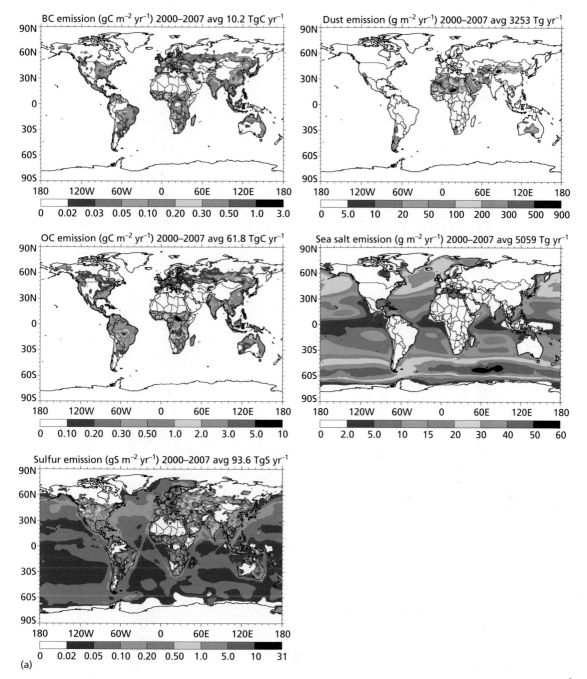

**Plate 1 (Fig. 4.3)** (a) Major emission source regions for sulfur $g\,S\,m^{-2}\,yr^{-1}$, organic and black (soot) carbon (OC + BC) $g\,C\,m^{-2}\,yr^{-1}$, soil dust $g\,m^{-2}\,yr^{-1}$ and seasalt $g\,m^{-2}\,yr^{-1}$, for year 1990, based on Chin *et al.* (2009). (b) Global total aerosol distribution based on satellite imagery.

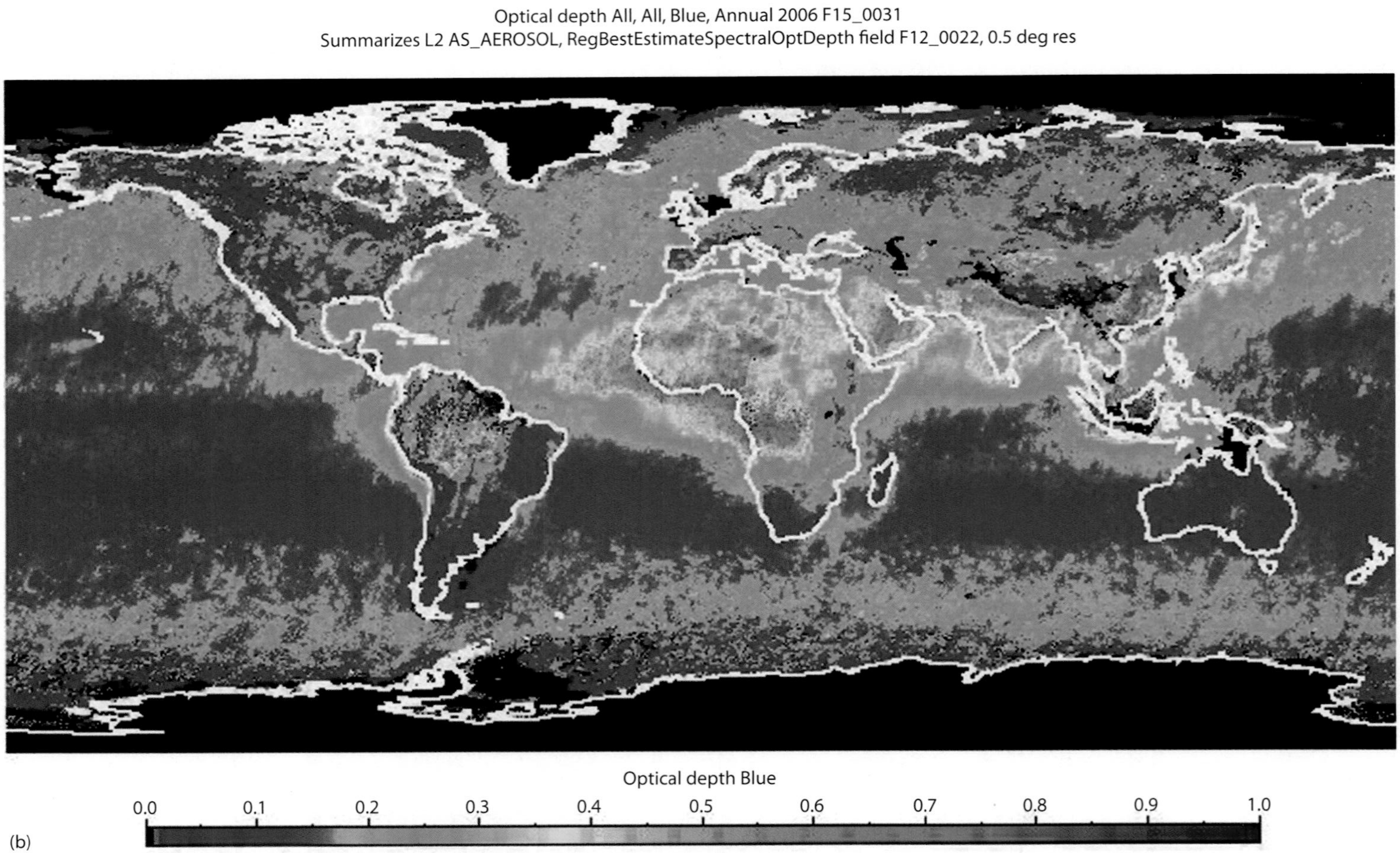

Optical depth All, All, Blue, Annual 2006 F15_0031
Summarizes L2 AS_AEROSOL, RegBestEstimateSpectralOptDepth field F12_0022, 0.5 deg res

Optical depth Blue

0.0   0.1   0.2   0.3   0.4   0.5   0.6   0.7   0.8   0.9   1.0

(b)

**Plate 1 (Fig 4.3)** *Continued*

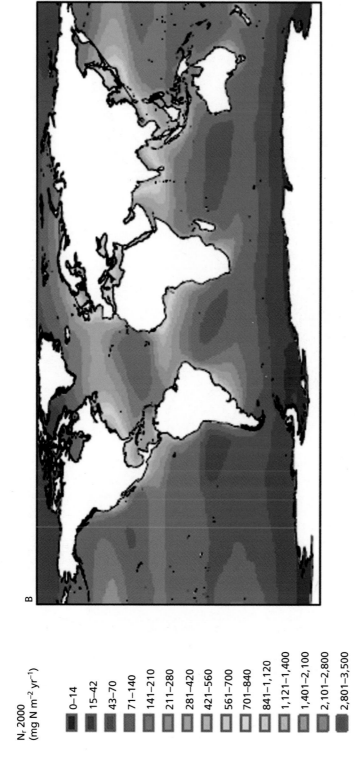

**N_r 2000
(mg N m⁻² yr⁻¹)**

0–14
15–42
43–70
71–140
141–210
211–280
281–420
421–560
561–700
701–840
841–1,120
1,121–1,400
1,401–2,100
2,101–2,800
2,801–3,500

**Plate 2 (Fig. 6.4)** Model estimates of total fixed nitrogen deposition to the oceans (Duce et al., 2008).

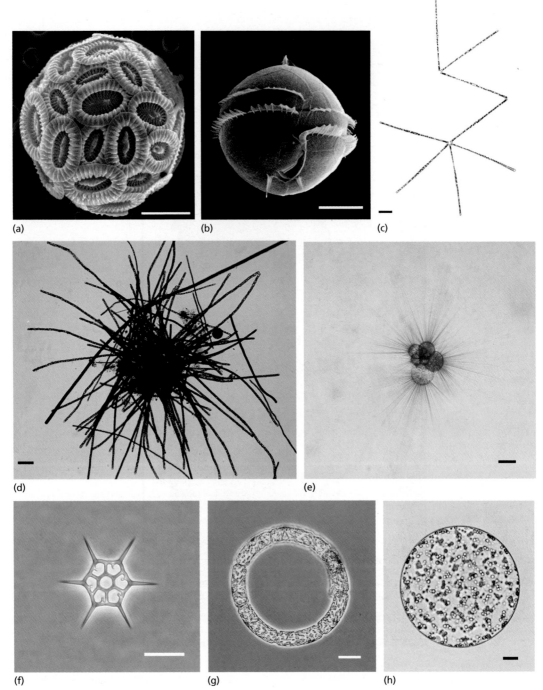

**Plate 3 (Fig. 9.4)** Micro-orgaisms within the marine plankton.
a) *Syracosphaera molischii*, a coccolithophorid;
b) *Protoperidinium cerasus*, a heterotrophic dinoflagellate;
c) a chain of cells of the pennate diatom *Thalassionema frauenfeldii*; d) a filamentous colony of filaments of the cyanobacterium *Oscillatoria*. The filaments form a low oxygen microclimate in the centre of the colony which allows the fixation of nitrogen; e) the foraminiferan *Globigerina bulloides*; f) half of the siliceous skeleton of the silicoflagellate *Dictyocha speculum*; a living cell is sandwiched in between two of these plates; g) part of a coiled chain of cells of the diatom *Guinardia striata*. The chains can extend for several tens of cells in a long coil; h) the spherical non-motile stage of the prasinophyte *Halosphaera minor*. Scale bars: a) 1 μm; b) 10 μm; c), f), g), h) 20 μm; d), e) 200 μm. a), b) Scanning electron micrographs; c), d), e), h) Light microscope, bright field; f), g) Light microscope, phase contrast. a), b) from samples preserved in formalin; c), d), f), g) From samples preserved in Lugol's Iodine; e), h) Live specimens. Picture credits: a) Pauhla McGrane; b)–h) Robin Raine.

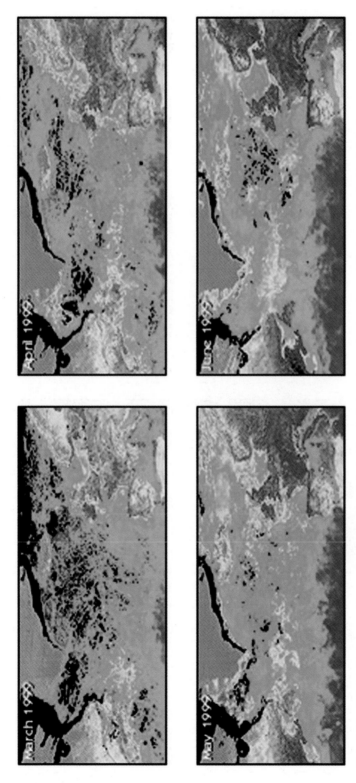

**Plate 4 (Fig. 9.5)** Satellite images of the seasonality of global phytoplankton stocks in surface waters based on satellite imagery for the North Atlantic.

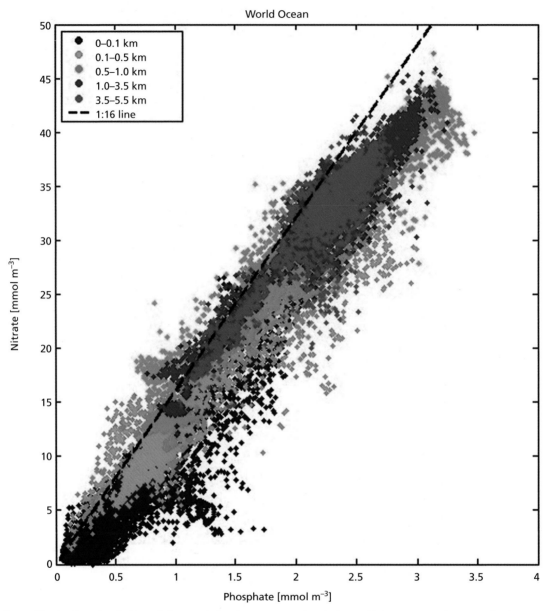

**Plate 5 (Fig. 9.10)** A plot of observed nitrate and phosphate concentrations in the world oceans arranged at water depth producing the so called Redfield ocean N:P ratio, courtesy of Dr T. Tyrrell and based on Tyrrell (1999).

# Annual silicate [μmol l⁻¹] at 3000 m. depth.

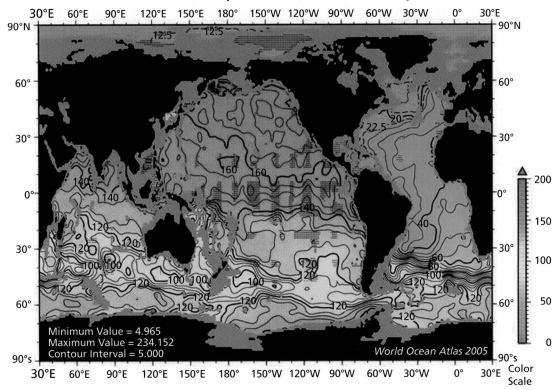

**Plate 6 (Fig. 9.11)** Map of nutrient concentrations (in the world oceans μmol l⁻¹) at 3000 m depth from *NOAA World Ocean Atlas.*

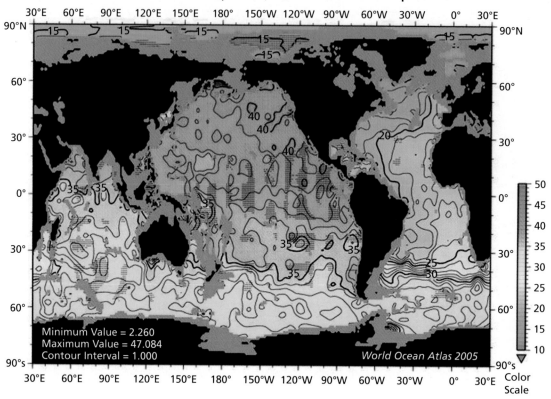

**Plate 6 (Fig. 9.11)** *Continued*

# Annual phosphate [umol l⁻¹] at 3000 m. depth.

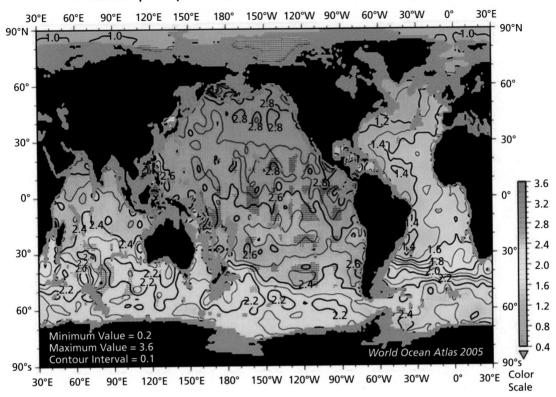

**Plate 6 (Fig. 9.11)** *Continued*

**Plate 7 (Fig. 9.13)** Map of oxygen concentrations (ml l⁻¹) in the world oceans at 3000 m depth from. Garcia *et al.*, (2006) *World Ocean Atlas 2005, Volume 3: Dissolved Oxygen, Apparent Oxygen Utilization, and Oxygen Saturation.* S. Levitus (ed.) NOAA Atlas NESDIS 63, U.S. Government Printing Office, Washington, D.C.

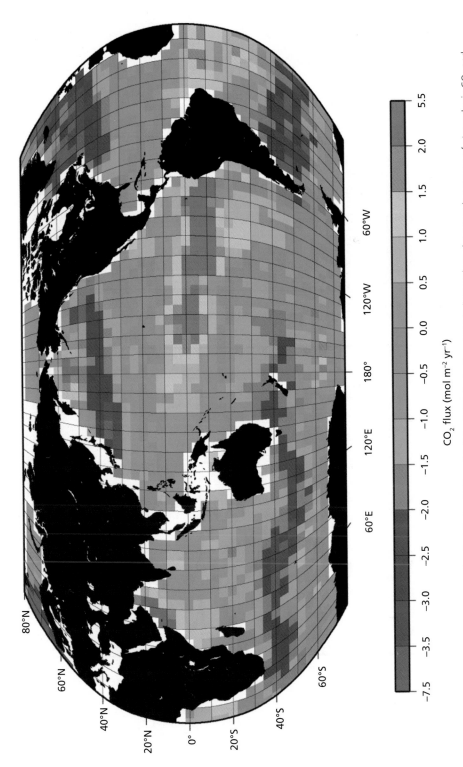

**Plate 8 (Fig. 9.17)** Mean annual net air-sea flux for $CO_2$ (mol $CO_2$ m$^{-2}$ yr$^{-1}$) for 1995. Red-yellow areas indicate where the ocean is a source of atmospheric $CO_2$ and blue-purple areas where it is a sink (Takahashi *et al.*, 2002). Available from IPCC. Reprinted with permission from Elsevier.

CO$_2$ flux (mol m$^{-2}$ yr$^{-1}$)

-7.5  -3.5  -3.0  -2.5  -2.0  -1.5  -1.0  -0.5  0.0  0.5  1.0  1.5  2.0  5.5

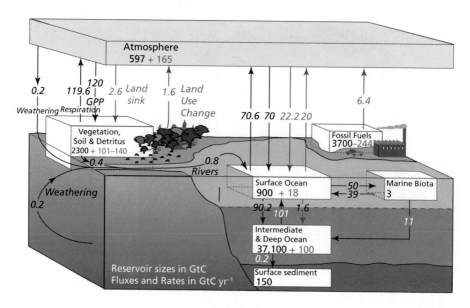

**Plate 9 (Fig. 9.18)** Global carbon cycle (Denman *et al.*, 2007). Available from IPCC.

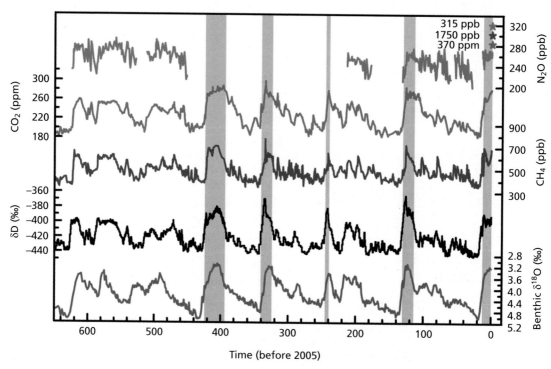

**Plate 10 (Fig. 9.20)** Changes in concentrations of the gases N₂O, CH₄ and CO₂ in bubbles in Antarctic ice over the last 600 000 years, the temperature of the ice as formed which is recorded as changes in the hydrogen isotope (δD) composition of the ice and the estimate of global ice volume based on the oxygen isotope composition of benthic foraminifera δ¹⁸O. Available from the IPCC, modern day concentrations of N₂O (315 ppb), CH₄ (1750 ppb) and CO₂ (370 ppm) are indicated by star symbols for comparison.

polysaccharides released from bacteria and phytoplankton which induce the aggregation of smaller particles into larger amorphous particles (Passow *et al.*, 2001). Shanks and Trent (1980) concluded that, like faecal pellets, marine snow can accelerate the transport of material to the deep ocean and may in fact be the main package in which the vertical flux takes place. It is apparent, therefore, that the main flux of particulate material from the surface ocean to the sediment surface is driven by large organic aggregates, which form the predominant carriers of the global carbon flux; that is, the organic matter that escapes recycling and is replaced in 'new' production (see Chapter 9). As they descend through the water column, these large aggregates drag down small-sized inorganic particles. Some of these inorganic particles are lithogenous in origin, and although they make up only a small fraction (usually less than ~5%) of the total TSM in surface water, they form the main contributors of material to lithogenous pelagic clays. McCave (1984) therefore suggested the intriguing idea that 'the processes of pelagic sedimentation of lithogenic matter may be viewed as a side effect of excretion and disposal of other waste products of the ocean's biological system'. It has also been suggested that the denser mineral particles trapped within these organic aggregates play an important role as 'ballast' encouraging the more rapid sinking of these particles, although the significance of this is still rather uncertain (e.g. Thomalla *et al.*, 2008).

The concept of a two-particle population in seawater has considerably advanced our understanding of how TSM is transported through the ocean reservoir. However, it has become apparent that the picture is more complicated than this simple twofold particle classification would suggest. The measurement of the natural series radionuclides of Pu, Pb, Th and U in the water column, which will be discussed further in Chapter 11, demonstrates that there is continued exchange between large and small particles during their transit through the water column to the sea bed, implying that the aggregation of small particles into larger particles and the reverse disaggregation occurs throughout the entire water column (Anderson, 2003).

The down-column flux is driven by CPM, which is composed mainly of biogenic aggregates. As a result, primary production in the euphotic layer exerts a fundamental control on the initiation of this down-column flux, via export production. An important thrust in marine biogeochemistry over the past few years has been to model the relationship between primary production at the surface and the flux of material through the interior of the ocean, and this is considered in Chapter 12. At present, it is sufficient to point out that it does appear that the primary down-column flux is closely related to surface productivity, and that the flux often varies in magnitude in relation to seasonal changes in photosynthetic activity in the euphotic zone. Further, it is generally recognized that around 95% of the organic matter produced in the surface ocean is recycled in the upper few hundred metres of the water column. However, the 5% or so that does escape from surface waters, that is, the export production that is replaced by new production (see Chapter 9), is composed largely of aggregates and it is these that dominate the down-column material flux. For example, Bishop *et al.* (1977) demonstrated that ~95% of the down-column flux of particulate material at a site in the equatorial Atlantic is associated with large faecal aggregates, although these only account for <5% of the total POC. These authors estimated that the transit times for the faecal material through a 4 km water column would be ~10–15 days. During a transit of this relatively short duration, lateral displacement resulting from deep-ocean advection in the region would be only ~40 km. As a result, material deposited in the underlying sediments would reflect the oceanic variability in the surface water TSM, thus offering an explanation for the source-related clay mineral distribution patterns found in deep-sea sediments (see Section 15.1.2). Such a relationship could not exist, given the speed of deep ocean currents, if settling rates were only a few metres per day. It has generally been considered that the *primary* flux of particulate material through the water column is in a downward direction; however, Smith *et al.* (1989) showed that upward fluxes of particulate organic matter composed of positively buoyant particles can also be important.

## 10.5 Down-column changes in the composition of oceanic TSM and the three-layer distribution model

The vertical transport of TSM in the water column is driven mainly by the CPM flux, that is, the flux is dominated by the larger particles. As the particles settle they undergo considerable modification via processes such as decomposition, aggregation–disaggregation, zooplankton grazing and dissolution.

The major components of the sinking particulate mass flux are biogenic (organic matter, calcium carbonate and opal) and lithogenic (see Section 12.1.1.2). The proportions of these components change as the particles descend the water column. The most notable change is that affecting organic matter, which undergoes bacterial degradation in subsurface waters. For example, Martin *et al.* (1987) concluded that 50% of the organic carbon removed from surface waters is regenerated at depths <300 m, 75% is regenerated by 500 m, and 90% by 1500 m (see Section 12.1). The exact values will vary from place to place and are generally poorly known, but the pattern of extensive loss of organic matter during sinking is very clear. The biogenic shell components change less dramatically with depth, but the overall effect is to increase the proportions of mineral material in the sinking mass flux. For example, Masuzawa *et al.* (1989) presented data on the mass flux composition of material collected by five sediment traps deployed between 890 m and 3420 m in the Japan Sea. A summary of the data from the traps is given in Table 10.2, from which it is apparent that the contribution of organic matter to the total flux decreases from ~35% at 890 m to ~5% in the deepest trap. In contrast, the contribution made by lithogenic material to the total flux increases from ~15% at 890 m to ~57% at 3420 m. As noted earlier, the role of mineral phases as 'ballast' to increase the sinking rate of material, means that the organic and inorganic fluxes are coupled.

In general, therefore, it may be concluded that internally produced biogenic components decrease, and lithogenic components increase, in importance as the CPM flux carries TSM down the water column. There are a number of reasons for this that can be related to the three-layer TSM water-column distribution model, which combines downward and upward particulate signals, but which also can be affected by lateral signals.

*The downward signal.* Part of the decrease in the proportion of the biogenic components results from their loss due to the bacterial decomposition of POM and the dissolution of shell material, processes that occur both in the euphotic zone and at depth during the descent of the TSM through the water column. It is these processes that lead to the setting up of the clear-water minimum zone, and they have two effects on the bulk TSM.

1 Because biogenic components make up a large percentage of the surface water TSM, their destruction leads to a decrease in the absolute concentration of particles towards the clear-water minimum zone.

2 As these components are removed, there is an increase in the relative proportions of the non-biogenic material that survives to reach mid-water depths.

*The upward signal.* The organic carbon content of most deep-sea sediments is generally quite low (<5%; see Section 14.2), with the result that the resuspended particulate population is relatively rich in aluminosilicates and sometimes also in shell material. The in situ resuspended flux of bottom sediments, which results in the production of nepheloid layers, therefore will increase the proportions of non-biogenic material in bottom water TSM.

**Table 10.2** Changes in the composition of the particle mass flux composition with depth (data from Masuzawa *et al.*, 1989).

| Trap depth | Mass flux | Flux composition (%) | | | |
|---|---|---|---|---|---|
| (m) | (mg m$^{-2}$ day$^{-1}$) | CaCO$_3$ | Opal | Organic matter | Lithogenic matter |
| 890 | 139 | 8.7 | 41.0 | 34.3 | 16.0 |
| 1100 | 116 | 8.2 | 41.0 | 33.2 | 17.6 |
| 1870 | 50.4 | 9.6 | 36.8 | 23.8 | 29.8 |
| 2720 | 49.4 | 8.3 | 37.4 | 9.4 | 44.8 |
| 3420 | 60.0 | 9.8 | 27.2 | 6.0 | 57.0 |

*The lateral signal.* The mixing of bottom sediment into the water column in the erosive boundary currents on the western edges of the ocean basins, and its advective transport, is now known to be a major pathway for the introduction of small-sized refractory particles into the basin interiors (see e.g. Brewer *et al.*, 1980; Honjo *et al.*, 1982; Spencer, 1984). The increase in the proportions of non-biogenic components at mid-water depths below the clear-water minimum can therefore also result from a direct addition of fine aluminosilicate material that is transported laterally from the continental margins and is then transferred into the vertical CPM flux.

The general relationships in the processes that drive the down-column TSM flux are illustrated schematically in Fig. 10.3.

Up to this point we have concentrated on TSM, and the distinction between TSM and dissolved components is usually operationally defined by a ~0.45 μm cut-off, using a membrane filter. There is, however, a population of material that bridges the 'dissolved–particulate' classification, but which is usually described as 'dissolved' by conventional filtration techniques. This population consists of *colloids*, that is, submicron particles between ~1 nm and ~1 μm. Isao *et al.* (1990) identified colloids in the size range 0.38–1 μm in the North Pacific. They reported that 95% of the colloidal particles were *non-living* and occurred in the upper 50 m of the water column at concentrations in the order of $5 \times 10^7$–$8 \times 10^7$ particles ml$^{-1}$. The concentrations decreased to $2 \times 10^6$ ml$^{-1}$ at 200 m. In the upper 200 m, ~95% of the colloidal

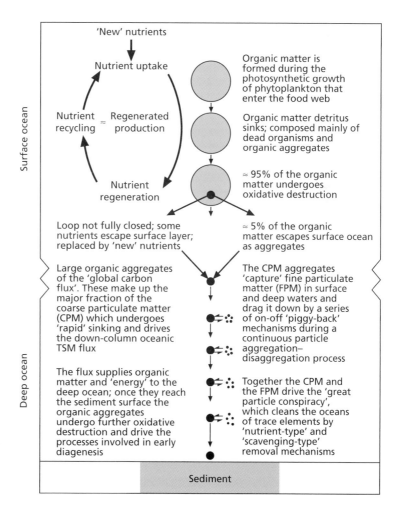

**Fig. 10.3** A schematic representation of the processes that drive the down-column flux of oceanic TSM. Reprinted with permission from Elsevier.

material measured was <0.6 μm. Wells and Goldberg (1991) presented data on the concentration and vertical distribution of colloids (<120 nm) in waters ~15 km off-shore in the Santa Monica Basin. They found that the colloids were highly stratified, that is, (i) concentrations were below detection (<$10^4$ particles ml$^{-1}$) at the surface; (ii) they increased sharply near the lower thermocline (~40–100 m) to concentrations of >$10^9$ particles ml$^{-1}$; and (iii) decreased again to <$10^4$ particles ml$^{-1}$ below the thermocline. This distribution is different from that of the larger colloids (0.38–1 μm) described by Isao *et al.* (1990).

Organic matter appears to be the most important constituent of the colloids reported in both studies, although Wells and Goldberg (1992) also identified inorganic material such as iron colloids and clay minerals, together with trace metals, in oceanic colloids. The mainly organic nature of the colloids suggests that they are derived primarily from biological processes, and Isao *et al.* (1990) concluded that a significant proportion of the material defined by conventional techniques as dissolved organic matter (DOM) in the upper ocean may be in the form of these colloids (see also Chapter 9.4). According to Wells and Goldberg (1991) the vertical stratification of the colloids they described indicates that they are reactive and have short residence times in seawater. In this context, Moran and Buesseler (1992) estimated that the residence time of colloidal $^{234}$Th (<0.2 μm) with respect to aggregation into small particles (>0.2–53 μm) in the upper ocean was around 10 days. The $^{234}$Th activity of the colloidal particles was the same, on average, as that of small particles (>0.2–53 μm), implying that the colloidal matter may be as important as traditionally defined TSM in the cycling of Th, and possibly other reactive elements, between dissolved and particulate forms in seawater (see Section 11.6.3.1).

It may be concluded, therefore, that macromolecular colloidal matter in the upper ocean has a short residence time and a rapid turnover. Colloids may act as a reactive intermediary in the marine geochemistry of trace metals (see Chapter 11), and a biologically labile pool of colloidal matter would also affect carbon cycling in the oceans. The colloids are not explicitly represented in Fig. 10.3 but are present throughout the water column exchanging with the fine particulate matter which is in turn exchanging with the coarse particulate matter.

## 10.6 Particulate material in the oceans: summary

1 The vertical distribution of TSM in the water column can be described in terms of a three-layer model in which a surface layer, a clear-water minimum layer and a deep-water layer are distinguished. The principal mechanisms that contribute to the setting up of this model are a primary downward signal from the ocean surface, transporting mainly biogenic aggregates, and a secondary (or resuspended) upward signal from the bottom sediment, carrying mainly aluminosilicates. The two end-members in the system are therefore the *surface ocean* and the *surface sediment*. The surface ocean is the zone in which biological cycles are especially active, and it is here that biogenic solids interact with the dissolved and particulate elements transported by river run-off, atmospheric deposition and vertical mixing, and so enable them to enter the oceanic biogeochemical cycles.

2 Oceanic TSM consists of a variety of components from both external sources (e.g. aluminosilicates and quartz) and internal sources (e.g. biological material, such as tissues and shells). The internal sources usually dominate.

3 There is a continual cycle of aggregation and disaggregation of oceanic TSM, which leads to the production of a particle population that has no specific size classes within it. From the point of view of biogeochemical reactivity, however, it is convenient to distinguish between colloidal material, fine particulate matter (FPM, diameters <5–10 μm) and coarse particulate matter (CPM, diameters >50 μm).

4 Much of the particulate flux to deep waters and sediments is driven by the CPM, which consists mainly of large organic aggregates. The CPM can also carry FPM to deep waters by 'on–off' piggyback-type mechanisms, in which particles can join the sinking flux and be lost from it back to the water column.

In the present chapter we have described the distribution, composition and sinking characteristics of the oceanic TSM population. It is this microcosm of particles that plays a vital role in regulating the chemical composition of seawater via *particulate–dissolved* equilibria. In the next chapter the distributions of trace elements in the oceans will be discussed, and in doing this the TSM story will be used as a background to assess the factors that control the down-column transport of these elements. It is this vertical transport to the benthic boundary layer that eventually results in the ultimate removal of the elements into the sediment reservoir, thus initiating the final, but still complex, act in the 'great particle conspiracy', which was used as the starting point in the present chapter.

# References

Anderson, R.F. (2003) Chemical tracers of particle transport, in *Treatise on Geochemistry*, H.D. Holland and K.K. Turekian (eds), Vol. 6, The Oceans and Marine Geochemistry, H. Elderfield (ed.) pp.247–291. Oxford: Elsevier Pergamon.

Biscaye, P.E. and Eittreim, S.T. (1977) Suspended particulate loads and transports in the nepheloid layer of the abyssal Atlantic Ocean. *Mar. Geol.*, **23**, 155–172.

Bishop, J.K.B., Edmond, J.M., Kellen, D.R., Bacon, M.P. and Silker, W.B. (1977) The chemistry, biology, and vertical flux of particulate matter from the upper 400 m of the equatorial Atlantic Ocean. *Deep Sea Res.*, **24**, 511–548.

Brewer, P.G., Nozaki, Y., Spencer, D.W. and Fleer, A.P. (1980) Sediment trap experiments in the deep North Atlantic: isotopic and elemental fluxes. *J. Mar. Res.*, **38**, 703–728.

Brun-Cottan, J.C. (1976) Stokes settling and dissolution rate model for marine particles as a function of size distribution. *J. Geophys. Res.*, **81**, 1601–1605.

Chester, R. (1982) The concentration, mineralogy, and chemistry of total suspended matter in sea water, in *Pollutant Transfer and Transport in the Sea*, G. Kullenberg (ed.), pp.67–99. Boca Raton: CRC Press.

Chester, R. and Aston, S.R. (1976) The geochemistry of deep-sea sediments, in *Chemical Oceanography*, Vol. 6, J.P. Riley and R. Chester (eds), pp.281–390. London: Academic Press.

Chester, R. and Stoner, J.H. (1972) Concentration of suspended particulate matter in surface seawater. *Nature*, **240**, 552–553.

Chester, R. and Stoner, J.H. (1974) The distribution of Mn, Fe, Cu, Ni, Co, Ga, Cr, V, Ba, Sr, Sn, Zn and Pb in some soil-sized particulates from the lower troposphere over the World Ocean. *Mar. Chem.*, **2**, 157–188.

Cho, B.C. and Azam, F. (1988) Major role of bacteria in biogeochemical fluxes in the ocean's interior. *Nature*, **322**, 441–443.

Collier, R.W. and Edmond, J.M. (1983) Plankton compositions and trace element fluxes from the surface ocean, in *Trace Metals in Sea Water*, C.S. Wong, E. Boyle, K.W. Bruland, J.D. Burton and E.D. Goldberg (eds), pp.789–809. New York: Plenum.

Connary, S.C. and Ewing, M. (1972) The nepheloid layer and bottom circulation in the Guinea and Angola Basins, in *Studies in Physical Oceanography*, A.L. Gordon (ed.), pp.169–184. London: Gordon and Breach.

Eittreim, S., Thorndike, E.M. and Sullivan, L. (1976) Turbidity distribution in the Atlantic Ocean. *Deep Sea Res.*, **23**, 1115–1127.

Ewing, M. and Thorndike, E. (1965) Suspended matter in deep ocean water. *Science*, **147**, 1291–1294.

Forchhammer, G. (1865) On the composition of sea water in the different parts of the ocean. *Philos. Trans. R. Soc. London*, **155**, 203–262.

Fowler, S.W. (1977) Trace elements in zooplankton particulate products. *Nature*, **269**, 51–53.

Fowler, S.W., Buat-Menard, P., Yokoyama, Y., Ballestra, S., Holm, E. and Nguyen, H.V. (1987) Rapid removal of Chernobyl fallout from Mediterranean surface waters by biological activity. *Nature*, **329**, 56–58.

Gardner, W.D., Southard, J.B. and Hollister, C.D. (1985) Sedimentation, resuspension and chemistry of particles in the northwest Atlantic. *Mar. Geol.*, **65**, 199–242.

Gross, T.F., Williams, A.J. and Nowell, A.R.M. (1988) A deep-sea sediment transport storm. *Nature*, **331**, 518–521.

Honjo, S. (1980) Material fluxes and modes of sedimentation in the mesopelagic and bathypelagic zones. *J. Mar. Res.*, **38**, 53–97.

Honjo, S., Spencer, D.W. and Farrington, J.W. (1982) Deep advective transport of lithogenic particles in Panama Basin. *Science*, **216**, 516–518.

Honjo, S., Doherty, K.W., Agrawal, Y.C. and Asper, V.L. (1984) Direct optical assessment of large amorphous aggregates (marine snow) in the deep ocean. *Deep Sea Res. I*, **31**, 67–76.

Isao, K., Hara, S., Terauchi, K. and Kogure, K. (1990) Role of sub-micrometre particles in the ocean. *Nature*, **345**, 242–244.

Jerlov, N.G. (1953) Particle distribution in the ocean. *Rep. Swed. Deep Sea Exped.*, **3**, 71–97.

Jickells, T.D., Deuser, W.G. and Belastock, R.A. (1990) Temporal variations in the concentrations of some particulate elements in the surface waters of the Sargasso Sea and their relationship to deep-sea fluxes. *Mar. Chem.*, **29**, 203–219.

Krishnaswami, S. and Sarin, M.M. (1976) Atlantic surface particulate: composition, settling rates and dissolution in the deep-sea. *Earth Planet. Sci. Lett.*, **32**, 430–440.

Lal, D. (1977) The oceanic microcosm of particles. *Science*, **198**, 997–1009.

Lampitt, R.S. and Antia, A.N. (1997) Particle flux in deep seas: regional characteristics and temporal variability. *Deep-Sea Res. I*, **44**, 1377–1403.

Lisitzin, A.P. (1972) *Sedimentation in the World Ocean*. Tulsa, OK: Society of Economic Paleontologists and Mineralogists, Special Publication, **17**.

McCave, I.N. (1975) Vertical flux of particles in the ocean. *Deep Sea Res. I*, **22**, 491–502.

McCave, I.N. (1984) Size spectra and aggregation of suspended particles in the deep ocean. *Deep Sea Res. I*, **31**, 329–352.

McCave, I.N., Hall, I.R., Antia, A.N., *et al.* (2001) Distribution, composition and flux of particulate material over the European margin at 47°–50°N. *Deep-Sea Res. II*, **48**, 3107–3139.

Martin, J.H. (1970) The possible transport of trace metals via moulted copepod exoskeletons. *Limnol. Oceanogr.*, **15**, 756–761.

Martin, J.H. and Knauer, G.A. (1973) The elemental composition of plankton. *Geochim. Cosmochim. Acta*, **37**, 1639–1653.

Martin, J.M. and Whitfield, M. (1983) The significance of the river input of chemical elements to the ocean, in *Trace Elements in Sea Water*, C.S. Wong, E. Boyle, K.W. Bruland, J.D. Burton and E.D. Goldberg (eds), pp.265–296. New York: Plenum.

Martin, J.H., Knauer, G.A., Karl, D.M. and Broenkow, W.W. (1987) VERTEX: carbon cycling in the Northeast Pacific. *Deep Sea Res. I*, **34**, 267–286.

Masuzawa, T., Noriki, S., Kurosaki, T. and Tsunogai, S. (1989) Compositional changes of settling particles with water depth in the Japan Sea. *Mar. Chem.*, **27**, 61–78.

Moran, S.B. and Buesseler, K.O. (1992) Short residence time of colloids in the upper ocean estimated from 238U–234Th disequilibria. *Nature*, **359**, 221–223.

Nittrouer, C.A., Curtin, T.B. and DeMaster, D.J. (1986) Concentration and flux of suspended sediment on the Amazon continental shelf. *Cont. Shelf Res.*, **6**, 151–174.

Passow, U., Shipe, R.F., Murray, A., Pak, D.K., Brezinski, M.A. and Alldredge, A.L. (2001) The origin of transparent exopolymer particles (TEP) and their role in the sedimentation of particulate matter. *Cont. Shelf Res.*, **21**, 327–346.

Shanks, A.L. and Trent, J.D. (1980) Marine snow: sinking rates and potential role in vertical flux. *Deep Sea Res. I*, **27**, 137–143.

Sherrell, R.M., Field, M.P. and Gao, Y. (1998) Temporal variability of suspended mass and composition in the Northeast Pacific water column: relation to sinking flux and advection. *Deep-Sea Res. II*, **45**, 733–761.

Smith, C.R., Mincks, S. and DeMaster, D. (1989) A synthesis of bentho-pelagic coupling on the Antarctic shelf: Food banks, ecosystem inertia and global climate change. *Deep-Sea Res. II*, **53**, 875–894.

Spencer, D.W. (1984) Aluminium concentrations and fluxes in the ocean, in *Global Ocean Flux Study*, pp.206–220. Washington, DC: National Academy Press.

Spencer, D.W., Brewer, P.G., Fleer, A., Honjo, S., Krishnaswami, S. and Nozaki, Y. (1978) Chemical fluxes from a sediment trap experiment in the deep Sargasso Sea. *J. Mar. Res.*, **36**, 493–523.

Stemmann, L., Jackson, G.A. and Ianson, D. (2004) A vertical model of particle size distributions and fluxes in the midwater column that includes biological and physical processes – Part 1: model formulation. *Deep Sea Res. I*, **51**, 865–884.

Thomalla, S.J., Poulton, A.J., Sanders, R., Turnewitch, R., Holligan, P.M. and Lucas, M.I. (2008) Variable export fluxes and efficiencies for calcite, opal, and organic carbon in the Atlantic Ocean: A ballast effect in action? *Global Biogeochem. Cycl.*, **22**, GB1010, doi:10.1029/2007GB002982,

Turekian, K.K. (1977) The fate of metals in the oceans. *Geochim. Cosmochim. Acta*, **41**, 1139–1144.

Wells, M.L. and Goldberg, E.D. (1991) Occurrence of small colloids in sea water. *Nature*, **353**, 342–344.

Wells, M.L. and Goldberg, E.D. (1992) Marine submicron particles. *Mar. Chem.*, **40**, 5–18.

# 11     Trace elements in the oceans

Trace elements are present in seawater at concentrations that range down to picomoles per litre (p mol l$^{-1}$) and even lower. Such small concentrations pose extreme analytical problems, and it is only relatively recently that these have been fully overcome. It is now known, for example, that contamination and the lack of sufficiently precise analytical techniques have led in the past to concentration data reported that is, for some trace elements, too high by factors as much as 10$^3$. With recent improvements in sampling and analysis techniques, chemical oceanographers can now collect and analyse seawater for essentially all the elements in the periodic table and the results demonstrate distributions that can be understood in terms of well established biogeochemical processes. A selection of the new trace-element concentration data in seawater is given in Table 11.1. At the same time that trace-element concentration data were being refined, there were also advances in our understanding of the speciation of the elements in seawater.

## 11.1 Introduction

The definition of trace elements is inevitably subjective. There are a small number of major ions that dominate the chemical composition of seawater – sodium, potassium, calcium, magnesium, chloride, sulfate and bicarbonate. All the other elements in the periodic table are present in seawater, usually at concentrations of μmol l$^{-1}$ or often very much less. Most of the elements in the periodic table are metals and so the terms trace elements and trace metals are often used almost interchangeably, but here the term trace elements will be used mostly. In this chapter the primary focus will be on these trace elements, although comparisons to the major ions will be made

where this is useful. The behaviour of the major nutrients nitrogen, phosphorus and silicon has been considered in detail in Chapter 9 and will not be repeated here. However, the behaviour of some trace elements closely resembles that of the nutrients and hence the behaviour of these major nutrients will be noted where appropriate.

The predicted inorganic chemical speciation of selected elements is also presented in Table 11.1. These predictions are based on fundamental physicchemical data such as equilibrium constants (Byrne, 2002). They are valid in the absence of organic matter as a competing trace element ligand, but such organic complexation is in fact widespread as we shall see later (Section 11.6).

The extremely low concentrations of many trace elements make their accurate analysis challenging. Much of the earlier data collected prior to 1970 is now known to be inaccurate and the problems were primarily one of sample contamination, rather than analytical detection limits. The pioneering work of Claire Patterson and others (Flegal, 1998) demonstrated that rigorous precautions are necessary to avoid the contamination of samples during collection, both from sampling equipment or the ship itself, and subsequently in the laboratory. We also now understand the procedures necessary to avoid the loss of analytes which can, for instance, occur by volatilization or adsorption to sample bottle walls. The latter problem can be greatly reduced for many elements by acidifying, and/or freezing, the samples during storage. It is now common practice to filter samples upon collection, but in open ocean waters, the dissolved concentrations tend to greatly exceed the particulate concentration, so this step is sometimes excluded.

*Marine Geochemistry*, Third Edition. Roy Chester and Tim Jickells.
© 2012 by Roy Chester and Tim Jickells. Published 2012 by Blackwell Publishing Ltd.

**Table 11.1** Selected data on some trace and major elements in seawater. Major species and average concentration along with oceanic distribution from Bruland and Lohan (2003), Nozaki (http://www.agu.org/eos_elec/97025e.html) and Johnson (http://www.mbari/org/chemsensor/pteo.htm), residence time based on net dissolved + particulate fluvial inputs Table 6.13 and mean oceanic concentrations, using an ocean volume of $1.4 \times 10^{21}$ l. Details of more element distributions are available from these references. The behaviour of many elements can be simply described as conservative, nutrient or scavenged but some deviate from the ideal and are marked *, for instance Ca, V and Cr are almost conservative but show a slight surface depletion, and so have some nutrient-like behaviour.

| Element | Dominant Inorganic Form | Mean Ocean Concentration | Residence Time years | Behaviour |
|---|---|---|---|---|
| Li | $Li^+$ | $26 \mu mol\,l^{-1}$ | $2 \times 10^6$ | conservative |
| B | $H_3BO_3$ | $416 \mu mol\,l^{-1}$ | $14 \times 10^6$ | conservative |
| Na | $Na^+$ | $468\,mmol\,l^{-1}$ | $68 \times 10^6$ | conservative |
| Mg | $Mg^{2+}$ | $53\,mmol\,l^{-1}$ | $7 \times 10^6$ | conservative |
| Al | $Al(OH)_3\ Al(OH)_4^-$ | $2\,nmol\,l^{-1}$ | <200 | scavenged |
| S | $SO_4^{2-}$ | $28\,mmol\,l^{-1}$ | $6 \times 10^6$ | conservative |
| Cl | $Cl^-$ | $546\,mmol\,l^{-1}$ | $119 \times 10^6$ | conservative |
| K | $K^+$ | $10.2\,mmol\,l^{-1}$ | $6 \times 10^6$ | conservative |
| Ca | $Ca^{2+}$ | $10.3\,mmol\,l^{-1}$ | $0.6 \times 10^6$ | conservative* |
| V | $HVO_4^{2-}$ | $36\,nmol\,l^{-1}$ | 10000 | conservative* |
| Cr | $CrO_4^{2-}$ | $5\,nmol\,l^{-1}$ | 2000 | conservative* |
| Mn | $Mn^{2+}$ | $0.3\,nmol\,l^{-1}$ | 12 | scavenged |
| Fe | $Fe(OH)_2^+, Fe(OH)_3$ | $0.5\,nmol\,l^{-1}$ | 0.6 | see text |
| Co | $Co^{2+}$ | $0.02\,nmol\,l^{-1}$ | 62 | scavenged |
| Ni | $Ni^{2+}$ | $8\,nmol\,l^{-1}$ | 4700 | nutrient |
| Cu | $CuCO_3$ | $3\,nmol\,l^{-1}$ | 1400 | nutrient/scavenged |
| Zn | $Zn^{2+}$ | $5\,nmol\,l^{-1}$ | 1000 | nutrient |
| Cd | $CdCl_2$ | $0.6\,nmol\,l^{-1}$ | 6200 | nutrient |
| Pb | $PbCO_3$ | $0.01\,nmol\,l^{-1}$ | 9 | scavenged |
| Total N | $NO_3^-$ | $30 \mu mol\,l^{-1}$ | 28000[@] | nutrient |
| DIP | $HPO_4^{2-}$ | $2.3 \mu mol\,l^{-1}$ | 23000[b] | nutrient |
| DSi | $H_4SiO_4$ | $100 \mu mol\,l^{-1}$ | 20000[c] | nutrient |

[@](based on fluvial TNinput); [b](based on fluvial TPinput); [c](based on fluvial dissolved Si).

It is now routine to rigorously pre-clean sampling and storage equipment, to use sampling equipment that avoids waters influenced by the ship, to subsequently handle samples in 'clean rooms' where the air is filtered to avoid contamination and for all participants to wear clothing that minimizes the risk of contaminating samples (Howard and Statham, 1993). These procedures coupled to modern high sensitivity analytical methods have delivered a revolution in our understanding of trace element behaviour This improved confidence in the data has also illustrated the fundamental analytical challenge of demonstrating that the results obtained are reliable when conventional laboratory analytical blank procedures alone are an inadequate method to demonstrate that samples have not been contaminated or indeed that analytes have not been lost by inadequate storage procedures. Boyle et al., (1977) laid out three criteria that can be used in such circumstances to build confidence in the data:

1 The new trace element concentrations must be confirmed by inter-laboratory agreement.

2 The vertical distribution profiles obtained from the new data must show smooth variations that can be related to hydrographic and chemical features displayed by conventionally measured properties, which themselves have well established distributions.

**3** The regional variations derived on the basis of the new data should be compatible with the large-scale physical and chemical circulation patterns known to operate in the ocean system. In other words, the new trace element data must be oceanographically consistent.

As with all components within the marine environment, the distribution of trace elements reflects:
• their inputs from atmospheric, fluvial, sedimentary and mid ocean ridge sources,
• their internal biogeochemical cycling within the ocean,
• ocean circulation.

The major improvement in the quality and quantity of oceanic trace element data is now ushering a new era where there is sufficient high quality information to map the oceanographic distribution of many trace elements and understand the processes controlling this distribution (SCOR, 2007). The relationship between these controls on trace elements and their oceanic distributions reflects, in part, the residence time of the trace element in the oceans in comparison to the timescales of circulation and mixing of water within the ocean. The concept of residence time was briefly introduced in Chapter 9 for the nutrients and it is now explored in more detail.

## 11.2 Oceanic residence times

The residence time of an element in the oceans is the average time it spends in the sea before being removed into the sediment sink. In a steady-state system it is assumed that the input of an element (mainly via river run-off, atmospheric deposition and hydrothermal inputs) per unit time is balanced by its output (mainly via sedimentation). There are a number of ways in which equations can be written to describe this relationship. For example, the residence times ($\tau$) of an element is often calculated from the Equation

$$\tau = A/(\mathrm{d}A/\mathrm{d}t) \tag{11.1}$$

where $A$ is the total amount of the element in suspension or solution in seawater, and $\mathrm{d}A/\mathrm{d}t$ is the amount introduced or removed per unit time. In making this type of calculation it is assumed that the element is completely mixed in the system in a time that is short compared with its residence time, and that neither $A$

nor $\mathrm{d}A/\mathrm{d}t$ change appreciably in three to four times this period – the steady state assumption. This equation can be simplified to

$$\tau = \text{Inventory/Input,} \tag{11.2}$$
$$\text{where Input = Removal at steady state}$$

Whitfield (1979) introduced the concept of a *mean oceanic residence time* (MORT), which is defined as the total quantity of an element present in the oceans divided by its input rate (from rivers) or its output rate (to the sediment). The MORT values, which assume the whole ocean to be a well-stirred system, are only approximate quantities because they ignore a number of important input (e.g. atmospheric deposition, hydrothermal venting) and output (e.g. atmospheric exchange, mid ocean ridge processes) terms. The MORT approach is used to provide the residence times in Table 11.1. The fluvial fluxes used are taken from Table 6.13, and the ocean volume is taken as $1.4 \times 10^{21}$ l. Ignoring atmospheric will mean that the residence times presented are over estimated. Ignoring mid ocean ridge fluxes will in general also mean that residence times are overestimated, but in a few cases the mid ocean ridges represent net sinks not sources of trace elements (see Chapters 5 and 6). As discussed in Chapters 2–6 all the input fluxes are rather uncertain. A particular concern for some elements with low solubility such as Fe and Al, is that the total flux includes particulate material from which the trace elements may not be released into solution. Such an overestimate of the soluble and bioavailable flux will lead to an underestimate of residence times. A final concern is that the contemporary fluxes of some elements such as nitrate have been greatly increased by human activity and the oceans have yet to reach steady state with respect to these fluxes. All of these caveats suggest that the mean ocean residence times should be considered as having quite large uncertainties, possibly of an order of magnitude. These uncertainties can be put into perspective by considering the very large range of residence times in Table 11.1 which go from <200 to $10^9$ years.

The elements can be divided into three groups based on their residence times, and these distinctions are clearly robust despite the uncertainties in the exact values of the residence times.

Group 1    Residence times of the order of $10^6$ years or longer including Li, B, Na, Mg, $SO_4^{2-}$, $Cl^-$, Ca, K.

Group 2    Residence times of the order of $10^3$ years including Cs, V, Cr, Cu, Zn, N (as nitrate), P and Si.

Group 3    Residence times less than $10^3$ years including Al, Mn, Fe, Co and Pb.

This grouping reflects patterns of behaviour of trace elements in the oceans which are the subject of the remainder of this chapter. These patterns can be classified into three broad types of behaviour; conservative, nutrient like and scavenged as will be discussed later. These classifications are independent of, but completely consistent with, the residence time groupings.

The wider problems involved in the overall concept of oceanic residence times have been discussed by a number of authors.

Bruland (1980) attempted to re-evaluate the residence time concept. He pointed out that for elements that have very long residence times in seawater (Group 1), the residence time can be defined as outlined above, that is $\tau = A/(dA/dt)$. The concept of oceanic residence times, however, becomes more complex for the other elements with shorter residence times. The major nutrients nitrate, phosphorus and silicon (see Chapter 10), and some metals such as Cd and Zn, which show similar behaviour to the nutrients as we will see later, are involved in active biological removal mechanisms, and show surface depletion and a subsurface regeneration. These elements therefore undergo an extensive process of internal cycling within the oceans between surface and deepwater mediated by biological and physical processes prior to their final removal in the sediments. Their residence times in the ocean are long with respect to these internal cycles and to the ocean water internal circulation timescales of a few thousand years (Section 7.3.3.), but much shorter than the residence times of elements in Group 1. Once residence times become less than the ocean mixing times (~1000 years) the assumption behind the residence time calculations break down so the actual residence times reported for the elements in Group 3 are not reliable, but it is clear that the residence times are very much shorter than the elements in Group 1 and 2. For the elements, in Group 3 atmos-

pheric and mid-ocean ridges inputs are also important in addition to river inputs, further complicating the issue.

## 11.3 The vertical distribution of dissolved trace elements in the oceans

### 11.3.1 Conservative-type distributions

This term is used for elements that have essentially constant vertical distributions in the ocean with depth (e.g. Fig. 11.1). Such distributions imply very limited interactions with particulate matter or biological processes relative to their abundance, and hence they have very long residence times. All the elements classified within Group 1 with residence times of the order of $10^6$ years above show such a conservative distribution (Bruland and Lohan, 2003). The conservative elements include all of the Group A and most of the B elements of the periodic Table (Li, Na, K, Cs, Mg, Sr). Ca is also included although it has a significant biological role, being used to build skeletons for some phytoplankton and zooplankton. In this case the biological cycling is small relative to the available concentrations and hence the concentrations approximate to conservative behaviour. Anions in seawater which have little biological importance such as fluoride, chloride and bromide also show conservative behaviour. Bicarbonate and sulfate do have important biological roles but, as with calcium, the biological requirement is small compared to the available supply so the behaviour again approximates to the conservative type behaviour.

Some transition metals in high oxidation states exist in seawater as oxyanions including V ($HVO_4^{2-}$), Cr ($CrO_4^{2-}$) and Mo($MoO_4^{2-}$) (Bruland and Lohan, 2003). Since most particles in the ocean are positively charged, such anionic species have rather limited interactions with particles and hence, if there is also little biological demand compared to the available supply, these elements will also approximate to conservative behaviour in the sea, although their residence times tend to be shorter than that of the elements discussed previously.

### 11.3.2 Nutrient-type distribution

We have noted before in Chapter 9, the distribution of the major nutrients (nitrate phosphate and sili-

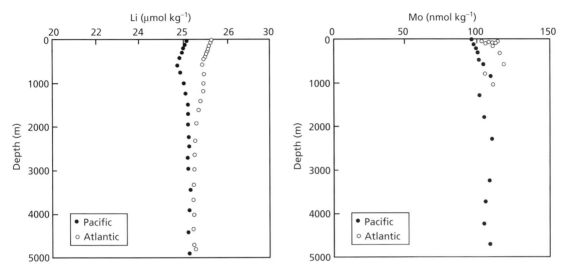

**Fig. 11.1** Vertical distribution of Li and Mo in the Atlantic and Pacific Oceans as examples of conservative type behaviour. Lithium was estimated from the formula 1323*Salinity/1.80655/(1000-1.81578*Salinity/1.80655) where 1.80655 is the Salinity/Chlorinity ratio and 1323 is the number of μmoles of Li/Chlorinity – Stoffyn-Egli and Mackenzie (1984) see http://www.mbari.org/chemsensor/pteo.htm Mo based on Sohrin *et al.* (1987) and Morris *et al.* (1975).

cate) in ocean waters is characterised by vertical distribution profiles with low concentrations in surface waters, due to biological uptake, and higher concentrations at depth due to the degradation and dissolution of the products of surface water production at depth. Concentrations of these elements also increase in deep waters along the global thermohaline circulation from the Atlantic to the Pacific Oceans (Chapter 9). Some trace elements (e.g. Cd and Zn) are now known to show similar distributions to the macronutrients (Fig. 11.2). Such a distribution implies a close involvement of the elements with the biological cycling of carbon and major nutrients within the oceans. The residence times of these elements are of the order of tens of thousands of years. This is much shorter than that of the conservative elements reflecting a relatively more efficient removal mechanism via sedimentation of biological particles, but still much longer than the ocean mixing times. Other elements whose profiles suggest nutrient like cycling include Sc, Ni, Ba, Cd, Ge, As and Se (Bruland and Lohan, 2003).

The relationships between some trace elements and nutrients is so strong that it is possible to demonstrate extremely high correlations between the elements and nutrient concentrations in linear *x*, *y*

plots. This has been demonstrated most impressively for Cd which has been studied quite extensively. (see e.g. Boyle *et al.*, 1976; Martin *et al.*, 1976; Bruland, 1980; Bruland and Franks, 1983; Danielsson and Westerlund, 1983; Moore, 1983; Boyle and Huested, 1983; Burton *et al.*, 1983; Boulegue, 1983; Spivack *et al.*, 1983; Kremling, 1983; 1985). In the North Pacific, for example, Bruland (1980) reported that dissolved Cd and phosphate are linearly correlated, at phosphate concentrations in excess of $0.2\,\mu mol\,l^{-1}$, a relationship described by the regression:

$$[Cd] = (0.347 \pm 0.007)[P] - (0.068 \pm 0.017)$$
$$(mean \pm SD; r = 0.992)$$

where [Cd] is in units of $n\,mol\,l^{-1}$ and [P] is in units of $\mu mol\,l^{-1}$. However, further sampling and careful inspection of the global data base has shown that this apparently simple linear correlation is better represented by two straight lines with rather different slopes at higher and lower concentrations reflecting the Pacific and Atlantic deep waters respectively (Frew and Hunter, 1992). Thus classification of nutrient like behaviour does indeed reflect behaviour by a trace element that is similar to that of the nutrients with surface water uptake and deep water

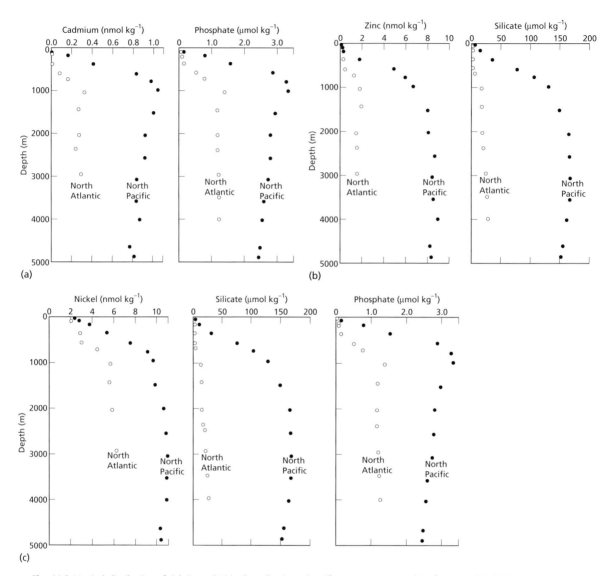

**Fig. 11.2** Vertical distribution of Cd, Zn and Ni in the Atlantic and Pacific Oceans as examples of nutrient like behaviour. DIP (phosphate) and dissolved Si (silicate are also shown) based on Bruland (1983). Reprinted with permission from Elsevier.

regeneration. However, it does not imply that the biogeochemical cycling of these elements are identical during both uptake and more particularly during deepwater regeneration, the process that dominates the observed relationships when these components are plotted one against the other.

The association between phytoplankton and the trace elements may reflect a direct requirement for the element by the organism; zinc for instance is an essential element required in the enzymes responsible for carbon uptake in phytoplankton. However, trace element uptake may occur by simple adsorption to cells or by essentially accidental uptake. It has been argued for instance that Cd may be taken up via Zn uptake pathways that are not able to adequately discriminate between the two elements at very low zinc concentrations. More recently, however, it has become clear that Cd can substitute for Zn in some enzyme systems, so it may also have a direct biological role (Morel and Price, 2003).

It was noted in Chapter 9 that the behaviour of nitrate and silicate in the oceans were subtly different, with silicate regenerated at greater depth than nitrate (or phosphate which behaves in a very similar manner to nitrate). This difference can also be identified in the behaviour of trace elements. Cadmium for instance tends to be well correlated to phosphate, while zinc correlates better with silicate. The uptake of both elements into the soft tissue associated with organic carbon cycling would be predicted if they both have an active role within the cell. Careful studies have shown that zinc concentrations within the opal skeletons is very small and cannot explain the observed relationship between Si and Zn (Ellwood and Hunter, 2000) which suggests that the observed relationship between Si and Zn in the oceans may be in part a reflection of processes during the degradation of organic matter and hence nutrient regeneration at depth rather than simply being controlled by phytoplankton uptake.

There is considerable interest in estimating the concentrations of nutrients in the oceans in the past as an indicator of past ocean circulation patterns which are related to climate. Sediment cores offer an archive of ocean processes, but the cycling of nutrients in the sediments is too complex to allow the measured concentrations of nitrogen or phosphorus to be used to estimate water column concentrations. It has, however, now been established that the concentration of Cd in the shells of calcareous foraminifera (invertebrates which live as zooplankton and as benthic species) are directly proportional to the water column concentrations. This has allowed the measurements of Cd concentrations (or to be more precise the Cd/Ca ratios) in the skeletons of these organisms to be used to estimate past water column Cd concentrations and from these past water column phosphate concentrations. This approach has, for instance, been used to demonstrate that during glacial periods, changes in the ocean thermohaline circulation changed the ocean distribution of cadmium and presumably phosphate, but not the overall ocean inventory (Boyle, 1988).

Morel and Price (2003) suggest that Mn, Fe, Co, Ni, Cu and Zn are all important phytoplankton micronutrients, but not all show nutrient like behaviour. In the case of Mn and Co this is because particle reactivity is sufficiently high to overwhelm the nutrient like profile and result in a scavenged type profile, as will be discussed below. Ni clearly does show a nutrient like profile (Fig. 11.2), but concentrations in surface waters do not fall to extremely low concentrations. This illustrates that the scale of surface water depletion represents a balance between supply, by mixing from below augmented by atmospheric or lateral inputs to the surface ocean mixed layer, and rates of uptake by phytoplankton. In the case of Ni, the biological uptake rate is relatively slow compared to supply, so concentrations are not extremely depleted in surface waters, in contrast to Zn and Cd.

### 11.3.3 Scavenged-type trace metal distributions

Note that the behaviour of iron is a complex and important example of a scavenged type element and is discussed in a separate section later.

The profiles of elements showing scavenged type distributions display surface water enrichments and depletions at depth. Elements showing such distributions include Al, Mn, Co and Hg (Bruland and Lohan, 2003). These elements also have very short oceanic residence times of hundreds of years or less. These residence times are therefore less than ocean mixing times and invalidate the simple calculations of residence times. Concentrations within the oceans will therefore vary considerably reflecting local inputs. These complications aside, it is clear that the members of this group of elements are highly reactive in seawater resulting in their rapid removal compared to other elements. The surface maximum in concentrations for these elements reflects a surface input, usually from atmospheric dust, and the decreasing concentrations with depth reflect the relatively rapid and effective net scavenging of the dissolved element onto sinking particulate matter and subsequent removal from the oceans. This pattern is illustrated for Al in Fig. 11.3. Concentrations of Al in surface waters reflect atmospheric inputs (and indeed can be used to estimate the atmospheric

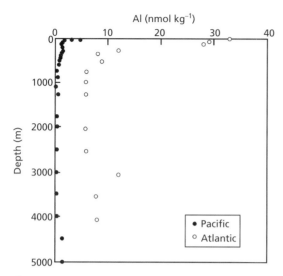

**Fig. 11.3** Vertical distribution of Al in the Atlantic and Pacific Oceans as an example of scavenged type behaviour. Based on Orians and Bruland (1986) and Hydes (1983) and http://www.mbari.org/chemsensor/pteo.htm.

inputs) with much higher concentrations seen in the surface waters of the Mediterranean and off the coast of the Sahara desert (of the order of 100 nmol l$^{-1}$) where dust deposition rates are high, compared to waters of the Southern Ocean remote from atmospheric inputs, where concentrations are about 1 nmol l$^{-1}$ (Measures and Vink, 2000)

The short residence time of the scavenged elements reflects their high reactivity in the aqueous environment and their tendency to stick to particles (Bruland and Lohan, 2003), a pattern also seen in their relative insolubility through the weathering cycle (see Chapter 2). This pattern of behaviour reflects their basic underlying physico-chemical properties as will be discussed later (see Section 11.5). The short residence time also means that this group of elements show considerable spatial and even temporal variability, but in general if there are no additional inputs, concentrations tend to decrease along deep water flow paths, (e.g. Atlantic concentrations > Pacific: Fig. 11.3), the opposite pattern to that seen for nutrient elements. The decrease in concentrations reflects continuing removal of the elements with time along the deep water flow paths.

A particularly interesting example of scavenged element cycling is provided by lead. Lead was exten-

sively used as a car fuel additive from the 1930s until about 1990 (see also Chapter 4) with resultant massive increases in lead emissions to the atmosphere relative to natural fluxes. These emissions peaked in the 1970s. This rapid increase in emissions followed by the rapid phasing out provides a fascinating biogeochemical tracer experiment (described by amongst others; Wu and Boyle, 1997; and Kelly et al., 2009), but also complicates any interpretation of Pb cycling because steady state assumptions are no longer valid.

Fortunately geochemists can use the cycling of $^{210}$Pb as a steady state analogue. The radioactive decay of uranium in soils releases radon gas into the atmosphere which decays to $^{210}$Pb and then adheres to particles and deposits. This is a natural cycle unaltered by human activity and one that very closely mirrors the emissions of pollutant lead from high temperature sources into the atmosphere. We can estimate the residence time of $^{210}$Pb to be about two years in the surface ocean and 100 years in the deep ocean, entirely consistent with scavenged type behaviour. We can assume the behaviour of other forms of lead to be the same as that of $^{210}$Pb, and hence use our knowledge of $^{210}$Pb to help interpret the behaviour of lead.

Monitoring of the lead distribution in the ocean has been undertaken repeatedly in the Sargasso Sea near Bermuda since 1979. In 1979 itself, the lead profile showed a classic scavenged-like shape with a surface maximum in concentrations. The rapid decline in atmospheric inputs since that time together with the rapid removal of the lead from the surface waters and slower removal at depth, where particle concentrations and scavenging are less, resulted in profiles that in the midst of the decline showed a maximum at depths of a few hundred meters (Fig. 11.4a). This time series also shows that surface water concentrations of lead have declined from a peak of about 150 pmol l$^{-1}$ in 1979 to <40 pmol l$^{-1}$ by 2000.

The Pb$^{2+}$ cation substitutes for Ca$^{2+}$ in the calcium carbonate of coral skeletons at a proportion that reflects the ratio of these ions in seawater. Measurement of lead concentrations from corals in reefs near Bermuda have been used to provide a record of lead concentrations in surface waters of this region that predates any actual analytical measurements of ocean lead concentrations (Fig. 11.4b). These coral samples can also be analysed for their lead isotopic composition which can provide a tracer of the source

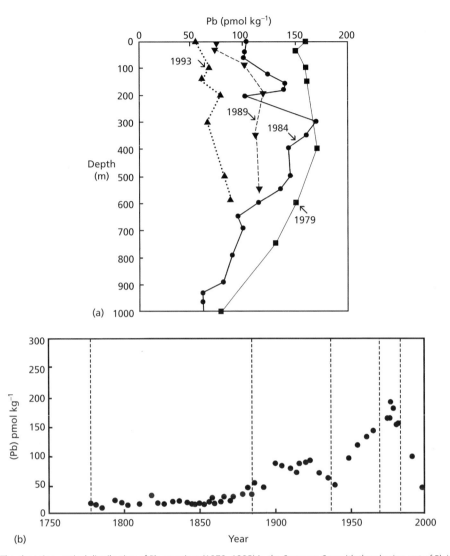

**Fig. 11.4** (a) The changing vertical distribution of Pb over time (1979–1993) in the Sargasso Sea with the phasing out of Pb in car fuels (based on Wu and Boyle, 1997). Reprinted with permission from Elsevier. (b) The change in the Pb concentration over time (1750–2000) recorded in a coral skeleton collected from Bermuda (based on Kelly *et al.*, 2009). Reprinted with permission from Elsevier.

of the lead. The results show lead concentrations began to increase in the mid 1800s associated primarily at that time with emissions from coal combustion and ore processing. Lead concentrations then increased rapidly from 1945 reflecting Pb use in car fuels and concentrations then rapidly declined after the 1970s. It is even possible to see in the coral lead isotopic record changes that reflect car fuel emissions in the USA declining before those from Europe. This coral record can also be used to estimate the pre-

anthropogenic baseline of surface water Pb concentrations to be about 15 pml l⁻¹. Thus concentrations have now declined more than four-fold from their peak concentrations but remain more than double the natural background.

### 11.3.4 Intermediate type behaviour

In the discussion above, elements have been classified into three distinct groups, conservative, nutrient-like

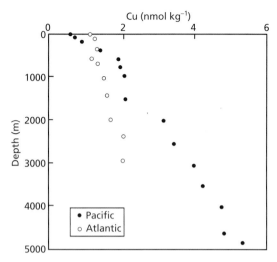

**Fig. 11.5** Vertical distribution of Cu in the Atlantic and Pacific Oceans as an example of intermediate type behaviour between scavenged and nutrient. (Bruland, 1980; Bruland and Franks 1983).

and scavenged on the basis of their vertical profiles and their residence times. However, all elements are subject to the same processes of scavenging, biological cycling and physical mixing, it is just that for some elements the distribution seen is dominated by one type of process as in the examples above. However, there are inevitably elements whose behaviour falls between these idealized end member types of behaviour. An example is copper whose vertical distribution (Fig. 11.5) shows lower concentrations in surface waters which then increase steadily with depth. This behaviour has been interpreted as reflecting very important roles for both biological cycling and scavenging in regulating the distribution of this element (Bruland and Lohan, 2003) together with a source from benthic sediments (see Chapter 14).

### 11.3.5 Iron

Another very important example of an element showing intermediate type behaviour is iron (see Boyd and Ellwood, 2010; Bruland and Lohan, 2003; Bergquist and Boyle, 2006; Morel and Price 2003). As noted in Chapter 9, iron is now recognized as a key nutrient limiting phytoplankton primary productivity over of the order of 25% of the oceans. Iron supply can also influence the rate of oceanic bacterial

nitrogen fixation (Section 9.1) and hence indirectly further affect ocean productivity. Iron has a very short oceanic residence time reflecting its rapid scavenging. The interplay between the major nutrient requirement for iron and the intense scavenging of Fe(III), the thermodynamically stable form of inorganic Fe in seawater, results in a nutrient like profile for iron Fig. 11.6, but a short ocean residence time of ~200 years similar to that of scavenged elements. The concentrations of dissolved iron range from <0.1 nmol l$^{-1}$ in surface waters to about 0.5 nmol l$^{-1}$ in deep water. Iron is taken up rapidly in surface waters by phytoplankton and possibly bacteria and can be regenerated at depth during the subsequent degradation of sinking organic matter within the water column, with dissolved iron released by this regeneration process subject to relatively rapid scavenging. Dissolved iron concentrations are very low, particularly given the abundance of iron in the crust, but even so these concentrations are above those that would be predicted for inorganic Fe$^{3+}$, and it is now known that essentially all of the dissolved iron in ocean waters is strongly bound to or complexed with organic matter, as is also seen for several other metals (see Section 11.6). Some of the organic compounds involved in this complexation appear to be specific iron binding ligands or siderophores and these seem particularly evident in surface waters, with weaker and less specific, but still important, ligands in deepwater. Much of organically complexed iron is probably colloidal rather than truly dissolved, and phytoplankton are believed to have active biochemical mechanisms to take up Fe from these organic complexes (Boyd and Ellwood, 2010).

Surface water iron concentrations are higher in regions subject to high dust inputs, demonstrating the importance of atmospheric dust as an iron source, despite dissolved iron residence times in surface waters that are <1 year. Iron has an active redox chemistry and can be reduced from Fe(III) to Fe(II) relatively easily. Fe(II) is the more soluble reflecting its greater size and lower charge (see Section 11.5) so that reduction of iron in sediments, mid ocean ridges and in water column hypoxic zones can locally increase concentrations – although the increased iron is often rapidly removed by scavenging upon oxidation. The overall residence time of iron therefore reflects inputs from several sources including rivers, mid ocean ridges, margin sediments, the atmosphere

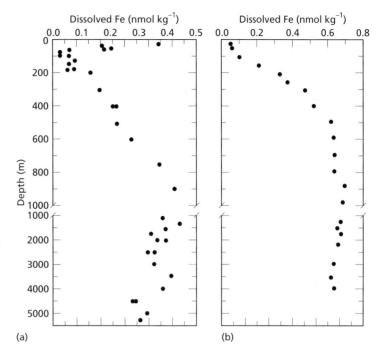

**Fig. 11.6** Vertical water-column profiles of dissolved iron. (a) A profile from the North Pacific central gyre, typical of a scavenged-type trace metal having a significant external source (after Bruland *et al.*, 1991). Reprinted with permission from Elsevier. (b) A profile from a remote subarctic Pacific HNLC region with the lack of a strong external iron source (after Martin *et al.*, 1989).

and glaciers with subsequent organic complexation, intense biological cycling and rapid scavenging producing a nutrient-like profile, very low dissolved iron concentrations and a short residence time. This short residence time reflects the very rapid loss of inorganic iron present at low concentrations and the slower loss of organically complexed iron.

## 11.4 Surface water distributions

### 11.4.1 Open ocean distributions

Open ocean surface water distributions of elements are clearly related to their biogeochemistry. The distribution of conservative-type elements in surface ocean waters essentially follows that of salinity with very minor horizontal gradients in concentrations within the open ocean. The distribution of nutrient-like elements in surface ocean waters is relatively uniform showing very low concentrations in surface waters, but concentrations do increase with increasing nutrient values associated with upwelling as predicted from the nutrient/trace element relationships (e.g. Kremling and Streu, 2001) However, detailed

studies suggest that as with the deepwater cycling discussed above, the exact stoichiometric coupling of nutrients and trace elements such as cadmium and phosphorus can break down within upwelling systems because of difference in the detailed cycling of the elements or because of additional inputs from coastal sources (e.g. Takesue and van Geen, 2002).

The distribution of scavenged-type elements in surface waters is more complex because the more rapid cycling (as shown by the shorter residence time) allows more local variability associated with particular inputs. The relatively close relationship between surface water dissolved aluminium concentrations and atmospheric dust inputs (Measures and Vink, 2000) was noted above and a similar relationship can be seen for Mn (e.g. Statham and Burton, 1986; Morley *et al.*, 1993).

### 11.4.2 Coastal waters

Coastal seas are regions of importance for both their economic (e.g. mineral and oil exploration, fishing, waste disposal) and social (e.g. leisure activities, wildlife preservation) value. The external inputs

**Table 11.2** Trace-metal fluxes to the North Sea; units, t yr$^{-1}$ (after Chester *et al.*, 1994).

| Trace metal | Atmospheric fluxes | Fluvial fluxes | Discharges and dumping fluxes |
| --- | --- | --- | --- |
| Ni | 869 | 240–270 | 903 |
| Cr | 277 | 590–630 | 3382 |
| Cu | 544–1067 | 1290–1330 | 1580 |
| Zn | 3891–4145 | 7360–7370 | 9850 |
| Pb | 1058–2042 | 920–980 | 2470 |
| Cd | 49 | 46–52 | 43 |

(mainly fluvial run-off and atmospheric deposition) that supply trace metals to oceanic surface waters from continental sources are stronger to coastal than open-ocean waters. Further, there are a number of additional pollutant inputs to the coastal zone. These include: (i) *disposal dumping*: for example sewage sludge, radioactive waste, dredge spoils, military hardware, off-shore structures; (ii) *deliberate discharges*: for example from sewage works, power stations, industrial plants, oil refineries, radioactive reprocessing plants; and (iii) *accidental discharges*: for example from oil tanker wrecks. An indication of the importance of these additional contaminant sources to a coastal zone is given by the data in Table 11.2, from which it can be seen that for the North Sea in the 1990s, when a particularly thorough set of flux estimates were prepared, inputs from discharges and dumping are comparable to those from fluvial and atmospheric sources for some trace metals. In Europe and many other areas considerable efforts have now been to reduce the discharge of harmful contaminants to the coastal seas. More recent data from the North Sea indicates that total inputs of these trace elements have declined by 2–10-fold, and direct discharges in particular have been greatly reduced (OSPAR, 2010). It may be concluded that shallow coastal seas are a major receiving zone for both natural and pollutant inputs to the ocean, and the health of coastal waters has given rise to much concern over recent years. However, technological control of the emission of contaminants can be achieved, and in general this is easier for point source inputs such as those from industry or sewage discharge, than from diffuse contaminant sources such as nitrate from agricultural fertilizer use.

One effect of the enhanced inputs to the coastal zone is that the surface-water concentrations of some trace elements are higher in coastal and shelf waters than they are in open-ocean waters. Examples of this have been provided for a number of individual trace metals.

1 *Iron.* Martin and Gordon (1988) reported a 300-fold decrease in surface-water total Fe concentrations on a transect from the North Pacific central gyre (~0.3 nmol l$^{-1}$) to the California coast (~100 nmol l$^{-1}$), and Symes and Kester (1985) found a 100-fold decrease in total Fe from offshore (~3 nmol l$^{-1}$) to inshore waters (~300 nmol l$^{-1}$) on a transect across the USA Atlantic continental shelf.

2 *Manganese.* Landing and Bruland (1987) measured a 10-fold decrease in dissolved Mn in surface waters on a California Current (~10 nmol l$^{-1}$) to North Pacific central gyre (~1 nmol l$^{-1}$) transect.

3 *Zinc.* Bruland and Franks (1983) reported a 40-fold decrease in Zn concentrations in Atlantic surface water from the USA east coast (~2.6 nmol l$^{-1}$) to the Sargasso Sea (~0.06 nmol l$^{-1}$).

In the coastal zone trace elements are delivered to surface or, for some pollutant inputs, subsurface waters and the coastal → open-ocean trace-metal concentration decreases reported above largely reflect differences in external source inputs between the two environments. However, the coastal zone is not simply a body of water that acts as the receiving reservoir for the *external* inputs of trace elements. Within it, the trace elements are distributed between the water column, biota and sediment reservoirs, and the zone provides an environment of intense *internal* reactivity involving a complex interplay between physical, chemical and biological processes (see Fig. 11.7). Together, these processes constrain the fates of trace elements, and may result in their permanent, or temporary, retention within the coastal zone, or their export out of it.

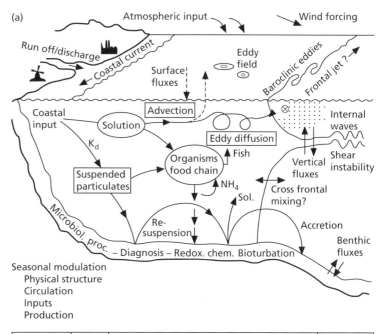

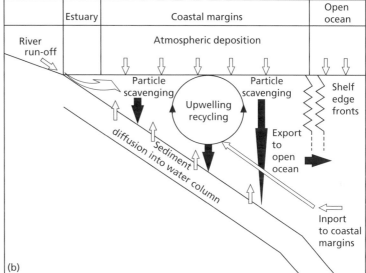

**Fig. 11.7** Trace metals in the coastal zone. (a) General processes influencing the fates of trace metals and pollutants in coastal margins (from Simpson, 1994). (b) Schematic representation of trace-metal pathways in the coastal margins.

The export of trace elements out of the coastal zone involves the physical transfer of either dissolved or particulate trace elements by advective water transport, particle transport and sedimentation. The retention of trace elements within the zone occurs via a number of often interlinked physical, biological and chemical processes. The exchange of water between the open ocean and the shelf is a key determinant of shelf biogeochemistry as was also noted in Section 6.5. In some areas there is rapid and efficient exchange. For example, Bruland and Franks (1983) were able to describe the distribution of Mn, Ni, Cu, Zn and Cd on a transect from the east coast of the USA out into the North Atlantic gyre and the

Sargasso Sea as simple linear mixing between a high concentration river source and a low concentration open ocean surface water end-member. In other coastal areas oceanographic fronts represent rather clear boundaries between the two water masses. Where they occur, these sharp boundaries are shown by sharp boundaries in the distribution of trace metals as shown by Kremling (1983; 1985) in a study of dissolved concentrations of Cd, Cu, Ni and Mn on an 'open-ocean North Atlantic to northern Scottish coast to European North Sea coast' transect (Fig. 11.8).

Chemical processes can also influence the behaviour of trace metals on the shelf with both biological and abiological uptake of metals to particles which then sink to the sediments. Both the suspended particle concentrations and rates of phytoplankton growth tend to be higher in coastal waters than in the open ocean, so the impact of these processes in shelf waters can be very important. However, incorporation into sediments need not be a 'one-step' trapping process. Coastal and shelf sediments have a relatively high concentration of organic matter (see Section 14.2), and are thus susceptible to suboxic as well as oxic diagenesis (see Section 14.1). As a result, elements that have been incorporated into reducing coastal sediments can be released by diagenetically mediated processes (see Chapter 14), and some frac-

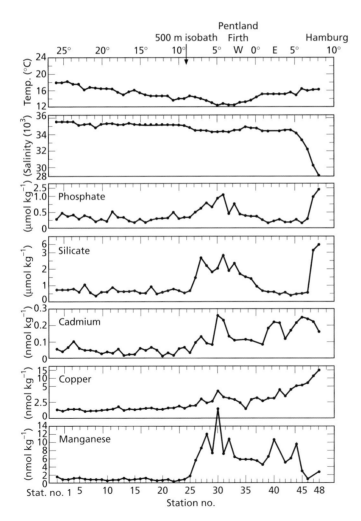

**Fig. 11.8** Surface-water distributions of Cd, Cu, Ni, Mn and Al on an open-ocean Atlantic–European coast transect (from Kremling, 1983): the shelf edge is indicated by the 500m isobath.

tion of these can escape back into the overlying waters; for example from direct diffusion across the sediment–water interface, or by the sweeping out of interstitial waters during sediment resuspension. This effect has now been shown to be of considerable importance for oceanic iron biogeochemistry as discussed earlier (see Section 11.3.5).

The complexity of the relationships that can develop at the shelf boundary can be illustrated by a study of trace metals in surface ocean waters of the North Pacific on a transect that extended from off the coast of California out to the central gyre (Schaule and Patterson, 1981). The transect passed through a variety of oceanic environments, which included: (i) the nutrient-rich waters of the biologically productive outer shelf region; (ii) an intermediate open-ocean region; and (iii) the biologically non-productive centre of the North Pacific gyre. Generalized surface-water distributions of the trace metals determined in the investigation are illustrated in Fig. 11.9. On the basis of their distributions along this transect the trace metals can be divided into four broad groups.

**1** *Copper and nickel*: The distributions of Cu and Ni exhibit a negative, that is, decreasing, horizontal surface-concentration gradient away from the coastal upwelling area out towards the open-ocean, and reach their lowest values around the central gyre.

This type of negative coastal → open-ocean surface-water concentration profile is to be expected:

(a) if trace metals are supplied to the coastal receiving zones by processes such as river run-off, atmospheric deposition, diffusion from sediments and shallow-depth coastal upwelling; and

(b) if only a fraction of the elements from these inputs subsequently escapes the coastal zone and is transported horizontally to the open ocean by advection–diffusion processes.

**2** *Lead*: The concentrations of Pb in the Pacific transect exhibit a gradient that is the opposite to that shown by both Cu and Ni, that is, the lowest Pb concentrations are found in the biologically active waters of the outer shelf region and there is a general increase out towards the central gyre, in which the surface waters have about three times as much dissolved Pb as those of the shelf region. This work was carried out at a time when the atmospheric emissions of lead were still very high (see Section 11.3). The concentrations of dissolved Pb at depth in the water column were lower than those in surface waters (see Section 11.3), which precludes the possibility that vertical redistribution by upwelling can be the dominant supply mechanism for Pb in the mixed layer. Schaule and Patterson (1981) concluded, therefore, that the surface-water distribution of dissolved Pb

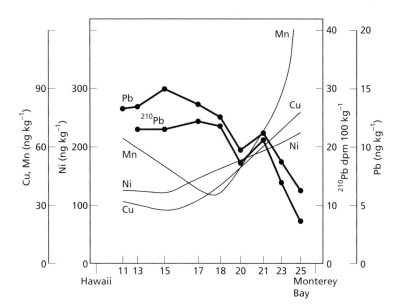

**Fig. 11.9** Trace-element surface-water distribution patterns on a North Pacific transect (from Schaule and Patterson, 1981). Reprinted with permission from Elsevier.

had resulted from strong external inputs to the surface waters. However, the surface-water Pb distribution pattern, with lower values in the coastal areas and higher values in the central gyre, has a concentration gradient which means that the external input into coastal waters cannot all subsequently undergo off-shelf transport by advection–diffusion to the open-ocean. The distribution of stable Pb along the Pacific transect is very similar to that of $^{210}$Pb, which has a predominantly aeolian input to the ocean (see Section 11.3). Both forms of Pb are subject to removal processes from surface waters, and Schaule and Patterson (1981) proposed that the distribution of Pb in the surface waters of the Pacific transect was maintained by the element having a strong atmospheric input to both coastal and open-ocean areas, coupled with a higher rate of removal (scavenging) from surface waters in the coastal areas, where the rates of biological activity, and so the concentrations of suspended particles, are higher than they are in the central gyres.

3 *Manganese*: The surface-water distribution of Mn along the Pacific transect exhibits a pattern that is intermediate between those of Cu and Ni on the one hand, and Pb on the other. That is, the highest concentrations of Mn are found in the coastal waters and there is a general decrease out into the open ocean; thus, Mn exhibits a negative concentration gradient, which is similar to those of Cu and Ni: see Fig. 11.9. This type of profile suggests that there must be a significant input of Mn to the coastal receiving zone (e.g. by river run-off or diffusion from bottom sediments), where much of it is removed from the surface waters, so leading to a diminished lateral transport of the element to the open ocean. In the coastal zone, therefore, it would appear that the dominant input of Mn is by fluvial transport and/ or sediment diffusion, whereas that for Pb is largely atmospheric. In the open-ocean waters, however, the surface distribution of Mn differs from those of Cu and Ni in that its concentrations increase to reach higher values in the central gyre. As with Pb, this type of distribution may indicate an atmospheric input to the open-ocean North Pacific, with the lowest rate of removal from surface waters occurring in the biologically inactive central gyre. Unlike Pb, however, the input of Mn from river run-off and/or sediment diffusion in the coastal regions overshad-

ows that from the atmosphere (see also Landing and Bruland, 1980; Bruland, 1983).

## 11.5 Scavenging mechanisms

The adsorptive removal of trace metals occurs on to particles that sink down the water column, and Goldberg (1954) gave this removal process the general name of *scavenging*. The dissolved down-column concentration profiles for scavenged elements are concave in deep water when plotted against a conservative tracer, indicating loss from the water column. The extent of this scavenging removal can be expressed in terms of a deep-water scavenging residence time; a selection of these residence times is given in Table 11.3. Much of our knowledge of the scavenging of dissolved elements from seawater has been derived from the unique chemistry of some radionuclide elements, and their distributions in the oceans have allowed scavenging models to be constructed. In this respect, members of three natural radioactive decay series have proved to be useful tools for the interpretation of scavenging reactions; these are the $^{238}$U series, the $^{232}$Th series and the $^{235}$U series.

The fundamental concept behind the use of these radionuclides is the parent–daughter relationship which can be used for a number of radioactive decay series and is illustrated here for uranium and thorium. The distribution of uranium (U the parent isotope) in seawater is essentially conservative while that of the thorium (Th) daughter isotope produced from

**Table 11.3** The deep-water scavenging residence times of some trace elements in the oceans (data from Balistrieri *et al.*, 1981; Orians and Bruland, 1986; Whitfield and Turner, 1987).

| Element | Scavenging residence time (years) | Element | Scavenging residence time (years) |
|---------|-----------------------------------|---------|-----------------------------------|
| Sn | 10 | Mn | 51–65 |
| Th | 22–33 | Al | 50–150 |
| Fe | 40–77 | Sc | 230 |
| Co | 40 | Cu | 385–650 |
| Po | 27–40 | Be | 3 700 |
| Ce | 50 | Ni | 15 850 |
| Pa | 31–67 | Cd | 177 800 |
| Pb | 47–54 | Particles | 0.365 |

the radioactive decay of the uranium is dominated by scavenging. The rate of radioactive decay of all naturally occurring radionuclides is known with great accuracy. Hence, knowing the uranium concentration it is possible to predict precisely the concentration of Th expected from the rate at which it is produced from uranium decay and in turn lost by radioactive decay. In reality less Th is usually found in the water column because some is removed by scavenging in addition to the loss by radioactive decay. The deficit of Th arising from scavenging and the scavenging rate can therefore be estimated from measurements of Th abundance or radioactivity.

Hence:

Rate of Production of daughter (D) = Rate of decay of parent (P) = Ap radioactivity of parent = $\lambda p[P]$ where P is the parent concentration and $\lambda$ the radioactive decay rate

And Rate of loss of daughter = rate of decay of daughter Ad + S

where S = scavenging rate

Ad = $\lambda d[D]$ and assuming scavenging rate S can be described by a rate constant $k_s[D]$ then at steady state

$$\lambda p[P] = \lambda d[D] + k_s[D] \qquad (11.2)$$

If the activities of the parent and daughter can be measured or estimated, the scavenging rate can therefore be calculated.

The different rates of radioactive decay of parent and daughters within different decay series can be used to study processes with different timescales. Thus $^{234}$Th which has a radioactive half life of 24.1 days is very valuable for looking at surface water processes with timescales of days. In contrast, $^{210}$Pb with a half life of ~22 years is more suitable for investigating deep ocean scavenging processes on timescales of decades and $^{230}$Th has a half life of 75 000 years making it suitable to investigate longer scale processes.

Extensive modelling of measurements of thorium isotopes in ocean profiles have allowed various scavenging hypotheses to be tested. It has proven impossible to explain the observed data in terms a simple hypothesis of irreversible uptake of scavenging to particles that are then lost directly to sediments. As was noted in Chapter 10, the ocean particle population can be described in terms of a background of small particles, which therefore have a relatively large surface area, and a population of larger fast sinking particles. Adsorption to particles depends on surface area, so a second hypothesis that can be tested is that scavenging occurs to the smaller particles which are then become irreversibly associated with the fast sinking particles, but again this mechanism cannot explain the observed data well. The only mechanism that can adequately explain the observed profiles involves reversible scavenging onto fine particles (and colloids) which are then packaged into larger particles which sink and disaggregate at depth only to be repackaged. Thus within a thorium ions residence time in the ocean of <100 years it appears to be exchanged between the dissolved and particulate phase numerous times before finally reaching the sediments (Anderson, 2003).

The process can be illustrated (Fig. 11.10) for the decay sequence of $^{234}$U to $^{230}$Th to $^{226}$Ra where $\lambda$ represents radioactive decay, k represents reversible scavenging and d and p represent dissolved and particulate phases respectively.

An important conclusion that can be drawn from this work is that, at least to a first approximation, the suspended particulate matter in deep water exists in a state of equilibrium with respect to the exchange of Th by adsorption–desorption reactions. This can be described in terms of an equilibrium model. In the framework of models of this type, the residence time of an element with respect to its removal from the oceans by scavenging is controlled by its equilibrium partitioning between dissolved and adsorbed forms,

**Fig. 11.10** An illustration of part of the *uranium thorium radium* radioactive decay series and water column scavenging. The subscripts d and p indicate the dissolved and particulate phases respectively.

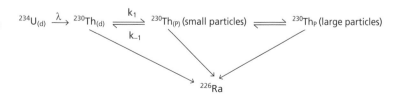

and by the residence time of the particles with which it reacts. Thus, a knowledge of the equilibrium distribution of the element, combined with an independently determined value for the residence time of the particulate material, will yield an estimate of the scavenging rate of the element itself. This relationship can be expressed in the following manner:

$$\frac{1}{\tau_{Me}} = \frac{1}{\tau_P} \frac{[Me]_p}{[Me]_t} \qquad (11.3)$$

where $\tau_{Me}$ is the residence time of the element (metal), $\tau_p$ is the residence time of the particulate matter, and $[Me]_p$ and $[Me]_t$ are the particulate (adsorbed) and total concentrations of the element, respectively.

The quantity $[Me]_p/[Me]_t$ can be calculated from direct field measurements, thus allowing the scavenging residence time of an element to be determined. This approach was adopted by Bacon and Anderson (1982), who concluded that on the basis of their equilibrium model the removal of Mn, Cu, Pb, Th and Pa from seawater appears to be controlled by a single population of particles with a residence time of about 5–10 yr in the water column. It was shown in Chapter 10 that most of the vertical flux of particulate matter from the surface to deep water is controlled by large aggregates (CPM; diameter >50 μm), which have much faster sinking times than 5–10 yr. However, scavenging is dominated by the small-sized population (FPM; diameter <<10 μm), which has a long residence time but can undergo accelerated settling by 'piggy-backing' on to and off the large aggregates. This conclusion, combined with reversible scavenging, yields scavenging residence times for trace metals and isotopes that lie between the residence times of the individual small- and large-particle populations. The deep-water scavenging residence times of a number of metals in seawater derived using this approach are listed in Table 11.3.

The concept that dissolved–particulate reactions in the water column may be controlled by an equilibrium mechanism, and so lead to the possibility that marine chemists may be able to describe the reactions involved in terms of classic surface chemistry theory, was not a new one. Equilibrium models, that is those involving reversible uptake, had in fact been proposed by other workers to explain trace-metal scavenging within the water column. For example,

Balistrieri *et al.* (1981) had defined an equilibrium scavenging model, based on the principles of surface chemistry that could be combined with field data to determine the scavenging fate of trace metals in the oceans. To do this, the quantity $[Me]_p/[Me]_t$ was calculated on the basis of equilibrium constants. Such an approach is intellectually appealing because it allows the oceanic residence times of elements to be predicted without a prior knowledge of their seawater concentrations. The results obtained by Balistrieri *et al.* (1981) strongly indicated that the scavenging component of marine particulate matter has an *organic* nature, that is, the trace-metal scavenging is controlled by organic coatings on particles. The approach adopted by Balistrieri *et al.* (1981) is open to criticism because it may have oversimplified particle–metal interactions in the ocean. Nonetheless, it represented an important step towards the stage when trace element scavenging in seawater could be modelled adequately.

It is now known that the rate of transfer of dissolved Th to the particulate state is a function of particle concentration as well as particle residence time (see e.g. Nyffeler *et al.*, 1984), that is, the rate constants of adsorption vary with the particle concentration. Hence scavenging rates will increase in more particle rich environments. The key to understanding the removal processes, therefore, lies in the fact that the adsorption of Th varies with the concentration of particles in the system. Honeyman *et al.* (1988) pointed out that this has important implications on Th fractionation because in near-shore surface ocean waters the particle residence time is equal to, or less than, the residence time of the dissolved Th. As particles are leaving the system the fraction of Th in the particulate phase will be less than it would be if the particles had an infinite residence time. Under these conditions, therefore, the system should not be viewed as an equilibrium system.

Honeyman *et al.* (1988) derived a series of equations to describe Th uptake and then plotted the log value for the forward Th sorption rate constant, that is, uptake, against the log value of the particle concentration, using data for a wide variety of oceanic environments. These included the highly productive California coastal zone, with medium particle concentrations; the low productive China shelf zone, with high particle concentrations; and the deep

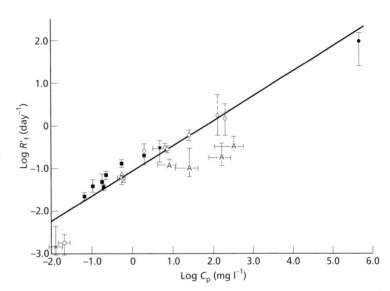

**Fig. 11.11** A plot of the forward sorption rate constant (log $R_f$) for Th versus particle concentration using field data from various oceanic regions (from Honeyman et al., 1988). The oceanic regions cover a wide variety of particle concentrations, particle types and primary productivity intensities: (h and ×) deep sea; (j) surface water, California Current; (s) coastal waters, Yangtze River; (A) coastal waters, Amazon River; (n) surface waters, Funka Bay; (.) surface waters, Narragansett Bay; (d) sediment porewaters, Buzzards Bay. The full line represents the linear regression excluding the Amazon data. Reprinted with permission from Elsevier.

ocean, which is low in both biological productivity and particle concentrations. The plot is reproduced in Fig. 11.11, and shows that there is a strong linear relationship between the log value of the forward sorption rate constant ($k_1$ in Fig. 11.10) and the log value of the particle concentration. This particle dependence of the scavenging rate holds over seven orders of magnitude variation in particle concentration. The results from very different systems fall close to the main regression line although results for the Amazon seem to fall systematically below the best fit line. This overall pattern suggests that the main determinant of scavenging rates is the particle concentration, but other factors such as particle type may also have an effect. Linear regression analysis of the data (excluding that for the Amazon) gave a slope of 0.58, and the authors described the partitioning of Th between dissolved and particulate phases by the equation:

$$\frac{A_{Th,part}}{A_{Th,diss}} = \frac{R_f (C_p)^{0.58}}{R_r + \lambda_{Th} + \lambda_{part}} \qquad (11.4)$$

where $A_{Th, part}$ and $A_{Th, diss}$ are the activities of particulate and dissolved Th, respectively; $R_f$ (= 0.079 l$^{-0.58}$ mg$^{-0.58}$ day$^{-1}$) is the rate constant for the transfer of Th from the dissolved to the particulate phase; $R_r$ (= 0.007 day$^{-1}$) is the rate constant for the reverse reaction; $C_p$ is the mass concentration of particles

(expressed in units of mg l$^{-1}$); and $\lambda_{Th}$ and $\lambda_{part}$ are the decay constant for the thorium isotope (day$^{-1}$) and the rate constant for particle removal (day$^{-1}$), respectively. Thus, the partitioning of Th depends on an interplay between the reaction rates, the particle concentration, the decay characteristics of the particular Th isotope, and the residence time of the particles in the system, which depends on the water column height, the particle concentration and the particle flux. It is generally assumed that these insights into scavenging behaviour gained from studies of thorium are applicable to scavenging processes in general.

Trace-metal removal times based on adsorption reactions are thought to be rapid, that is, of the order of seconds, but in natural systems they appear to be of the same order as physical mixing processes, that is, days to weeks. In order, therefore, to combine all the various features of the Th removal processes, Honeyman et al. (1988) suggested that the aggregation or coagulation of colloids, which also is a function of particle concentration, may play a part in controlling these slow rates. They proposed, therefore, that the constancy of the 'Th removal rate constant–particle concentration' relationship can be attributed to a Th removal mechanism that involves a combination of surface coordination reactions (based on surface coordination chemistry involving surface complex formation: see previous) and colloidal particle-to-particle aggregation. This surface coordination–colloidal aggregation model implies that although particles in a highly productive system may be of biological origin and may be supplied at a rate that is controlled by primary production, they simply serve as new adsorptive passive substrates for the surface-active Th. It must be stressed, however, that Honeyman et al. (1988) based their model on Th, and other metals may have different controlling properties. Thus, the extent to which Th can act as an analogue for other surface-active metals requires further research. Nonetheless, the model proposed by Honeyman et al. (1988) links micro-level adsorption reactions to macro-level oceanic scavenging processes, and offers an explanation for the correlations found between dissolved Th removal rate constants and biological productivity.

The effect of particle concentration on scavenging rates means that the Th removal rate in surface waters will be related to primary productivity as the main source of particles in surface waters. The deficit of Th relative to the predicted concentration in surface waters will therefore reflect the efficient scavenging of Th onto biogenic particles, and the subsequent removal of those particles. This removal of biogenic particles is related to the export of organic matter from surface waters which on a long enough time scale is equivalent to 'new' primary production as discussed in Chapter 9 and indeed the measurement of thorium isotope distributions is now one of the main ways that 'new' production is measured (e.g. Coale and Bruland, 1985; 1987; Buessler et al., 2006).

The vertical seawater profiles of some trace metals can be perturbed by the effects of processes that occur at the sediment–water interface in both coastal and open-ocean areas. For example, Bacon and Anderson (1982) showed that there was a sharp decrease in the concentration of dissolved $^{230}$Th towards the sea bed, which they interpreted as resulting from an accelerated uptake of the isotope from solution on to resuspended sediment particles, which act as a sink for dissolved elements. Similar perturbations to the dissolved down-column profiles also have been reported for $^{210}$Pb (see e.g. Bacon et al., 1976). This arises because the higher concentrations of fine particles at the ocean boundaries lead to zones of enhanced scavenging (e.g. turbidity maxima in

estuaries: see Section 3.2.4). Thus, processes at the ocean boundaries may act as a *source* for fine particles and as a *sink* for some dissolved elements via enhanced scavenging. Although the vertical transport of elements down the water column by incorporation into the sinking particle flux offers a useful insight into how trace elements are removed from seawater, it is evident that lateral transport into and out of the boundary regions (e.g. along isopycnals) must also be considered in assessing the processes that control the marine cycles of the trace elements.

To summarize, the scavenging of trace elements from seawater is a complex process. The scavenging removal itself is dominated by the fine-particle population, but is then complicated by an association of the fine particles with the large-particle flux by reversible physical mechanisms (e.g. 'piggy-backing') and by reversible chemical adsorption that may tend towards an equilibrium state.

Scavenging rates therefore depend in part on the concentrations of particles, but it is also clear that rates vary for different elements, as illustrated by their different residence times. The uptake of elements on to the surfaces of particulate matter involves a distribution ratio ($K_d$), which is the ratio of the concentration of the element in the solid phase to that in the aqueous phase. A variety of theoretical models have been proposed to describe the inorganic adsorption of trace elements on to particulate surfaces. The general aim of such models is to predict, or at least rationalise, the behaviour of elements from their fundamental chemical properties. The surface complexation model (see e.g. Schindler, 1975) has been most commonly used in marine chemistry, and this will serve to illustrate the general mechanisms that are thought to be involved in the ultimate removal of elements reaching the sediment surface. In this model, it is assumed that metal ions are removed by adsorption on to inorganic particulate matter at oxide surfaces covered with OH groups, which act as ligands. The hydrolysis of the oxide surfaces produces hydrous oxide surface groups such as $\equiv Si-OH$, $\equiv Mn-OH$ and $=Fe-OH$. The extent of adsorption is fundamentally related to the Gibbs Free Energy of adsorption onto the surface and the bonding of the ion in question to the surface. These terms can be parameterised in several different ways based on physicochemical data, although many of these parameterisations are closely related to one

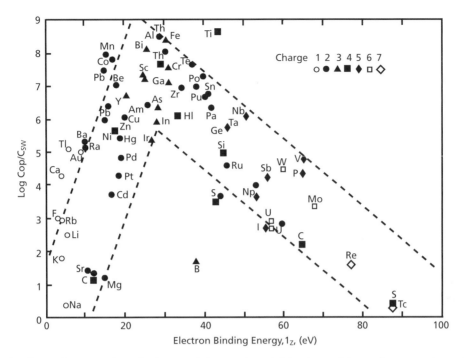

**Fig. 11.12** The elemental concentrations of various metals in oceanic pelagic clays divided by those in seawater (log $C_{op}/C_{sw}$) as a measure of the tendency of elements to partition to the dissolved or particulate phase plotted against the electron biding energies to cations Iz, based on Li (1981). Note that elements showing conservative type behaviours plot in the lower left hand part of the graph, scavenged type elements in the upper central part of the graph with elements showing nutrient like behaviour plot between the two. Elements plotting in the lower right hand portion of the graph show conservative type behaviour because they form oxyanions. Reprinted with permission from Elsevier.

another (Li, 1991). This arises because the charge and size of ions can be shown to be a fundamental control on these particle water interactions, because the removal of the water of hydration from the ion is a fundamental precursor to any such interaction. Li (1991) has therefore used a parameter Iz which is related to electron binding energy of the element to characterize this. In Fig. 11.12 Iz is plotted against the ratio of the concentrations of elements in marine clays and dissolved in seawater, representing the extent to which the elements partition between the dissolved and particulate phase. Li also considers the distribution with other solid phases and derives essentially the same result as that in Fig. 11.12. This figure shows that scavenged type behaviour is seen for elements forming small highly charged cations such as $Mn^{4+}$, $Al^{3+}$ and $Th^{4+}$. By contrast conservative behaviour is seen for anions and large and low charge cations (e.g. $Na^+$, $Mg^{2+}$), and nutrient like

elements plot in between the two. As noted earlier, this separation of behaviours between conservative, nutrient and scavenged-like behaviour is also reflected in the residence time of the elements. Thus, the classification of behaviour presented earlier, based on empirical data on water column distributions and residence times, can be related to the fundamental chemical behaviour of the elements. While encouraging, it should be noted in Fig. 11.12 that a wide range of behaviour is seen for elements with rather similar Iz values such as Mg, Cd and Mn. This, in part, reflects issues around the speciation of elements in seawater which will be discussed next.

## 11.6 Trace-element speciation

The chemical speciation of components in multi-electrolyte solutions such as seawater is notoriously difficult to assess, and a full treatment of the subject will not be attempted here. Instead, attention will be focused on two key questions.

1 What is speciation?

2 Why is speciation important to our understanding of the chemical dynamics of trace elements in seawater?

For a general discussion of speciation in the environment, the reader is referred to the volume edited by Ure and Davidson (2002).

### 11.6.1 What is speciation?

*Speciation* can be defined as the individual physicochemical forms of an element that together make up its total concentration. The full water-column speciation of an element therefore involves its distribution between free ions, ion pairs, complexes (both inorganic and organic), colloids and particles. The speciation of elements influences all their chemical reactions in the ocean including uptake by organisms and particle-water interactions by altering their chemical reactivity (Byrne, 2002). A number of generalities can be identified with respect to the speciation of elements in seawater.

1 For a few elements, mainly the non-metallic and metalloid elements, the formation of covalent bonds is important in controlling species distribution in seawater.

2 Coordination bonding chemistry (acid–base reactions, precipitation–dissolution, complex formation) dominates the species distribution of the metallic elements.

3 Redox reactions lead to changes in electron configuration that are reflected in both covalent and coordination bonding characteristics.

Because both proton exchange (acid–base reactions) and electron exchange (redox) processes are extremely important in controlling speciation distribution, pH and redox potential (see Chapter 14 for a full discussion) are uniquely important parameters in aqueous chemistry, and are often termed the 'master variables' of the system.

The basic concepts underlying trace-element speciation in seawater have been described by Stumm and Brauner (1975). Atoms, molecules and ions will tend to increase the stability of their outer-shell electron configurations by undergoing changes in their coordinative relationships, for example, by acid–base, precipitation and complex formation reactions.

Any combination of cations (the *central atom*) with molecules or anions (the *ligand*) containing unshared electron pairs (*bases*) is termed coordination (or complex formation); this can be either electrostatic, covalent, or a mixture of both. In an aqueous solution, cations are coordinated with water molecules, and Stumm and Brauner (1975) distinguish between two types of complex species in seawater.

1 In an *ion pair* (or outer-sphere species), the metal ion, the ligand or both retain the coordinated water when the complex is formed, so that the metal ion and the ligand are separated by water molecules. In an ion pair, the association between the cation and anion is largely the result of long-range electrostatic attraction.

2 In a *complex* (or inner-sphere species), the interacting ligand is immediately next to the metal cation, and a dehydration step must precede the association reaction. In this type of species, short-range or covalent forces contribute towards the bonding. The number of linkages attaching ligands to a central group is known as the coordination number. When a compound contains more than one ligand, and can thus occupy more than one coordination position in a complex, it is termed a multidentate (as opposed to a unidentate) ligand; complex formation with these ligands is termed chelation, and the complexes are referred to as *chelates*. Chelates are usually much more stable relative to the corresponding complexes with unidentate ligands. The order of stability of the transition metal chelates formed with ligands (especially organic ligands) follows the so-called Irving–Williams order $Mn^{2+} < Fe^{2+} < Co^{2+} < Ni^{2+} < Cu^{2+} < Zn^{2+}$.

### 11.6.2 Important trace-element species in seawater

#### 11.6.2.1 Ligand exchange: complex chemistry

Following Bailey *et al.* (1978) the formation of a complex in solution can be represented by a number of steps, each of which can be described by an *equilibrium constant* (formation constant). The complexes usually undergo continuous breaking and remaking of the metal–ligand bonds, and this introduces a fundamental concept in speciation chemistry. If the bond-breaking step is rapid (time scales of minutes or often much less), the ligand exchange reactions are also rapid, and the complex is classified as being *labile*. In contrast, if the reactions are slow (time scales of days or potentially much longer), the complex is classified as *non-labile*, or inert (Filella *et al.*, 2002). Considerable advances in trace-metal speciation chemistry have been made using electrochemical techniques. In some analytical configurations these techniques measure only labile forms of the elements, in some one particular redox state and in others total concentrations, so the lability of complexes can be directly measured. Electrochemical techniques can also be used to directly measure organically complexed forms of metals and even determine the equilibrium constants for naturally occurring complexes. (Bruland and Lohan, 2003; Luther *et al.*, 2001).

According to Nurnberg and Valenta (1983) the inorganic ligands normally present in seawater usually form labile mononuclear complexes with trace metals. Because of factors such as their high rate constants of formation and disassociation, these complexes are very mobile and will undergo reversible electrode processes in electrochemical determinations. This also is the situation for certain weak complexes formed between trace metals and some ligands of the dissolved organic matter (DOM) in seawater. In contrast, other DOM ligands form more stable non-labile complexes with trace metals.

In terms of their association–disassociation rate constants it is therefore convenient from a biogeochemical viewpoint to consider some aspects of both the inorganic and organic trace-metal complex formation in seawater within the framework of a classification that distinguishes between two general species types.

1 Those that form kinetically very *labile* complexes. These species can be considered to be in thermodynamic equilibrium, and species distributions can be predicted from mathematical models.

2 Those that are involved in the formation of very *non-labile* (or inert) complexes. For these, species distribution can be identified using experimental techniques, which can isolate the inert complexes.

1 *Speciation involving labile complexes.* This form of trace-metal speciation can occur with inorganic, and also some organic, ligands. The determination of the labile complexes uses theoretical equilibrium models based on thermodynamic compositions. There are a number of problems involved in both the

setting up of these models and in the manner in which the data they generate are interpreted; for example, factors such as the use of different suites of stability constants often yield different speciation pictures. In seawater, the abundant inorganic ligands that are significant for trace-metal speciation are $Cl^-$, $OH^-$, $CO_3^{2-}$ and $SO_4^{2-}$. The processes involved in this speciation can be influenced by the formation of ion pairs (electrostatic attraction only) between cations and anions, and by mixed ligand complexes. Other effects that must be considered include side reactions and mixed ligand complexes. Byrne (2002) has thoroughly reviewed the speciation of elements in seawater, focusing particularly on the labile complexes.

A list of predicted trace-metal species in oxygenated seawater is given in Table 11.1 and two main features can be identified.

1. The alkali and alkaline earth metals do not form strong complexes, and even their tendency to form ion pairs is limited, with the result that these elements exist in seawater largely as simple cations.

2. For the trace elements, the species are distributed between the free ion and complexes with the various ligands.

It must be pointed out, however, that the data in the table disregard the formation of complexes with organic ligands, which are often dominant as discussed in the next section.

2 *Speciation involving non-labile complexes with organic matter.* Some organic ligands can form stable complexes, or even chelates, with trace metals in seawater and electrochemical techniques have demonstrated the significance of these strong complexes. More than 90% of the Fe(III), Cu, Co, Zn, 80% of the Cd (Bruland and Lohan, 2003) and about 40% of the lead (Copodaglio *et al.*, 1990) in seawater is now known to be organically complexed, and it seems likely that detailed studies of other transition metals will reveal that for them too organic complexation is probably important. We know very little about the chemical structure of organic ligands in seawater, although we have direct estimates of the equilibrium constants for their binding to some metals from electrochemical measurements. It seems likely that many different organic compounds are involved. The poorly characterized chemistry of dissolved organic matter (DOM) is discussed in Chapter 9. It is likely that at least some of this highly degraded organic matter is involved in the complexation of

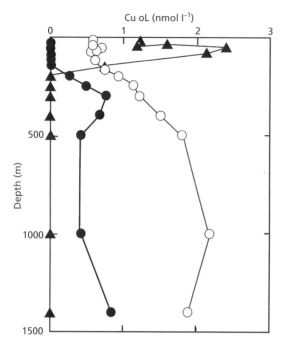

**Fig. 11.13** Vertical distribution of total dissolved Cu (○), inorganic free copper (●) and the copper binding ligand L₁ oL (▲) at a site in the Pacific Ocean (based on Coale and Bruland, 1988).

metals, and in addition there is evidence for the release of specific binding ligands which, for instance, complex copper to reduce its toxicity and others that complex iron (see Sections 11.3.4 and 11.3.5), probably to retain it in solution (Morel and Price, 2003).

Free copper is known to be toxic to many species of phytoplankton, so the organic complexation is of direct biological importance. Coale and Bruland (1988) for instance showed that organic complexation reduces the inorganic copper ($Cu^{2+}$ and $CuCO_3$) concentrations from about $10^{-9}\,mol\,l^{-1}$ to $10^{-13}\,mol\,l^{-1}$ in surface waters Fig. 11.13, thereby greatly reducing the potential for deleterious biological impacts of free inorganic $Cu^{2+}$. In this study two distinct ligands were detected (Fig. 11.13), one (L1) dominant in surface waters and a weaker but more abundant complex (L2) in deeper waters.

### 11.6.2.2 Electron exchange: redox chemistry

There are regions of the oceans where oxygen concentrations fall to low levels (Section 9.3), and

this also occurs at depth in sediments. Under such conditions the reduction/oxidation potential (redox, Chapter 14) changes, with less oxidized forms of the elements coming to dominate. Redox speciation affects some elements (e.g. Mn, Fe, I, Cr, As and Se) that can exist in variable oxidation states in seawater. For example, the thermodynamically unstable Mn(II) is soluble in seawater but oxidizes to form rather insoluble Mn(IV) oxyhydroxides. This change in behaviour reflects the efficiency of scavenging for small highly charged cations such as MnIV compared to larger less highly charged ions such as Mn II (see Section 11.5). As a result, Mn is solubilized by reduction and precipitated by oxidation, and cycling between the two species of the metal can occur throughout the water column and in bottom sediments (see also Section 14.6). In subsurface waters Mn oxidation changes can occur at the $O_2$–$H_2S$ boundary and in the dissolved oxygen minimum zone. Thus Mn concentrations increase dramatically in moving from well oxygenated to deoxygenated seawater in areas such as the Cariaco Trench, in the Caribbean Sea where the deep waters below a few hundred metres are isolated from exchange with the surface waters (Fig. 11.14) and the Black Sea. The relative ease of conversion between Mn oxidation states and their strikingly different solubility (or tendency to be scavenged) means that Mn has a particularly rich ocean chemistry. In addition to the dramatic changes in dissolved Mn associated with changes from oxygenated to reduced environments discussed above, there are two other environments where Mn concentrations can be increased by redox processes. The first is in hydrothermal plumes (see Chapter 5) in which acidic and reducing condition within hydrothermal systems result in very high concentrations of many metals in the vent fluids – black and white smokers (Fig. 11.15). The oxidation of MnII to Mn IV requires two electrons and the simultaneous provision of two electrons is much more difficult than providing one, so the oxidation rate of Mn is much slower than that of FeII which is also released in hydrothermal plumes but which is rather rapidly oxidized to FeIII and removed to particles. The detection of enriched Mn plumes has often been used as a means of mapping and discovering hydrothermal vent systems (Chapter 5) since these signals persist over tens or even hundreds of km away from the source. Mn can also be

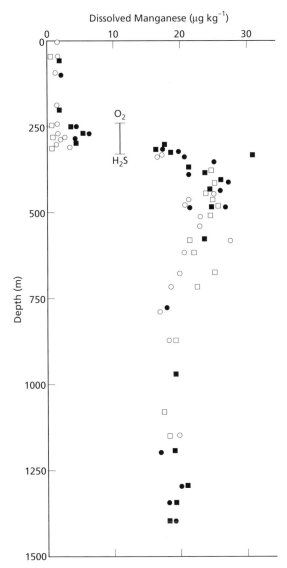

Fig. 11.14 Vertical distribution of dissolved Mn from the Cariaco Trench in the Caribbean Sea showing the very marked increase in concentrations across the boundary between oxygenated ($O_2$) and deoxygenated ($H_2S$)waters (based on Bacon *et al.*, 1980). Reprinted with permission from Elsevier.

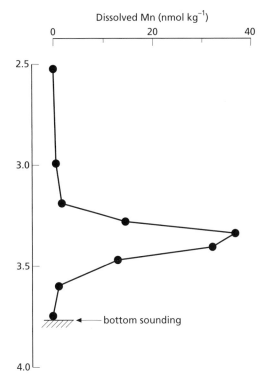

Fig. 11.15 Vertical distribution of dissolved Mn over the mid-Atlantic ridge (26°N) showing the effects of hydrothermal inputs (based on Klinkhammer *et al.*, 1986). Reprinted with permission from Elsevier.

reduced by photochemical reactions by which energy is first transferred to dissolved organic matter and this then in turn acts as a reducing agent for MnIV (Sunda and Huntsman, 1994).

Chromium is another trace metal that has a highly soluble redox state ([Cr(VI)] as the chromate ion) and a very insoluble state [Cr(III)] in seawater; in contrast to Mn, however, Cr is solubilized by oxidation and precipitated by reduction.

### 11.6.2.3 The hydride elements and methylation

Another area in which trace-element speciation is important has been highlighted by the work on the 'hydride' elements in seawater. Andreae (1983) has pointed out that metals and metalloids in the fourth, fifth and sixth main groups of the periodic table including Ge, Sn, As, Sb and Se, can be present in seawater in organometallic forms. For example, Froelich *et al.* (1983; 1985) have identified at least three forms of Ge in marine and estuarine waters, inorganic germanic acid ($Ge_i$), monomethylgermanium (MMGe) and dimethylgermanium (DMGe), and the authors showed that a large fraction of the oceanic Ge apparently exists in the methyl forms. The geochemical importance of these findings is that the different forms of Ge follow different geochemical pathways, for instance in the estuarine–marine environment. In general, $Ge_i$ behaves like silica; it has a high concentration in river water, where its source is the weathering of crustal rocks, and low concentrations in seawater. In estuaries, the behaviour of $Ge_i$ follows that of silica, and it can exhibit non-conservative behaviour following uptake by diatoms. In the oceans, the horizontal and vertical distributions of $Ge_i$ also mimic those of silica and reflect uptake on to, and dissolution from, siliceous organisms; that is, it is a nutrient-type element displaying a non-conservative behaviour pattern. In contrast, both MMGe and DMGe behave in a conservative manner in estuaries and in the open ocean; both species are barely detectable in river water, and in estuaries they probably have an oceanic source.

Selenium is another element that has organic and inorganic species that exhibit different biogeochemical behaviour patterns in the oceans. Cutter and Bruland (1984) reviewed the oceanic chemistry of the element, and demonstrated the existence of three dissolved Se species: selenite (Se(IV)), selenate (Se(VI)) and an operationally defined organic selenide. The water-column distribution of these species is illustrated in Fig. 11.16. In surface waters of the North and South Pacific, organic selenide comprised ~80% of the total dissolved Se, with the maximum being associated with primary production. In deepwaters, however, organic selenide was undetectable. In contrast to the organic Se, the selenite and selenate species exhibited nutrient-type vertical distributions and were enriched in deep-waters. On the basis of this data, the authors summarised the marine biogeochemical cycle of Se as illustrated in Fig. 11.16. This involves the selective uptake of the element, its reductive incorporation into biogenic material, its delivery to the deep sea as particulate organic selenide via sinking detritus, and a regeneration back into the dissolved state. While the cycle of Se is of interest in its own right, it also serves to illustrate the complex and rich ocean chemistry of elements that with modern techniques can now be described.

Methylation is important in the marine chemistries of As, Sb, Sn and Hg. Arsenic is present in seawater as As(V), as As(III) and methylated As species. As(V) (arsenate) is the predominant dissolved As species, especially in deep-water. In the surface layer, however, arsenate is taken up by phytoplankton, together with phosphate, and is excreted as As(III) (arsenite) and as the methylated species methylarsenate and dimethylarsenate. The methylated species making up ~10% of the total As in the euphotic zone in some oceanic regions (Andreae, 1983). Thus, plankton (and bacteria) can influence the speciation of As in seawater, and the speciation transformations in the euphotic zone lead to As(V) having a nutrient-type distribution in the water column. In some respects, the distribution of Sb in the water column resembles that of As, with Sb(V) predominating over Sb(III), and methylated Sb making up ~10% of the total antimony in surface water. Iodine also shows a redox chemistry with iodate ($IO_3^-$) dominant in deep water but significant iodide ($I^-$) present in surface waters due to uptake of iodate and release of iodide associated with primary production in surface waters (Campos et al., 1996). Note that the reduced species of Sb, As and I are found almost exclusively in surface waters, while reduced forms of Se persist deep into the oceans, indicating that oxidation of reduced selenium is very slow compared to these other elements, although the reasons for this are not well understood (Measures et al., 1980). Although numerous elements in seawater show redox-like behaviour, their detailed individual distributions will respond differently to changes in redox conditions reflecting the different redox potentials of the individual oxidation states of the element as well as kinetic factors, and the role of organic complexation (Rue et al., 1997).

Mercury cycling within the oceans also involves methylated forms in addition to inorganic forms of mercury including $Hg^{2+}$ and $Hg^0$, the latter potentially available for gaseous air–sea exchange, a unique feature of the Hg cycle amongst all the metals. The mercury cycle has been considerably perturbed by human activity and as discussed earlier for lead, this means that the measured profiles in the ocean of this scavenged element may not represent steady state distributions. The mercury can be bioaccumulated at higher trophic levels and this has led to health concerns associated with eating fish such as tuna which are top predators (Fitzgerald et al., 2007).

Methylated species of Sn have been identified in polluted estuarine and coastal waters; however, they are at very low concentrations in the open ocean. Toxic butyltin, like methyltin, is produced in indus-

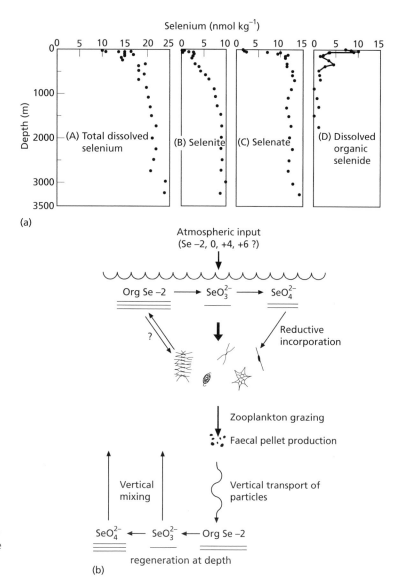

**Fig. 11.16** Selenium in seawater.
(a) Vertical water-column profiles of
total dissolved selenium, selenite,
selenate and organic selenide in the
Pacific Ocean (from Cutter and Bruland,
1984). (b) Schematic representation of
the marine biogeochemical cycle of
selenium (from Cutter and Bruland,
1984). The underlining indicates the
relative concentrations of selenium
species in surface and deep waters. The
preferential uptake of selenite in surface
waters is shown by the largest of the
dissolved-to-particulate arrows.

trial processes and it is also used as an antifouling agent, and these can be sources of the compound to coastal waters with demonstrable deleterious effects on some organisms (Hoch, 2001).

It may be concluded that speciation is important because different species of the same element can behave in different ways with respect to their entry into marine biogeochemical cycles. In particular, the speciation of a trace element has a strong influence on its bioavailability; that is,. uptake into biota, and its particle reactivity; that is, uptake on to inorganic particles. Some speciation forms, especially those involving organic complexation, can stabilize metals in solution. However, it is the transition from the dissolved to the particulate state that ultimately controls the residence time of an element in seawater and mediates its delivery to the sediment sink and its journey out of the seawater reservoir.

## 11.7 Trace elements in seawater: summary

1 Trace-element distributions in seawater are controlled by a complex interaction between their source strengths, their removal rates and water circulation patterns.

2 The vertical distributions of trace elements in the water column can be related to a number of analogues. On this basis, the profiles can be divided into the following classes: (i) conservative type (unreactive); (ii) nutrient surface-depletion–depth-enrichment type; (iii) scavenging surface-enrichment–depth-depletion type. Some elements show behaviour intermediate between these classes and in some other cases, such as Mn, the behaviour of an element is very different in different redox environments.

3 Dissolved trace elements are removed from seawater by the oceanic microcosm of particles, via scavenging-type and nutrient-type mechanisms, and the subsequent settling of these particles down the water column, either directly from the surface water or indirectly following subsurface lateral advection. Scavenging efficiency can be related to the fundamental chemical properties of the elements, particularly the charge and size of the ions they form.

4 Relatively large fractions (>50%) of the dissolved concentrations of some trace metals (e.g. Cu, Zn, Cd, Pb) in open-ocean surface waters are bound in organic complexes by ligands, which are often metal-specific. This organic speciation plays an important role in the biogeochemical cycles of the metals.

The processes involved in the down-column transport of the trace elements are part of the 'great particle conspiracy', and in the next chapter an attempt will be made to assess the rate at which the conspiracy operates.

## References

Anderson, R.F. (2003) Chemical tracers of particle transport, in *Treatise on Geochemistry*, H.D. Holland and K.K. Turekian (eds) Vol. 6, *The Oceans and Marine Geochemistry*, H. Elderfield (ed.), pp.247–291. Oxford: Elsevier Pergamon.

Andreae, M.O. (1983) The determination of the chemical species of some of the 'hydride elements' (arsenic, antimony, tin and germanium) in seawater: methodology and results, in *Trace Metals in Sea Water*, C.S. Wong, E. Boyle, K.W. Bruland, J.D. Burton and E.D. Goldberg (eds), pp.1–19. New York: Plenum.

Bacon, M.P. and Anderson, R.F. (1982) Distribution of thorium isotopes between dissolved and particulate forms in the deep sea. *J. Geophys. Res.*, 87, 2045–2056.

Bacon, M.P., Spencer, D.W. and Brewer, P.G. (1976) 210Pb/226Ra and 210Po/210Pb disequilibria in seawater and suspended particulate matter. *Earth Planet. Sci. Lett.*, 32, 277–296.

Bacon, M.P., Brewer, P.G., Spencer, D.W., Murray, J.W. and Goddard, J. (1980) Lead-210, polonium-210, manganese and iron from the Cariaco Trench. *Deep Sea Res. A*, 27, 119–135.

Bailey, R.A., Clarke, H.M., Ferris, J.P., Krause, S. and Strong, R.L. (1978) *Chemistry of the Environment*. New York: Academic Press.

Balistrieri, L., Brewer, P.G. and Murray, J.W. (1981) Scavenging residence time of trace metals and surface chemistry of sinking particles in the deep ocean. *Deep Sea Res. A*, 28, 101–121.

Bergquist, B.A. and Boyle, E.A. (2006) Dissolved iron in the tropical and subtropical Atlantic Ocean, *Global Biogeochem. Cycl.*, 20, GB1015, doi:10.1029/2005 GB002505.

Boulegue, J. (1983) Trace metals (Fe, Zn, Cd) in anoxic environments, in *Trace Metals in Sea Water*, C.S. Wong, E. Boyle, K.W. Bruland, J.D. Burton and E.D. Goldberg (eds), pp.563–577. New York: Plenum.

Boyd, P.W. and Ellwood, M.J. (2010) The biogeochemical cycle of iron in the ocean. *Nature Geosciences*, 3, 675–682.

Boyle, E.A. (1988) Cadmium: chemical tracer of deep water oceanography. *Palaeooceanography*, 3, 471–489.

Boyle, E.A. and Huested, S. (1983) Aspects of the surface distributions of copper, nickel, cadmium and lead in the North Atlantic and North Pacific, in *Trace Metals in Sea Water*, C.S. Wong, E. Boyle, K.W. Bruland, J.D. Burton and E.D. Goldberg (eds), pp.379–394. New York: Plenum.

Boyle, E.A., Sclater, F. and Edmond, J.M. (1976) On the marine geochemistry of cadmium. *Nature*, 263, 42–44.

Boyle, E.A., Sclater, F. and Edmond, J.M. (1977) The distribution of dissolved copper in the Pacific. *Earth Planet. Sci. Lett.*, 37, 38–54.

Bruland, K.W. (1980) Oceanographic distributions of cadmium, zinc, nickel and copper in the North Pacific. *Earth Planet. Sci. Lett.*, 47, 176–198.

Bruland, K.W. (1983) Trace elements in sea water, in *Chemical Oceanography*, J.P. Riley and R. Chester (eds), Vol. 8, pp.157–220. London: Academic Press.

Bruland, K.W. and Franks, R.P. (1983) Mn, Ni, Zn and Cd in the western North Atlantic, in *Trace Metals in Sea Water*, C.S. Wong, E. Boyle, K.W. Bruland, J.D. Burton and E.D. Goldberg (eds), pp.395–414. New York: Plenum.

Bruland, K.W., Donat, J.R. and Hutchins, D.A. (1991) Interactive influences of bioactive trace metals on biological production in oceanic waters. *Limnol. Oceanogr.*, **36**, 1555–1577.

Bruland, K.W. and Lohan, M.C. (2003) Controls of trace metals in seawater, in K.W. Bruland, and Lohan, M.C. (2004) Controls on trace metals in seawater, in the *Oceans and Marine Geochemistry* H. Elderfield (ed.) Vol. 6, *Treatise on Geochemistry* H.D. Holland and K.K. Turekian (eds.), pp. 23–49. London: Elsevier.

Buessler K.O., Benitez, C.R., Moran, S.B., *et al.* (2006) An assessment of particulate organic carbon to thorium-234 ratios in the ocean and their impact on the application of 234Th as a POC flux proxy, *Marine Chemistry*, **100**, 213–233.

Burton, J.D., Maher, W.A. and Statham, P.J. (1983) Some recent measurements of trace metals in Atlantic Ocean waters, in *Trace Metals in Sea Water*, C.S. Wong, E. Boyle, K.W. Bruland, J.D. Burton and E.D. Goldberg (eds), pp.415–426. New York: Plenum.

Byrne, R.H. (2002) Speciation in seawater, in *Chemical Speciation in the Environment*, A.M. Ure and C.M. Davidson (eds), pp.322–357. Blackwell Publishing Ltd.

Campos, M.A.M., Farrenkopf, A.M., Jickells, T.D. and Luther, G.W. (1996) A comparison of dissolved iodine cycling at the Bermuda Atlantic Time-series and Hawaii Ocean time series station. *Deep-Sea Res. II*, **43**, 445–466.

Capodaglio, G., Coale, K. H. and Bruland, K. W. (1990). Lead speciation in surface waters of the Eastern North Pacific. *Mar.Chem.* **29**, 221–233.

Chester, R., Bradshaw, G.F., Ottley, C.J. *et al.* (1994) The atmospheric distributions of trace metals, trace organics and nitrogen species over the North Sea, in *Understanding the North Sea System*, H. Charnock, K.R. Dyer, J.M. Huthnance, P.S. Liss, J.H. Simpson and P.B. Tett (eds). *Philos. Trans. R. Soc. London, Ser. A*, **343**, 545–556.

Coale, K.H. and Bruland, K.W. (1985) 234Th:238U disequilibria within the California Current. *Limnol. Oceanogr.*, **30**, 22–33.

Coale, K.H. and Bruland, K.W. (1987) Oceanic stratified euphotic zone as elucidated by 234Th:238U disequilibria. *Limnol. Oceanogr.*, **32**, 189–200.

Coale, K.H. and Bruland, K.W. (1988) Copper complexation in the northeast Pacific. *Limnol. Oceanogr.*, **33**, 1084–1101.

Cutter, G.A. and Bruland, K.W. (1984) The marine biogeochemistry of selenium: a re-evaluation. *Limnol. Oceanogr.*, **29**, 1179–1192.

Danielsson, L.G. and Westerlund, S. (1983) Trace metals in the Arctic Ocean, in *Trace Metals in Sea Water*, C.S. Wong, E. Boyle, K.W. Bruland, J.D. Burton and E.D. Goldberg (eds), pp.85–95. New York: Plenum.

Ellwood, M.J. and Hunter, K.A. (2000) The incorporation of zinc and iron into the frustule of the marine diatom Thalassiosira pseudonana. *Limnology and Oceanography*, **45**, 1517–1524.

Filella, M., Town, R.M. and Buffle, J. (2002) Speciation in freshwaters, in *Chemical Speciation in the Environment*, 2nd edn, A.M. Ure and C.M. Davidson (eds), pp.188–236, Blackwell Publishing Ltd.

Fitzgerald, W.F., Lamborg, C.H. and Hammerschmidt, C.R. (2007) Marine biogeochemical cycling of mercury, *Chem. Rev.*, **107**, 641–662.

Flegal, A. (1998) Clair Paterson's Influence on Environmental Research. *Environmental Research*, **78**, 65–70.

Frew, R.D. and Hunter, K.A. (1992) Influence of Southern Ocean waters on the cadmium–phosphate properties of the global ocean. *Nature*, **360**, 144–146.

Froelich, P.N., Hambrick, G.A. and Andreae, M.O. (1983) Geochemistry of inorganic and methyl germanium species in three estuaries. *Eos*, **45**, 715.

Froelich, P.N., Hambrick, G.A., Kaul, L.W., Byrd, J.T. and Lecointe, O. (1985) Geochemical behaviour of inorganic germanium in an unperturbed estuary. *Geochim. Cosmochim. Acta*, **49**, 519–524.

Goldberg, E.D. (1954) Marine geochemistry. I. Chemical scavengers of the sea. *J. Geol.*, **62**, 249–265.

Hoch, M., 2001. Organotin compounds in the environment an overview. *Applied Geochemistry*, **16**, 719–743.

Honeyman, B.D., Balistrieri, L.S. and Murray, J.W. (1988) Oceanic trace metal scavenging: the importance of particle concentration. *Deep Sea Res. A*, **35**, 227–246.

Howard, A.G. and Statham, P.J. (1993) *Inorganic Trace Analysis: Philosophy and Practice*. Chichester: John Wiley & Sons, Ltd.

Hydes, D.J. (1983) Distribution of aluminium in waters of the north east Atlantic 25°N to 35°N. *Geochim. Cosmochim. Acta*, **47**, 967–973.

Kelly, A.E., Reuer, M.K., Goodkin, N.F. and Boyle, E.A. (2009) Lead concentrations and isotopes in corals and water near Bermuda, 1780–2000. *Earth Planet. Sci. Lett.*, **283**, 93–100.

Klinkhammer, G., Elderfield, H., Greaves, M., Rona, P. and Nelson, T. (1986) Manganese geochemistry near high temperature vents in the Mid-Atlantic Rift valley. *Earth Planet. Sci. Lett.*, **80**, 230–240.

Kremling, K. (1983) Trace metal fronts in European shelf waters. *Nature*, **303**, 225–257.

Kremling, K. (1985) The distribution of cadmium, copper nickel, manganese and aluminium in surface waters of the open Atlantic and European shelf area. *Deep Sea Res. A*, **32**, 531–555.

Kremling, K. and Streu, P. (2001) The behavior of dissolved Cd, Co, Zn and Pb in North Atlantic near-surface waters (30°N/60°W–60°N/2°W). *Deep Sea Res. I*, **48**, 2541–2567.

Landing, W.M. and Bruland, K.W. (1980) Manganese in the North Atlantic. *Earth Planet. Sci. Lett.*, **49**, 45–56.

Landing, W.M. and Bruland, K.W. (1987) The contrasting biogeochemistry of iron and manganese in the Pacific Ocean. *Geochim. Cosmochim. Acta*, **51**, 29–43.

Li, Y.-H. (1991) Distribution patterns of the elements in the ocean: A synthesis. *Geochim. Cosmochim. Acta*, **55**, 3223–3240.

Luther, G.W., Rozan, T.F., Witter, A. and Lewis, B. (2001) Metal-organic complexation in the marine environment. *Geochem. Trans.*, **9**, doi:10.1186/1467-4866-2-65.

Martin, J.H., Bruland, K.W. and Broenkow, W.W. (1976) Cadmium transport in the California Current, in *Marine Pollutant Transfer*, H.L. Windom and R.A. Duce (eds), pp.159–184. Lexington, MA: Lexington Books.

Martin, J.H. and Gordon, R.M. (1988) Northeast Pacific iron distributions in relation to phytoplankton productivity. *Deep Sea Res. A*, **35**, 177–196.

Martin, J.H., Gordon, R.M. and Broenkow, W.W. (1989) VERTEX: phytoplankton/iron studies in the Gulf of Alaska. *Deep Sea Res. A*, **37**, 1639–1653.

Measures, C.I., McDuff, R.E. and Edmond, J.M. (1980) Selenium redox chemistry at GEOSEC I re-occupation. *Earth Planet Sci. Lett.*, **49**, 102–108.

Measures C I and Vink S, (2000) On the use of dissolved aluminum in surface waters to estimate dust deposition to the ocean, *Global Biogeochem. Cycl.*, **14**, 317–327.

Moore, R.M. (1983) The relationship between distributions of dissolved cadmium, iron and aluminium and hydrography in the central Arctic Ocean, in *Trace Metals in Sea Water*, C.S. Wong, E. Boyle, K.W. Bruland, J.D. Burton and E.D. Goldberg (eds), pp.131–142. New York: Plenum.

Morel, F.M.M. and Price, N.M. (2003) The biogeochemical cycles of trace metals in the oceans. *Science*, **300**, 944–947.

Morley, N.H., Statham, P.J. and Burton, J.D. (1993) Dissolved trace metals in the southwestern Indian Ocean, *Deep-Sea Res. I*, **40**, 1043–1062.

Morris, A.W. (1975) Dissolved molybdenum and vanadium in the northeast Atlantic Ocean. *Deep-Sea Res.*, **22**, 49–54.

Nurnberg, H.W. and Valenta, P. (1983) Potentialities and applications of voltammetry in chemical speciation of trace metals in the sea, in *Trace Metals in Sea Water*, C.S. Wong, E. Boyle, K.W. Bruland, J.D. Burton and E.D. Goldberg (eds), pp.671–697. New York: Plenum.

Nyffeler, U.P., Li, Y.-H. and Santschi, P.S. (1984) A kinetic approach to describe trace element distribution between particles and solution in natural aquatic systems. *Geochim. Cosmochim. Acta*, **48**, 1513–1522.

Orians, K.J. and Bruland, K.W. (1986) The biogeochemistry of aluminium in the Pacific Ocean. *Earth Planet. Sci. Lett.*, **78**, 397–410.

OSPAR Commission (2010) *Quality Status Report.* Available at: http://qsr2010.ospar.org/en/index.html (accessed 15 March, 2012).

Rue, E.L., Smith, G.J., Cutter, G.A. and Bruland, K.W. (1997) The response of trace element redox couples to suboxic conditions in the water column, *Deep-Sea Res. I*, **44**, 113–134.

Schaule, B.K. and Patterson, C.C. (1981) Lead concentrations in the Northeast Pacific: evidence for global anthropogenic perturbations. *Earth Planet. Sci. Lett.*, **54**, 97–116.

Schindler, P.W. (1975) Removal of trace metals from the oceans: a zero order model. *Thalassia Jugoslav.*, **11**, 101–111.

SCOR 2007 SCOR Working Group (2007) GEOTRACES – An international study of the global marine biogeochemical cycles of trace elements and their isotopes. *Chemie der Erde Geochemistry*, doi:10.1016/j.chemer.2007.02.001.

Simpson, J.H. (1994) Introduction to the North Sea Project, in *Understanding the North Sea System*, H. Charnock, K.R. Dyer, J.M. Huthnance, P.S. Liss, J.H. Simpson and P.B. Tett (eds). *Philos. Trans. R. Soc. London, Ser. A*, **343**, 1–4.

Sohrin, Y., Isshiki, K. and Kuwamoto, T. (1987) Tungsten in North Pacific waters. *Mar. Chem.*, **22**, 95–103.

Spivack, J., Huested, S.S. and Boyle, E.A. (1983) Copper, nickel and cadmium in the surface waters of the Mediterranean, in *Trace Metals in Sea Water*, C.S. Wong, E. Boyle, K.W. Bruland, J.D. Burton and E.D. Goldberg (eds), pp.505–12. New York: Plenum.

Statham, P.J. and Burton, J.D. (1986) Dissolved manganese in the North Atlantic Ocean, 0–35°N. *Earth Planet. Sci. Lett.*, **79**, 56–65.

Stoffyn-Egli, P. and MacKenzie, F.T. (1984) Mass balance of dissolved lithium in the oceans. *Geochim. Cosmochim. Acta*, **48**, 859–872.

Stumm, W. and Brauner, P.A. (1975) Chemical speciation, in *Chemical Oceanography*, J.P. Riley and G. Skirrow (eds), Vol. 1, pp.173–239. London: Academic Press.

Symes, J.L. and Kester, D.R. (1985) The distribution of iron in the northwest Atlantic. *Mar. Chem.*, **17**, 57–74.

Sunda W.G. and Huntsman, S.A. (1994) Photoreduction of manganese oxides in seawater. *Mar. Chem.*, **46**, 133–152.

Takesue, R.K. and van Geen, A. (2002) Nearshore circulation during upwelling inferred from the distribution of dissolved cadmium off the Oregon coast. *Limnol. Oceanogr.*, **47**, 176–185.

Ure, A.M. and Davidson, C.M. (eds) (2002) *Chemical Speciation in the Environment*, 2nd edn. London: Chapman and Hall.

Whitfield, M. (1979) The mean oceanic residence time (MORT) concept, a rationalization. *Mar. Chem.*, **8**, 101–123.

Whitfield, M. and Turner, D.R. (1987) The role of particles in regulating the composition of sea water, in *Aquatic Surface Chemistry Chemical Processes at the Particle–Water Interface*, W. Stumm (ed.), pp.457–493. New York: John Wiley & Sons, Inc.

Wu, J. and Boyle, E.A. (1997) Lead in the western North Atlantic Ocean: completed response to leaded gasoline phaseout. *Geochimica et Cosmochimica Acta*, **61**, 3279–3283.

# 12 Down-column fluxes and the benthic boundary layer

In the last three chapters we have described the distributions of particulate matter, nutrients, carbon and dissolved trace elements in the oceans, and have discussed the manner in which they interact in the water column to exert a control on the elemental composition of seawater. The principal mechanism underlying this control is the removal of dissolved trace elements by the oceanic microcosm of particles via scavenging-type and nutrient-type carrier-phase associations, and the subsequent settling of these particles down the water column. These transport processes are part of the *great particle conspiracy*. In the present chapter an attempt will be made to estimate the rate and processes by which material travels to deep waters via the particulate carriers. This sinking material is also a major component of the global carbon cycle and a source of food and energy to the deep ocean. Following this discussion, the journey undertaken by the particulate matter will be taken a stage further by tracking the sedimenting solids to the sediment–water interface.

## 12.1 Down-column fluxes

It was shown in Section 10.4 that the total suspended material in the oceans can be divided conveniently into two general populations: (i) a small-sized population, termed fine particulate matter (FPM), which sinks at a slow rate; and (ii) a large-sized population (CPM), which consists mainly of biogenic particle aggregates that undergo relatively fast vertical settling. The FPM sink mainly by association with the CPM, the coupling between the two particulate populations occurring by 'piggy-backing'. The total suspended matter (TSM) flux sinking vertically down the water column from the surface therefore can be defined as:

$$\text{Total Flux} = \text{FPM}_f + \text{CPM}_f \tag{12.1}$$

where $\text{FPM}_f$ and $\text{CPM}_f$ are the fluxes associated with the fine (FPM) and coarse (CPM) material respectively.

The treatment given in the following will focus on the total, or particle mass, flux down the water column. The FPM is especially important for the scavenging of trace elements. However, most of the mass flux of material is carried down the water column in association with large-sized aggregates, such as faecal pellets and 'marine snow'. Thus, the vertical transport is dominated by the global carbon flux, that is, the export flux that escapes surface water recycling and is balanced by new production (see Section 9.1). In general, the strength of the down-column flux varies with the degree of primary production in surface waters, although this relationship is not always simple as discussed later. The magnitude of the flux has been estimated by both sediment-trap and radioisotope techniques. The general use of radioisotopes in dissolved–particulate equilibrium studies has been reviewed in Section 11.5, and attention at this stage will be confined largely to sediment-trap techniques.

Much of our knowledge of the strengths of down-column, or more strictly down-column, plus laterally advected fluxes has come from the use of sediment traps, which according to Brewer *et al.* (1986)

*Marine Geochemistry*, Third Edition. Roy Chester and Tim Jickells.
© 2012 by Roy Chester and Tim Jickells. Published 2012 by Blackwell Publishing Ltd.

provide direct evidence of 'sediments in the making'. The traps consist of large cones approximately a metre tall and with an opening at the top with a diameter of the order of a metre or so. These can be deployed at a prescribed depth within the ocean to be recovered at a later time with sinking particles collected in sequentially opening bottles at the base of the cone. The down-column fluxes that carry the material forming the sediments are driven by particulate matter and in many of the field experiments the sediment traps have been deployed in relation to the three-layer model for the distribution of TSM in the water column (see Section 10.2); that is samples have been collected in the upper ocean, the clear-water minimum and the deep-water layer. This relationship between particle collection and the three-layer model is very important because it identifies a fundamental constraint that must be placed on the interpretation of sediment-trap data. That is, when attempts are made to estimate the strengths of down-column particulate fluxes it is necessary to make a distinction between the two sources of TSM to the water column, that is a primary *downward* flux from the surface layer and a resuspended *upward* flux from the sea-bed. The overall relationship between the three-layer TSM model and the various particle fluxes is illustrated in Fig. 12.1.

There are a number of difficulties associated with the design, calibration and water-column location of sediment traps, including hydrodynamic effects and the role of swimmers (Buesseler *et al.*, 2007a). Sinking particles in the oceans fall vertically relatively slowly compared to their horizontal transport speed so collecting them efficiently within the cone of a sediment trap is difficult, analogous to the problems of accurately collecting snow in rain gauge. These hydrodynamic effects can lead to over and underestimates of fluxes. Swimmers is the term used to describe the active entry of deep water animals, from zooplankton to fish, into the traps which can lead to an underestimate of flux if they escape with the food, or overestimation if they die within the bottle due to poisons used to preserve the flux. With care and good practice, however, it is possible to minimize these problems (Buesseler *et al.*, 2007a). The main advantage of the sediment traps is that they are capable of collecting both small and large-sized sinking material, and despite the inherent problems, they have played a vital role in mass-flux studies.

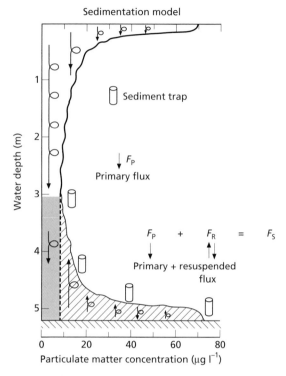

**Fig. 12.1** The relationship between the down-column location of sediment traps and the three-layer oceanic TSM flux model (from Gardner *et al.*, 1985). Reprinted with permission from Elsevier.

Many of the problems associated with sediment traps are minimized in deep water where current flows and animal density is low, so much of the most useful data available is from sediment trap deployments in the deep ocean, but at depths well above the effects of any benthic resuspension. Over the past decade or so, sediment-trap experiments have provided a considerable body of data on the magnitude and composition of the down-column particle fluxes to the interior of the ocean. This data includes deployments for periods of the order of a year or more (thus long enough to define the seasonal cycle) at sites around the world (e.g. see syntheses by Lampitt and Antia, 1997; Francois *et al.*, 2002; Lutz *et al.*, 2007), and at a very few sites, the maintenance of long term multi-year deployments of sediment traps, most notably that off Bermuda initiated in 1978 by Werner Deuser and still continuing (Conte *et al.*, 2001).

## 12.1.1 The magnitude and major component composition of the particle mass flux to the interior of the ocean

Compilations of the data from a series of mostly long-term (at least 1 yr) trap deployments have been produced by Lampitt and Antia (1997), Francois *et al.* (2002) and Lutz *et al.* (2007). Table 12.1 based on Lampitt and Antia (1977) provides a useful general average of the flux data. These authors noted that polar area show some of the most extreme flux variability so provided averages from all the available data and with averages polar regions excluded. A number of features of the particle mass flux to the interior of the ocean can be identified when the data in Table 12.1 are combined with other findings reported in the literature. These features are summarized in the table.

### 12.1.1.1 The magnitude of the particle mass flux

Mass fluxes to the interior of the ocean vary greatly, even for the non-polar regions there is a 10-fold range from $8–66\,\mathrm{g\,m^{-2}\,yr^{-1}}$ with greater variability still in Polar Regions. The fluxes are dominated by biogenic material and so a relationship to ocean primary production can be anticipated and indeed the highest fluxes are associated with regions of high productivity such as ocean upwelling zones off North Africa and in the Indian Ocean, with lower fluxes collected beneath oligotrophic gyres as illustrated in Table 12.2. However, the relationship between ocean primary productivity and even the sediment trap organic carbon fluxes, let alone the total mass fluxes, is not straightforward. Lampitt and Antia (1997) suggested there is an approximately linear relationship up to primary production rates of about $200\,\mathrm{g}$ organic $\mathrm{C\,m^{-2}\,yr^{-1}}$, but above this the relationship

flattened out and sediment trap fluxes did not increase even at much higher productivity. The authors suggested this may reflect particularly efficient recycling of organic matter within very high productivity ecosystems. Francois *et al.* (2002) suggested there may be a very important role for mineral material (which is dominated by calcium carbonate) acting as ballast and promoting rapid and efficient transfer of organic matter from the surface ocean to the depths (see also Chapter 9). Changes in ballast supply may therefore also affect the relationship between primary production and deep ocean fluxes and Francois suggested that a linear relationship exists between the proportion of export production from surface waters that reaches sediment traps and the flux of calcium carbonate. Francois also suggested that export from diatom blooms may be less efficiently transferred to deep water because of either the less efficient role of opal as ballast or because diatom aggregates from blooms are consumed or broken up more readily than those associated with calcium carbonate during sinking. Such a difference in flux characteristics from different regions may explain why high latitude systems, in which diatoms blooms are most common, appear to be less efficient at transferring organic matter formed during primary production to deep water (Lampitt and Antia, 1997; Francois *et al.*, 2002; Lutz *et al.*, 2007).

## 12.1.2 The major components of the particle mass flux

Based on Table 12.1 and the average proportion of C in organic matter and calcium carbonate and Si in opal, it is possible to estimate the average composition of the sinking flux. The residual material not contained in the analyses in Table 12.1 is assumed to be

**Table 12.1** Sediment trap fluxes for all stations, and for all stations excluding those in polar areas $(\mathrm{g\,m^{-2}\,yr^{-1}})$ for the major components (sd, standard deviation) normalized to 2000 m. (Lampitt and Antia, 1997).

| | All data | | | Non Polar | | |
|---|---|---|---|---|---|---|
| | Max | Min | Median ± sd | Max | Min | Median ± sd |
| Dry weight | 148 | 0.26 | 22 ± 22 | 66 | 7.8 | $23 ± 14\,\mathrm{g\,m^{-2}\,yr^{-1}}$ |
| Organic C | 5.9 | 0.01 | 1.4 ± 1.3 | 4.2 | 0.4 | $1.5 ± 1.1\,\mathrm{g\,C\,m^{-2}\,yr^{-1}}$ |
| Carbonate | 3.6 | 0.001 | 1.4 ± 0.9 | 3.6 | 0.6 | $1.7 ± 0.8\,\mathrm{g\,C\,m^{-2}\,yr^{-1}}$ |
| Opal Si | 8.9 | 0.1 | 1.6 ± 2.0 | 8.9 | 0.4 | $1.9 ± 1.9\,\mathrm{g\,Si\,m^{-2}\,yr^{-1}}$ |

**Table 12.2** Particle mass fluxes to the interior of the northeast Atlantic derived from sediment-trap deployments: data normalized to 4000 m for the loss of POC (data after Jickells *et al.*, 1996).

(a) Particle mass flux and total flux composition; mg m$^{-2}$ day1.

| Site | Mass flux | CaCO3 | Opal | Organic matter | Lithogenic matter |
|------|-----------|-------|------|----------------|-------------------|
| 48°N 21°W | 71.5 | 42.4 | 15.4 | 5.1 | 8.6 |
| 48°N 20°W | 59.0 | 31.7 | 11.3 | 8.4 | 7.6 |
| 34°N 21°W | 58.7 | 34.9 | 5.5 | 5.3 | 13.0 |
| 31°N 24°W | 33.6 | 24.0 | – | 3.1 | <6.5 |
| 28°N 22°W | 27.7 | 16.7 | 1.4 | 3.0 | 6.6 |
| 24°N 23°W | 41.1 | 20.3 | 2.9 | 3.9 | 14.3 |
| 21°N 20°W | 177.0 | 78.7 | 14.8 | 9.0 | 74.5 |
| 21°N 21°W | 150.9 | 79.0 | 6.5 | 8.1 | 57.3 |
| 19°N 20°W | 227.7 | 110.2 | 16.9 | 25.4 | 75.2 |
| 22°N 25°W | 51 | – | – | – | 18.2 |

(b) Percentage total flux composition.

| Site | CaCO$_3$ | Opal | Organic matter | Lithogenic matter |
|------|----------|------|----------------|-------------------|
| 48°N 21°W | 59 | 21.5 | 7 | 12 |
| 48°N 20°W | 54 | 19 | 14 | 13 |
| 34°N 21°W | 59 | 9 | 9 | 22 |
| 31°N 24°W | 71 | – | 9 | <19 |
| 28°N 22°W | 60 | 5 | 11 | 24 |
| 24°N 23°W | 49 | 7 | 9 | 34 |
| 21°N 20°W | 45 | 8 | 5 | 42 |
| 21°N 21°W | 52 | 4 | 5 | 38 |
| 19°N 20°W | 48 | 7 | 11 | 33 |
| 22°N 25°W | – | – | – | 36 |

(c) Percentage biogenic flux composition and biogenic flux.

| Site | CaCO$_3$ | Opal | Organic matter | Biogenic flux (mg m$^{-2}$ day$^{-1}$) |
|------|----------|------|----------------|-----------------------------------------|
| 48°N 21°W | 67 | 24 | 8 | 62.9 |
| 48°N 20°W | 62 | 22 | 16 | 51.4 |
| 34°N 21°W | 76 | 12 | 12 | 45.7 |
| 31°N 24°W | 71–89 | 0–9.3 | 9–11 | 27.1–33.6 |
| 28°N 22°W | 79 | 7 | 14 | 21.1 |
| 24°N 23°W | 75 | 11 | 14 | 27.1 |
| 21°N 20°W | 77 | 14 | 9 | 102.5 |
| 21°N 21°W | 84 | 7 | 9 | 93.6 |
| 19°N 20°W | 72 | 11 | 17 | 152.5 |

aluminosilicate soil or sediment material. This component is not always directly measured on sediment trap material, but there are sufficient measurements available to support the assumption that it dominates the residual matter. Using this approach the global average composition of sediment trap material is about 13% organic matter, 60% calcium carbonate, 17% opal and 10% aluminosilicate. This aluminosilicate proportion is also consistent with the estima-

tion of Francois *et al.* (2002). While it is possible to calculate such a global average flux, there is strong evidence to show the composition of sinking material does vary systematically within the oceans.

Lampitt and Antia (1997) showed that the composition of sediment trap material varies with latitude with the percentage of calcium carbonate being about four times higher in equatorial regions than in polar regions, while the opal flux shows a less dramatic trend in the opposite direction with higher proportions of opal at high latitude. Table 12.2 also illustrates such a trend in the composition of sediment trap material. These trends do, at least in part, reflect the dominant phytoplankton composition in the overlying waters with diatoms being a relatively more important component of the phytoplankton at higher latitudes (see Chapter 9). The relative proportion of the minor lithogenic component depends in part on the fluxes of the major components (organic matter, calcium carbonate, opal) which can act to dilute the background input from, for instance, atmospheric dust. However, in the Atlantic Ocean at least it is clear that the absolute fluxes, and indeed the relative proportions, of aluminosilicate material, increases into the region underlying the Saharan dust storm transport compared to areas further north (Table 12.2) showing that the sediment trap fluxes can also reflect atmospheric dust inputs.

### 12.1.3 The mass particle flux and primary production in surface waters

#### 12.1.3.1 Seasonality in the mass particle flux

Various authors have identified seasonal signals in mass flux data obtained from sediment traps. The data are illustrated in Fig. 12.2 for the long term programme of sediment trap measurements off Bermuda and show evidence of both seasonal and inter-annual variability in the magnitude of the particle fluxes. The seasonal variations exhibit a winter–spring maximum and a summer–autumn minimum,

a seasonality in phase with that of surface water phytoplankton growth in this region with a lag of only a few weeks (Deuser, 1986; 1987; Conte et al., 2001). This was dramatic confirmation of rapid particle sinking in the oceans and showed that even the deepest ocean abyss was fuelled by a seasonal food supply from the surface of the oceans. This sedimentation event and its subsequent consumption by the deep sea benthos was subsequently recorded on camera (Billet et al., 1983).

Recent syntheses of sediment trap data show that there is little seasonality in sediment trap fluxes in equatorial regions, except where there is a strong seasonal driver such as the monsoon. Seasonality increases in subtropical waters (e.g. about 3-fold range in the Sargasso sea: Fig. 12.2; Conte et al., 2001) with increasing seasonality and a steadily later seasonal flux event moving poleward toward a maximum seasonality (of the order of 100-fold range) in polar regions (Lutz et al., 2007). This pattern exactly mirrors the patterns and scale of seasonality in primary production in overlying waters and implies particle sinking rates of 100–200 m d$^{-1}$ (Lampitt and Antia, 1997). At sites where there are sediments traps deployed at several depths, such as off Bermuda, the lag in the arrival of periods of high flux between sediment traps floating 500 m and 3000 m below the sea surface is less than 15 days (the sampling intervals for individual samples collected by the traps) again implying sinking rates of 100–200 m d$^{-1}$. Such sinking rates cannot be achieved by individual sinking phytoplankton or other fine particulate matter (FPM), but rather can only be

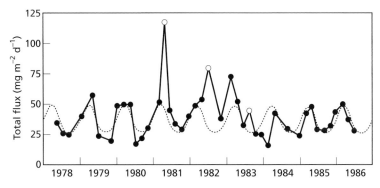

**Fig. 12.2** Temporal variations in particle fluxes in the deep ocean (from Deuser, 1987). An eight-year record of variations in total particle flux to a sediment trap located at 3200 m in the Sargasso Sea is shown. Points represent average fluxes over two-month trap deployment periods, plotted at the period mid-points. The dotted curve represents the average timing of the winter–spring flux maximum: the three measurements indicated by open circles were not included in the calculation of the average curve.

achieved by relatively large particles (CPM) associated with marine snow, aggregates or faecal pellets. These fast sinking rates mean that the material collected in a sediment trap will have originated from the surface waters relatively close to the sampling site, (within perhaps many 10s of square kilometres), rather than being dispersed over large distances by ocean currents during slow settling.

Although clearly closely coupled, the relationship between primary production and sediment trap fluxes does vary subtly with latitude. At high latitudes seasonality of primary production is generally greater than seasonality of sediment trap fluxes, with the reverse situation at lower latitudes (Lutz *et al.*, 2007). This suggests that the modification of organic matter fluxes within the euphotic zone and during sinking through the deep ocean is important in modifying the flux seasonality as well as the total flux. These processes probably involve bacteria and zooplankton grazing (Conte *et al.*, 2001) which consumes components of the sinking flux (see Section 12.2), but they are currently not well understood.

Although there is evidence of strong coupling of surface water primary production to deep water fluxes on time scales of weeks, the relationship is not always completely straightforward. There are regions of the ocean where lateral advection of material appears to greatly influence the magnitude and timing of fluxes, such as near to ocean margins (Lampitt and Antia, 1997). There are also now documented examples of short term high productivity 'events' in surface waters, associated for example with large eddies, which can lead to periods of high flux that do not conform to the classic seasonal cycle at any particular location (Conte *et al.*, 2003).

Given processes that can rapidly aggregate phytoplankton into large and relatively fast sinking particles, the observed seasonality of biogenic fluxes into the deep sea is readily understandable. Sediment traps also reveal that this process couples non-biological component fluxes to this seasonal biogenic flux. Figure 12.3 shows that the flux of aluminium, which is almost exclusively incorporated in clay particles, is very well correlated with the organic carbon flux to a sediment trap in the Sargasso Sea. This suggests that non-biological particles are therefore incorporated into sinking biological particles, possibly via indiscriminate feeding by zooplankton in surface waters resulting in clays and organic matter

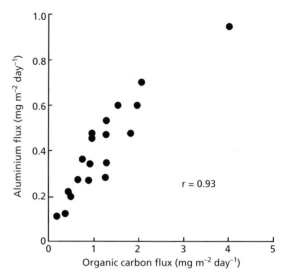

**Fig. 12.3** Correlation between fluxes of organic carbon and abiogenic Al (a tracer for clay) in a sediment trap at a depth of 3200 m in the Sargasso Sea over a four year period (based on Deuser *et al.*, 1983).

being transported together in faecal pellets. The amounts of clay have also now been shown to respond to changes in the dust load from year to year (Jickells *et al.*, 1998). Strong correlations between lithogenic and organic fluxes are seen at other sites even where dust loadings are much higher, such as underneath the Sahara dust plume off the coast of North Africa (Bory and Newton, 2000). This implies that the biological flux has the necessary capacity to carry all the available dust flux in most ocean areas. However, Bory and Newton found that the relationship between lithogenic and biogenic fluxes can become more complex when both dust and productivity are high and have different seasonal patterns such as in the upwelling region off North Africa. Hence, while most ocean regions show a close relationship between lithogenic and biogenic fluxes, this pattern can break down.

Subsequent analyses have shown that not only clays, but also contaminants such as trace metals and trace organic pollutants are incorporated into the sinking flux (Conte *et al.*, 2001). Sinking particulate matter showing enrichments of relatively insoluble trace metals relative to crustal abundance (see Section 4.2) that reflect atmospheric contamination by industrial emissions in the atmosphere above the sediment trap collection site (Huang and Conte, 2009). A striking example of such contaminant incorporation was the rapid transfer of radioactive material from the Chernobyl nuclear accident to the deep sea on timescales of days as shown by sediment trap measurements in the Mediterranean (Fowler *et al.*, 1987).

## 12.2 Evolution of particle flux with depth

Chapters 9 and 11 demonstrated that the dissolved concentrations of nutrients and trace metals vary with depth in the oceans due to the breakdown of sinking organic matter, coupled to the global ocean circulation. The data from sediment traps allows us to begin to quantify the scale of these processes. Lampitt and Antia (1997) attempted this on a global scale (Table 12.3). As discussed in Chapter 9, estimates of ocean productivity and export productivity are uncertain and different compilations arrive at somewhat different average fluxes, but the estimates in Table 12.3 are broadly consistent with estimates presented earlier in Chapter 9. The results in Table 12.3 make the key point that only about 1% of the organic carbon produced in the surface ocean sinks to the ocean depths.

Degradation during sinking systematically alters the composition of sinking organic matter as bacteria and zooplankton utilize the organic matter. Conte *et al.* (2001) compared the average composition of

the organic matter collected in sediment traps at 500 m and 3200 m in the Sargasso sea off Bermuda to describe the these changes in flux with depth (Table 12.4). Although this data is for only one site, it seems likely that similar trends will occur throughout the oceans. At this site there is evidence of lateral transport of sediment at depth from the ocean margins and this may explain the small increase in calcium carbonate, but the most striking effect is the large scale loss of organic matter during sinking. This means that the sinking flux becomes systematically depleted in organic carbon relative to inorganic components. The nutrients N and P also appear to be lost slightly more than organic carbon. To take account of these effects of changes during sinking, the data in Tables 12.1, 12.3 and 12.4 have all been converted to equivalent fluxes at a fixed depth of 2000 m based on relationships that have been developed to mathematically describe these loss processes.

The relationship between sediment trap measured organic matter fluxes and primary productivity has been described as a non-linear function of primary production or the flux leaving the surface layer and depth. A commonly used formulation is that originally proposed by Martin *et al.* (1987).

$$F = F_{100}(z/100)^b \qquad (12.2)$$

where F is the flux at depth, $F_{100}$ the flux exiting the base of the euphotic zone at 100 m, z the depth and b is a unit-less parameter describing the attenuation. Buesseler *et al.* (2007b) have shown that the parameter b varies with location and surface water productivity regime in the oceans. Buesseler *et al.* (2007b) and Lee *et al.* (2004) both emphasise that much of the degradation of the sinking organic matter occurs just below the euphotic zone (a region sometimes referred to as the twilight zone) and also in the benthic boundary layer, a region which is discussed in the following section. An example of the declining organic matter flux is shown in Fig. 12.4. The

**Table 12.3** Estimates of global average C fluxes through the ocean (Lampitt and Antia, 1997).

| Flux | Gt C yr$^{-1}$ |
|---|---|
| ocean primary productivity | 36 |
| export production from the euphotic zone | 4.5 |
| sediment trap flux at 2000 m | 0.34 |

**Table 12.4** Changes in fluxes of selected major components and C : N : P ratio of sinking material collected in sediment traps with depth in the Sargasso Sea (Conte *et al.*, 2001). Mass g m$^{-2}$ yr$^{-1}$ other fluxes as mM m$^{-2}$ yr$^{-1}$.

| Sediment trap | Mass | CaCO$_3$ | Organic C | Nitrogen | Phosphorus | C : N : P |
|---|---|---|---|---|---|---|
| 500 m | 13 | 65 | 125 | 15 | 0.75 | 166 : 20 : 1 |
| 3000 m | 13 | 77 | 52 | 5 | 0.3 | 182 : 18 : 1 |

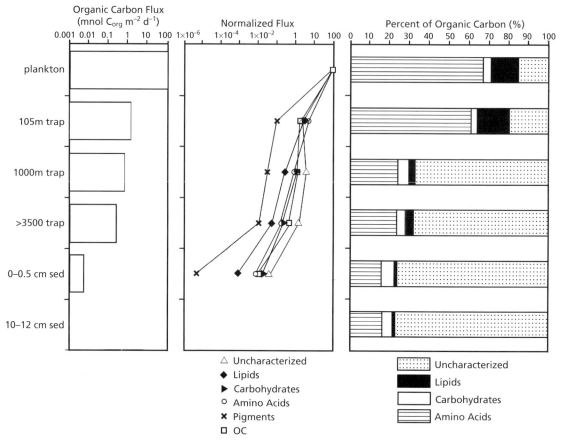

**Fig. 12.4** Changes with depth in the sinking organic C flux and the relative importance of major organic carbon compositional groups for sites in the equatorial Pacific Ocean (based on Lee *et al.*, 2004). Left hand panel is flux of organic carbon for plankton (productivity), sediment traps at 105, 1000 and > 3500 m depth plus in upper (0–0.5 cm) sediment layer, based on dated sediment material. Central panel shows fluxes of individual chemical components of the same samples. Note log scale for x axis. OC is organic carbon. Uncharacterized is chemically uncharacterized material. Right hand column shows the percentage composition.

chemical characterization of this material is discussed further in Section 12.1.1.

In a subsequent study, Huang and Conte (2009) considered a wide range of trace metals and showed that the sinking flux at 1500 m at this site in the Sargasso Sea was systematically depleted in trace metals such as Cd and Zn, which are cycled in association with organic carbon as discussed in Chapter 11. In contrast, the fluxes of Mn, Fe and Co increased, consistent with their scavenged-type behaviour discussed in Chapter 11. The solid state speciation of the metals was also shown to change with depth as they were lost from organic phases and trapped onto

inorganic carrier phases. This kind of redistribution is illustrated in Fig. 12.5 based on earlier work by Chester *et al.* (1988). These results are broadly consistent with independent measurements of particulate trace metal concentrations also carried out in the Sargasso Sea (Sherrell and Boyle, 1992) which show increases of the concentrations of scavenged metals (see Chapter 11) like Mn and Al with depth, while nutrient like metals such as Cd decrease in concentrations (Fig. 12.6). Note also that as discussed in Chapter 11, the particulate metal concentrations are much smaller than the dissolved concentrations. The increase in concentrations close to the seabed reflect

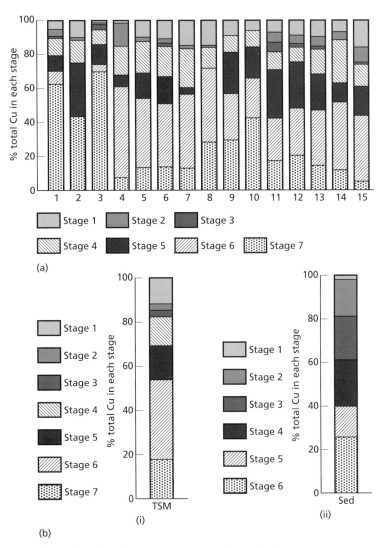

**Fig. 12.5** The solid-state speciation of Cu in surface water particulates (TSM) and deep-sea sediments (from Chester *et al.*, 1988). (a) The average partitioning signatures for ΣCu in Atlantic surface seawater particulates. Samples 1–15 were collected along a north–south Atlantic transect. The Cu host associations for the various stages are as follows: stage 1, loosely held or exchangeable associations; stage 2, carbonate and surface oxide associations; stage 3, easily reducible associations, mainly with 'new' oxides and oxyhydroxides of manganese and amorphous iron oxides; stage 4, humic organic associations; stage 5, moderately reducible associations, mainly with 'aged' manganese oxides and crystalline iron oxides; stage 6, refractory organic associations; stage 7, detrital or residual associations. (b) The average partitioning signatures for ΣCu in Atlantic deep-sea sediments. For the deep-sea sediment samples the humic and refractory organic associations were combined into one organic-associated stage. The Cu host associations are as follows: stage 1, loosely held or exchangeable associations; stage 2, carbonate and surface oxide associations; stage 3, easily reducible associations, mainly with 'new' oxides and oxyhydroxides of manganese and amorphous iron oxides; stage 4, moderately reducible associations, mainly with 'aged' manganese oxides and crystalline iron oxides; stage 5, organic associations; stage 6, detrital or residual associations. Note the differences in the solid-state speciation of Cu between the open-ocean surface water TSM (samples 4–15), in which 50% of the ΣCu is held in some form of organic association, and the deep-sea sediments, in which and10% of the ΣCu is associated with organic hosts. Reprinted with permission from Elsevier.

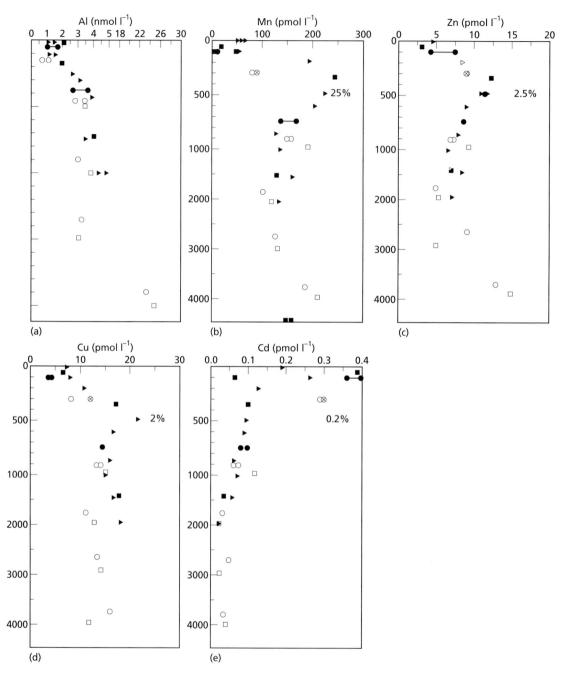

**Fig. 12.6** Particulate trace-metal profiles in the Sargasso Sea (after Sherrell and Boyle, 1992). Depths are in metres. Samples were collected using a two-stage filter holder with a 53 μm prefilter and a 1.0 μm primary filter. Symbols indicate different sampling sites, although all are within the same region and can be merged to create a single profile. Symbols connected by a line represent analyses of replicate subsamples of a single filter. Circle with line indicates replicate sample collected using a 0.4 μm filter. Where indicated, the percentages are the particulate fraction of the total metal (dissolved + particulate) at 500 m, the region of the particulate metal maximum for Mn, Zn and Cu. Reprinted with permission from Elsevier.

the resuspension of deep sea sediments which will be discussed later.

These results illustrate a very consistent pattern of behaviour shown by metals in the dissolved and particulate phases reflecting their very active cycling during rapid sinking through the water column, over timescales of a few weeks.

Earlier in this chapter and in Chapter 11 the interaction between fine slow sinking (FPM) and large fast sinking (CPM) particles was discussed. In Chapter 11 it was shown that the analysis of natural series radionuclides has shown that there must be continuous interactions between the CPM and FPM. Sherrell and Boyle (1992) used the data in Fig. 12.3 to investigate further how this coupling operates. In setting up the model it is assumed that the CPM 'sinking' population is formed in surface waters by biological activity and that the FPM 'suspended' population becomes associated with it by 'piggy-backing'. In this context, the scavenging of trace metals can be regarded as the sum of two processes: (i) incorporation into the surface-water particulates which form the CPM that sinks rapidly to the deep ocean; and (ii) association with deep suspended particles of the FPM.

The 'Sherrell–Boyle' model uses a one-dimensional two-box flux approach, and the particulates are divided into CPM and FPM populations. The CPM is free-sinking but the suspended FPM, which makes up the bulk of the TSM, sinks only by association with the CPM, and is maintained by disaggregation of the larger particles (CPM). Thus, the large particulate flux regulates the removal of the suspended particulates, but it is only the suspended FPM which, because of its greater concentration, larger surface areas and longer residence times, is assumed to exchange metals with the dissolved pool.

The 'Sherrell–Boyle' model is illustrated in Fig. 12.7. The total metal flux leaving the deep box is equal to the sum of (i) the surface-derived flux ($F_S$) and (ii) the flux associated with the removal of suspended particulate material, that is, the 'repackaging flux' ($F_R$); thus

$$F_T = F_S + F_R \tag{12.3}$$

where $F_T$ was taken as the mean flux determined from sediment-trap data, and the 'repackaging' flux was calculated from the suspended metal data, where

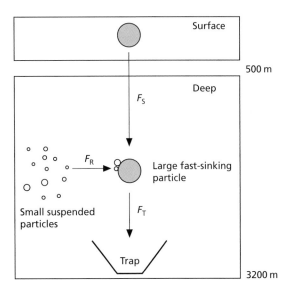

**Fig. 12.7** The 'Sherrell–Boyle' model to describe the coupling between suspended (FPM) and sinking (CPM) particulates in the down-column transport of trace metals in the Sargasso Sea (from Sherrell and Boyle, 1992). $F_s$ is the sinking flux formed in surface waters by the association of FPM with CPM. $F_R$ is the 'repackaging' flux associated with the removal of suspended (FPM) particulate metals. $F_T$ is the total flux, that is, $F_T = F_s + F_R$. The flux arrows indicate only the processes of interest in the model and are not intended to demonstrate all fluxes in the system. For a full description of the 'Sherrell–Boyle' model, see text. Reprinted with permission from Elsevier.

$$F_R = (Me_P D)/\tau_P \tag{12.4}$$

where $Me_P$ is the mean deep-ocean suspended metal concentration per volume of seawater, $D$ is the depth of the deep box and $\tau_P$ is the estimated suspended particle residence time. The value of $\tau_P$ was estimated by Sherrell and Boyle (1992) using measurements of particulate [230]Th in the region. The model requires knowledge of the relative contributions made to the total flux by surface scavenging and deep-water removal at a single location, and Sherrell and Boyle (1992) used data for both contributions at the Sargasso Sea site, utilizing sediment-trap data and their own suspended-particulate data.

Using the model, Sherrell and Boyle (1992) demonstrated that the repacking flux ($F_R$) provided less than about half (29–54%) of the total flux ($F_T$) for Al, Fe, Mn, Ni, Cu, Zn and Cd in the Sargasso Sea region. Further, the fraction yielded by the repackaged flux was remarkably constant: see Table 12.5,

**Table 12.5** The 'Sherrell–Boyle' down-column particulate flux model applied to the Sargasso Sea water column; units, $ng\,cm^{-2}\,yr^{-1}$ (data from Sherrell and Boyle, 1992). For explanation of the model, see text.

| Trace metal | 'Repackaging' flux (FR) | Total flux (FT) | FR/FT |
|---|---|---|---|
| Al | 4045 | 17530 | 0.23 |
| Fe | 2792 | 9494 | 0.30 |
| Mn | 335 | 1044 | 0.32 |
| Ni | 14.7 | 45 | 0.32 |
| Cu | 39.5 | 95 | 0.41 |
| Zn | 20 | 189.5 | 0.11 |
| Cd | 0.225 | 0.337 | 0.67 |
| Pb | 20 | 66 | 0.30 |

column 4. Thus, the repackaging and vertical removal of deep suspended particles plays a relatively minor role in governing the total flux of trace metals in the deep ocean, and processes taking place in the surface waters (<500 m) contribute about two-thirds of the deep flux.

The overall scenario to emerge from applying the 'Sherrell–Boyle' model to the Sargasso Sea depicts large particles (CPM) that, having interacted with small particles (FPM) in the upper water column, sink through the deep ocean exchanging, on average, only a minor portion of their mass and trace-metal content with the suspended particulates (FPM).

If these Sargasso Sea results can be applied to other central ocean gyres, it becomes apparent that the removal of suspended particles in the deep ocean interior is not the dominant sink for trace metals from the ocean. Rather, as suggested by a growing body of evidence, the control is exerted at the *ocean boundaries*: that is, (i) the surface ocean, and (ii) the continental margins. Thus, the whole oceanic residence times of trace metals may be strongly influenced by processes occurring at the ocean boundaries rather than in dissolved–particulate fractionation in deep waters. Hence although for many purposes it is useful to think of the ocean particulate flux transport as a relatively simple vertical process, the reality is a three-dimensional ocean where water transport, processes of resuspension and even inputs from major rivers that can transport material across the continental shelf (as in the case of the Amazon and Orinoco; Jickells *et al.*, 1990) can also play a role.

## 12.2.1 Organic components of the particle mass flux to the interior of the ocean

Organic carbon decreases in concentration as the particle mass flux sinks from surface waters with only about 1% of primary production reaching 4000 m depth (Eglinton and Repeta, 2003). Bulk analysis (Table 12.4) shows that this bacterial degradation of the sinking organic matter selectively remineralizes nitrogen and phosphorus compared to carbon (Conte *et al.*, 2001). Other authors have considered how the chemistry of specific groups of organic compounds change during the degradation of sinking matter which begins life with a composition similar to that of surface water particulate matter (see Chapter 9). However, the complexity of such a task should not be underestimated; Conte *et al.* (2003) report that over 100 extractable lipid compounds are present in oceanic particulate matter and these lipids represent only a part of the total organic matter. The complexity of identifying individual compounds means that it is often more useful to consider groups of compounds (e.g. Fig. 12.4). The bulk of the organic matter is composed of organic compounds with amino acid (mainly as proteins), lipid or carbohydrate (mainly as polysaccharides) character, plus a fraction that cannot currently be chemically characterized. Analysis of sinking material shows that some particularly reactive groups such as lipids and particularly chlorophyll pigments are lost relatively faster than the other groups. Detailed analysis of individual compounds shows inevitably that some compounds are selectively broken down faster than others. However, the relative proportions of the three main groups of compounds (amino acids, lipids, carbohydrates) remains rather similar, but the proportion of the organic carbon that is uncharacterized increases markedly from <20% in surface waters to >70% in deep water traps (Lee *et al.*, 2004). Lee *et al.* (2004) stress the point that our inability to chemically characterize this material does not necessarily imply it is chemically complex and non biodegradable, indeed the evidence is that it is degradable by bacteria. The process by which the characterized organic matter is transformed into the uncharacterized fraction is not understood but probably involve bacterial degradation and chemical reactions between compounds to produce larger polymeric material (Lee *et al.*, 2004).

These processes continue into the sediments, see Section 14.2.

The general similarity of the characterized organic matter composition with depth has led to the suggestion that a fraction of the organic matter formed in the surface undergoes little degradation during sinking through the water column, possibly because it is intimately bound up with inorganic mineral material which makes bacterial and enzymic breakdown difficult (Lee *et al.*, 2004). The importance of mineral material as 'ballast' to enhance sinking rates was noted earlier and here we see another role as also potentially protecting the sinking organic matter from degradation.

## 12.3 The benthic boundary layer: the sediment–water interface

The benthic boundary layer (BBL) is a zone in which the deep oceanic circulation interacts with the ocean bottom, evidenced for example in the formation of nepheloid layers (see Chapter 10). In addition to these physical forces the BBL, which includes the sediment–water interface, is the site of relatively large gradients in chemical and biological properties, and the processes associated with these gradients are involved in the active cycling of elements between the water and sediment reservoirs. Overall, therefore, the BBL is one of the most dynamic environments in the entire ocean system.

The down-column transport of TSM transfers nutrients and organic carbon from the surface ocean to deep waters. The transfer of minerals and contaminants occurs in association with this organic flux. The remineralization of the carbon provides energy to the bottom benthos. Burial in the sediments removes bioactive material from the system and therefore regulates oceanic fertility (Brewer *et al.*, 1986). In addition, the composition of the TSM and the magnitude of its flux play an important role in controlling the chemical environment in the bottom sediments (Dymond, 1984). The reason for this is that organic carbon is the determining factor controlling the type of redox conditions that are set up in the sediments (see Chapter 14). Because these redox conditions largely control the diagenetic reactions at both the sediment–seawater and the sediment–interstitial-water interfaces, the POC flux

to the sediments may be regarded as the driving force behind most of the diagenetic reactions occurring in marine sediments (Emmerson and Dymond, 1984).

In Section 10.5 it was pointed out that in order to assess the net downward flux to the BBL it is necessary to distinguish between the primary and the resuspended fluxes that operate in this deep-ocean reservoir. According to Dymond and Lyle (1994) it also is necessary to distinguish between *distal* and *local* resuspension fluxes. Distal resuspension, for example, by nepheloid transport or turbidity currents, can be part of the primary, or net, flux of material to a sediment site. However, local resuspension represents an artificial flux as it merely recycles material already deposited.

The mass particle flux that descends to the sea bed supplies material to marine sediments. This flux may be termed the rain rate, and the extent to which the flux material is incorporated into bottom sediments may be referred to as the burial rate. The difference between the rain rate (material supply) and the burial rate (material preservation) is then a measure of the recycling of the flux material that reaches the sea floor: that is, the benthic regeneration flux. This flux indicates the extent of the exchange of material between the sediment and the overlying water column and is a key component in the process by which the particulate rain is transformed into the sedimentary record. In its simplest form Dymond and Lyle (1994) expressed this 'benthic regeneration equation' for a particular component of the flux as:

$$
\begin{aligned}
\text{the rain rate} = &\ \text{the burial rate} \\
&+ \text{benthic regeneration flux}
\end{aligned} \tag{12.5}
$$

Dymond (1984) proposed an elegant conceptual model to assess how much of the net mass particle flux is buried in the underlying sediments. The model is illustrated in Fig. 12.8 and offers an excellent framework within which to address the problem of the fate of down-column transported elements once they reach the BBL. In designing the model, Dymond (1984) made a distinction between (i) *refractory* elements (e.g. Al, Fe, Ti), and (ii) *labile* elements (e.g. organic C, P, N). The primary particulate flux of the refractory elements can increase with depth as a result of either particle scavenging of dissolved elements, or additions from horizontal advection (e.g. from the continental margins and distal resuspension). Close

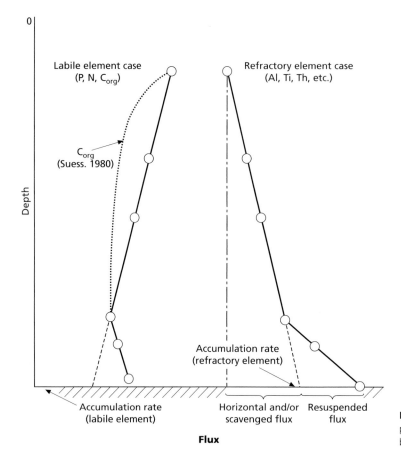

**Fig. 12.8** Conceptual model of particle-associated fluxes to the benthic boundary layer (from Dymond, 1984).

to the sea bed the refractory elements can also exhibit a sharp increase in flux from the resuspended flux, because the sediments are enriched in refractory material relative to the sinking flux. In contrast, the particulate flux of the labile elements will decrease with depth, owing to processes such as particle decomposition, dissolution and disaggregation.

Local resuspension of sedimentary material does not appear to be significant at elevations greater than a few hundred metres above the sea bed (the resuspension zone), and in the Dymond (1984) model net, or primary, particulate flux measurements are extrapolated downwards through the resuspension zone from measurements taken above the zone itself: see Fig. 12.8. Dymond and Lyle (1994) used this model to compare the primary particle *rain rate*, determined from sediment-trap data, to the sediment

burial rate at 11 sites in the Atlantic and Pacific oceans that covered a number of different biogeochemical oceanic environments. The burial flux inevitably averages over timescales of hundreds to thousands of years so it is difficult to directly compare this to a single year of sediment trap data. Despite this caveat Dymond and Lyle (1994) found that the rain and burial rates for terrigenous components such as Al, Fe and Ti primarily associated with clay minerals agreed well. The flux of biogenic material (consisting mainly of organic carbon, calcium carbonate and opal) dominates the particle mass flux. Organic carbon preservation is relatively low at the pelagic sites, with less than ~5% of the down-column rain being buried in the sediments. In contrast, at the sites closer to the continental margins there is a higher preservation of organic carbon, and the burial

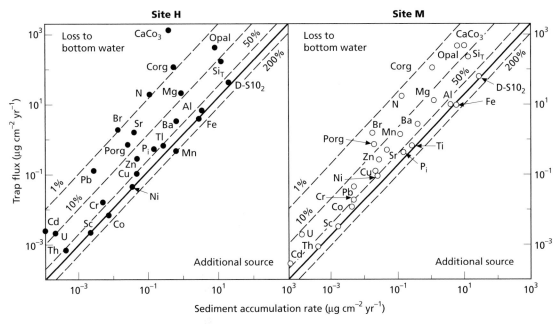

**Fig. 12.9** Comparison between sediment accumulation rates and measured particle fluxes at two sites in the eastern equatorial Pacific (from Dymond, 1984). The heavy line indicates equal particulate and burial fluxes; the broken lines indicate the percentage of the particle flux preserved in the sediment. $P_i$ is inorganic P, Si is total Si, D-SiO$_2$ is detrital silica.

rate ranges between ~10% and ~30% of the rain rate. This is consistent with the overall preservation of organic matter in marine sediments and indicates the importance of the continental margins as major sites for the removal of organic carbon from the ocean; this topic is discussed in Chapter 14. Calcium carbonate preservation in marine sediments is a function of water depth (see Section 15.2) but in general less than ~30%, and in many cases ~5%, of the carbonate rain is preserved. Overall, the percentage of opal preserved in the sediments ranges from ~1 to ~20% of the rain rate. The preservation of opal in marine sediments is more complex than that of calcium carbonate (see Section 15.2), but the rain rate of opal provides a constraint on its dissolution. The data provided by Dymond and Lyle (1994) indicate that there is a relationship between the magnitude of the opal rain rate and its sediment burial rate, with, in general, the highest rain rate resulting in a higher burial rate in the sediments. The processes of organic matter degradation, as well as the compositional trends to an increasing proportion of material whose organic chemical composition cannot be readily characterized, continues through this region

of the water column and on into the sediments (Lee *et al.*, 2004).

Dymond (1984) also provided data on the particle-flux–sedimentation-rate relationships for a wide range of trace metals at the H and M sites. The author concluded that a number of particle reactive elements are accumulated in the sediments at about the same rates at which they are delivered down the water column: see Fig. 12.9. Lead was an exception to this, with only less than ~20% of its flux to the BBL being preserved in the depositing sediment; however, this may have been because much of the Pb in the water column is of a recent anthropogenic origin, and so has not yet been integrated into the sediments, which accumulate only at slow rates.

The resuspension of the bottom sediment has an important effect on the down-column trace-metal flux profiles. This resuspended trace-metal flux can be generated in a number of ways. For example, it can result from the addition of metal-rich sediment particles directly to the water column from the underlying sediments. However, the resuspended particles can themselves scavenge additional dissolved metals from the bottom-water column, thus leading to a region of 'enhanced scavenging' in the resuspension zone. This could arise, for example, as metals released by the regeneration of organic particles are made available for scavenging by the resuspended material. An alternative mechanism to explain the increases in deep-water particulate fluxes was postulated by Dymond (1984), who introduced the concept of a rebound flux. This rebound flux was thought to result from the resuspension not of well-deposited sediment components, but rather of recently laid down primary flux material (see e.g. Billet *et al.*, 1983). Dymond (1984) suggested that the rebound material may be similar to the organic-rich 'fluff' found over deep-sea sediments and believed to be derived from recently deposited material. The concept was later modified, however, by Walsh *et al.* (1988) to exclude 'fluff' and in order to distinguish it from the resuspension of surface sediment, the rebound flux was defined as particles that have reached the sediment surface by settling out of the water column but which have not yet become incorporated into the sediment itself. Using the modified terminology, therefore, the material in the BBL can be divided into: (i) a local primary down-column flux; (ii) a local rebound flux; and (iii) a local, or distal, resuspended sediment flux. The relative contribution of these various components can be estimated by identifying suitable tracer for each. Using this approach Lampitt *et al.* (2000) showed that both the recently deposited 'fluff' and local resuspended sediments contributed to the enhanced flux measured in deep waters at a site in the north east Atlantic, and that the amount of resuspension can vary significantly with time.

## 12.4 Down-column fluxes and the benthic boundary layer: summary

1 There is a 'rain' of particulate material to the sea floor over the World Ocean, with sinking speeds in the range $\sim$50–200m day$^{-1}$. The magnitude of the particle mass flux varies in both space and time but it appears to be generally in the range $\sim$20–250m gm$^{-2}$ day$^{-1}$, with the highest values being found in areas of coastal upwelling and the smallest in the central ocean gyres and in some polar regions.

2 The particle mass flux to the interior of the ocean is composed primarily of organic carbon, calcium carbonate, opal and lithogenic material, with calcium carbonate dominating in most open-ocean regions. As it sinks down the water column the relative proportions of the major components change, and the percentage of organic matter decreases from as much as $\sim$90% in surface waters to usually less than $\sim$10% at depth. The composition of the organic matter changes as it sinks and the proportion we cannot readily characterize increases markedly with depth.

3 The down-column fluxes, which are driven by export production from the euphotic zone, are related to primary production in the surface waters and often show seasonal variations. Thus, the down-column mass particle flux can be related to major oceanographic parameters in surface waters and 'oceanographic consistency' can be injected into the system. However, the relationship between primary production and deep fluxes is not simple or linear.

4 The down-column fluxes of trace metals generally vary in relation to the total particle mass fluxes, being highest in coastal and marginal areas, and lowest in the open-ocean. The total particulate down-column transport in most regions involves a coupling between biogenic and lithogenic components, with the result that most lithogenic elements are transported to depth in association with organic components. The lithogenic material, however, can be decoupled from the biogenic material, and sink down the water column independently, when there is a relatively large input of terrigenous components: for example. during the sporadic delivery of aeolian material in the form of dust 'pulses', which perturb the steady-state conditions in surface waters.

5 As it approaches the sea bed the sinking particle mass flux enters the benthic boundary layer (BBL). This layer is an oceanic region of considerable reactivity through which down-column and laterally transported TSM must pass, and perhaps be recycled, before entering the sediment reservoir.

6 The difference between the down-column rain rate and the sediment burial rate is a measure of the extent to which elements undergo benthic recycling. In open-ocean regions in which the bottom deposits are not affected by distal transport (e.g. turbidity currents) the rain rates and burial rates of refractory elements, such as Al, Fe, Ti, are generally in good agreement. In contrast, for labile elements, such as C and P, rain rates exceed burial rates, indicating their recycling in the BBL. Some trace metals can also be recycled across the BBL.

7 It has become clear, therefore, that although particulate matter is the driving force behind the reactions that control the elemental composition of seawater, the movement of both particulate and dissolved elements through the ocean system is not a simple one-stage journey, but rather involves a series of complex interactive recycling stages.

This complicated journey has now been followed down to the sediment–seawater interface. It must be stressed, however, that even when the material has crossed this interface and has been incorporated into the bottom deposits it has not simply entered a static reservoir for the permanent storage of components extracted from the water column. On the contrary, diagenetic reactions occur within the sediment–interstitial-water complex, and these can severely modify the mineral and elemental compositions of the sediments themselves. Before looking at the reactions that occur in the sediment reservoir, however, it is necessary to know something of the fundamental characteristics of the oceanic sediments themselves and these are described in the next chapter.

# References

Billet, D.S.M., Lampitt, R.S., Rice, A.L. and Mantoura, R.F.C. (1983) Seasonal sedimentation of phytoplankton to the deep-sea benthos. *Nature*, **302**, 520–522.

Bory, A.J.-M. and Newton, P.P. (2000) Transport of airborne lithogenic material down through the water column in two contrasting regions of the eastern subtropical North Atlantic Ocean. *Global Biogeochem. Cycl.*, **14**, 297–315.

Brewer, P.G., Bruland, K.W., Eppley, R.W. and McCarthy, J.J. (1986) The Global Ocean Flux Study (GOFS): status of the U.S. GOFS Program. *Eos*, **67**, 827–832.

Buesseler, K.O. *et al.* (2007a) An assessment of the use of sediment traps for estimating upper ocean particle fluxes. *J. Mar Res.*, **65**, 345–416.

Buesseler, K.O. *et al.* (2007b) Revisiting carbon flux through the twilight zone. *Science*, **316**, 567–570.

Chester, R., Thomas, A., Lin, F.J., Basaham, A.S. and Jacinto, G. (1988) The solid state speciation of copper in surface water particulates and oceanic sediments. *Mar. Chem.*, **24**, 261–292.

Conte, M.H., Ralph, N. and Ross, E.H. (2001) Seasonal and interannual variability in deep ocean particle fluxes at the Oceanic Flux Program (OFP/Bermuda Atlantic Time Series (BATS) site in the western Sargasso Sea near Bermuda. *Deep-Sea Res. II*, **48**, 1471–1505.

Conte, M.H., Dickey, T.D., Weber, J.C., Johnson, R.J. and Knap, A.H. (2003) Transient physical forcing of pulsed export of bioreactive material to the deep Sargasso sea. *Deep-Sea Res. I*, **50**, 1157–1187.

Deuser, W.G. (1986) Seasonal and interannual variations in deep-water particulate fluxes in the Sargasso Sea and their relation to surface hydrography. *Deep Sea Res. A*, **33**, 225–291.

Deuser, W.G. (1987) Variability of hydrography and particle flux: transient and long-term relationships, in *Particle Flux in the Ocean*, E. Izdar and S. Honjo (eds), pp.179–193. Hamburg: Mitt. Geol.-Palont. Inst. Univ. Hamburg, SCOPE/UNEP, Sonderband 62.

Deuser, W.G., Brewer, P.G., Jickells, T.D. and Commeau, R.F. (1983) Biological control of the removal of biogenic particles from the surface ocean. *Science*, **219**, 388–391.

Dymond, J. (1984) Sediment traps, particle fluxes, and benthic boundary layer processes, in *Global Ocean Flux Study*, pp.261–284. Washington, DC: National Academy Press.

Dymond, J. and Lyle, M. (1994) Particle fluxes in the ocean and implications for sources and preservation of ocean sediments, in *Material Fluxes on the Surface of the Earth*, pp.125–142. Washington, DC: National Academy Press.

Eglinton, T.I. and Repeta, D.J. (2003) Organic matter in the contemporary ocean, in H. Elderfield (ed.), *Treatise on Geochemistry*, Vol. 6, H.D. Holland and K.K. Turekian (eds), pp.145–180. London: Elsevier.

Emmerson, S. and Dymond, J. (1984) Benthic organic carbon cycles: toward a balance of fluxes from particle settling and pore water gradients, in *Global Ocean Flux Study*, pp.205–384. Washington, DC: National Academy Press.

Fowler, S.W. *et al.* (1987) Rapid removal of Chernobyl fallout from Mediterranean surface waters by biological activity. *Nature*, **329**, 56–58.

Francois, R., Honjo, S., Krishfield, R. and Manganini, S. (2002) Factors controlling the flux of organic carbon to the bathypelagic zone of the ocean. *Global Biogeochem. Cycl.*, **16**, doi:10.1029/2001GB001722.

Gardner, W.D., Souchard, J.B. and Hollister, C.D. (1985) Sedimentation, resuspension and chemistry of particles in the northwest Atlantic. *Mar. Geol.*, **65**, 199–242.

Huang, S. and Conte, M.H. (2009) Source/Process Apportionment of major and trace elements in sinking particles in the Sargasso Sea. *Geochim. Cosmochim. Acta*, **73**, 65–90.

Jickells, T.D., Deuser, W.D., Fleer, A. and Hembleben, C. (1990) Variability of some elemental fluxes in the western tropical Atlantic Ocean. *Oceanologica Acta*, **13**, 291–298.

Jickells, T.D., Newton, P.P., King, P., Lampitt, R.S. and Boutle, C. (1996) A comparison of sediment trap records of particle fluxes from 19°N to 48°N in the northeast Atlantic and their relation to surface water productivity. *Deep Sea Res. I*, **43**, 971–986.

Jickells, T.D., Dorling, S., Deuser, W.G., Church, T.N., Arimoto, R. and Prospero, J. (1998) Air-borne dust fluxes to a deep water sediment trap in the Sargasso Sea. *Global Biogeochem. Cycl.*, **12**, 311–320.

Lampitt, R.S. and Antia, A.N. (1997) Particle flux in deep sea: regional characteristics and temporal variability. *Deep-Sea Res. I*, **44**, 1377–1403.

Lampitt, R.S., Newton, P.P., Jickells, T.D., Thomson, J. and King, P. (2000) Near bottom particle flux in the abyssal northeast Atlantic. *Deep-Sea Res. II*, **47**, 2051–2071.

Lee, C., Wakeham, S. and Arnosti, C. (2004) Particulate organic matter: the compositional conundrum. *Ambio*, **33**, 565–575.

Lutz, M.J., Caldeira, K., Dunbar, R.B. and Behrenfeld, M.J. (2007) Seasonal rhythms of net primary production and particulate organic carbon flux describe biological pump efficiency in the global ocean. *J. Geophys. Res.*, **112**, doi: 10.1029/2006JC003706.

Martin, J.H., Knauer, G.A., Karl, D.M. and Broenkow, W.W. (1987) VERTEX: carbon cycling in the northeast Pacific. *Deep Sea Res. A*, **34**, 267–286.

Sherrell, R.M. and Boyle, E.A. (1992) The trace metal composition of suspended particles in the oceanic water column near Bermuda. *Earth Planet. Sci. Lett.*, **111**, 155–174.

Walsh, I., Fischer, K., Murray, D. and Dymond, J. (1988) Evidence for resuspension of rebound particles from near-bottom sediment traps. *Deep Sea Res. A*, **35**, 59–70.

# Part III
# The Global Journey: Material Sinks

# 13 Marine sediments

Marine sediments represent the major sink for material that leaves the seawater reservoir. Before attempting to understand how this sink operates, however, it is necessary briefly to describe the sediments within a global ocean context.

## 13.1 Introduction

The sea floor can be divided into three major topological regions: the continental margins, the ocean-basin floor and the mid-ocean ridge system.

*The continental margins.* These include the continental shelf, the continental slope and the continental rise. The continental shelf is the seaward extension of the land masses, and its outer limit is defined by the *shelf edge*, or break, beyond which there is usually a sharp change in gradient as the continental slope is encountered. The continental rise lies at the base of this slope. In many parts of the world both the continental slope and rise are cut by submarine canyons. These are steep-sided, V-shaped valleys, which are extremely important features from the point of view of the transport of material from the continents to the oceans because they act as conduits for the passage of terrigenous sediment from the shelves to deep-sea regions by processes such as turbidity flows. Trenches are found at the edges of all the major oceans, but are concentrated mainly in the Pacific, where they form an interrupted ring around the edges of some of the ocean basins. The trenches are long (up to ~4500 km in length), narrow (usually <100 km wide) features that form the deepest parts of the oceans and are often associated with island arcs; both features are related to the tectonic generation of the oceans. The trenches are important in the oceanic sedimentary regime because they can act as traps for material carried down the continental shelf. In the absence of trenches, however, much of the bottom-transported material is carried away from the continental rise into the deep sea.

*The ocean-basin floor.* The ocean-basin floor lies beyond the continental margins. In the Atlantic, Indian and northeast Pacific Oceans abyssal *plains* cover a major part of the deep-sea floor. These plains are relatively flat and underlain by thick (>1000 m) layers of sediment, and have been formed by the slow accumulation of biogenic and abiogenic material settling through the water column and surviving diagenesis and the transport of material from the continental margins by turbidity currents and other off shelf transport processes. Turbidity currents (see later) spread their loads out on the deep-sea floor to form thick turbidite sequences. Thus, large amounts of sediment transported from the continental margins are laid down in these abyssal plains, which fringe the ocean margins in the so-called 'hemi-pelagic' deep-sea areas along with the material (predominantly calcium carbonate, opal and organic carbon) which sinks through the water column. The plains are found in all the major oceans, but because the Pacific is partially ringed by ocean trenches that act as a sediment trap, they are more common in the Atlantic and Indian Oceans. Because the Pacific has fewer abyssal plains than the other major oceans, abyssal hills are more common here, covering up to ~80% of the deep-sea floor in some areas. Seamounts are volcanic hills rising above the sea floor, which may be present either as individual features or in chains. Seamounts are especially abundant in the Pacific Ocean.

*Marine Geochemistry*, Third Edition. Roy Chester and Tim Jickells.
© 2012 by Roy Chester and Tim Jickells. Published 2012 by Blackwell Publishing Ltd.

*The mid-ocean ridge system.* This ridge system is one of the major topographic features on the surface of the planet. It is an essentially continuous feature, which extends through the Atlantic, Antarctic, Indian and Pacific Oceans for more than 60 000 km, and the 'mountains' forming it rise to over 3000 m at the ridge crest above the sea bed in some areas. The topography of the ridge system is complicated by a series of large semi-parallel fracture zones, which cut across it in many areas. It is usual to divide the ridge system into crestal and flank regions. The flanks lead away from ocean basins, with a general increase in height as the crestal areas are approached. In the Mid-Atlantic Ridge the crestal regions have an extremely rugged topography with a central rift valley (~1–2 km deep) that is surrounded by rift mountains. The geochemical cycling processes within ridge systems are discussed in Chapter 5.

The distribution of these various topographic features is illustrated for the North Atlantic in Fig. 13.1 and for the World Ocean in Fig. 13.2(a). The way in which these sea-bed features were formed can be related to the tectonic history of the oceans in terms of the theory of sea-floor spreading. In essence, this theory can be summarized as follows. The mid-ocean ridges are associated with the rising limbs of convection cells in the mantle and the sea floor 'cracks apart' at the crest regions. New crust is formed here, and as it is generated the previous crust is moved away on either side of the crests and is lost at the edges of the oceans under zones of trenches, island arcs and young mountains associated with the descending limbs of convection cells where the ocean floor is carried back down into the mantle. Thus, the ocean floor is continually being created and destroyed in response to convection in the mantle. According to the evidence available it would appear that the spreading takes place at rates of a few centimetres per year. However, some ridges spread faster than others, allowing fast and slow spreading centres to be identified, although the spreading rates do not appear to have been constant with time. The slowest rates are found along parts of the Mid-Atlantic and Mid-Indian Ocean Ridges (~1–2.5 cm yr$^{-1}$), and the fastest around the East Pacific Rise (~5 cm yr$^{-1}$).

One of the most important findings to emerge from these tectonic studies is that the ocean basins are relatively young on the geological time-scale, and magnetic evidence, combined with sediment age data

from the international deep ocean drilling projects (now IODP: www.iodp.org), indicates that the oldest oceanic crust existing in the oceans today is around Middle Jurassic in age; that is, it was formed ~170 Ma ago. The concept of a dynamic ocean floor that is being generated at the ridge crests and resorbed into the mantle at the ocean edges is intimately related to the overall pattern of global tectonics. The unifying theory which brings together the various aspects of modern thinking on global structure is the plate tectonic concept. Put simply, it is now thought that the surface of the Earth consists of a series of thin (~100–150 km) plates that are continuously in motion. Six (or sometimes seven) major plates were originally identified, but additional plates have since been included (see Fig. 13.2b). The plates have boundaries between them, and Jones (1978) has classified these into three principal kinds.

1 Conservative boundaries, at which the plates slide past each other without any creation or destruction of oceanic crust.

2 Constructive boundaries, which are marked by the crests of the active spreading oceanic ridges where new crust is formed as the plates diverge. Thus, constructive boundaries are crust *sources*.

3 Destructive boundaries, at which the plates converge and one plate moves beneath another one. These are marked by the presence of trenches, island arcs or young mountain belts. Thus, destructive margins are crust *sinks*, and the crust is lost by subduction under the trenches, where it eventually becomes sorbed back into the mantle. This is an extremely important process, because when the crust is sorbed back into the mantle it takes with it the sediment that has accumulated on the ocean floor. This has important implications for marine geochemistry because it means that the sediments deposited at the bottom of the sea, which are a principal sink for much of the material poured into seawater, are eventually returned back to the land surface and participate further in the global geochemical cycle. Given that the oldest ocean sediments are about 170 Ma, this cycle must have been repeated numerous times during the almost 4 billion year long history of the Earth.

A large part of the floor of the World Ocean is covered by sediments. The thickness of this sediment blanket varies from place to place; for example, in the Atlantic it averages >1 km, in the Pacific it is

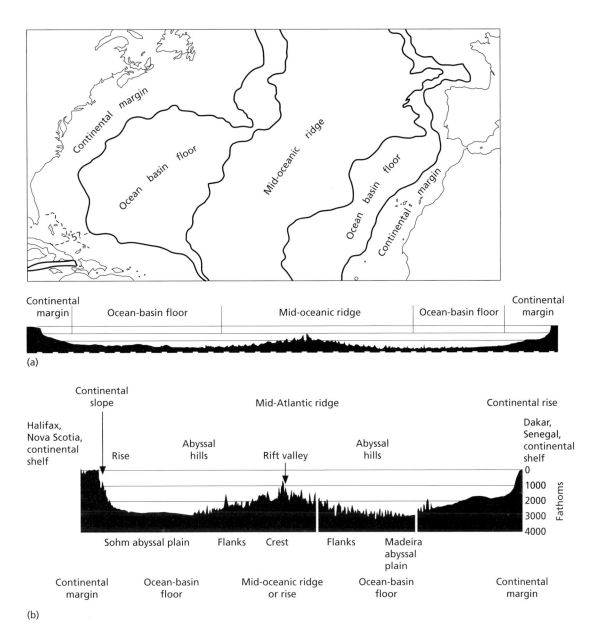

**Fig. 13.1** The topography of the North Atlantic ocean floor (a) from Heezen *et al.* (1959) (b) from Kennet (1982) after Holcombe (1977).

<1 km, and over the whole ocean it averages ~500 m. The sediments forming this blanket are the ultimate marine sink for all the particulate components that survive destruction within the ocean reservoir. The sediments are also the major sink for dissolved elements, although for some of these elements basement rocks can also act as a sink (see Chapter 5). Marine sediments therefore form one of the principal oceanic reservoirs and are ultimately subducted at oceanic trenches.

In terms of a very simplistic model, the initial sediment in a newly formed ocean basin will be

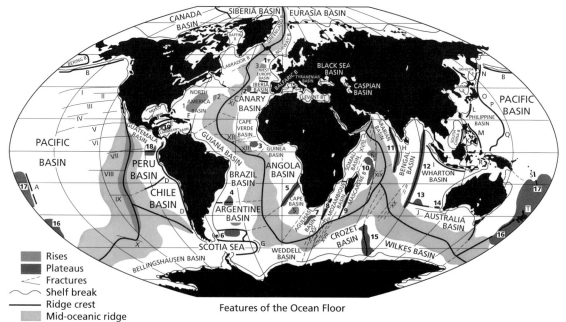

**Features of the Ocean Floor**

Rises
Plateaus
Fractures
Shelf break
Ridge crest
Mid-oceanic ridge

**Fracture zones:**

| | |
|---|---|
| I. Mendocino | XI. Atlantis |
| II. Pioneer | XII. Vema |
| III. Murray | XIII. Romanche |
| IV. Molokai | XIV. Chain |
| V. Clarion | XV. Mozambique |
| VI. Clipperton | XVI. Prince Edward |
| VII. Galapagos | XVII. Malagasy |
| VIII. Marquesas | XVIII. Owen |
| IX. Easter | XIX. Rodriguez |
| X. Eltanin | XX. Amsterdam |

**Plateaus:**

| | |
|---|---|
| 1. Rockall Plateau | 11. Chagos-Laccadive Plateau |
| 2. Azores Plateau | 12. Ninetyeast Ridge |
| 3. Sierra Leone Plateau | 13. Broken Ridge |
| 4. Rio Grande Plateau | 14. Naturaliste Plateau |
| 5. Walfisch Ridge | 15. Kerguelen Plateau |
| 6. Falkland Ridge | 16. New Zealand Plateau |
| 7. Agulhas Plateau | 17. Melanesia Plateau |
| 8. Mozambique Ridge | 18. Galapagos Plateau |
| 9. Madagascar Ridge | 19. Jamaica Plateau |
| 10. Mascarene Plateau | |

**Trenches:**

| | |
|---|---|
| A. Kermedec-Tonga | K. Manila |
| B. Aleutian | L. Nansei-Shoto |
| C. Middle America | M. Philippine |
| D. Peru-Chile | N. Kurile-Kamchatka |
| E. Cayman | O. Japan |
| F. Puerto Rico | P. Bonin |
| G. South Sandwich | Q. Mariana |
| H. Chagos | R. Banda |
| I. Java | S. New Hebrides |
| J. Diamantina | T. Hikurangi |

**Rises**

1. Bermuda Rise
2. Corner Rise
3. Rockall Rise
4. Argentine Rise
5. Schmidt-Ott Rise
6. Madingley Rise

(a)

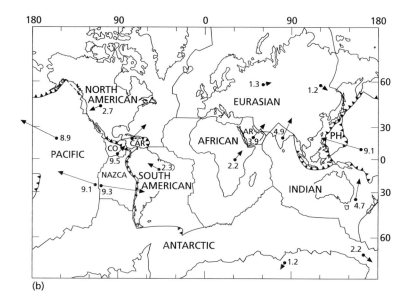

(b)

**Fig. 13.2** Topographic features in the oceans. (a) The distribution of the major oceans and their principal topographic features (from Heezen and Hollister, 1971). (b) The distribution of the principal lithospheric plates (from Parsons and Richter, 1981). The principal plates are identified; the abbreviations CO = Cocos, CAR = Caribbean, AR = Arabian, PH = Philippine. Absolute plate velocities (cm yr$^{-1}$) are indicated at selected points. Convergent boundaries are indicated by arrow-heads that point from the subducting plate towards the non-subducting plate.

←

deposited on to the basalt basement as fresh crust is generated at the ridge spreading centres. This initial sediment is a hydrothermal deposit composed of metal-rich precipitates (see Sections 15.3.6 and 16.5.2), and as the ridge lies above the carbonate compensation depth (CCD; see Section 15.2.4) carbonate sediments begin to accumulate. As the basin continues to open up, the sediment surface away from the ridge falls below the CCD and clays are deposited (see Fig. 16.7 later). As sedimentation continues, the composition of the deposits is controlled by a variety of interacting chemical, physical and biological factors. Because of changes in the relative importance of these various factors as the plate on which the sediment is deposited moves through different oceanic environments (e.g. high- and low-productivity zones), the nature of the deposits forming the marine 'sediment blanket' has varied through geological times as the ocean basins evolved to their present state in response to sea-floor spreading. As a result of this sea-floor spreading, or continental drift, changes have occurred in the sizes, shapes, latitudinal distributions, water circulation patterns and depths of the ocean basins, all of which affected the patterns of sedimentation.

Much of our current knowledge of the deeper parts of the oceanic sediment column has come from material collected during the ocean drilling programmes. Detailed examination of the sediment cores from these programmes using sedimentological, chemical, and palaeontological evidence has shown how much the Earth has changed over time and also how the ocean system we describe in this book has evolved. We now know for instance that 50 Myr ago the Arctic was almost 20°C warmer than today, was ice free and had an almost freshwater surface layer (Moran *et al.*, 2006). This ocean core material has also documented the regular glacial interglacial cycles that have characterized the Quaternary Epoch of the last approximately 2 Ma (see e.g. Shackleton *et al.*, 1983), and the dramatic short (hundreds of years) cooling and warming events

known as Heinrich or Dansgaard-Oeschger events (Hemming, 2004).

The importance of changes in the type of deposit with depth in the oceanic sediment column will be identified where necessary in the text. In the present volume, however, we are concerned chiefly with the role played by the sediments as marine sinks for components that have flowed through the seawater reservoir. Diagenetic changes, which have their most immediate effect on the composition of seawater, take place in the upper few metres of the sediment column, that is, during early diagenesis (see Section 14.1), and for this reason attention here will be focused mainly on the present-day sediment distribution patterns, which reflect contemporary trends in climatic conditions, oceanic current patterns and sea-floor topography.

Marine sediments are deposited under a wide variety of depositional environments. However, it is useful at this stage to make a fundamental distinction between nearshore and deep-sea deposits, a distinction that recognizes the importance of the shelf break in dividing two very different oceanic depositional regimes.

Nearshore sediments are deposited mainly on the shelf regions under a wide variety of regimes that are strongly influenced by the adjacent land masses. As a result, physical, chemical and biological conditions in nearshore areas are much more variable than in deep-sea regions. Nearshore depositional environments include estuaries, fjords, bays, lagoons, deltas, tidal flats, the continental terrace and marginal basins.

Deep-sea sediments are usually deposited in depths of water >500 m, and factors such as remoteness from the land-mass sources, reactivity between particulate and dissolved components within the oceanic water column, and the presence of a distinctive biomass lead to the setting up of a deep-sea environment that is unique on the planet. Because of this, deep-sea sediments, which cover more than 50% of the surface of the Earth, have very different characteristics from those found in continental or nearshore environments. Two of the most distinctive characteristics of these deep-sea sediments are (i) the particle size, and (ii) the rate of accumulation, of their land-derived non-biogenic components.

1 The land-derived fractions of deep-sea sediments are dominated by clay-sized, that is, <2 μm diameter, components, which usually account for ~60–70% of the non-biogenic material in them. In contrast, material having a wide variety of particle sizes is found in nearshore sediments, and in general clay-sized components constitute a much smaller fraction of the land-derived solids: see Table 13.1.

2 Various techniques are available for the measurement of the accumulation rates of marine sediments. These include dating by magnetic reversals, fossil assemblages and the decay of radionuclides. The radionuclide decay methods are the most commonly used, and normally involve either $^{14}C$ or members of the uranium, thorium and actino-uranium decay series. There are differences between the accumulation rates derived by the various techniques, and sometimes even between those obtained using the same technique, probably reflecting complex chemical and physical processes within the sediments, but they results available do provide a reasonable indication of sediment accumulation rates such as that in Fig. 13.3. Rates of sediment accumulation vary with diagenesis, particularly the dissolution of calcium carbonate below the CCD (Section 15.2), and with the supply of material from the continent. Rates vary from about 0.1–2.5 g cm$^{-2}$ 10$^3$ yr$^{-1}$ (Jahnke, 1996) approximately equivalent to 0.2–5 cm per 1000 years (Libes, 2009) with the highest rates associated with large terrestrial inputs (e.g. North Indian and tropical Atlantic Oceans) and lowest rates in the north and south Pacific (Fig. 13.3).

It is apparent, therefore, that under present-day conditions, land-derived material is accumulating in deep-sea sediments at a rate of the order of a few

**Table 13.1** The average <2 μm fraction content in marine sediments and suspended river particulates (data from Griffin et al., 1968).

| Sediment type | Location | Wt% <2 μm fraction |
|---|---|---|
| Pelagic sediment | Atlantic Ocean | 58 |
| | Pacific Ocean | 61 |
| | Indian Ocean | 64 |
| Shelf sediment | USA Atlantic coast | 2 |
| | Gulf of Mexico | 27 |
| | Gulf of California | 19 |
| | Sahul Shelf, northwest Australia | 72 |
| Suspended river particulates | 33 USA rivers | 37 |

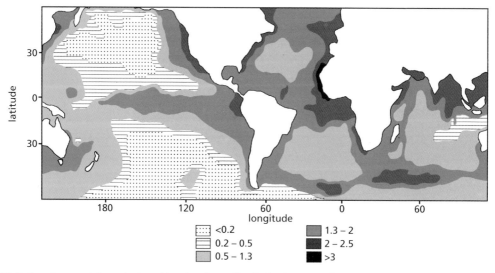

**Fig. 13.3** Sediment accumulation rates based in units of g cm$^{-2}$ kyr$^{-1}$, that is g per cm$^2$ per 1000 years based on Jahnke (1996).

millimetres per 1000 years. In contrast to deep-sea deposits, those deposited in nearshore areas have land-derived fractions that can accumulate at rates much greater than a few millimetres per year.

## 13.2 The formation of deep-sea sediments

The processes that are involved in the formation of deep-sea sediments can be linked to the manner in which material is transported to, and distributed within, the World Ocean, and by considerably simplifying the situation it is possible to distinguish between five general types of sediment transport mechanisms that operate in the deep sea.

*Gravity currents.* Gravity currents transport material to the deep sea from the shelf regions by slides, slumps and gravity flows. Of the various types of gravity flows, it is the turbidity currents that have the greatest influence on the movement of material from the shelf regions to the open ocean. These turbidity currents are short-lived, high-velocity, density currents that can carry vast quantities of suspended sediment off the shelves, often via submarine canyon conduits. The deep-sea deposits generated by these currents are termed turbidites, and usually they are made up of sand layers interbedded with pelagic deposits of smaller grain size. *Proximal* turbidites have been deposited relatively close to the source of the transported sediment, whereas *distal* types have been carried for much greater distances. Turbidite deposition is an extremely important marine sedimentary process and is thought to be responsible for the formation of features such as submarine fans on the continental rise, and the abyssal plains on the deep-sea floor.

*Geostrophic deep-ocean or bottom currents.* Bottom currents have a significant influence on the distribution of sediment on the deep-sea floor. These currents transport material that has a finer grain size than that carried by turbidity currents, and result in the formation of features such as sediment piles and ridges. Bottom currents are most strongly developed along the western boundaries of the oceans (see Section 7.3.3), and it is here that nepheloid layers (layers of suspended sediment) are best developed. These deep ocean currents therefore can be important for redistributing material within the oceans.

*Mid-depth currents.* Various kinds of suspended sediments can be transported by advection via mid-depth oceanic circulation patterns. These include material released directly from nearshore sediments, material resuspended from deep-sea sediments at basin edges, and hydrothermal components. As well as transporting material away from the basin margins, mid-depth currents can also carry material laterally from the centres to the edges of the oceanic basins.

*Surface and near-surface currents.* Surface current movements, which are dominated by the gyre circulation patterns (see Section 7.3.2), transport the oceanic biomass, together with fine-grained land-derived material introduced to the surface ocean by river run-off, atmospheric deposition and glacial transport. As discussed in Chapter 12, rapid packaging of particulate matter into large fast sinking aggregates means that the time available for transport of material by these currents prior to settling may be rather short.

*Vertical or down-column transport.* This is the great ocean-wide carbon-driven transport process (or global ocean flux), which carries material from the surface ocean to the sea bed. During the process, the material becomes incorporated into the major oceanic biogeochemical cycles, which are involved in the down-column flux of particulate material, and play such a vital role in controlling the chemistry of the oceans (see Sections 11.5, 12.1 and 12.2). The down-column flux can also incorporate the advective mid-depth flux (see Section 10.5).

On the basis of these different mechanisms, important distinctions therefore can be made between the transport vectors involved in: (i) *lateral* off-shelf movements; (ii) *lateral* sea-bed movements; (iii) *lateral* mid-depth movements; (iv) *lateral* sea surface movements; and (v) *vertical* down-column movements.

Each of these transport vectors has a material flux associated with it that contributes material to deep-sea sediments, leading to a complicated mixture of processes contributing to the final sediment assemblage. Using various geochemical tracers it is often possible to identify the various sources involved. This is illustrated in the case study that follows.

### A case study of sediment sources to the North Atlantic

A study reported by Grousset and Chesselet (1986) can be used both to illustrate how the transport vectors identified above operate on a quasi-global scale and to assess the extent of any coupling between them. These authors carried out an investigation into the Holocene (10000 yr BP) mid-ocean ridge sedimentary regime that prevailed in the North Atlantic between ~45°N and ~65°N. The major sources of sediment to the region in the Holocene were the North American mainland and Iceland, with local mid-ocean ridge sources being generally of only minor importance. By using a number of tracers to identify the sediment source materials, and to elucidate the mechanisms by which they were transported, the authors were able to construct a first-order model for the Holocene sedimentary regime in the region. This model is illustrated in Fig. 13.4, and involves the following source–flux relationships.

**1** Source 1 was the North American mainland from which material was supplied by surface currents ($\Phi_s$), turbidity currents ($\Phi_t$), and aeolian transport ($\Phi_e$). Material from this source decreased in the underlying sediments in a west-to-east direction, and it was suggested that a coupling of the aeolian transport flux into the down-column flux ($\Phi_v$) was the principal mechanism responsible for driving this gradient.

**2** Source 2 was Iceland, from which material was supplied via the turbidity current flux ($\Phi_t$) and the

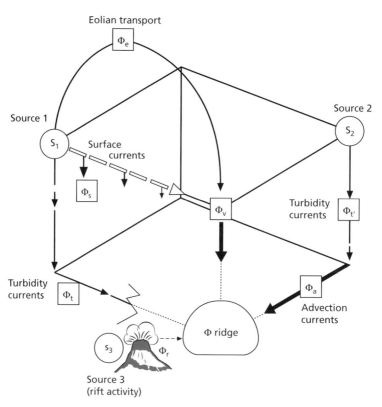

**Fig. 13.4** A first-order model for the mid-ocean ridge Holocene sedimentary regime in the North Atlantic (from Grousset and Chesselet, 1986). The three major sediment sources are $S_1$, $S_2$ and $S_3$, which supply the following specific flux materials: $\Phi_t$ and $\Phi_{t'}$ which are fluxes associated with downslope processes; $\Phi_s$, which is the surface current flux; $\Phi_a$ which is the advective bottom flux; $\Phi_e$ which is the aeolian flux; $\Phi_r$, which is the rift flux. These are the main components of $\Phi_v$ (the vertical flux) and $\Phi_a$ (the advective flux): see text. Reprinted with permission from Elsevier.

geostrophic current flux ($\Phi_a$). Material from this source decreased in concentration in a north-to-south direction in the underlying sediments, and it was proposed that this gradient arose from a coupling of the two fluxes that transported material from Iceland.

**3** Source 3 was the Mid-Atlantic Ridge, but this supplied only a minor amount of sedimentary material to the region.

The two thick arrows in the flux model (Fig. 13.4), representing the down-column ($\Phi_v$) and the geostrophic ($\Phi_a$) fluxes, are the main pathways by which continental material was transported from the North America and Icelandic sources, respectively, and these combine to form the ridge flux ($\Phi_{ridge}$), the hydrothermal flux being negligible.

---

The case study based on the work of Grousset and Chesselet (1986) clearly demonstrates how the various mechanisms that transport land-derived material to the ocean reservoir interact on a quasi-global oceanic scale. This kind of interaction is extremely important in controlling the chemical composition of the bottom sediments. There are a number of reasons for this, and three of the more important of these are identified:

**1** Solids derived from different continental sources may have different chemical compositions, and the extent to which they are mixed will affect the overall composition of the sediments to which they contribute.

**2** Material can be transported at varying rates by the different transport mechanisms; for example, turbidity currents deposit material at a much faster rate than down-column settling. The accumulation rate of sediments determines the length of time their surface is in contact with the overlying water column, and therefore influences both the degree to which the sediment components react with material dissolved in seawater and the extent to which the diagenetic sequence proceeds within the sediment column.

**3** The relative magnitudes of the sea bottom and the vertical down-column particle fluxes are extremely important in constraining the chemical composition of non-biogenous deep-sea sediments. There are two main reasons for this. First, the material that is initially dumped on the shelf regions has a coarser particle size, and is poorer in trace metals, than the material that escapes the coastal zone via the surface ocean; this concept was developed in Chapters 3 and 6. Secondly, the material that undergoes down-column transport is involved in the dissolved–particulate reactions that remove trace elements from seawater and so becomes a sink for additional trace

elements, which are not picked up by bottom-transported material. This bottom-transport–down-column vertical transport trace-metal fractionation is considered in more detail in Section 16.4.

## 13.3 A general scheme for the classification of marine sediments

A number of the parameters discussed above can be combined together to outline a general scheme for the classification of marine sediments. It must be stressed, however, that this is by no means a rigorous sediment classification, and several attempts have been made to produce much more detailed and lithologically consistent classifications of deep-sea sediments. Nonetheless, the simplified classification will serve its purpose, which is merely to act as a framework within which to describe the geochemistry of marine sediments. This general scheme is outlined in the following and is based on the major components of the various sediment types.

### 13.3.1 Nearshore sediments

Nearshore or coastal sediments are deposited on the margins of the continents under a wide variety of conditions in chemical environments that range from oxic to fully anoxic in character. The sediments include gravels, sands, silts and muds and they are composed of mixtures of terrigenous, authigenic and biogenous components; the latter is mainly shell material, but relatively high concentrations of organic carbon are found in sediments deposited under reducing conditions. Nearshore sediments contain material having a wide variety of grain sizes, fine-grained material being found in low-energy environments and sands in high-energy environments.

Over many shelves terrigenous sediments (e.g. terrigenous muds, which accumulate at relatively high rates, greater than several millimetres per year) are the prevalent type of deposit, but on some shelf regions carbonate deposition can predominate, for example on broad shallow shelves where the supply of terrigenous debris is small. In specialized hot arid near coastal environments, evaporite deposits of minerals such halite and gypsum can form from evaporation of seawater. Such deposits are rare in the current ocean but relatively abundant in the geological record (Libes, 2009).

### 13.3.2 Deep-sea sediments

On the basis of the major transport mechanisms that supply the sediment-forming material, deep-sea deposits can be subdivided into hemi-pelagic and pelagic types.

#### 13.3.2.1 Hemi-pelagic deep-sea sediments

These sediments are deposited in areas that fringe the continents, for example, on the abyssal plains. The land-derived material in hemi-pelagic sediments has been transported mainly by bottom processes, that is, by the turbidity current and geostrophic bottom-current fluxes described above, and much of it originated on the shelf regions. The inorganic hemi-pelagic sediments include lithogenous clays (or muds), glacial marine sediments, turbidites and mineral sands. All of these sediments can contain varying proportions of biogenous shell material. The rates of deposition of the land-derived material in hemi-pelagic deep-sea sediments can be $>10\,mm\ 10^3\,yr^{-1}$, and they often contain as much as ~1–5% organic carbon. The preservation of this organic carbon is a function of the extent to which the diagenetic sequence has proceeded (see Sections 14.2), and hemi-pelagic clays are often grey-green in colour, indicating reducing conditions, below a thin oxidized red layer.

#### 13.3.2.2 Pelagic deep-sea sediments

These sediments accumulate in open-ocean areas. They are generally deposited in the absence of effective bottom currents, and Davies and Laughton (1972) have defined pelagic sediments as those 'laid down in deep-water under quiet current conditions'. The bulk of the material forming these deposits has settled down the water column to blanket the bottom topography with a sediment cover. Thus, pelagic sediments are formed via the vertical down-column flux identified above, and so can be differentiated from hemi-pelagic deposits which, although they also have a vertical flux, have been formed largely from bottom transport processes. It is common practice to subdivide pelagic (and hemi-pelagic) deep-sea sediments into inorganic and biogenous categories.

*Inorganic pelagic deep-sea sediments.* These are defined as containing <30% biogenous skeletal remains, and as having a large fraction (>60%) of their non-biogenous material in the <2 μm (clay) size class. Traditionally, therefore, it is these sediments that have been known as the pelagic clays, or simply the deep-sea clays. These sediments usually have slow accumulation rates of the order of a few millimetres per 1000 years. As a result, much of the organic carbon reaching the sediment surface is destroyed in the early stages of the diagenetic sequence and before burial (see Section 14.1), and the sediments usually contain only ~0.1–0.2% organic carbon. Pelagic clays are therefore oxidizing to a considerable depth, and often have a red colour owing to the presence of ferric iron. Because of this, they have often been termed red clays. Some authors have further subdivided pelagic clays on the basis of the origin of their principal components, for example, into lithogenous and hydrogenous types; for a description of these categories, see Chapter 15.

*Biogenous pelagic deep-sea sediments.* These sediments have been defined traditionally as containing >30% biogenous shell remains. They are usually referred to by the term *oozes*, and are subdivided into calcareous and siliceous types. They accumulate more quickly than the deep sea clays with sedimentation rates of the order of a few centimetres per thousand years

The calcareous oozes contain >30% skeletal carbonates, and are classified on the basis of the predominant organisms present into foraminiferal ooze (sometimes termed *Globigerina* ooze after the most common of the forams), nanofossil ooze (or coccolith ooze) and pteropod ooze.

The siliceous oozes contain >30% opaline silica skeletal remains, and are subdivided into diatom oozes and radiolarian oozes, depending on the principal silica-secreting organism present.

## 13.4 The distribution of marine sediments

The distribution of sediments in the World Ocean is illustrated in Fig. 13.5. In Fig. 13.5(a), the sediments are classified on a general basis in which individual types of deep-sea clay are not specified. In contrast, Fig. 13.5(b) presents an example of a classification in which the deep-sea clays are subdivided into lithogenous and hydrogenous types. A number of the principal features in the distributions of deep-sea sediments can be identified from these diagrams.

1 Deep-sea clays and calcareous oozes are the predominant type of deep-sea deposit.

2 The calcareous oozes cover large tracts of the open-ocean floor (~48% OU, 1989) at water depths <3–4 km (see also Section 15.2).

3 The siliceous oozes cover about 14% of the ocean floor (OU, 1989) forming a ring around the high-latitude ocean margins in the Antarctic and North Pacific (both diatom oozes), and are also found in a band in the equatorial Pacific (radiolarian oozes). This distribution reflects at least in part that of the regions where siliceous organisms form an important part of the plankton community.

4 Extensive deposits of glacial marine sediments are confined to a band fringing Antarctica, and to the high-latitude North Atlantic.

5 Lithogenous clays cover a large area of the North Pacific, whereas the clays in the South Pacific are mainly hydrogenous in character. Pelagic clays overall cover about 38% of the ocean floor (OU, 1989). They are the dominant component of the sediments in regions where other components are absent, such as below the carbonate compensation depth CCD (see Section 15.2).

## 13.5 The chemical composition of marine sediments

The material that is finally deposited in the marine sediment reservoir has undergone a complex journey before reaching its sea-bed sink. The sediments forming this reservoir represent, if not the ultimate end-point, at least a major geological time-scale halt in the global mobilization–transportation cycle. A knowledge of the chemical composition of this sediment reservoir is therefore important for an understanding of the global cycles of many elements. However, before attempting to synthesize the various factors that act together to control the chemistry of these sediments it is necessary to establish a compositional database from which to work. For this purpose, three data compilations have been prepared.

1 The first compilation is given in Table 13.2, and lists the overall elemental compositions of the principal types of marine sediments, that is, nearshore muds, deep-sea clays and deep-sea carbonates. The table also includes chemical data for the continental crust, soils, suspended river particulates and crustal aerosols, which are given so that the compositions of marine sediments can be evaluated within the global context of their principal terrestrial feeder materials.

2 The second compilation is given in Table 13.3, and lists the *major element compositions* of the main types of deep-sea sediments.

3 The third compilation is given in Table 13.4, and presents data showing a number of overall trends in the elemental compositions of a wide range of deep-sea deposits. The aim of this table is to permit compositional differences in the deposits to be evaluated within an ocean-wide framework.

From these combined databases a number of gross features in the geochemistry of marine sediments can be identified, and for this purpose it is convenient to treat the major and trace elements separately.

*Major elements.* The major element composition of marine sediments is largely controlled by the relative proportions of the sediment-forming minerals. The principal minerals are the clays, and biogenic carbonates and opal. On the basis of the mutual proportions of these minerals the sediments can be divided into three broad types, that is, clays, carbonate oozes and siliceous oozes. Average major element analyses for the three sediment types are given in Table 13.3, from which it can be seen that aluminium is concentrated in the clays, calcium in the carbonates and silicon in the siliceous oozes. Manganese and Fe are included in Table 13.3, and although they are usually present as major elements they will also be included

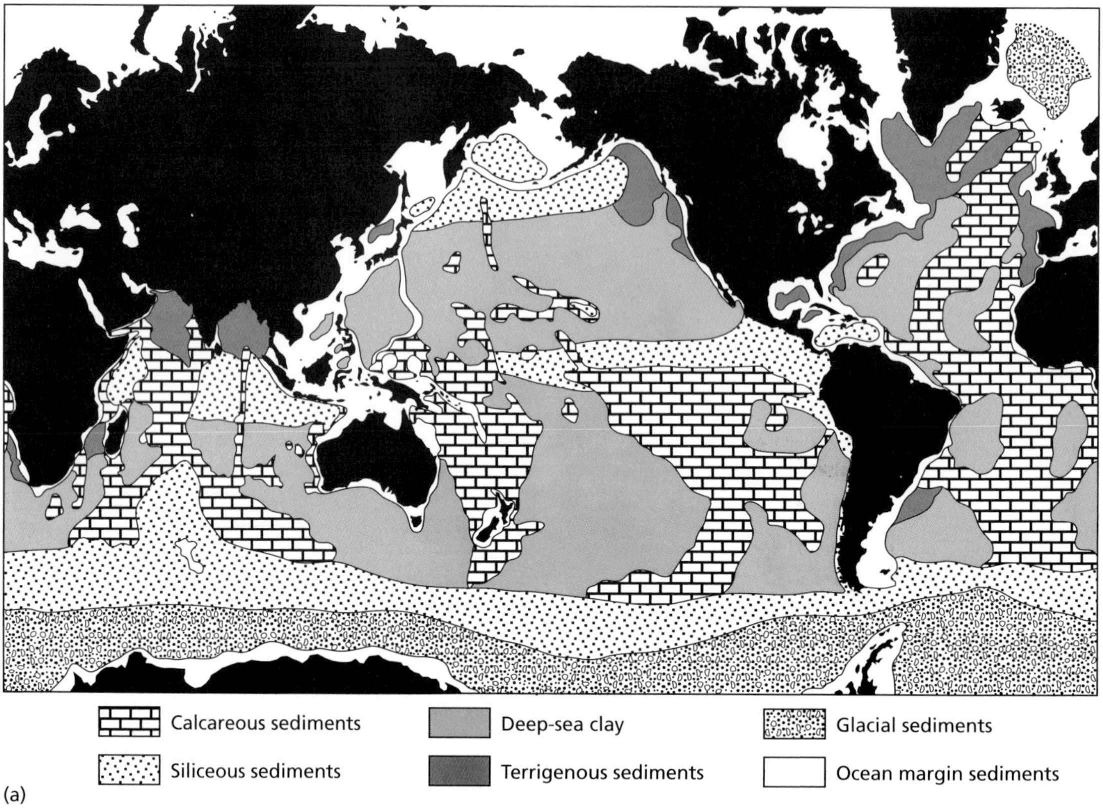

**Fig. 13.5** The present-day distributions of the principal types of marine sediments. (a) The general distribution of marine sediments (from Davies and Gorsline, 1976). Reprinted with permission from Elsevier. (b) The distribution of deep-sea sediments classified on the basis of their components (from Riley and Chester, 1971). Reprinted with permission from Elsevier.

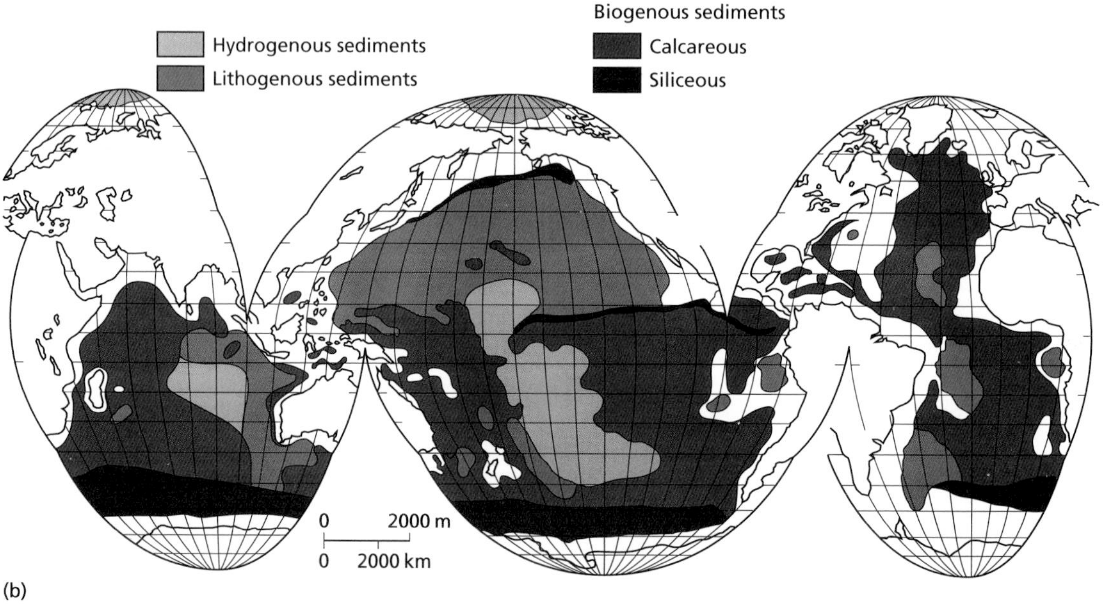

**Table 13.2** Elemental composition of marine sediments and some continental material (units, $\mu g\,g^{-1}$). Estimated values are given in brackets.

| Element | Continental crust* | Continental soil* | River particulate material* | Crustal dust† | Near-shore mud‡ | Deep-sea clay* | Deep-sea carbonate§ |
|---|---|---|---|---|---|---|---|
| Ag | 0.07 | 0.05 | 0.07 | – | – | 0.1 | 0.0X |
| Al | 69 300 | 71 000 | 94 000 | 82 000 | 84 000 | 95 000 | 20 000 |
| As | 7.9 | 6 | 5 | – | 5 | 13 | 1.0 |
| Au | 0.01 | 0.001 | 0.05 | – | – | 0.003 | 0.00X |
| B | 65 | 10 | 70 | – | – | 220 | 55 |
| Ba | 445 | 500 | 600 | – | – | 1500 | 190 |
| Br | 4 | 10 | 5 | – | – | 100 | 70 |
| Ca | 45 000 | 15 000 | 21 500 | – | 29 000 | 10 000 | 312 400 |
| Cd | 0.2 | 0.35 | (1) | – | – | 0.23 | 0.23 |
| Ce | 86 | 50 | 95 | – | – | 100 | 35 |
| Co | 13 | 8 | 20 | 23 | 13 | 55 | 7 |
| Cr | 71 | 70 | 100 | 79 | 60 | 100 | 11 |
| Cs | 3.6 | 4 | 6 | – | – | 5 | 0.4 |
| Cu | 32 | 30 | 100 | 47 | 56 | 200 | 50 |
| Er | 3.7 | 2 | (3) | – | – | 2.7 | 1.5 |
| Eu | 1.2 | 1 | 1.5 | – | – | 1.5 | 0.6 |
| Fe | 35 000 | 40 000 | 48 000 | 48 000 | 65 000 | 60 000 | 9000 |
| Ga | 16 | 20 | 25 | – | – | 20 | 13 |
| Ge | 1.5 | X | – | – | – | 1.2 | 0.1 |
| Gd | 6.5 | 4 | (5) | – | – | 7.8 | 3.8 |
| Hf | 5 | – | 6 | – | – | 4.5 | 0.41 |
| Ho | 1.6 | 0.6 | (1) | – | – | 1.0 | 0.8 |
| In | 0.1 | – | – | – | – | 0.08 | 0.02 |
| K | 24 000 | 14 000 | 20 000 | 20 000 | 25 000 | 28 000 | 2900 |
| La | 41 | 40 | 45 | – | – | 45 | 10 |
| Li | 42 | 25 | 25 | – | 79 | 45 | 5 |
| Lu | 0.45 | 0.40 | 0.50 | – | – | 0.50 | 0.50 |
| Mg | 16 400 | 5000 | 11 800 | – | 21 000 | 18 000 | – |
| Mn | 720 | 1000 | 1050 | 865 | 850 | 6000 | 1000 |
| Mo | 1.7 | 1.2 | 3 | 1.8 | 1 | 8 | 3 |
| Na | 14 200 | 5000 | 5300 | 5100 | 40 000 | 40 000 | 20 000 |
| Nd | 37 | 35 | 35 | – | – | 40 | 14 |
| Ni | 49 | 50 | 90 | 73 | 35 | 200 | 35 |
| P | 610 | 800 | 1150 | – | 550 | 1400 | 350 |
| Pb | 16 | 35 | 100 | 52 | 22 | 200 | 9 |
| Pr | 9.6 | – | (8) | – | – | 9 | 3.3 |
| Rb | 112 | 150 | 100 | – | – | 110 | 10 |
| Se | 0.05 | 0.01 | – | – | – | 4.5 | 0.5 |
| Sb | 0.9 | 1 | 2.5 | – | – | 0.8 | 0.15 |
| Sc | 10 | 7 | 18 | – | 12 | 20 | 2 |
| Si | 275 000 | 330 000 | 285 000 | – | 250 000 | 283 000 | 32 000 |
| Sm | 7.1 | 4.5 | 7.0 | – | – | 7.0 | 3.8 |
| Sn | 2 | (0.1) | – | – | 2 | 1.5 | 0.X |
| Sr | 278 | 250 | 150 | – | 160 | 250 | 2000 |
| Ta | 0.8 | 2 | 1.2 | – | – | 1 | 0.0X |
| Tb | 1.05 | 0.7 | 1 | – | – | 1 | 0.6 |
| Th | 9.3 | 9 | 14 | – | – | 10 | – |
| Ti | 3800 | 5000 | 5600 | 5700 | 5000 | 5700 | 770 |

*Continued*

**Table 13.2** *Continued*

| Element | Continental crust* | Continental soil* | River particulate material* | Crustal dust† | Near-shore mud‡ | Deep-sea clay* | Deep-sea carbonate§ |
|---|---|---|---|---|---|---|---|
| Tm | 0.5 | 0.6 | (0.4) | – | – | 0.4 | 0.1 |
| U | 3 | 2 | 3 | – | – | 2 | – |
| V | 97 | 90 | 170 | 120 | 145 | 150 | 20 |
| Y | 33 | 40 | 30 | – | – | 32 | 42 |
| Yb | 3.5 | – | 3.5 | – | – | 3 | 1.5 |
| Zn | 127 | 90 | 250 | 75 | 92 | 120 | 35 |
| Zr | 165 | 300 | – | – | 240 | 150 | 20 |

X, order of magnitude.
*Data for Ge, In, Se, Sn and Zr from a variety of sources; data for all other elements from Martin and Whitfield (1983).
†A Saharan dust population from the Atlantic Northeast Trades; data from Murphy (1985).
‡Data from Wedepohl (1960).
§Turekian and Wedepohl (1961).

**Table 13.3** The major element composition of the principal types of deep-sea sediments* (units, wt% oxides).

| Major element | Calcareous | Lithogenous clay | Siliceous | Oceanic average† |
|---|---|---|---|---|
| $SiO_2$ | 26.96 | 55.34 | 63.91 | 42.72 |
| $TiO_2$ | 0.38 | 0.84 | 0.65 | 0.59 |
| $Al_2O_3$ | 7.97 | 17.84 | 13.30 | 12.29 |
| $Fe_2O_3$ | 3.00 | 7.04 | 5.66 | 4.89 |
| FeO | 0.87 | 1.13 | 0.67 | 0.94 |
| MnO | 0.33 | 0.48 | 0.50 | 0.41 |
| CaO | 0.30 | 0.93 | 0.75 | 0.60 |
| MgO | 1.29 | 3.42 | 1.95 | 2.18 |
| $Na_2O$ | 0.80 | 1.53 | 0.94 | 1.10 |
| $K_2O$ | 1.48 | 3.26 | 1.90 | 2.10 |
| $P_2O_5$ | 0.15 | 0.14 | 0.27 | 0.16 |
| $H_2O$ | 3.91 | 6.54 | 7.13 | 5.35 |
| $CaCO_3$ | 50.09 | 0.79 | 1.09 | 24.87 |
| $MgCO_3$ | 2.16 | 0.83 | 1.04 | 1.51 |
| Organic C | 0.31 | 0.24 | 0.22 | 0.27 |
| Organic N | – | 0.016 | 0.016 | 0.015 |
| Total | 100.0 | 100.0 | 100.0 | 100.0 |
| Total $Fe_2O_3$ | 3.89 | 8.23 | 6.42 | – |

*Data from El Wakeel and Riley (1961).
†Weighted mean calculated on the basis of the areal coverage of the sea floor by each type of sediment: calcareous, 48.7%; lithogenous, 37.8%; siliceous, 13.5%.

with the trace elements because both Fe and Mn phases can act as scavenging agents for dissolved metals.

*Trace elements.* From the data given in Table 13.4, it is possible to establish a number of trends in the distributions of some trace elements in deep-sea sedi-ments (see e.g. Chester and Aston, 1976). Li and Schoonmaker (2004) present a more recent compila-tion of pelagic clay composition, but not of the other components in Tables 13.2 to 13.4 but this recent compilation confirms and extends the conclu-sions below. These trends can be summarized as follows.

**Table 13.4** The concentrations of some trace elements in deep-sea deposits* (units, $\mu g\, g^{-1}$).

| Trace element | Nearshore muds | Deep-sea carbonate | Atlantic deep-sea clay | Pacific deep-sea clay | Active ridge sediment | Ferromanganese nodules |
|---|---|---|---|---|---|---|
| Cr | 100 | 11 | 86 | 77 | 55 | 10 |
| V | 130 | 20 | 140 | 130 | 450 | 590 |
| Ga | 19 | 13 | 21 | 19 | – | 17 |
| Cu | 48 | 30 | 130 | 570 | 730 | 3 300 |
| Ni | 55 | 30 | 79 | 293 | 430 | 5 700 |
| Co | 13 | 7 | 38 | 116 | 105 | 3 400 |
| Pb | 20 | 9 | 45 | 162 | – | 1 500 |
| Zn | 95 | 35 | 130 | – | 380 | 3 500 |
| Mn | 850 | 1000 | 4 000 | 12 500 | 60 000 | 220 000 |
| Fe | 69 900 | 9000 | 82 000 | 65 000 | 180 000 | 140 580 |

*From Chester and Aston (1976).

1 In general, deep-sea carbonates are impoverished in most trace elements relative to deep-sea clays, Sr being an exception to this.

2 Certain trace elements, for example, Cr, V and Ga, have similar concentrations in both nearshore muds and deep-sea clays (DSC).

3 In contrast, other trace elements, for example, Mn, Cu, Ni, Co and Pb, are enhanced in the deep sea clays (DSC) relative to nearshore muds. Thus, a fundamental oceanic fractionation between nearshore muds and deep-sea clays can be introduced for these enriched or excess elements. Li and Schoonmaker (2004) argue that this fractionation occurs as a result of the formation of ferromanganese phases (see below). Raiswell and Anderson (2005) propose a global scale cycling of reactive (as opposed to that bound within relatively inert aluminosilicate structures) sedimentary iron from shelf seas to the open ocean, associated with sediment diagenesis in shelf sediments, the formation of particulate phases in the water column and their subsequent transport to deep ocean sites of sediment accumulation.

4 The excess trace elements are enhanced to a greater extent in Pacific than in Atlantic deep-sea clays.

5 Ferromanganese nodules (see Chapter 15) have particularly high concentrations of the elements that are enhanced in deep-sea clays (point 3 previously), but only small concentrations of the elements that are not enriched in DSC (point 2 previously).

6 Metalliferous, active ridge, sediments are enhanced in elements such as Fe, Mn, Cu, Zn, Ni and Co, relative to normal DSC.

7 Although it is not apparent from the data given in Table 13.4, some of the early surveys carried out on the distributions of trace elements in deep-sea sediments revealed that the excess trace elements reached their highest values in deposits that had accumulated at very slow rates in areas remote from the land masses.

These various trends offer a skeleton around which to build a discussion of the factors that combine together to control the elemental composition of marine sediments. Even at this stage a number of elemental fractionations can be identified in the marine sediment complex. For example, there is a major fractionation of some trace metals between nearshore muds and deep-sea clays. In addition, further fractionation stages occur within the various kinds of deep-sea deposits themselves, for example, between the ridge-crest metalliferous deposits and the deep-sea clays. Thus, it begins to appear as if some kind of sequential enhancement is occurring for certain elements within marine sediments. The following chapters will be devoted to an attempt to understand how these elemental fractionations, and sequential enrichments, may have arisen. To provide a framework for this, further use will be made of the concept of chemical signals.

## 13.6 Chemical signals to marine sediments

Marine sediments may be thought of as having received a variety of chemical signals, or fluxes, which, in a number of combinations, have resulted in them acquiring their present composition. Two principal questions therefore must be asked in order to understand how the sediments attained this composition.

1 What is the chemical composition of the dissolved and particulate material carried by the signals themselves?

2 How can the effects of the individual signals be unscrambled in order to provide a reasonably coherent explanation for the geochemical characteristics of individual sediments, or sediment suites?

One way of addressing these questions is to identify the individual components that combine together to form marine sediments, and then to establish whether or not the *processes* by which they are formed can be related to specific individual chemical signals, that is, to adopt a process-orientated approach to the problem of describing the chemical compositions of the sediments. It is possible to classify the components of marine sediments into a series of genetically different types, and a number of schemes have been proposed for this purpose. The scheme adopted in the present volume is a modification of that outlined by Goldberg (1954), which classifies the components in terms of their geospheres of origin. In the original scheme the sediment components were subdivided into a single aqueous phase, that is, interstitial waters, and four solid phases, which were classified according to the origin of their component elements as lithogenous, biogenous, hydrogenous and cosmogenous. In the modified scheme used here, however, the hydrogenous material will be subdivided into a number of different types (see Section 15.3).

## 13.7 Marine sediments: summary

1 A large part of the floor of the World Ocean is covered by a blanket of sediment, which has an average thickness of ~500 m.

2 Nearshore sediments are deposited on the shelf region, under a wide variety of depositional environments. Deep-sea sediments are deposited seaward of the shelf, under conditions of slow accumulation, and cover more than 50% of the surface of the Earth. The deep-sea sediments are subdivided into hemipelagic types, which are deposited in areas fringing the continents mainly via bottom transport mecha-

nisms, and pelagic types, which are deposited in open-ocean areas mainly via down-column transport mechanisms.

3 Calcareous oozes cover large tracts of the open-ocean floor at water depths <<3–4 km. Siliceous oozes form a ring around the high-latitude ocean margins.

4 The components of marine sediments can be subdivided into an aqueous phase (interstitial water) and four solid phases, which, on the basis of their geospheres of origin, are classed as lithogenous, hydrogenous, biogenous and cosmogenous.

5 Relative to nearshore muds, deep-sea clays contain enhanced concentrations of some trace elements, for example, Mn, Cu, Ni, Co and Pb; these are often referred to as excess elements.

The solid sediment-forming components that make up marine sediments can be thought of as building blocks, which are stacked together in various proportions to form an individual sediment, or a suite of sediment types. However, at this stage an extremely important concept must be introduced; this is that sediments are *not* an inert reservoir, and as a result the building blocks are not simply stacked together in a way that retains their original compositions. Rather, they can be subjected to a series of *diagenetic* reactions following their deposition. Further, it is important to understand that these diagenetic reactions, which take place mainly via the medium of the interstitial waters, not only modify the compositions of pre-existing building blocks but also can supply elements that result in the formation of new blocks. In order, therefore, to be able to evaluate fully the processes involved in the formation of the sediment components, diagenesis will be described before the components themselves are considered. To do this, diagenetic reactions will be discussed in terms of the aqueous, that is, the interstitial water, sediment phase. This will be followed by a description of the individual sediment-forming components themselves, and finally an attempt will be made to identify, and unscramble, the chemical signals that are transmitted to marine sediments.

## References

Chester, R. and Aston, S.R. (1976) The geochemistry of deep-sea sediments, in *Chemical Oceanography*, J.P.

Riley and R. Chester (eds), Vol. 6, pp.281–390. London: Academic Press.

Davies, T.A. and Laughton, A.S. (1972) Sedimentary processes in the North Atlantic, in *Initial Reports of the Deep Sea Drilling Project*, Vol. 12, pp.905–934. Washington, DC: US Government Printing Office.

Davies, T.A. and Gorsline, D.S. (1976) Oceanic sediments and sedimentary processes, in *Chemical Oceanography*, J.P. Riley and R. Chester (eds), Vol. 5, pp.1–80. London: Academic Press.

El Wakeel, S.K. and Riley, J.P. (1961) Chemical and mineralogical studies of deep-sea sediments. *Geochim. Cosmochim. Acta*, **25**, 110–146.

Goldberg, E.D. (1954) Marine geochemistry. Chemical scavengers of the sea. *J. Geol.*, **62**, 249–255.

Griffin, J.J., Windom, H. and Goldberg, E.D. (1968) The distribution of clay minerals in the World Ocean. *Deep Sea Res.*, **15**, 433–459.

Grousset, F.E. and Chesselet, R. (1986) The Holocene sedimentary regime in the northern Mid-Atlantic Ridge region. *Earth Planet. Sci. Lett.*, **78**, 271–287.

Heezen, B.C. and Hollister, C.D. (1971) *The Face of the Deep*. New York: Oxford University Press.

Heezen, B.C., Tharp, M. and Ewing, M. (1959) The floors of the oceans. *Geol. Soc. Am. Spec. Pap.*, **65**, 1–122.

Hemming, S.R. (2004) Heinrich events: massive late pleistocene detritus layers of the North Atlantic and their global imprint. *Rev. Geophys.*, **42**, 1–43.

Holcombe, T.L. (1977) Ocean bottom features – terminology and nomenclature. *Geojournal*, **6**, 25–48.

Jahnke, R.A. (1996) The global ocean flux of particulate organic carbon: areal distribution and magnitude. *Global Biogeochem. Cycl.*, **10**, 71–88.

Jones, E.J.W. (1978) Sea-floor spreading and the evolution of the ocean basins, in *Chemical Oceanography*, J.P. Riley and R. Chester (eds), Vol. 7, pp.1–74. London: Academic Press.

Kennet, J.P. (1982) *Marine Geology*. Englewood Cliffs, NJ: Prentice Hall.

Li, Y-H and Schoonmaker, J.E. (2004) Chemical composition and mineralogy of marine sediments, in *Treatise on Geochemistry*, H.D. Holland and K.K. Turekian (eds), Vol. 7, *Sediments, Diagenesis and Sedimentary Rocks*, F.T. Mackenzie (ed.), pp.1–35. Oxford: Elsevier Pergamon.

Libes, S.M. (2009) *Introduction to Marine Biogeochemistry*, 2nd edn, Academic Press.

Martin, J.-M. and Whitfield, M. (1983) The significance of the river input of chemical elements to the ocean, in *Trace Metals in Sea Water*, C.S. Wong, E.A. Boyle, K.W. Bruland, J.D. Burton and E.D. Goldberg (eds), pp.265–296. New York: Plenum.

Moran, K. *et al.* (2006) The Cenozoic palaeoenvironment of the Arctic Ocean. *Nature*, **441**, 601–605.

Murphy, K.J.T. (1985) The trace metal chemistry of the Atlantic aerosol. PhD thesis, University of Liverpool.

OU (1989) *Ocean Chemistry and Deep-Sea Sediments*, Open University, Elsevier.

Parsons, B. and Richter, F.M. (1981) *Mantle convection and the oceanic lithosphere, in The Sea*, Vol. 7, C. Emiliani (ed.), pp.73–117. New York: John Wiley & Sons, Inc.

Raiswell, R. and Anderson, T.F. (2005) Reactive iron enrichment in sediments deposited beneath euxinic bottom waters: Constraints on supply by shelf recycling. *Geol. Soc. Special Pub.*, **248**, 179–194.

Riley, J.P. and Chester, R. (1971) *Introduction to Marine Chemistry*. London: Academic Press.

Shackleton, N.J., Imbrie, J. and Hall, M.A. (1983) Oxygen and carbon isotope record of East Pacific core V19–30: implications for the formation of deep water in the late Pleistocene North Atlantic. *Earth Planet. Sci. Lett.*, **65**, 233–244.

Turekian, K.K. and Wedepohl, K.H. (1961) Distribution of the elements in some major units of the Earth's crust. *Bull. Geol. Soc. Amer.*, **72**, 175–192.

Wedepohl, K.H. (1960) Spurenanalytische Untersuchungen an Tiefseetonen aus dem Atlantik. *Geochim. Cosmochim. Acta*, **18**, 200–231.

Interstitial waters are aqueous solutions that occupy the pore spaces between particles in rocks and sediments. In some nearshore deposits groundwater seepages can occur, and around the ridge-crest areas circulating hydrothermal solutions (i.e. modified seawater) can enter the sediment column. For most marine sediments, however, the interstitial fluids originated as seawater trapped from the overlying water column. The interstitial-water–sediment complex is a site of intense chemical, physical and biological reactions, which can lead both to the formation of new and altered mineral phases and to changes in the composition of waters themselves. These changes may be grouped together under the term *diagenesis*, which has been defined by Berner (1980) as 'the sum total of processes that bring about changes in a sediment or sedimentary rock subsequent to its deposition in water'. Many of the important diagenetic changes that affect marine sediments take place during *early diagenesis*, which occurs during the burial of the deposits. The highest rates of biological activity occur in the top tens of cm of the sediments but biological activity is now known to continue to depths in excess of 1 km (Roussel *et al.*, 2008).

## 14.1 Early diagenesis: the diagenetic sequence and redox environments

### 14.1.1 Introduction

Many of the chemical changes that take place during early diagenesis are redox-mediated, that is, they depend on the redox (chemical reduction/oxidation reactions) environment in the sediment–interstitial-water–seawater system. In turn, this redox environ-ment is largely controlled by the degree to which organic carbon is preserved, or undergoes decomposition, in the sediment complex. Most of the ocean water column is oxic, and hence the degradation in the water column occurs mainly with oxygen as the oxidizing agent. However, within the ocean sediments oxygen can become depleted and alternative oxidizing agents become important. Oxidation of organic carbon can use alternative electron acceptor and is now chemically defined as the loss of electrons (See Worksheet 14.1).

### 14.1.2 The diagenetic sequence

Diagenetic processes in sediments are driven by redox reactions that are mediated by the decomposition of organic carbon, and some of the basic concepts involved in sedimentary redox processes are described in Worksheet 14.1.

It is now generally recognized that there is a diagenetic sequence of catabolic processes in sediments, the nature of which depends on the particular oxidizing agent that 'burns' the organic matter. As sedimentary organic matter is metabolized it donates electrons to various oxidized components in the interstitial-water–sediment complex, and when oxygen is present it is the preferred electron acceptor. During the diagenetic sequence, however, the terminal electron-accepting species alter as the oxidants are consumed in order of decreasing energy production per mole of organic carbon oxidized. Thus, as oxygen is exhausted, microbial communities switch to a succession of alternative terminal electron acceptors in order of decreasing thermodynamic advantage (see e.g. Froelich *et al.*, 1979; Galoway and Bender, 1982; Wilson *et al.*, 1985). Using the schemes outlined by,

*Marine Geochemistry*, Third Edition. Roy Chester and Tim Jickells.
© 2012 by Roy Chester and Tim Jickells. Published 2012 by Blackwell Publishing Ltd.

## Worksheet 14.1: Redox reactions in sediments

Aqueous solutions do not contain free protons and free electrons. However, according to Stumm and Morgan (1996) it is possible to define the relative proton and electron activities in these solutions. Acid–base processes involve the transfer of protons, and pH, which can be written

$$pH = -\log_{10} a_{H+} \qquad \text{(WS14.1.1)}$$

which measures the relative tendency of a solution to accept or transfer protons. The activity of a hypothetical hydrogen ion is high at low pH and low at high pH, and pH is a master variable in acid–base equilibria.

In a similar manner, it is also possible to define a convenient parameter to describe redox intensity. Redox reactions involve the transfer of electrons, and $p\varepsilon$, which can be written

$$p\varepsilon = -\log_{10} a_{e-} \qquad \text{(WS14.1.2)}$$

which measures the relative tendency of a solution to accept or transfer electrons. A high $p\varepsilon$ indicates a relatively high tendency for oxidation, and $p\varepsilon$ is a master variable in redox equilibria.

An oxidation–reduction reaction is termed a *redox* reaction, and can be written as two half-reactions in which a reduction is accompanied by an oxidation in terms of a redox couple. To illustrate this, Drever (1982) used the reduction of $Fe^{3+}$ by organic matter, represented by (C). Thus:

$$4Fe^{3+} + (C) + 2H_2O \rightarrow 4Fe^{2+} + CO_2 + 4H^+. \qquad \text{(WS14.1.3)}$$

In this equation neither molecular oxygen nor electrons are shown explicitly. The equation can be broken down into two half-reactions, one involving only Fe and the other only C. Thus:

$$4Fe^{3+} + 4e^- \rightarrow 4Fe^{2+} \qquad \text{(WS14.1.4)}$$

in which $Fe^{3+}$ undergoes reduction to $Fe^{2+}$, and

$$(C) + 2H_2O \rightarrow CO_2 + 4H^+ + 4e^- \qquad \text{(WS14.1.5)}$$

in which the organic matter undergoes oxidative destruction to yield $CO_2$. It must be remembered, however, that these half reactions do not represent complete chemical reactions because aqueous solutions do not contain free electrons.

The half-reaction concept can be related to measurements in electrochemical half-cells, and allows another parameter to be introduced into redox chemistry. This parameter is $E_h$, in which the electron activity is expressed in volts, the h subscript indicating that the $E_h$ value is expressed relative to the standard hydrogen electrode, which is used as a zero reference. The relative activity of electrons in a solution can therefore be expressed in units of electron activity ($p\varepsilon$), which is a dimensionless quantity, or in volts ($E_h$), and the relation between $p\varepsilon$ and $E_h$ is given by:

$$p\varepsilon = \frac{F}{2.3RT} E_h \qquad \text{(WS14.1.6)}$$

where $F$ is Faraday's constant, $R$ is the gas constant and $T$ is the absolute temperature; at 25°C $E_h = 0.059p\varepsilon$. It is possible to measure $E_h$ using electrodes, analogous to the measurement of pH. However, electrode-measured $E_h$ values in oxidizing natural waters are difficult to relate to a specific redox pair, and both Stumm and Morgan (1996) and Drever (1982) have pointed out that it is important to distinguish between electrode-measured $E_h$ and $E_h$ calculated from the activities of a redox pair.

*continued on p. 292*

In the present text the general concept of *redox conditions* will be used, in which positive $E_h$ (redox potential) values indicate oxidizing conditions and negative values indicate reducing conditions; that is, half-reactions of high $E_h$ are oxidizing, and those of low $E_h$ are reducing. Thus, a half-reaction with a lower $E_h$ will undergo oxidation when combined with a half-reaction of higher $E_h$. This reaction combination can be used to describe redox-mediated diagenetic reactions in sediments. $E_h$ can be converted to energy units (e.g. Fig. WS14.1.1) and hence describes the thermodynamic, but not necessarily the kinetic, tendency for a reaction to proceed.

$E_h$ conditions in sediments are controlled mainly by the decomposition of photosynthetically produced organic matter by non-photosynthetic bacteria, and are constrained by the rate of supply of the organic matter (primary production), the rate at which it accumulates (sedimentation rate) and the rate of supply of oxidizing agents. The breakdown of organic matter is carried out almost exclusively by bacteria. However, this bacterial decomposition of organic matter is driven by a sequence of reactions that switch to a successive series of oxidants, or electron acceptors, which represent lower $p\varepsilon$ levels and hence lower energy yields. During the reaction sequence, in which the organic matter is decomposed by micro-organisms, the organisms acquire energy for their metabolic requirements and communities able to carry out the highest energy yielding reactions will always dominate.

Only relatively abundant elements with oxidation states that can change under conditions found within the environment will play an important role in sedimentary diagenetic redox reactions. Hence, only a relatively few elements (C, N, O, S, Fe and Mn) are predominant participants in aquatic redox processes, although the redox state of other trace components may be altered. The overall relationships that involve these elements in the microbially mediated redox sequence have been summarized diagrammatically by Stumm and Morgan (1996); their scheme is reproduced in Fig. WS14.1.1 in which the energy yields associated with the various half reaction processes in the diagenetic sequence are given in the form of reaction combinations that are initiated at various $E_h$ and $p\varepsilon$ values. These are illustrated in Fig. WS14.1.1 and Table WS14.1.1. For example, the first stage in the sequence involves the oxidation of organic matter by dissolved oxygen (A + L), with successive reactions following the decreased $p\varepsilon$ and $E_h$ levels. The full 'diagenetic sequence', and the sedimentary environments associated with the various stages in the sequence, are discussed in detail in the text. Examples

**Table WS14.1.1** Environmentally Important Reduction and Oxidation Reactions (Stumm and Morgan, 1996) at pH7, $p\varepsilon$ values are for reactions proceeding left to right, but all are theoretically reversible. Letters refer to Fig WS14.1.1. Note M, N, O and P are the reverse of reactions G, E, D and C respectively. The energy of organic matter (simplified here to $CH_2O$) oxidation depends on the products for example, F and L. e represents an electron transferred in the reaction.

|   |   | $p\varepsilon$ |
|---|---|---|
| A | $1/4 O_{2(g)} + H^+_{(aq)} + e \rightarrow 1/2\, H_2O$ | +13.75 |
| B | $1/5 NO^-_{3(aq)} + 6/5 H^+_{(aq)} + e \rightarrow 1/10 N_{2(g)} + 3/5 H_2O$ | +12.65 |
| C | $1/2 Mn(IV)O_{2(s)} + 1/2 HCO^-_{3(aq)} + 3/2 H^+_{(aq)} + e \rightarrow 1/2 Mn(II)CO_{3(s)} + H_2O$ | +8.9 |
| D | $1/8 NO^-_{3(aq)} + 5/4 H^+_{(aq)} + e \rightarrow 1/8 NH^+_{4(aq)} + 3/8 H_2O$ | +6.15 |
| E | $FeOOH_{(s)} + HCO^-_{3(aq)} + 2H^+_{(aq)} + e \rightarrow FeCO_{3(s)} + 2H_2O$ | −0.8 |
| F | $1/2 CH_2O + H^+_{(aq)} + e \rightarrow 1/2 CH_3OH$ | −3.01 |
| G | $1/8 SO_4^{2-} + 9/8 H^+_{(aq)} + e \rightarrow 1/8 HS^-_{(aq)} + 1/2 H_2O$ | −3.75 |
| H | $1/8 CO_{2(g)} + H^+_{(aq)} + e \rightarrow 1/8 CH_{4(g)} + 1/4 H_2O$ | −4.13 |
| J | $1/6 N_2 + 4/3 H^+_{(aq)} + e \rightarrow 1/3 NH^+_{4(aq)}$ | −4.68 |
| L | $1/4 CH_2O_{(aq)} + 1/4 H_2O \rightarrow 1/4 CO_{2(g)} + H^+_{(aq)} + e$ | −8.2 |

*continued*

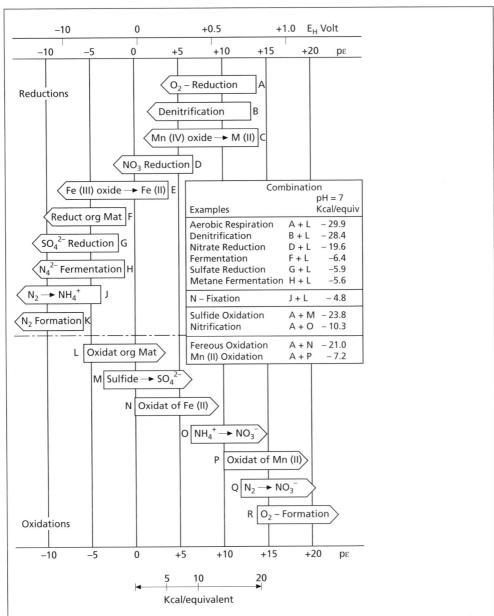

**Fig. WS14.1.1** The microbially mediated diagenetic sequence in sediments (from Stumm and Morgan, 1996). The relative energy yield of the various redox half reactions are presented in term of decreasing energy yield (as Eh, pε, and ΔG). Arrows point in the direction of the spontaneous reaction with the starting point representing the energy for that particular half reaction. The half reactions A to P are listed in the box.

of the diagenetic succession are given in the box, in Fig. WS14.1.1, where the overall energy yield of the combined reactions are given. From this it can be seen, for example, that there is a tendency for the more energy-yielding reactions to take precedence over those that are less energy-yielding. Thus, the sequence begins with aerobic respiration (A + L), followed by denitrification (B + L), and so on.

among others, Froelich *et al.* (1979) and Berner (1980), the general *diagenetic sequence* in marine sediments can be outlined in the following general way.

*Aerobic metabolism.* Aerobic organisms can use dissolved oxygen from the overlying or interstitial waters to 'burn' organic matter. The organic matter that undergoes early diagenesis can be considered to have the Redfield composition of $(CH_2O)_{106}(NH_3)_{16}$ $(H_3PO_4)$ (see Section 9.2). The oxidation of organic matter by aerobic organisms therefore can be represented by a general equation such as that proposed by Galoway and Bender (1982):

$$5(CH_2O)_{106}(HN_3)_{16}(H_3PO_4) + 690O_2$$
$$\rightarrow 530CO_2 + 80HNO_3 + 5H_3PO_4 + 610H_2O. \quad (14.1)$$

The $CO_2$ released during this reaction can lead to carbonate dissolution, and the ammonia can be oxidized to nitrate, a process termed *nitrification*. Under oxic conditions most of the remains of dead animals and plankton are apparently destroyed at this stage in the diagenetic sequence. For example, according to Bender and Heggie (1984), >90% of the organic carbon that reaches the deep-sea floor is oxidized by $O_2$. Oxygen therefore may be regarded as the *primary* oxidant involved in the destruction of organic matter, and in a closed system reaction (reaction 14.1) (oxic diagenesis) will continue until sufficient oxygen has been consumed to drive the redox potential low enough to favour the next most efficient oxidant. Thus, as dissolved oxygen becomes depleted, organic matter decomposition can continue using oxygen from *secondary* oxidant sources (suboxic diagenesis).

*Anaerobic metabolism.* Anaerobic metabolism takes over when the content of dissolved oxygen falls to very low levels, or becomes entirely exhausted, and a series of secondary oxidants are utilized depending on their energy yields (see Worksheet 14.1 where a simpler representation of organic composition is used). These secondary oxidants include nitrate, $MnO_2$, $Fe_2O_3$ or $FeOOH$ and sulfate.

**1** *Nitrate.* According to Berner (1980), when the dissolved oxygen levels fall to ~5% of their concentration in aerated waters nitrate becomes the preferred electron acceptor, a reaction that can be represented as follows:

$$5(CH_2O)_{106}(NH_3)_{16}(H_3PO_4) + 472HNO_3$$
$$\rightarrow 276N_2 + 520CO_2 + 5H_3PO_4 + 886H_2O$$
$$(14.2a)$$

This process is termed *denitrification*. The greenhouse gas $N_2O$ can be produced a by-product of this reaction as discussed in Section 9.1 In the reaction given above it is assumed that all organic nitrogen released is in the form of ammonia, which is then oxidized to molecular nitrogen by the reaction:

$$5NH_3 + 3HNO_3 \rightarrow 4N_2 + 9H_2O \quad (14.2b)$$

However, this is not the only possible pathway, and Froelich *et al.* (1979) have pointed out that the fate of the nitrogen has important consequences in diagenesis with respect to the sequence in which the secondary oxidants are used. These authors suggested that if all the nitrogen goes to $N_2$ then the use of nitrate as a secondary oxidant overlaps with that of $MnO_2$, but that if the nitrogen is released as ammonia and is not oxidized to $N_2$ then $MnO_2$ should be reduced before nitrate. This emphasises that the complexities of the ocean often confound simplistic chemical representation.

The use of the other secondary oxidants can be illustrated by reactions of the type described by Froelich *et al.* (1979) for the use of Mn(IV), Fe(III) and sulfate as alternative electron acceptors for the degradation of organic matter with Redfield stoichiometry.

**2** *Manganese oxides*

$$(CH_2O)_{106}(NH_3)_{16}(H_3PO_4) + 236MnO_2 + 472H^+$$
$$\rightarrow 236Mn^{2+} + 106CO_2 + 8N_2 + H_3PO_4 + 336H_2O$$
$$(14.3)$$

**3** *Iron oxides*

$$(CH_2O)_{106}(NH_3)_{16}(H_3PO_4) + 212Fe_2O_3 + 848H^+$$
$$\rightarrow 424Fe^{2+} + 106CO_2 + 16NH_3 + H_3PO_4 + 530H_2$$
$$(14.4)$$

**4** *Sulfate*

$$(CH_2O)_{106}(NH_3)_{16}(H_3PO_4) + 55SO_4$$
$$\rightarrow 106CO_2 + 16NH_3 + 55S^{2-} + H_3PO_4$$
$$+ 106H_2O \quad (14.5)$$

5 Following sulfate reduction, biogenic methane can be formed by two possible reaction pathways, which may be generalized as:

$$CH_3COOH \rightarrow CH_4 + CO_2 \qquad (14.6a)$$

or

$$CO_2 + 8H_2 \rightarrow CH_4 + 2H_2O \qquad (14.6b)$$

Thus, diagenesis proceeds in a general sequence in which the oxidants are utilized in the order: oxygen > nitrate > manganese oxides > iron oxides > sulfate. In this diagenetic sequence it is assumed that in marine sediments $O_2$, $NO_3^-$, $MnO_2$, $Fe_2O_3$ (or FeOOH) and $SO_4^{2-}$ are the only electron acceptors, and that organic matter (represented by the Redfield composition) is the only electron donor. Furthermore, it is assumed that the oxidants are limiting, that is, each reaction proceeds to completion before the next one starts. However, the diagenetic processes are not always sequential; for example, although it is usually thought that sulfate reduction precedes methane formation, Oremland and Taylor (1978) have suggested that the two processes can occur simultaneously, that is, they are not mutually exclusive.

## 14.1.3 Diagenetic and redox environments

As diagenesis proceeds a number of end-member sedimentary environments are set up, which Berner (1981) was able to relate to a diagenetic zone sequence.

1 *Oxic environments* are those in which the interstitial waters of the sediments contain measurable dissolved oxygen, and diagenesis occurs via aerobic metabolism. Under these conditions in deep ocean sediments often little organic matter is preserved in the sediments, and in terms of the reactions given above the diagenetic sequence has proceeded only to the reaction in Equation 14.1 in oxic environments. Thus, the oxygen supply exceeds the degradable organic matter on, or in, the sediments.

2 *Anoxic environments* are those in which the sediment interstitial waters contain no measurable dissolved oxygen, that is, diagenesis here has to proceed via the secondary oxidants through anaerobic metabolism. The anoxic environments were subdivided into a number of types.

3 *Non-sulfidic post-oxic environments*. These environments, which contain no measurable dissolved sulfides, are common in many deep-sea sediments, and are perhaps more often referred to in the literature as *suboxic* environments. The condition necessary to set up this type of sedimentary environment is a supply of organic carbon sufficient that diagenesis can proceed beyond the oxic stage. Under these conditions, nitrate, manganese oxides and iron oxides are used as secondary oxidants, but the sequence does not reach the stage at which sulfate is utilized for this purpose. In suboxic sediments, therefore, there is a relatively large, but still limited, supply of metabolizable organic matter, and the diagenetic sequence has proceeded to the reactions in Equations 14.2–14.4 given previously.

4 *Sulfidic environments*. These result when the diagenetic sequence has reached the stage at which the bacterial reduction of dissolved sulfate takes place with the production of $H_2S$ and $HS^-$. If a sufficient supply of metabolizable organic matter is available, sulfate reduction can be a common feature in marine sediments as a result of the relatively high concentration of sulfate in both seawater and marine sediment interstitial waters. In practice, however, constraints on the supply and preservation of organic matter mean that sulfate reduction is largely restricted to nearshore sediments where organic carbon supply is relatively high. In sulfidic environments the diagenetic sequence has now proceeded to Equation 14.5. The production of sulfide is particularly important because it is toxic to many larger organisms and also because it will react with many metals to form very insoluble sulfides. These include: iron sulfides, such as the polysulfide pyrite ($FeS_2$) and its diamorph marcasite ($FeS_2$); the metastable iron sulfides mackenawite and greigite; glauconite; chamosite; manganese sulfide (MnS). For a detailed description of these minerals the reader is referred to Calvert (1976).

5 *Non-sulfidic methanic environments*. In some sediments that contain a relatively large amount of metabolizable organic matter the diagenetic reactions can pass through the stage at which oxygen, nitrate, manganese oxides, iron oxides and sulfate are sequentially utilized. Continued decomposition of organic matter results in the formation of dissolved methane, for example by reactions in Equations 14.6a and 14.6b with methane released as bubbles from the sediment.

As noted in Section 11.6 such suboxic and even sulfidic environments can occur within the water column where deep waters are isolated from exchange with surface waters, such as in the Black Sea and Cariaco Trench.

The general diagenetic sequence outlined earlier, in which the various oxidants are consumed in the order

oxygen → nitrate → manganese oxides
→ iron oxides → sulfate

leads to the setting up of a series of diagenetic zones in sediments. These zones give rise to an environmental succession in the processes of organic matter decomposition, which involves oxygen consumption (respiration), nitrate (and/or manganese and iron) reduction, sulfate reduction and methane formation, and depending on the amount of available organic matter, any sediment can pass through each of these zones during deposition and burial. Thus, a vertical diagenetic zone sequence can be set-up in a sediment. This zonation can be elegantly illustrated by a study by Froelich *et al.* (1979) of early diagenesis in suboxic hemi-pelagic sediments from the eastern equatorial Atlantic. The cores investigated typically had a light tan-coloured surface layer ~35 cm in thickness, with a low organic carbon content (0.2–0.5% organic carbon by weight), underlain by a dark olive green terrigenous sediment, which had a higher organic carbon content (~0.5 to >1%). The authors gave data on a number of constituents in the interstitial waters of the sediments, and their findings can be summarized as follows.

**1** Dissolved nitrate concentrations increased from those of the ambient bottom water to a maximum, then decreased linearly to approach zero at approximately the depth of the tan–olive green lithological transition.

**2** Dissolved $Mn^{2+}$ concentrations were very low at the surface but began to increase at a depth lying between the nitrate maximum and the depth at which nitrate falls to zero.

**3** Dissolved $Fe^{2+}$ concentrations were below the detection limit to a depth below that at which the nitrate concentration fell to zero, then began to increase.

**4** Sulfate concentrations never differed detectably from those of the ambient bottom water, that is,

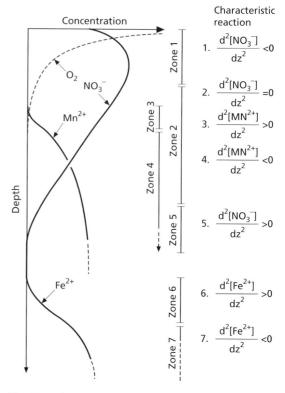

**Fig. 14.1** Schematic representation of diagenetic zones and trends in interstitial-water profiles during the suboxic diagenesis of marine sediments, z represents depth (from Froelich *et al.*, 1979). Reprinted with permission from Elsevier.

there was no indication that the sediments had entered the sulfate reduction zone.

Froelich *et al.* (1979) interpreted their data in terms of the general vertical-depth-zone diagenetic model, which is related to the sequential use of oxidants for the destruction of organic carbon, and they were able to identify a number of distinct zones in the sediments. The sequence is illustrated in Fig. 14.1, and the individual zones are described in the following.

*Zone 1.* This is the interval over which oxygen is being consumed during the destruction of organic matter (diagenetic reaction Equation 14.1). During this process the ammonia derived from the degradation of organic matter is oxidized to nitrate.

*Zones 2 and 5.* Below the nitrate maximum, nitrate diffuses downwards, the linearity of the gradient suggesting that nitrate is neither consumed nor produced in this zone. The downward diffusing nitrate is reduced by denitrification (diagenetic reaction Equation 14.2) at the depth of the nitrate zero in zone 5.

*Zones 3 and 4.* These overlap with zones 2 and 5. Zone 4 is the interval over which organic carbon is oxidized by manganese oxides (diagenetic reaction Equation 14.3) to release dissolved $Mn^{2+}$ into the interstitial waters. As discussed in Chapter 11, Mn is relatively soluble in the reduced $Mn^{2+}$ form but relatively insoluble as oxidized Mn(IV). This $Mn^{2+}$ then diffuses upwards to be oxidatively converted to solid $MnO_2$ at the top of the diffusion gradient in zone 3. This reduction–oxidation cycling results in the setting up of a 'sedimentary manganese trap' (see Section 14.4.3 and Worksheet 14.4).

*Zones 6 and 7.* Zone 7 is the region over which organic carbon oxidation takes place via the reduction of ferric oxides (diagenetic reaction 14.4). $Fe^{2+}$ is released into solution (again the reduced form of Fe is more soluble than the oxidized form see Chapter 11), and diffuses upwards to be consumed near the top of zone 7 and in zone 6.

This study therefore provided clear evidence of how the diagenetic sequence operates in hemi-pelagic sediments, and demonstrates that the oxidants are, in fact, consumed in the predicted sequence, that is, oxygen > nitrate $\cong$ manganese oxides > iron oxides > sulfate (although the sulfate reduction stage was not reached in these sediments).

Truly anoxic waters, where sediments are *initially* deposited under anoxic conditions, prevail over only a small area of the oceans (see Section 9.3 and 11.6). The vast majority of environments at the sea floor are therefore oxidizing, and there is usually a layer of oxic material at the sediment surface. While the organic matter is at, or very close, to the sediment surface it will usually have access to a very large supply of oxygen from within the water column. As sediments become buried, the degradation of the organic matter becomes isolated from water column oxygen due to the slow diffusion of water and gas in and out of sediments and oxic decomposition depends on the oxygen dissolved within the sediment interstitial water. As a result of the consumption of

dissolved oxygen in the interstitial waters the sediments can become reducing, and ultimately anoxic, at depth as the diagenetic sequence proceeds. The depth at which the oxic–anoxic change occurs depends largely on a combination of the magnitude of the down-column carbon flux, by which the carbon is *supplied*, and the sediment accumulation rate, by which it is *buried* since For example, according to Muller and Mangini (1980) a bulk sedimentation rate of $\leq 1$–$4\,cm\ 10^3\,yr^{-1}$ is necessary for the deposition of an oxygenated sedimentary column. Thus, the thickness of the surface sediment oxic layer will tend to increase from nearshore to pelagic regions as the accumulation rate decreases. The organic carbon content of sediments ranges from <0.25 to 20% organic C dry weight (Burdige, 2007). Seiter *et al.* (2004) compiled a massive data base of the organic carbon content of surface (top 5 cm) marine sediments. They estimated the average organic carbon content of deep sea (>4000 m) sediments to be 0.5% dry weight and that of shallower sediments to be about 1.5%. Thus the organic carbon content of shallower sediments is markedly higher than deep seas sediments. The organic carbon contents are highest under water of relatively high primary production such as the upwelling zones off Namibia, California and in the Arabian Sea. Despite their low organic carbon content, deep sea sediments cover a vast area and so represent an important component of the global carbon system. The depth of the oxic layer (which can now be measured using thin oxygen electrodes but which is often readily identified as a sharp change in sediment colour) in ocean sediments varies from less than 1 cm in shallow water sediments to greater than 20 cm in the deep ocean (Soetart *et al.* 1996)

An example of how the thickness of the oxic surface layer in deep-sea sediments varies has been provided by Lyle (1983) for a series of hemi-pelagic deposits from the eastern Pacific. The redox boundary, which is indicative of the change between oxidizing (oxic–positive redox potential) and reducing (anoxic–negative redox potential) conditions in sediments is often accompanied by a colour change, which generally is from red-brown (oxidized) to grey-green (reduced). Morford and Emerson (1999) created a model to estimate the depth of oxygen penetration for ocean sediments (those in waters deeper than 1 km) globally (Fig. 14.2).

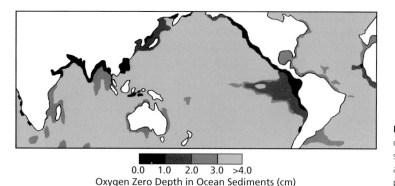

**Fig. 14.2** Variations in the thickness (in cm) of the surface oxic layer in sediments in ocean sediments (Morford and Emerson, 1999). Reprinted with permission from Elsevier.

0.0   1.0   2.0   3.0   >4.0
Oxygen Zero Depth in Ocean Sediments (cm)

This global map is consistent with the regional field data of Lyle (1983) and shows that shallow oxygen penetration in sediments is primarily a function of regions of high productivity associated with upwellings and areas of low deep water oxygen concentrations.

It is apparent, therefore, that there are a range of redox environments in marine sediments, which can be expressed, on the basis of the increasing thickness of the surface oxic layer, in the following sequence.

1 *Anoxic sediments.* These are usually found in coastal areas, or in isolated basins and deep-sea trenches. They have organic carbon contents in the range ~5–≥10%, and are reducing throughout the sediment column if the redox boundary is found in the overlying waters.

2 *Nearshore sediments.* These sediments, which usually have organic carbon contents of ≤5%, accumulate at a relatively fast rate and become anoxic at shallow depths so that the brown oxic layer is usually no more than a few centimetres in thickness.

3 *Hemi-pelagic sediments.* These have intermediate sedimentation rates, and organic carbon contents that are typically around 2%. The thickness of the oxic layer in these deposits ranges from a few centimetres up to around a metre.

4 *Pelagic sediments.* These are deposited at very slow rates and have organic carbon contents that usually are only ~0.5 and usually 0.1–0.2% (Burdige, 2007). In these sediments the oxic layer extends to depths well below 1 m, and often to several tens of metres.

It was pointed out above that early diagenesis in marine sediments follows a *vertical* zone sequence.

It is also apparent that there is a *lateral*, that is, nearshore → hemi-pelagic → pelagic, diagenetic zone sequence in the oceanic environment.

The concentration of organic matter in a sediment is a critical parameter in determining how far the diagenetic sequence progresses, and the factors controlling the distribution, protection and preservation of organic matter in marine sediments are discussed in the following sections.

## 14.2 Organic matter in marine sediments

### 14.2.1 Introduction

The organic matter in marine sediments is important, not only because the sediments provide a significant reservoir in the global carbon cycle, but also because organic matter drives early diagenesis and thus plays a major role in the chemistry of the oceans.

Burdige (2007) has reviewed the organic carbon burial in ocean sediments and his summary has been used to compile Table 14.1. The author notes these data are rather uncertain; firstly because of inadequate sampling of the ocean sediments (and the associated sedimentation rates) and also because of uncertainties in our knowledge of the efficiency of carbon burial, particularly for shelf sediments where porous sandy sediments are common. These sandy sediments have low carbon content and are not important carbon sinks. However, with their porous nature allowing water to flow through them and providing a surface for bacteria to grow on, they are probably sites of very efficient organic carbon recy-

**Table 14.1** Ocean Organic Carbon Burial $10^{12}$ g organic C $yr^{-1}$. Based on Burdige (2007).

| Depth Range (km) | % ocean area | Organic Carbon Burial (% of total in brackets) |
|---|---|---|
| 0–0.2 | 7 | 152 (50%) |
| 0.2–2 | 9 | 96 (31%) |
| >2 | 84 | 61 (20%) |

cling. Regardless of the uncertainties in the actual values two striking features are apparent from Table 14.1.

Firstly, considering the open ocean waters (>2 km), in Table 12.3 the flux of organic matter through 2000 m was estimated at $340 \times 10^{12}$ g C $yr^{-1}$, so less than 20% of the carbon passing this depth and reaching the sediments is finally buried, illustrating again the efficiency of carbon recycling throughout the ocean system. Indeed Burdige suggests that the burial efficiency in waters deeper than 4 km (63% of the ocean area) is <5%.

The second point from Table 14.1 is the dominance of shallow water systems, and in fact deltas in particular, in ocean carbon burial.

## 14.2.2 The sources and distribution of organic matter in marine sediments

The organic matter in marine sediments is derived from terrestrial, marine and anthropogenic sources.

In Chapter 6 the fluvial and atmospheric fluxes of organic carbon to the oceans were estimated at 216 and $180 \times 10^{12}$ g $yr^{-1}$ respectively, a sum that is only about 1% of the organic carbon produced within the oceans (see Chapter 9) and hence the overall carbon cycle in the oceans is dominated by internal production by photosynthesis. However, as we have seen only about 1% of ocean primary production reaches the sediments, the rest being degraded while sinking through the water column. By contrast, the external inputs to the oceans, particularly those arriving through the fluvial system, have already been extensively degraded during transport and include material rather more resistant to degradation. In particular, fluvial material includes lignin, a phenolic compound forming a major structural element in terrestrial

plants, including wood in trees. As a unique tracer of land plant material which is relatively resistant to degradation, lignin along with carbon isotope composition, has been used to follow the fate of terrestrial organic matter in the oceans which includes other plant material plus some man-made compounds and soot (black) carbon (Burdige, 2007). Most of the fluvial terrestrial organic matter is buried on the continental shelf, particular in delta sediments where it may form a major component of the organic matter and be preserved with a much higher burial efficiency (>10%) than seen for organic matter in open ocean sediments (Burdige, 2007).

The extent to which organic matter is preserved in a sediment is critical in determining how far the diagenetic sequence (Section 14.1) progresses. There are considerable variations in the organic matter content of marine sediments. On the basis of the information summarized above, however, it is apparent that two major marine sedimentary organic matter reservoirs can be identified, and it is important to distinguish between them because they have very different preservation characteristics.

*Nearshore sediments.* Fluvial inputs are delivered initially to nearshore regions, which are also the sites of much of the oceanic primary productivity. Nearshore deposits, which accumulate at relatively fast rates, usually contain ~1–5% organic carbon, but the concentrations can be considerably higher in sediments deposited in some anoxic basins and under areas of high primary production; for example, Calvert and Price (1970) reported that organic-rich diatomaceous muds on the Namibian shelf contained up to ~25% organic carbon.

*Deep-sea sediments.* Here the organic carbon input is dominated by marine carbon sources and very efficient recycling in the water column and in the sediments leads to very low organic average carbon contents (Seiter *et al.*, 2004) of ~0.5%.

Overall organic matter *burial efficiency*, that is, how much of the rain of organic matter to the sediments escapes early diagenesis varies with sediment accumulation rate from <1% in slow sedimentation open ocean regimes to >50% in fast sedimentation delta regimes (Fig. 14.3).

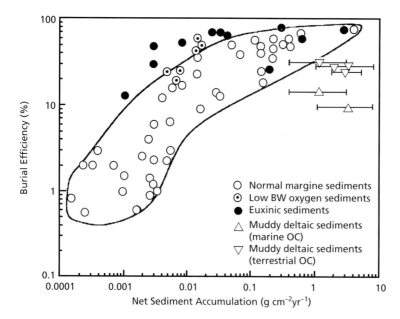

**Fig. 14.3** Burial efficiency of organic matter (defined as the fraction of organic matter arriving at the sediment which escapes early diagenesis and is buried) as a function of sediment accumulation rate. Note also that there is little relationship between the burial efficiency and the water column oxygen status (Burdige, 2007). The envelope shown here defines the commonly observed pattern in normal marine sediments of burial efficiency increasing with increasing sedimentation rate. Redrawn from Burdige 2007.

### 14.2.3 The classification of organic matter in marine sediments

The organic matter in marine sediments is composed of a wide range of organic molecules. These can be grouped into three main classes of compounds, amino acids (including proteins), carbohydrates and lipids plus lignin-like compounds from terrestrial plants as discussed above. However, much of the organic material in sediments cannot be characterized by current chemical techniques and this proportion increases from a few percent in fresh phytoplankton material to >60% in sedimentary material (Burdige, 2007; Lee *et al.*, 2004) an extension of the trend toward increasing proportion of uncharacterized material discussed in Section 12.1.3, Fig. 12.4.

The organic matter in marine sediments is ultimately converted to kerogen a term used for the poorly characterized material remaining after early diagenesis representing amorphous, high molecular weight and insoluble material (Burdige, 2007). This material continues to undergo chemical transformations as it is buried deeper in the sediments where, at depths >1 km, temperatures increase. Degradation processes continue and evolve throughout this process (Fig. 14.4) and the organic material enters

the geosphere possibly forming hydrocarbon reserves, or the residual organic material is uplifted with the sediments onto land to be degraded as the rock cycle continues. The full sequence is in the order: diagenesis (down to ~1000 m) → catagenesis (several kilometres depth, with an increase in temperature and pressure) → metagenesis or metamorphism (Tissot and Welte, 1984).

### 14.2.4 The overall behaviour of organic matter during diagenesis

Both the organic carbon content, its composition and its burial efficiency upon arriving at the sediments varies. Understanding the controls on these processes of carbon preservation is important for a wide range of issues on many times scales, from the long term global carbon and oxygen cycle to food supplies for benthic organisms. There has been a good deal of research on the controls on organic matter burial which has also shown that the chemical composition of organic matter changes during burial. This trend is an extension of that already described in the water column (Section 12.1.3) with the proportion of chemically uncharacterized organic matter increasing from <20% in surface water phytoplankton to >60%

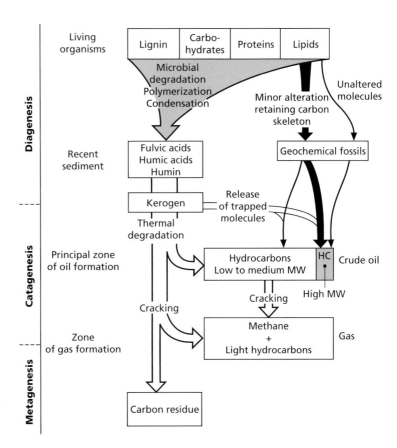

**Fig. 14.4** Stages in the diagenesis, catagenesis and metagenesis of organic matter in sediments (from Tissot and Welte, 1984).

in the sediments. As Lee *et al.* (2004) and Burdige (2007) note, this increase in the uncharacterized fraction does not necessarily imply the material is any less biodegradable. Burdige (2007) identifies three processes that play a role in determining burial efficiency and in this chemical evolution to an organic composition dominated by this uncharacterized fraction.

**1** Degradation of organic matter produces reactive products that can then react to form this complex uncharacterized material.

**2** The bacterial degradation may selectively remove the characterized fraction.

**3** The organic matter may be physically protected within, or on, inorganic/organic substrates. The binding of organic matter to surfaces such as clay particles may in itself protect organic matter from degradation, or offer physical protection by trapping material in pore spaces that are sufficiently small to prevent entry by bacteria and or enzymes (Hedges

and Keil, 1995; Lee *et al.*, 2004). This mechanism may be particularly important in protecting organic matter from degradation (see also Section 12.1.3), but reactions on the surfaces of the bound organic matter may change its chemical character.

These mechanisms are not mutually exclusive and all may occur together, or in sequence, during the breakdown of organic matter and the relative importance of each of them may vary between different environments or sediment types (Burdige, 2007).

The role of oxygen concentrations in influencing organic matter preservation has been extensively discussed. The very limited organic matter preservation in well oxygenated oceanic abyssal sediments and the organic rich sediments underlying an anoxic environment, such as the Black Sea, suggest an important role for oxygen concentration in regulating sediment organic carbon preservation. However, the data in Fig. 14.3 show that burial efficiency is not particularly closely related to oxygen concentrations.

However, the oxygen concentrations in bottom waters are closely related to the organic carbon content of sediments because the breakdown of that organic matter can consume oxygen (Middleburg and Levin, 2009). In most sedimentary environments, oxygen is available for the initial degradation of organic matter (Middleburg and Levin, 2009), and hence the organic matter degraded by alternative electron acceptors has already been modified from that degraded using oxygen (Middleburg and Levin, 2009). Burdige (2007) concludes from an extensive review of the literature that;

1 Fresh labile organic matter is degraded under oxic and anoxic conditions, although the rates may be different.

2 Refractory organic matter may be degraded more slowly, and/or less efficiently, under anoxic or low oxygen conditions. This may be because the breakdown of such refractory material requires oxygen (Middleburg and Levin, 2009).

Burdige identifies two other factors that may be potentially important in regulating organic matter burial. The first is the time that organic matter is exposed to oxygen, a factor that explains the very low organic carbon burial in deep ocean sediments and the effect of burial rate (Fig. 14.3). The second factor is that a temporally variable redox environment tends to promote organic matter degradation. Such temporal variability is most likely in shallow water systems and in sediments subject to extensive bioturbation.

The discussion to date has assumed a relatively simple conceptual model in which organic matter falls onto the sediment surface continuously and is steadily and slowly buried. This model is good approximation for most of the oceans most of the time. However, these system may not always be in steady state, and Burdidge (2007) notes that on glacial/interglacial timescales, the nature of continental shelves have changed fundamentally and with that change an important change in the ocean carbon cycle can be anticipated given the current dominance of the shelf seas in organic carbon sedimentation. Wilson *et al.* (1985) provide another example of non-steady state conditions in a study of the interstitial water chemistry of a mixed-layer pelagic–turbidite sediment in the north east Atlantic. This sediment (station 10554) consisted of a thin (≤10 cm) upper light brownish grey layer of pelagic material,

underlain by a turbidite sequence made up of an upper light brownish grey unit and a lower green unit. The sediment had an overall accumulation rate of ~10 cm $10^3$ yr$^{-1}$. The organic carbon content was ~0.6% in the surface layer and rose to ~1.6% around the top of the green layer. The interstitial water chemistry of the sediment at this site was considerably more complex than that of the exclusively pelagic deposit reflecting the adjustment of the system to the introduction and subsequent burial of the organic rich turbidite sequence. Wilson *et al.* (1985) suggested that the sequence could be interpreted in terms of two zones. Zone 1 is close to the sediment surface reflecting normal pelagic sedimentation in which almost all the organic matter is consumed by oxic processes. Below this zone there is a second zone of high diagenetic activity reflecting the presence of the organic rich turbidite sediments which are oxidized by oxidants diffusing from above. This creates a downward-moving oxidation front, the rate of movement of which is controlled by the rates of diffusion of oxygen and nitrate from the bottom water to the front itself. The progressive *downward* migration of the turbidite oxidation front is not a steady-state process; rather it implies that there is a continuing readjustment of the redox profile *after* the deposition of an organic-rich unit, which initially was at the original pelagic sediment surface.

## 14.3 Diagenesis: summary

1 Early diagenesis in marine sediments follows a general pattern in which a series of oxidants are utilized for the destruction of organic carbon in the following general sequence:

$$oxygen \rightarrow nitrate \cong manganese\ oxides$$
$$\rightarrow iron\ oxides \rightarrow sulfate.$$

The consumption of these oxidants represents an important component of their global biogeochemical cycles. Thus, sedimentary denitrification, particularly in shelf, slope and hemipelagic sediments, is responsible for about 60% of the loss of nitrogen from the oceans, with the remainder lost by water column denitrification in low oxygen regions (~30%, see Chapter 9) and burial of organic matter (~10%) (Brandes *et al.*, 2007). Estimated sulfide burial in sediments is equivalent to almost half of the input of sulfate to the oceans (Berner, 1982).

**2** The diagenetic sequence passes through each of the oxidant utilization stages successively in a down-column direction, thus setting up a *vertical* gradient.

**3** Most organic matter delivered to the oceans and produced within the oceans is degraded within the ocean water column or else on or within the sediments and is not ultimately buried. The degree to which organic matter suffers degradation in marine sediments depends on the rate at which it is degraded versus the rate at which is it buried. Organic carbon burial in the oceans occurs predominantly in deltas and on the continental shelf as illustrated in Fig. 14.5.

**4** The extent to which the diagenetic sequence itself proceeds depends largely on the rate of supply of organic matter to the sediment surface and the rate at which it is buried. Both of these decrease away from the continental margins towards the open-

ocean, thus setting up a *lateral* gradient in the diagenetic sequence. As a consequence of this, the diagenetic stage reached drops progressively from (i) nearshore sediments with an anoxic layer close to the sediment surface sometimes only mm from the surface), to (ii) nearshore sediments having a slightly thicker oxic layer, to (iii) hemi-pelagic deep-sea sediments having an oxic layer of a few centimetres followed by zones of nitrate, manganese oxide and iron oxide reduction, to (iv) truly pelagic deep-sea sediments, in which diagenesis does not progress beyond the oxic stage at which oxygen consumes virtually all the organic carbon brought to the sediment surface. In shelf and slope sediments organic matter can be protected at least to some extent by association with sediment surfaces.

During diagenesis elements are mobilized into solution and so can migrate through the interstitial waters. Some of these elements (together with those already present in the interstitial waters) are incorporated into newly formed or altered minerals.

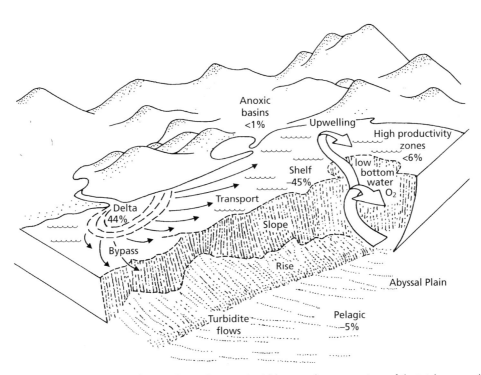

**Fig. 14.5** Idealized diagram depicting estimates of organic matter burial (expressed as a percentage of the total ocean sediment organic carbon burial), in a number of marine environments types (after Hedges and Keil, 1995). Reprinted with permission from Elsevier.

However, a fraction of the elements can escape capture in this way, and so be released into the overlying seawater. In the next section, therefore, the wider aspects of diagenetic mobilization will be considered in relation both to the depletion-enrichment of elements in interstitial waters and to the potential fluxes of the elements to the oceanic water column.

## 14.4 Interstitial water inputs to the oceans

### 14.4.1 Introduction

Early diagenesis takes place generally over the upper tens of centimetres of marine sediments involving the oxidative destruction of organic matter. Reactions continue as the sediment is buried deeper and may involve interactions with underlying basalts. However, rates of exchange of the sediment porewaters with the overlying seawater decrease with depth and hence the impacts on ocean water chemistry become less. These early diagentic processes can play an important role in the interstitial water chemistries and oceanic cycling of a number of components. These include the following.

1 The *bioactive*, or *labile*, elements such as C, N, P, together with Ca (calcium carbonate) and Si (opal). A characteristic of these elements is that only a relatively small fraction of their down-column rain rates are preserved in marine sediments, with most of them being recycled. The preservation of organic carbon was discussed earlier and the preservation of carbonate and silica are discussed in Chapter 15 since they profoundly affect the major solid components of ocean basin sediments.

2 The *oxidants* used to destroy organic matter are: oxygen, nitrate, Mn and Fe oxides, and sulfate (see Section 14.1.2). The consumption of nitrate and sulfate within sediments by these reactions is a major term in the global biogeochemical cycles of these ions.

3 *Trace metals.* Diagenetic processes involved in the oxidative destruction of organic matter are intimately related to the interstitial water chemistries of many trace metals, including those transported down the water column by organic carriers and those associated with the oxidants (e.g. Mn oxides). For metals in associations such as these, the diagenetic destruction of organic matter acts as a *recycling* term.

4 *Refractory elements.* Changes also occur during early diagenesis that are not related to the oxidative destruction of organic matter. These changes affect the non-bioactive, or refractory, elements (e.g. Na, K, Mg) and involve reactions with aluminosilicacte mineral phases which create biogeochemical sources for some elements and sinks for others.

The elemental composition of interstitial waters therefore is controlled by a number of interrelated factors, which include:

1 the nature of the original trapped fluid, usually seawater;

2 the nature of the water transport processes, that is, convection or diffusion;

3 reactions in the underlying basement, including both high- and low-temperature basalt–seawater interactions;

4 reactions in the sediment column;

5 reactions across the sediment–seawater interface.

As a result of these reactions, changes can be produced in the composition of the interstitial waters relative to the parent seawater, and diffusion gradients can be set up under which the components will migrate from high- to low-concentration regions. In contrast, under some conditions the compositions of the interstitial water and seawater will not differ significantly and concentration gradients will be absent.

The transport by diffusional processes can be described mathematically and this has allowed the development of some chemical models of sediment pore water processes that are much more sophisticated than can be used to describe water column processes. These are illustrated in the worksheets in this chapter.

The sampling of ocean sediment pore waters is usually done by the collection of sediment cores which can be divided into section of a cm in length or less from which pore water can be extracted by filtration or centrifugation. Some of the key chemical reactions are very sensitive to temperature and pressure and others to the presence or absence of oxygen, so care is required during sampling of pore waters to minimize the effect of changes due to sampling. For a small number of sediment components it is now possible to measure the pore water chemistry in situ using for example electrodes which can be automatically lowered into the sediment to increasing depths (Martin and Sayles, 2003).

### 14.4.2 Major elements

Diagenesis associated with the destruction of organic matter, involving C, N and P, has been described in Section 14.1, and with respect to the *major* interstitial water constituents, attention in the present section will be confined largely to the non-biogenic elements in porewaters. Interest in these components arises from trying to understand the role of chemical processes in sediment burial in creating large scale removal routes for the major ions $Na^+$, $K^+$ and $Mg^{2+}$ from the oceans. This process is often referred to as reverse weathering and involves the formation within ocean sediments of metal-rich clay material from degraded clays and silica formed during weathering (Mackenzie and Kump, 1995). Such processes have now been detected in at least coastal delta sediments and may occur more widely (Michalopoulis and Aller, 1995). Reverse weathering was proposed to explain imbalances in global major ion budgets. However, the discovery of the large scale hydrothermal processes at mid ocean ridges and the role of these systems as sources and sinks for many major ions, suggests that a major role for reverse weathering in major ion budgets may not be required, although this issue is far from settled (Mackenzie and Kump, 1995).

Analyses of interstitial waters were first carried out more than 100 years ago (see e.g. Murray and Irvine, 1895). Until recently, however, data for the chemistry of interstitial water have suffered from a number of major uncertainties. According to Sayles (1979) these arose from:

1 sampling procedures, such as temperature-induced artefacts inherent in the water extraction techniques;

2 imprecise analytical techniques (especially for trace elements); and

3 a lack of detail close to the sediment–seawater interface, a region where a number of important reactions take place.

Because of factors such as these, much of the early interstitial-water data must be regarded as being unreliable. In order to rectify some of these uncertainties, Sayles (1979) carried out a study of the composition of interstitial waters collected using in situ techniques from the upper 1–2 m of a series of sediments from the North and South Atlantic on a marginal–central ocean transect. A number of trends could be identified from the data obtained on this transect.

1 The interstitial waters were almost always enriched in $Na^+$, $Ca^{2+}$ and $HCO_3^-$, and *depleted* in $K^+$ and $Mg^{2+}$, relative to seawater. In addition, $SO_4^{2-}$ was slightly enriched at most stations, probably reflecting the role of sulfate as an alternative electron acceptor in organic matter oxidation.

2 The extent of these depletions or enrichments varied from element to element. For example, the enrichments in $Na^+$ were relatively small, and although the pore water concentration increased with depth the gradient was only gradual. In contrast, the other major cations had interstitial-water distribution profiles that were characterized by sharp gradients in the upper 15–30 cm of the sediments with only a limited change at greater depths.

3 The concentrations of $Mg^{2+}$, $K^+$, $Ca^{2+}$ and $HCO_3^-$ in the interstitial waters all exhibited a pronounced geographical variability, with the highest concentrations being found in waters from the marginal sediments and the lowest in those from the central ocean areas.

It may be concluded therefore that, relative to seawater, the interstitial fluids of the upper 1–2 m of oceanic sediments are generally enriched in calcium, sodium and bicarbonate and are depleted in potassium and magnesium. Some of the processes causing these interstitial-water enrichments and depletions are discussed below, and a number of the basic concepts relating to the behaviour of chemical species in interstitial water are discussed in Worksheet 14.2.

The depletion of $K^+$ is consistent with its incorporation into clay minerals as proposed in the reverse weathering concept (Mackenzie and Kump, 1995). The formation of interstitial-water gradients can be illustrated with respect to calcium and magnesium. Concentration gradients, showing increases in calcium and decreases in magnesium with depth, have been reported in the interstitial waters of many deep-sea sediments. The theories advanced to explain the existence of these gradients include:

1 the formation of dolomite or high-magnesium calcite during the dissolution and recrystallization of shell carbonates, which would account for the interstitial-water gains in calcium and losses in magnesium;

**Worksheet 14.2: Some basic concepts relating to the behaviour of chemical species in the sediment–interstitial-water complex**

Interstitial waters are the medium through which elements migrate during diagenetic reactions. In sediments the interstitial-water properties change very much more rapidly in the vertical than in the horizontal direction, with the result that the changes can often be described by one-dimensional models. The transport of solutes through interstitial waters takes place by convection, advection and diffusion. In the context used here advection refers to transport by the physical movement of the water phase, and diffusion refers to migration of a chemical species through the water as a result of a gradient in its concentration (or chemical potential). Diffusion in an aqueous solution can be described mathematically by Fick's laws, which, for one dimension, may be written as follows (see e.g. Berner, 1980):

1 First law

$$J_i = -D_i \frac{\partial c_i}{\partial x} \qquad \qquad \text{(WS14.2.1)}$$

2 Second law

$$\frac{\partial c_i}{\partial t} = D_i \frac{\partial^2 c_i}{\partial x^2} \qquad \qquad \text{(WS14.2.2)}$$

Here $J_i$ is the diffusion flux of component $i$ in mass per unit area per unit time, $c_i$ is the concentration of component $i$ in mass per unit volume, $D_i$ is the diffusion coefficient of $i$ in area per unit time, and $x$ is the direction of maximum concentration gradient; the minus sign in the first law indicates that the flux is in the opposite direction to the concentration gradient. Fick's first law is applied to calculations that involve steady-state systems, and the second law is applied to non-steady-state systems. Before Fick's laws can be applied directly to sediments, however, it is necessary to take account of the nature of the sediment–interstitial-water complex. Interstitial waters are dispersed throughout a sediment and the rate of diffusion of a solute through them is less than that in water alone, that is, as predicted by Fick's laws, because of the solids present (the porosity effect) and because the diffusion path has to move around the grains; the term *tortuosity* ($\theta$) is used to describe the ratio of the length of the sinuous diffusion path to its straight-line distance (Berner, 1980). Tortuosity is usually determined indirectly from measurements of electrical resistivity of sediments and of the interstitial waters separated from them, using the relationship

$$\theta^2 = \phi F \qquad \qquad \text{(WS14.2.3)}$$

where $\phi$ is the porosity and $F$ is the *formation factor* ($F = R/R_0$, where $R$ is the electrical resistivity of the sediment and $R_0$ is the resistivity of the interstitial water alone). Formation factors in marine sediments usually appear to lie in the range around 1 to around 10, so that the *effective diffusion coefficient* ($D'$) in sediment will be less than in the solution alone by a factor of up to around 10. The effective diffusion coefficient can be calculated from the relationship

$$D' = \frac{D_\phi}{\theta^2} \qquad \qquad \text{(WS14.2.4)}$$

*continued*

where $D$ is the diffusion coefficient in solution, $\phi$ is the porosity and $\theta$ is the tortuosity.

A detailed treatment of how to apply Fick's laws directly to sediments is given in Berner (1980), and for a comprehensive mathematical treatment of migrational processes and chemical reactions in interstitial waters the reader is referred to the 'benchmark' publication by Lerman (1977).

To illustrate this approach with an example, Gieskes (1983) has pointed out that if it is assumed that only vertical transport through interstitial waters is important, then the flux of a chemical constituent can be described by the equation

$$J_b = -pD_b \frac{\partial c}{\partial z} + puc \qquad \text{(WS14.2.5)}$$

where $J_b$ is the mass flux, $p$ is the porosity, $z$ is the depth coordinate in centimetres (positive downwards), $u$ is the interstitial-water velocity relative to the sediment–water interface in $cm_b\,s^{-1}$ (i.e. the advection rate), $c$ is the mass concentration in $mol\,cm_p^{-3}$ and $D_b$ is the diffusion coefficient in the bulk sediment (the subscript b indicates that concentrations and distances are measured over the bulk sediment (i.e. solids and interstitial waters) and the subscript p indicates the interstitial water phase only).

The mass balance of the solute is given by

$$\frac{\partial pc}{\partial t} = \frac{\partial}{\partial z}(J_b) + R \qquad \text{(WS14.2.6)}$$

where $R$ is a chemical source–sink term, that is, the reaction rate ($mol\,cm_b^{-3}s^{-1}$).

If the interstitial-water density and the solid density do not change in a given depth horizon, then

$$\frac{\partial p}{\partial t} = \frac{\partial pu}{\partial z} \qquad \text{(WS14.2.7)}$$

and Equation WS14.2.6 becomes

$$p\frac{\partial c}{\partial t} = \frac{\partial}{\partial z}\left(pD_b\frac{\partial c}{\partial z}\right) - pu\frac{\partial c}{\partial z} + R \qquad \text{(WS14.2.8)}$$

and when steady state exists, this becomes

$$0 = \frac{\partial}{\partial z}\left(pD_b\frac{\partial c}{\partial z}\right) - pu\frac{\partial c}{\partial z} + R. \qquad \text{(WS14.2.9)}$$

According to Gieskes (1983) if conditions (e.g. sedimentation rates, temperature gradients) have been stable during relatively recent times (the last 10–12 Ma), then the steady-state assumption is valid for pelagic sediments, which have accumulation rates of ~20 m Ma$^{-1}$. The author then considered how a concentration–depth gradient, such as that illustrated in Fig. WS14.2.1, could be explained.

Gieskes (1983) considered the factors that might control the concentration–depth relationship in the dissolved Ca profile illustrated in Fig. WS14.2.1 and related them to changes in three variables. These variables were diffusion ($D_b$), reaction rate ($R$) and advection ($u$); that

*continued on p. 308*

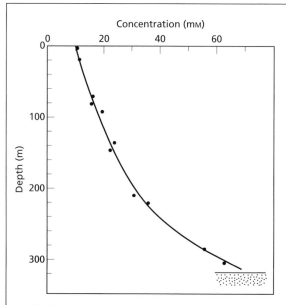

**Fig. WS14.2.1** Concentration–depth profile of dissolved Ca in interstitial waters of a DSDP core (from Gieskes, 1983). Reprinted with permission from Elsevier.

is, the profiles were interpreted within a *diffusion–advection–reaction* framework. To illustrate this approach, three cases were considered.

*Case 1*, in which the rate of diffusion varies; that is, $R = 0$, $D_b = f(z)$ and $u = 0$. Under these conditions, there is no reaction and no significant contribution from advection. Thus, only a gradual decrease in $D_b$ with depth could then explain the increased curvature with depth in the otherwise conservative profile.

*Case 2*, in which the reaction rate varies; that is, $R \neq 0$, $D_b$ is constant and $u = 0$. Thus, the profile implies a removal of calcium from solution, notwithstanding the significant source term for dissolved Ca at the lower boundary.

*Case 3*, in which the advection rate is not zero; that is, $R = 0$, $D_b$ is constant and $u \neq 0$. Thus, under these conditions of no reaction and constant diffusion coefficient, the curvature in the profile would be caused by the relatively large advective term.

Gieskes (1983) then considered two types of Ca–Mg interstitial-water concentration–depth profiles. In the first type, there are linear correlations between $\Delta$Ca and $\Delta$Mg, that is, $R = 0$. Using data that included information on porosity, and diffusion coefficients (evaluated from a knowledge of formation factors; see above), a solution of Equation WS14.2.9 assuming $R = 0$ indicated that the depth profiles of Ca and Mg could be explained in terms of conservative behaviour (i.e. transport through the interstitial water column alone and no reaction), with the boundary conditions being fixed by concentrations in the underlying basalts and the overlying seawater. In the second type, there are non-linear correlations between $\Delta$Ca and $\Delta$Mg, which implies reaction in the sediment column; that is, $R \neq 0$. Under these conditions, derivatives of Equation WS14.2.9 must be evaluated geometrically from the concentration–depth profiles in order to model the data. These two types of Ca–Mg profiles are described in the text.

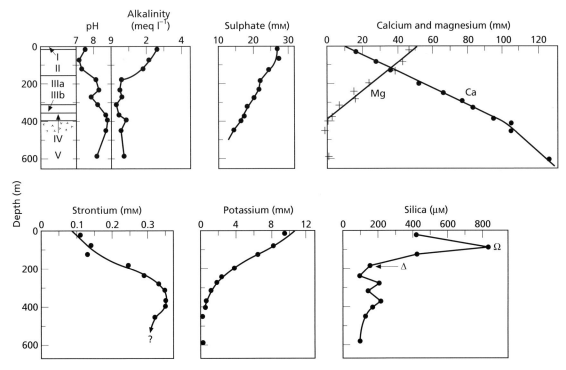

**Fig. 14.6** Interstitial water profiles at DSDP sites. Concentration–depth profiles at DSDP site 446 (24°42′N, 132°47′E). Lithology: I, brown terrigenous mud and clay; II, pelagic clay and ash, siliceous; IIIa, mudstone, clay stone, siltstone, sandstone; IIIb, calcareous clay and mudstones, turbidites; IV, calcareous claystones, glauconite, mudstones; V, basalt sills and intrusions. (From Gieskes 1983.) Reprinted with permission from Elsevier.

**2** the adsorption of magnesium on to opal phases, which would result in a decrease in magnesium in the interstitial waters of siliceous sediments.

However, in addition to changes in magnesium and calcium there is often a decrease in $\delta^{18}O$ values in the interstitial waters, and according to Lawrence *et al.* (1975) this cannot be explained by reactions involving either carbonate or opaline diagenesis. Instead, these authors proposed that the changes in $\delta^{18}O$ values result from reactions taking place during the alteration either of the basalts of the basement or of volcanic material dispersed throughout the sediment column. The problem was addressed by Gieskes (1983), who identified two types of calcium–magnesium profiles in the interstitial waters from 'long-core' oceanic sediments and linked them to reactions involving interstitial waters and volcanic rocks, either in the basalt basement or in the sediment column itself, or in both.

*Reactions involving basement rocks.* Reactions continue deep into the sediments and these can be sampled from sediments collected from the ocean drilling project. Such data is illustrated in Fig. 14.6. The profiles in Fig. 14.6 are interpreted as showing evidence of a range of geochemical processes.
**1** the maximum in the strontium concentrations probably results from the recrystallization of carbonates in the nanofossil ooze, during which new mineral phases are formed that have lower strontium concentrations than the parent material;
**2** potassium concentrations decrease with depth, probably as a result of the uptake of the element during the formation of potassium-rich clays;
**3** the sulfate profile shows a decrease with depth, indicating that there has been sulfate reduction within the sediment column;
**4** the calcium and magnesium profiles are interpreted as showing a sink for Mg at depth by reaction

with basalt and an associated source of calcium, although in other profiles detailed modelling suggests additional reactions of these ions occurring within the sediments as well as the deeper basalt interface.

### 14.4.3 Trace elements

The fate of a component following deposition in sediments is constrained by its post-depositional mobility, since in order to be added to the interstitial water from solid sediment phases it must first be solubilized to the dissolved state. For many trace metals this solubilization is intimately related to the oxidative destruction of organic matter during early diagenesis. The trace metals that are released in this process to become concentrated in interstitial waters relative to overlying seawater can follow one of two general pathways:

**1** they can be released (i.e. recycled) back into seawater via upward migration across the sediment–water interface;

**2** they can be reincorporated into sediment components following upward or downward migration through the interstitial waters.

Thus, it is necessary to introduce the concept of a *net* interstitial water flux, which in the present context is the flux that escapes reaccumulation and

is added directly to seawater. Careful sampling to avoid contamination is required for the analysis of trace metals in porewaters as with the water column (Chapter 11). The results of one study of trace metals in sediment pore waters is Fig. 14.7 and the distribution patterns seen here have been seen in other studies. Note that all the four metals considered are enriched in pore waters to some extent relative to seawater. The solubility of manganese is much greater in the Mn(II) form than Mn(IV) and the increase in Mn in porewaters below 15 cm in Fig. 14.7 reflects the onset of the use of Mn(IV) as an alternative electron acceptor in organic matter oxidation. The similar behaviour of Ni suggests that it is associated with manganese oxide phases in sediments and mobilized along with Mn. Cadmium appears to be unaffected by these processes so is presumably not subject to a net flux to the interstitial waters through organic matter breakdown in the upper part of the sediment core or during manganese reduction. Copper by contrast is strongly enriched in surface sediment porewaters probably through organic matter degradation and this generates a flux of copper into the overlying ocean waters (Boyle *et al.*; 1977, see also Chapter 11). The modelling of such sediment pore water profiles is considered in Worksheet 14.3.

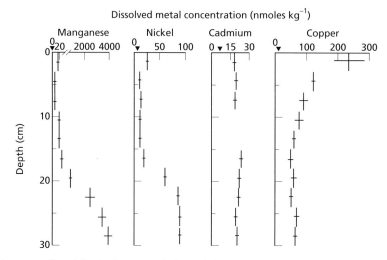

**Fig. 14.7** Interstitial-water profiles of dissolved trace metals (from Klinkhammer *et al.*, 1982) from a sediment core that was in a suboxic condition below an oxic surface layer. Arrowheads indicate the bottom water concentrations. Reprinted with permission from Elsevier.

**Worksheet 14.3: Models for the regeneration of trace metals in oxic sediments**

Klinkhammer *et al.* (1982) reported data on the concentrations of Ni, Cd and Cu in bottom waters and oxic sediment interstitial waters at MANOP site S in the central equatorial Pacific, and used these to set up diagenetic models. A characteristic feature of the metal interstitial-water profiles at the oxic site (S) is that the steepest concentration gradient is found across the sediment–seawater interface, which implies that most metal regeneration in these oxic sediments takes place across this interface. The concentration–depth profiles are maintained by a combination of the regeneration and the interaction between the dissolved metals and the sediment below the interface. Nickel and Cd exhibit little tendency to react with sediment components under these conditions, with the result that their interstitial water profiles are monotonic at site S and are generated by the burial of surficial pore water. In contrast, Cu is readily taken up by sediment components, which leads to an exponential decrease in interstitial-water dissolved Cu concentrations with depth: see Fig. WS14.3.1.

*Ni and Cd regeneration*
Under conditions of oxic diagenesis the degradation of organic matter in surface sediments utilizes dissolved oxygen, and Klinkhammer *et al.* (1982) assumed that in the simplest case early oxic diagenesis on the sea floor is analogous to oxidation in the overlying water column, that is, a 'continuum model'. Nickel and Cd are nutrient-type elements in seawater so that under oxic conditions it should be possible to predict the Ni and Cd concentrations in the surficial interstitial waters from the interstitial nutrient concentrations. To set up a model for

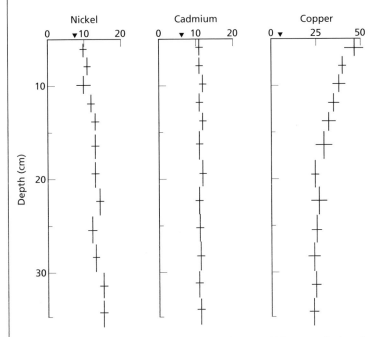

**Fig. WS14.3.1** Dissolved metal concentrations in the interstitial waters of oxic sediments at MANOP site S (nmol l⁻¹). Arrow heads indicate bottom-water concentrations.

*continued on p. 312*

this, Klinkhammer et al. (1982) assumed that both the metals and the nutrients are lost from the sediment–seawater interface by diffusion only, so that the flux of metal $M$ across the interface is related to the corresponding flux of a nutrient $N$ by a proportionality constant $M/N$. Thus

$$D_M \left( \frac{dM}{dz} \right)_{z=0} = \frac{M}{N} D_N \left( \frac{dN}{dz} \right)_{z=0}. \tag{WS14.3.1}$$

By assuming a linear concentration gradient across the interface, Equation (WS14.3.1) reduces to the relationship

$$\frac{M}{N} = \frac{(M_0 - M_{BW})D_M}{(N_0 - N_{BW})D_N}. \tag{WS14.3.2}$$

In the water column Ni mimics silica and Cd is related to nitrate (see Chapter 11), and using data from the literature the appropriate diffusion coefficient ratios ($D$) were calculated to be $D_{Cd}/D_{NO3} = 0.36$ and $D_{Ni}/D_{Si} = 0.68$. Variables $M_0$ and $M_{BW}$ are metal concentrations in the top interstitial-water interval and bottom seawater, and $N_0$ and $N_{BW}$ are the corresponding nutrient concentrations. A comparison of the results obtained from Equation WS14.3.2 with the ratios found in seawater is a test of the model. Klinkhammer et al. (1982) made such a comparison and the data from site S are reproduced in Table WS14.3.1, and strongly support the 'continuum model'.

*Cu regeneration*

Copper is readily taken up by sediment components, which leads to an exponential decrease in interstitial-water dissolved Cu concentrations with depth. In addition, dissolved Cu is released into seawater at the interface. The amount of Cu released in this way, however, is considerably greater than that which would be predicted from the decomposition of organic material consisting of plankton. Thus, the model developed for Ni and Cd is inappropriate for Cu. Klinkhammer et al. (1982) therefore assumed that the interstitial-water dissolved Cu profiles are sustained by scavenging from the water column combined with vigorous recycling at the interface and uptake into the sediment. The authors then attempted to model the interstitial-water dissolved Cu profile in the following manner.

The shape of the interstitial-water dissolved Cu profile at site S (see Fig. WS14.3.1) shows a concentration gradient that indicates diffusion upwards (into seawater) and downwards (into the sediment) from the interface, and the negative curve suggests an uptake by the sediment at depth. The authors concluded that the simplest model consistent with this type of profile

**Table WS14.3.1** Metal: nutrient ratios (Ni : Si and Cd : NO₃) at site S calculated from Equation WS14.3.2 compared with those observed in general sea water and bottom water at the site (from Klinkhammer et al., 1982).

| Element | $(M{:}N)_{\text{site S}} \times 10^5$ | $(M{:}N)_{\text{SW}} \times 10^5$ | $(M{:}N)_{\text{BW}} \times 10^5$ |
|---------|------------------|------------------|------------------|
| Ni | 3.2 | 3.3 | 5.3 |
| Cd | 3.3 | 2.3 | 1.8 |

*continued*

was a steady-state approach in which the Cu is controlled by diffusion and first-order removal, which can be represented by the Equation:

$$D\frac{\partial^2 C}{\partial z^2} - kC = 0. \tag{WS14.3.3}$$

The solution of Equation WS14.3.3 for a one-layer sediment is given by

$$C = \frac{C_1 \sinh[(h-z)] + C_2 \sinh(Rz)}{\sinh(Rh)} \tag{WS14.3.4}$$

where $R = (k/D)^{1/2}$, $C$ is the concentration predicted at depth $z$, $C_1$ and $C_2$ are the concentrations at the upper $(z = 0)$ and lower $(z = h)$ boundaries, $D$ is the diffusion coefficient corrected for porosity, tortuosity and temperature $(D = 1.4 \times 10^{-6}\,\text{cm}\,\text{s}^{-1})$, and $k$ is the reaction constant in $\text{s}^{-1}$. Values for $k$ were calculated using a best-fit approach. The authors modelled their data to a depth of ~15 cm in the sediment, below which there is little variation. The results for site S are illustrated in Fig. WS14.3.2. It is apparent from Fig. WS14.3.2 that there is very good agreement between the predicted and observed dissolved Cu interstitial-water profiles, that is, the profiles are adequately explained by the remobilization of Cu at the interface and its removal into the sediment below.

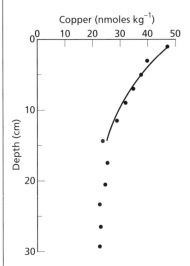

**Fig. WS14.3.2** Dissolved Cu interstitial-water profiles in the oxic sediments at site S. Filled circles are the measured concentrations in the interstitial waters. The full curve indicates the curve derived from Equation WS14.3.4 using the following parameters: $C_1 = 47\,\text{nmol}\,\text{kg}^{-1}$, $C_2 = 26\,\text{nmol}\,\text{kg}^{-1}$, $h = 13\,\text{cm}$ and $k = 0.17\,\text{yr}^{-1}$.

As is clearly evident from the Mn profile in Fig. 14.7, the redox environment can profoundly affect the cycling of trace metals in porewaters and these affects depend on the chemical characteristics of each element. Morford and Emerson (1999) suggest for instance that Mo and V are released from sediment solid phases along with Mn, whereas overall Re, Cd and U flux into the sediments because either they are less soluble in the reduced form (Re and U exist as anions in oxygenated seawater see Chapter 11) or because they form very insoluble sulfides.

Mobilization under reducing conditions within sediment porewaters will only result in release to the water column if the onset of reducing conditions is at or near the sediment surface.

*Manganese.* The sedimentary geochemistry of manganese has attracted considerable interest because of its role within the redox cycle and the impact of this cycle on the solubility of Mn in pore waters. The reactions involved in the diagenesis of Mn can be related to interstitial-water-phase–solid-phase changes, and can be expressed in a simple form as:

$$Mn^{2+}(soluble) \xleftrightarrow[oxidation]{reduction} MnO_x(solid\ hydrous\ oxide) \tag{14.7}$$

where $x$ generally is less than 2. This reaction governs the diagenetic mobility of Mn in sediments, and the general conditions that control both the solid-phase and dissolved Mn in the sediments. The mobilization of Mn at depth in the sediments results in increases in pore water concentrations and upward diffusion of Mn toward the oxic layer where it can reprecipitate resulting in a maximum in solid phase Mn within the sediments. This process has been described by a number of models. One of which is described in Worksheet 14.4.

## 14.5 Interstitial water inputs to the oceans: summary

1 The interstitial waters of most oceanic sediments originate as trapped seawater.
2 The external boundary conditions for the interstitial-water–sediment complex are set by the basalt basement below and the water column above.
3 Reactions take place in the sediment–interstitial-water sandwich, during which changes are produced in the composition of the interstitial waters relative to the parent seawater, and diffusion gradients are set up via which elements migrate from high- to low-concentration regions.
4 The most intense reactions usually take place in the upper ~30–50 cm of the sediment column. These reactions are in response to the diagenetic environment, the intensity of the reactions generally decreasing in the order highly reducing shelf sediments > hemi-pelagic sediments > pelagic sediments reflecting the supply of organic matter.
5 Data are now available on the diffusive interstitial water fluxes of a number of major and trace components. For some of these components interstitial waters act as sinks for their removal from seawater; for others they act as sources for their addition to seawater.
6 When considering interstitial water sources it is extremely important to make a distinction between primary and secondary fluxes. Interactions that take place between the basalt basement and interstitial waters, and between some types of dispersed volcanic material and interstitial waters, can lead to dissolved components being introduced into seawater for the first time. These are the primary fluxes in the same way as hydrothermal fluxes are primary in nature. In contrast, a large fraction of the trace metals that are mobilized into interstitial waters from particulate material via the diagenetic sequence has already been taken out of solution in seawater in the biogeochemical cycles that operate in the water column, and so is simply being returned to the dissolved state. These are the recycled or secondary fluxes. The dissolved trace metals associated with these fluxes can sometimes impose 'fingerprints' on the water column if they escape recapture at the sediment surface. However, the important point is that they have been recycled *within* the ocean system, and as such do not represent an additional primary external source. Despite this, the diagenetic remobilization and release of elements that are initially removed from solution by incorporation into estuarine and coastal sediments can sometimes reverse the processes that occur in estuaries, and Bruland (1983) has suggested that for these elements such a diagenetic release can be regarded as an 'additional dissolved river input' to the oceans.

The diagenetic reactions that take place both within the sediment–interstitial-water complex and

**Worksheet 14.4: The Burdige-Gieskes model for the steady-state diagenesis of Mn in marine sediments**

Burdige and Gieskes (1983) outlined a steady-state pore-water–solid-phase diagenetic model for Mn in marine sediments. The model has been described qualitatively in the text, and also can be used to illustrate how diagenetic equations are used. The Burdige-Gieskes steady-state diagenetic model, which involves a series of Mn reaction zones, is illustrated in Fig. WS14.4.1.

The derivation of the diagenetic equations used to translate this into a quantitative mathematical model is summarized in the following.

Burdige and Gieskes (1983) presented a mathematical treatment of their Mn 'zonation model' using the steady-state diagenetic equations given by Berner (1980):

$$D_b \frac{\partial^2 C_p}{\partial z^2} - w \frac{\partial C_p}{\partial z} + R(z) = 0 \qquad \text{(WS14.4.1)}$$

$$-w \frac{\partial C_s}{\partial z} - \frac{\phi}{1-\phi} R(z) = 0 \qquad \text{(WS14.4.2)}$$

where $D_b$ is the bulk sediment diffusion coefficient (identical with Berner's $D_s$ term), $C_p$ is the concentration of Mn in the pore waters, $C_s$ is the concentration of Mn in the solid phase, $w$ is the sedimentation rate, $\phi$ is the porosity and $R(z)$ is a rate expression for either oxidation or reduction. The term $\phi/(1 \times \phi)$, which has units cm$^3$ porewater / cm$^3$ dry sed, is used to convert Mn concentrations between the solid and liquid phases.

With respect to the Mn model the following assumptions are implicit in the diagenetic equations.

1 Steady-state diagenesis operates.
2 Vertical gradients are much more important than horizontal gradients, to the extent that the latter can be ignored.

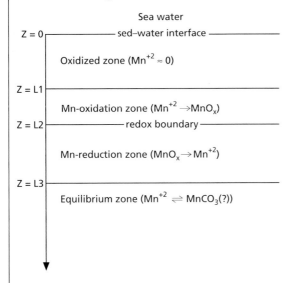

**Fig. WS14.4.1** Manganese steady-state diagenetic zonation model (from Burdige and Gieskes, 1983).

*continued on p. 316*

**3** Diffusion in pore waters occurs via molecular processes, that is, the diffusion follows Fick's laws.
**4** Porosity and diffusion coefficients are constant with depth.
**5** Advection is constant and equal to the sedimentation rate.
**6** Solid-phase diffusion can be neglected.
**7** The supply of solid Mn to the sediment surface is constant with time.
**8** Adsorption of $Mn^{2+}$ can be neglected.
**9** Bioturbation can be neglected because in most oceanic sediments it will be confined to the well oxygenated layer of sediments above the Mn reduction zone.
The authors also drew attention to two other factors, which are detailed in the following.

Variable $D_b$ differs from the free-ion diffusion in seawater ($D$) because of tortuosity effects resulting from the presence of sediment particles, and $D_b$ can be related to $D$ by the following equation:

$$D_b = D/\phi F \tag{WS14.4.3}$$

where $F$ is a 'formation factor', which is measured as the ratio of the bulk sediment resistivity to the pore-water resistivity (see Worksheet 14.3).

In Equations WS14.4.2 and WS14.4.3, $R(x)$ is the rate expression for either oxidation or reduction. In the presence of an abundant surface area (such as that of a sediment), Mn precipitation–oxidation is a pseudo-first-order process, so that

$$R_{ox}(z) = k_{ox}C_p \tag{WS14.4.4}$$

and Equation WS14.4.5 therefore should be an appropriate rate expression for oxidation under these conditions, in which the reaction product is assumed to be a hydrous oxide ($MnO_x$: see text). The authors assumed that whatever the mechanism involved, the rate of Mn reduction will be proportional to the amount of solid Mn available (which is presumed to be all hydrous oxide). Thus, $R_{red}(z)$ can be expressed as

$$R_{red}(z) = k_{red}C_s. \tag{WS14.4.5}$$

Combining the rate expressions with Equations WS14.4.1 and WS14.4.2, the following set of equations was obtained:
**1** For the oxidizing zone ($L_1 \leq z \leq L_2$)

$$D_b \frac{\partial^2 C_p^{ox}}{\partial z^2} - w \frac{\partial C_p^{ox}}{\partial z} - k_{ox}C_p^{ox} = 0 \tag{WS14.4.6}$$

$$-w \frac{\partial C_s^{ox}}{\partial z} + \frac{\phi}{1-\phi} k_{ox}C_p^{ox} = 0 \tag{WS14.4.7}$$

**2** For the reducing zone ($L_2 \leq z \leq L_3$)

$$D_b \frac{\partial^2 C_p^{red}}{\partial z^2} - w \frac{\partial C_p^{red}}{\partial z} + \frac{1-\phi}{\phi} k_{red}C_s^{red} = 0 \tag{WS14.4.8}$$

$$-w \frac{\partial C_s^{red}}{\partial z} - k_{red}C_s^{red} = 0 \tag{WS14.4.9}$$

By defining a series of boundary conditions the solutions to Equations WS14.4.6– WS14.4.9 were given as

*continued*

$$C_p^{ox} = A \sinh[\alpha(z - L_1)] \tag{WS14.4.10}$$

$$C_p^{red} = G - \frac{1-\phi}{\phi} \frac{Ew^2}{k_{red}D_b} \exp[-\beta(z - L_2)] \tag{WS14.4.11}$$

$$C_s^{ox} = C_s^\circ \cosh[\alpha(z - L_1)] \tag{WS14.4.12}$$

$$C_s^{red} = E \exp[-\beta(z - L_2)] \tag{WS14.4.13}$$

where

$$\alpha = (k_{ox}/D_b)^{1/2} \tag{WS14.4.14}$$

(assuming $4D_b k_{ox} \gg w^2$, i.e. advection in pore waters is negligible compared with diffusion) and

$$\beta = k_{red}/w \tag{WS14.4.15}$$

$$L_{ox} = L_2 - L_1 \tag{WS14.4.16}$$

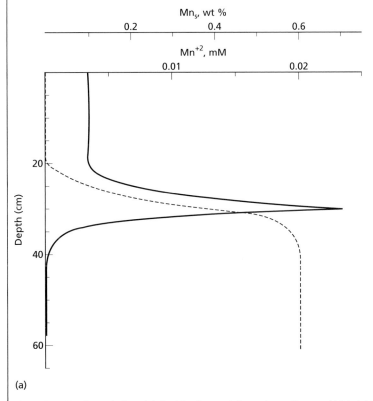

(a)

**Fig. WS14.4.2** Theoretical models for Mn diagenesis in marine sediments. (a) Model for porewater (broken curve) and solid-phase (full curve) Mn profiles (from Burdige and Gieskes, 1983). Profiles predicted from Equations WS14.4.10 to WS14.4.19 using the following parameters: $D_b = 71.6\,cm^2\,yr^{-1}$, $w = 3\,cm\,10^3\,yr^{-1}$, $\phi = 0.8$, $\rho = 2.6\,cm^{-3}$ of sediment, $L_1 = 20\,cm$, $L_2 = 30\,cm$, $C_s^\circ = 0.1$ wt%, $k_{ox} = 5\,yr^{-1}$, and $k_{red} = 1.50 \times 10^{-3}\,yr^{-1}$. (b) Schematic representation of model predicted by Froelich *et al.* (1979). Reprinted with permission from Elsevier.

*continued on p. 318*

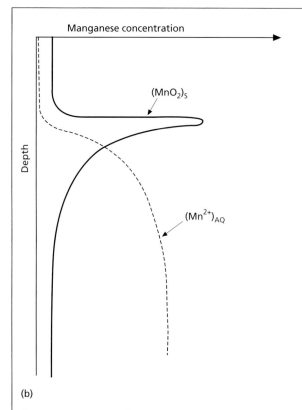

(b)

**Fig. WS14.4.2** *Continued*

$$A = \frac{C_s^o(1-\phi)w}{\phi \alpha D_b} \tag{WS14.4.17}$$

$$E = C_s^o \cosh(\alpha L_{ox}) \tag{WS14.4.18}$$

$$G = A[\sinh(\alpha L_{ox}) + \alpha/\beta \cosh(\alpha L_{ox})] \tag{WS14.4.19}$$

There are two unknowns ($k_{ox}$ and $k_{red}$) in the equation, which can be used for a least-squares fit of the equations to actual data.

By inserting typical parameters for marine sediments into the equations the authors derived model profiles for the diagenesis of dissolved and solid-phase Mn. These are illustrated in Fig. WS14.4.2(a), and show close agreement with the diagrammatic representation suggested by Froelich *et al.* (1979): see Fig. WS14.4.2(b).

The Mn post-depositional migration associated with this model may be summarized as follows (see text for detailed explanation). As the sediments are buried below the redox boundary, solid Mn oxides are reduced to $Mn^{2+}$, which diffuses upwards along the concentration gradient to be oxidized and reprecipitated. With further sedimentation, these reprecipitated oxides are again brought to the reducing zone to be redissolved, with the net result being the Mn trap at the narrow zone at the redox boundary.

Burdige and Gieskes (1983) also applied their equations to solid and porewater data from two cores in the eastern Equatorial Atlantic and obtained very good fits to the data.

at the seawater–sediment interface can strongly modify the composition of the material that reaches the sediment reservoir. The nature of these reactions has now been described, and in the following chapter we will consider the composition of the components that form the oceanic sediments in terms of their genetic histories, which, for some material, includes diagenetic modification.

# References

Bender, M.L. and Heggie, D.T. (1984) Fate of organic carbon reaching the deep sea flow: a status report. *Geochim. Cosmochim. Acta*, **48**, 977–986.

Berner, R.A. (1980) *Early Diagenesis: A Theoretical Approach*. Princeton, NJ: Princeton University Press.

Berner, R.A. (1981) A new geochemical classification of sedimentary environments. *J. Sed. Petrol.*, **51**, 359–365.

Berner, B.A. (1982) Burial of organic carbon and pyrite sulfur in the modern ocean: its geochemical and environmental significance. *Am J. Sci.*, **282**, 451–473.

Boyle, E.A., Sclater, F.R. and Edmond, J.M. (1977) The distribution of copper in the Pacific. *Earth Planet. Sci. Lett.*, **37**, 38–54.

Brandes, J.A., Devol, A.H. and Deutsch, C. (2007) New developments in the marine nitrogen cycle. *Chem. Rev.*, **107**, 577–589.

Bruland, K.W. (1983) Trace elements in sea water, in *Chemical Oceanography*, J.P. Riley and R. Chester (eds), Vol. 8, pp.156–220. London: Academic Press.

Burdige, D.J. and Gieskes, J.M. (1983) A pore water/solid phase diagenetic model for manganese in marine sediments. *Am. J. Sci.*, **283**, 29–47.

Burdige, D.J. (2007) Preservation of organic matter in marine sediments: controls, mechanisms and an imbalance in sediment organic carbon budgets? *Chem. Rev.*, **107**, 467–485.

Calvert, S.E. (1976) The mineralogy and geochemistry of near-shore sediments, in *Chemical Oceanography*, J.P. Riley and R. Chester (eds), Vol. 6, pp.187–280. London: Academic Press.

Calvert, S.E. and Price, N.B. (1970) Minor metal contents of Recent organic-rich sediments off South West Africa. *Nature*, **227**, 593–595.

Drever, J.I. (1982) *The Geochemistry of Natural Waters*. Englewood Cliffs, NJ: Prentice-Hall.

Froelich, P.N., Klinkhammer, G.P., Bender, M.L. *et al.* (1979) Early oxidation of organic matter in pelagic sediments of the eastern equatorial Atlantic: suboxic diagenesis. *Geochim. Cosmochim. Acta*, **43**, 1075–1090.

Galoway, F. and Bender, M. (1982) Diagenetic models of interstitial nitrate profiles in deep-sea suboxic sediments. *Limnol. Oceanogr.*, **27**, 624–638.

Gieskes, J.M. (1983) The chemistry of interstitial waters of deep sea sediments: interpretations of deep-sea drilling data, in *Chemical Oceanography*, J.P. Riley and R. Chester (eds), Vol. 8, pp.221–269. London: Academic Press.

Hedges, J.I. and Keil, R.G. (1995) Sedimentary organic matter preservation: an assessment and speculative synthesis. *Mar. Chem.*, **49**, 81–115.

Klinkhammer, G., Heggie, D.T. and Graham, D.W. (1982) Metal diagenesis in oxic marine sediments. *Earth Planet. Sci. Lett.*, **61**, 211–219.

Lawrence, J.R., Gieskes, J.M. and Broecker, W.S. (1975) Oxygen isotope and carbon composition of DSDP pore waters and the alteration of Layer II basalts. *Earth Planet. Sci. Lett.*, **27**, 1–10.

Lee, C., Wakeham, W. and Arnosti, C. (2004) Particulate organic matter in the sea: the composition conundrum. *Ambio*, **33**, 565–575.

Lerman, A. (1977) Migrational processes and chemical reactions in interstitial waters, in The Sea, E.D. Goldberg, I.N. McCave, J.J. O'Brien and J.H. Steele (eds), Vol. 6, pp.695–738. New York: John Wiley & Sons, Inc.

Lyle, M. (1983) The brown-green color transition in marine sediments: a marker of the Fe(II)–Fe(III) redox boundary. *Limnol. Oceanogr.*, **28**, 1026–1033.

Mackenzie, F.T. and Kump, L.R. (1995) Reverse weathering, clay mineral formation and oceanic element cycles. *Science*, **270**, 586–587.

Martin, W.R. and Sayles, F.L. (2003) The recycling of biogenic material at the seafloor, in *Treatise on Geochemistry*, H.D. Holland and K.K. Turekian (eds), Vol. 7, *Sediments, Diagenesis, and Sedimentary Rocks*, F.T. Mackenzie (ed.), pp.37–65. Oxford: Elsevier Pergamon.

Michalopoulis, P. and Aller, R.C. (1995) Rapid clay mineral formation in Amazon delta sediments: reverse weathering and oceanic elemental cycles. *Science*, **270**, 614–617.

Middleburg, J.J. and Levin, L.A. (2009) Coastal hypoxia and sediment biogeochemistry. *Biogeosciences*, **6**, 1273–1293.

Morford, J. and Emerson, S. (1999) The geochemistry of redox sensitive trace metals in sediments. *Geochim. Cosmochim. Acta*, **69**, 521–532.

Muller, P.J. and Mangini, A. (1980) Organic carbon decomposition ratio in sediments of the Pacific manganese nodule belt dated by [230]Th and [231]Pa. *Earth Planet. Sci. Lett.*, **51**, 96–114.

Murray, J. and Irvine, R. (1895) On the chemical changes which take place in the composition of the seawater associated with blue muds on the floor of the ocean. *Trans. R. Soc. Edinburgh*, **37**, 481–507.

Oremland, R.S. and Taylor, B.F. (1978) Sulfate reduction and methanogenesis in marine sediments. *Geochim. Cosmochim. Acta*, **42**, 209–214.

Roussel, E.G. *et al.* (2008) Extending the sub-sea floor biosphere. *Science*, **320**, 1046.

Sayles, F.L. (1979) The composition and diagenesis of interstitial solutions – I. Fluxes across the seawater–sediment interface in the Atlantic Ocean. *Geochim. Cosmochim. Acta*, **43**, 527–545.

Seiter, K., Hensen, C., Schröter, J. and Zabel, M. (2004) Organic carbon content in surface sediments – defining regional provinces. *Deep-Sea Res*. I, 2001–2026.

Soetart, K., Herman, P.M.J. and Middelburg, J.J. (1996) A model of diagenetic processes from the shelf to abyssal depths. *Geochim. Cosmochim. Acta.*, **60**, 1019–1040.

Stumm, W. and Morgan, J.J. (1996) *Aquatic Chemistry. Chemical Equilibria and Rates in Natural Waters*. New York: John Wiley & Sons, Inc.

Tissot, B.P. and Welte, D.H. (1984) *Petroleum Formation and Occurrence*. Berlin: Springer-Verlag.

Wilson, T.R.S., Thompson, J., Colley, S., Hydes, D.J. and Higgs, N.C. (1985) Early organic diagenesis: the significance of progressive subsurface oxidation fronts in pelagic sediments. *Geochim. Cosmochim. Acta*, **49**, 811–822.

# 15    The components of marine sediments

In this chapter the components of marine sediments themselves are discussed. The approach taken is not to try and describe all the constituents of the sediments, but rather to focus on the components that play important roles in the geochemistry of the sediments and oceans, or act as tracers of important processes. These sediment building-block components are described below in terms of a modification of the 'geosphere of origin' classification proposed by Goldberg (1954). The sediments of the ocean have been sampled and scientifically analysed for over 100 years. Techniques to characterize their gross composition have been available for much longer than that for the components in seawater itself, and hence our understanding of the sedimentary system is in some ways more advanced. The focus in this chapter is primarily on oceanic sediments, although some coastal areas are discussed. The issue of intertidal sediments and evaporites is not considered here.

## 15.1 Lithogenous components

### 15.1.1 Definition

Following Goldberg (1954), lithogenous components are defined as those which arise from land erosion, from submarine volcanoes or from underwater weathering where the solid phase undergoes no major change during its residence in seawater.

A wide variety of solid products are mobilized on the continents and transported to the oceans via river run-off, atmospheric deposition and glacial transport. These solid products have a wide range of composition and there is no reason why any type of continental mineral should not be found, albeit it at small concentrations, in oceanic sediments. Quantitatively, the most important land-derived minerals in the sediments are the clay minerals and quartz, together with smaller amounts of other minerals such as the feldspars. Some of these lithogenous minerals are deposited in nearshore sediments, but the finest fractions can reach open-ocean areas. The distributions of the fine-fraction lithogenous minerals in deep-sea sediments can therefore offer an insight into how material derived from the continents is dispersed throughout the marine environment. This dispersion process is illustrated in the following with respect to the distribution of the clay minerals.

### 15.1.2 The distributions of the principal clay minerals in deep-sea sediments

The detailed structure of the clay minerals is a vast subject which is widely discussed in many geology and soils text books. Here the focus is only on a simple description of their structure sufficient to understand their transport and deposition within the oceans. Clay mineral structures consist of a combination of tetrahedral layers, which are made up of $SiO_4$ tetrahedra linked to form a sheet, and octahedral layers, which are made up of two layers of oxygen atoms or hydroxyl ions with cations of Al, Fe or Mg between them. Riley and Chester (1971) divided the principal clay minerals into two general types. In type I clays, the basic structure consists of a two-layer sheet of one tetrahedral and one octahedral layer (1:1 clays); in type II clays, the basic unit is a sandwich of two tetrahedral layers (in which there is usually some substitution of $Si^{4+}$ by $Al^{3+}$ in the $SiO_4$ tetrahedra) with one octahedral layer between them (2:1 clays). The clay minerals can be formed in a number of ways; for example some are stable weathering residues, some are metamorphic, some

*Marine Geochemistry*, Third Edition. Roy Chester and Tim Jickells.
© 2012 by Roy Chester and Tim Jickells. Published 2012 by Blackwell Publishing Ltd.

are hydrothermal and some are reconstituted during reverse weathering processes (see Section 14.4). A range of these clay minerals have been found in marine sediments, but the most common varieties belong to the kaolinite, chlorite, illite and montmorillonite groups. These clay minerals make up a large proportion of the $<2\,\mu m$ land-derived (carbonate-free) fraction of deep-sea sediments, and as the members of the four major groups are sometimes formed under different geological conditions they are especially attractive for the study of the sources and dispersal patterns of solids in the oceans.

The studies reported by Biscaye (1965) and Griffin *et al.* (1968), on the $<2\,\mu m$ carbonate-free sediment fractions, provided a database that allowed major trends in the oceanic distributions of the clays to be established. The trends identified in these two studies are outlined below in terms of the principal clay mineral groups, and the distributions of the individual clays are illustrated in Fig. 15.1. It is well documented that most clay minerals can be formed under a variety of geological conditions. However, in looking for major trends in the distribution patterns of the clays only their most important general sources will be identified, and although this will of necessity produce an oversimplified picture it is still a useful approach to adopt on an ocean-wide scale.

*Kaolinite.* The formation of kaolinite, which is a type I clay, is characteristic of intense tropical and desert weathering, and it is clearly apparent from Fig. 15.1(a) that the highest concentrations of the mineral are found in deep-sea sediments from equatorial regions. The mineral is transported to these regions from the surrounding arid and semi-arid land masses via fluvial and atmospheric pathways, and because of its distribution, Griffin *et al.* (1968) referred to kaolinite as the *low-latitude* clay mineral.

*Chlorite.* Chlorite is a type II clay mineral and has a brucite layer between the basic three-layer sandwich. It can be seen from Fig. 15.1(b) that chlorite has its highest concentrations in sediments from the polar seas. The mineral is found in metamorphic and sedimentary rocks of the Arctic and Antarctic regions from which, in the general absence of chemical weathering, it is released by mechanical processes and is subsequently dispersed by ice rafting; the latter is confirmed by the sharp break-off in concentrations

around 50°N and 50°S, which corresponds roughly to the iceberg transport limit. Thus, this chlorite is a primary mineral and not a weathering residue. Chlorite is also formed under a variety of other geological–climatic conditions, but because of its high concentrations in sediments from polar regions, Griffin *et al.* (1968) termed it the *high-latitude* clay mineral. In general, the overall distribution of chlorite in deep-sea sediments is therefore the inverse of that of the low-latitude clay mineral kaolinite.

*Illite.* Members of the illite group are type II clays, which have $K^+$ ions lying between the three-layer basic sandwich. Illites are the most common of the clay minerals in deep-sea sediments. They are formed under a wide variety of geological conditions and, unlike kaolinites and chlorites, they are not confined to particular latitudinal bands. Most of the illites in deep-sea sediments are land-derived and their distributions are controlled by (i) the amount of land surrounding an oceanic area, and (ii) the extent to which they are diluted by clays that have specific source regions (e.g. kaolinite and chlorite). As a result of a combination of these two factors, the highest concentrations of illite are found in deep-sea sediments from *mid-latitudes* (where there is less dilution from chlorite and kaolinite) in the Northern Hemisphere (which has a higher proportion of surrounding land areas than the Southern Hemisphere). This is particularly apparent in the North Pacific where the sediments have an average of 40% illite, compared with only 26% in the South Pacific: see Table 15.1. The distribution of illite in deep-sea sediments is illustrated in Fig. 15.1(c).

*Montmorillonite.* This is sometimes termed smectite or 'expanding-lattice' clay. Montmorillonites are type II clays, in which water molecules are found lying between the basic three-layer sandwich units. According to Riley and Chester (1971) montmorillonites in deep-sea clays can be formed in at least three different ways:

**1** as primary detrital weathering residues formed on the continents;

**2** as secondary detrital degraded lattices of the illite or chlorite type which have had their intersheet $K^+$ ions replaced by water molecules;

**3** as in situ products of the submarine weathering of volcanic material.

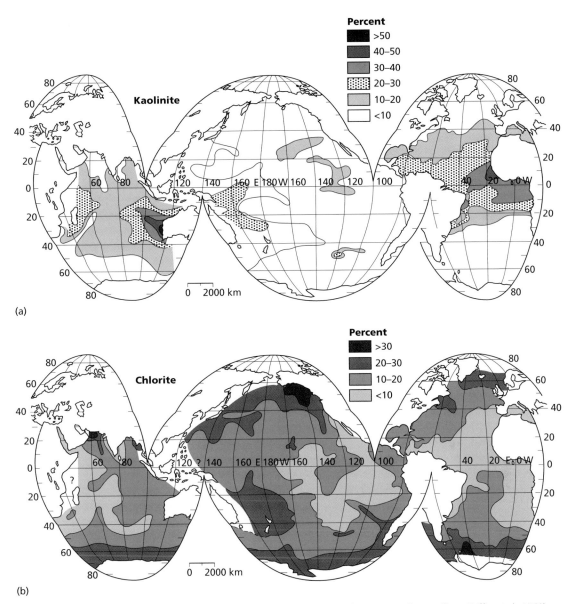

**Fig. 15.1** (a–d) The concentration of clay minerals in the <2 μm size fractions of deep-sea sediments (from Griffin *et al.*, 1968); values expressed as percentages of a 100% clay sample. Reprinted with permission from Elsevier.

It is the latter form that imposes specific patterns on the distribution of montmorillonite in deep-sea sediments. As a result, the highest concentrations of the clay are found in areas that have a plentiful supply of parent volcanic debris, and in which the sedimentation rates are low enough to allow the transformation reactions to occur before the volcanic debris is buried. This is reflected in the sequence of increasing montmorillonite concentrations in deep-sea sediments, which are in the order: North Atlantic (16%) < South Atlantic (26%) < North Pacific (35%) < Indian Ocean (41%) < South Pacific (53%): see Table 15.1. Thus, the deep-sea sediments from the South Pacific are richer in montmorillonite than

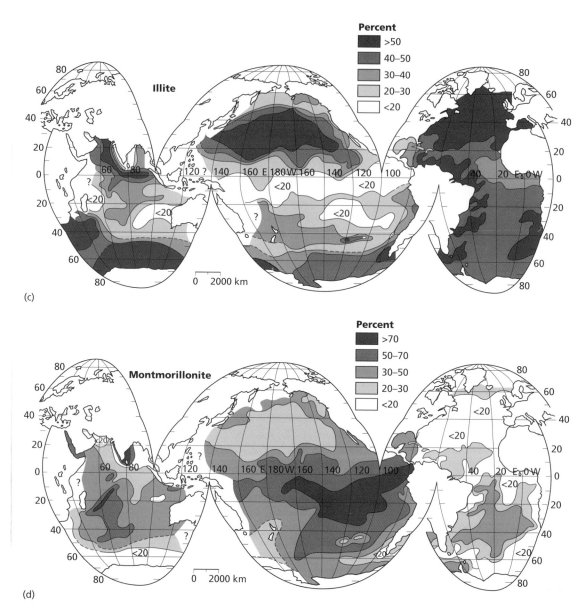

**Fig. 15.1** *Continued*

those from the other oceanic areas. According to Griffin *et al.* (1968), the highest concentrations of montmorillonite in the South Pacific itself are found in sediments from mid-ocean areas, where the mineral is found in association with volcanic material, with concentration gradients decreasing towards the continents. This led the authors to propose that much of the montmorillonite in the mid-ocean sedi-

ments has an in situ origin from the alteration of volcanic material, and on this basis they characterized montmorillonite as being indicative of a volcanic regime. The distribution of montmorillonite in deep-sea sediments is illustrated in Fig. 15.1(d).

The North Pacific has more rivers draining into it, and a greater area of surrounding land, than the South Pacific, and this is reflected in the distributions

**Table 15.1** Average concentrations of the principal clay minerals in the <2 μm carbonate-free fractions of sediments from the major oceans (data from Griffin *et al.*, 1968).

| Oceanic area | Average % clay minerals* | | | |
| --- | --- | --- | --- | --- |
| | Chlorite | Montmorillonite | Kaolinite | Illite |
| North Atlantic | 10 | 16 | 20 | 55 |
| South Atlantic | 11 | 26 | 17 | 47 |
| North Pacific | 18 | 35 | 8 | 40 |
| South Pacific | 13 | 53 | 8 | 26 |
| Indian | 12 | 41 | 17 | 33 |

*Individual clay mineral percentages are expressed in terms of a 100% clay sample.

of illite and montmorillonite in the deep-sea sediments from these two regions. That is, in the North Pacific, which receives a relatively large contribution of land-derived material, the sediments have some of the highest concentrations of the ubiquitous detrital clay mineral illite, whereas those in the South Pacific have the highest concentrations of the authigenic clay mineral montmorillonite. Griffin and Goldberg (1963) suggested therefore that the North Pacific is an area of mainly detrital deposition, but that in the South Pacific authigenic deposition is more important.

### 15.1.3 The chemical significance of the clay minerals in marine processes

Once they are brought into contact with saline waters the clay minerals can take part in ion-exchange reactions of the type

$$\text{cation A (clay)} + \text{cation B}^x \text{(solution)}$$
$$= \text{cation B (clay)} + \text{cation A}^x \text{(solution)} \quad (15.1)$$

where $x$ is the charge on the cation. These reactions are particularly important in estuarine environments where clay minerals which have equilibrated with river waters, where $Ca^{2+}$ is usually the major ion, encounter seawater where $Na^+$ is the major ion. Ion exchange processes drive a release of $Ca^{2+}$ ions from the clays balanced by an uptake of $Na^+$ ions. These ion exchange reactions occur relatively rapidly (timescales of minutes to days) and result in the exchange of geochemically significant amounts of major ions in the context of global budgets (e.g. Sayles and Mangelsdorf, 1977).

In addition to this normal rapid ion exchange, the clay minerals can remove major cations from seawater during the reconstitution of degraded clay minerals, that is, those that have had cations stripped from intersheet positions during weathering on land. A number of workers have attempted to model the composition of seawater on the basis of thermodynamic equilibria between components dissolved in the water and the mineral phases of marine sediments. This type of modelling was given a great impetus in the 1960s by Sillen. In his models, Sillen proposed that the composition of seawater was controlled by equilibria between individual minerals, which would fix the ratios of the cations in solution. The removal of major ions supplied to the oceans from river run-off was then assumed to take place by the *transformation* of one clay mineral phase into another, and the equilibria were considered to operate between phases such as seawater–quartz–kaolinite–illite–chlorite–montmorillonite–calcite–zeolites as part of the overall atmosphere–seawater–sediment system (see e.g. Sillen, 1961; 1963; 1965; 1967). In terms of these models, the coexistence of the various phases would fix the mutual ratios of all the cations, so that only the sum of their concentrations was variable, and this was assumed to be fixed by the chloride concentration. More recent studies demonstrate that the actions in the ocean are regulated by many other process, particularly biological ones. However Mackenzie and Garrels (1966) proposed that, rather than transformations between existing clay minerals, it was the transformation of either amorphous aluminosilicate material (formed during weathering) or degraded clays into new crystalline clay phases that controlled the reactions required by the equilibrium models. Thus

$$\text{X-ray amorphous Al silicate} + SiO_2 + HCO_3^- \\ + \text{cations} = \text{cation Al silicate} + CO_2 + H_2O \quad (15.2)$$

The normal weathering of aluminosilicates on land involves the uptake of $CO_2$ from the atmosphere and the release of bicarbonate. Thus

$$\text{silicate} + CO_2 + H_2O = \text{clay mineral} + HCO_3^- \\ + H_4SiO_4 + \text{cations} \quad (15.3)$$

In effect, therefore, the processes occurring in the sea (i.e. in which cations, dissolved silica and bicarbonate are removed from Equation solution 15.2) are the reverse of the weathering process on the continents; hence, they are often referred to as reverse weathering. These reactions were invoked as being major controls on seawater major ions chemistry, but the discovery of the key role of mid ocean ridge cycling (see Chapter 5) has meant that such reverse weathering reactions may not need to be invoked to balance major ion global biogeochemical budgets. However, more recently work on pore waters have confirmed reverse weathering type reactions do indeed take place, even if they are less important for global biogeochemical budgets than previously estimated (see Section 14.4.2).

### 15.1.4 Clay minerals in deep-sea sediments: summary

The average concentrations of the four principal clay minerals in deep-sea sediments from the major oceans are listed in Table 15.1, and their origin–distribution trends may be summarized as follows.
1 Illite, chlorite and kaolinite are largely land-derived, that is, lithogenous, clays.
2 Chlorite and kaolinite tend to be concentrated in surficial rocks and soils from particular latitudinal belts, and this is reflected in their distributions in oceanic sediments, chlorite being the *low-latitude* clay and kaolinite the *high-latitude* clay.
3 Illite is also mainly land-derived, but because it is formed under a wide variety of weathering conditions it is not diagnostic of any particular supply region.

4 The main features in the distribution of montmorillonite in deep-sea sediments arise from its transformation from volcanic debris. Thus, montmorillonite is characteristic of volcanic regimes.

## 15.2 Biogenous components

### 15.2.1 Definition

In the classification suggested by Goldberg (1954) biogenous components are defined as those produced in the biosphere, and as such include both organic matter and inorganic shell material. Organic matter in marine sediments has been discussed in Chapter 14, and will not be considered here. Other biogenous components include phosphates (e.g. skeletal apatite) and sulfates (e.g. barite). However, these usually make up only a few percent of marine sediments. They can be important as tracers of ocean processes in the paleo record and produce potentially valuable mineral resources. However, in this chapter attention will be focused on the major sediment-forming biogenic components, that is, carbonate and opaline silica shell material. The scales of the production of this biogenic skeletal material are very large. Milliman (1993) estimated the total calcium carbonate production in the oceans to be $5.3 \times 10^9$ tonnes $(10^{15}\,\text{g})$ $\text{yr}^{-1}$ and recent estimates (see next section) are even higher and Ragueneau *et al.* (2000) estimate opaline silica production rates in the ocean to be $0.2–0.3 \times 10^{15}\,\text{g Si yr}^{-1}$.

### 15.2.2 Carbonate and opaline skeletal material in marine sediments

The most important carbonate-secreting organisms in the oceans are foraminifera, coccolithophorids and pteropods (see Section 9.1.4). Most of these organisms produce calcium carbonate as calcite, although some produce a slightly different form of calcium carbonate known as aragonite. Calcite and aragonite are polymorphs of calcium carbonate with the same chemical structure, but different crystal geometries which mean they also have slightly different solubility. Foraminifera are grazers. The planktonic foraminifera have chambered shells (or tests) composed of calcite that range in size from ~30 μm to ~1 mm. These planktonic foraminifera, which are classified into a superfamily (the Globigerinacea), constitute the most common biogenous components in deep-sea sediments, evidenced by the fact that *Globigerina* oozes are the dominant type of pelagic sediment. There are also benthic foraminfera, but they are quantitatively much less abundant than planktonic forms and are not a significant source of calcium carbonate compared to other forms. Coccolithophorids are nanoplankton (phytoplankton) secreting calcite shells (generally around 10 μm in size), which after the death of the organism tend to break up into individual plates termed coccoliths. These coccoliths are a major component of many calcareous deep-sea sediments. Pteropods are pelagic molluscs that secrete large (millimetre-sized) shells of aragonite. These shells undergo dissolution at shallower depths than those composed of calcite (see the following), and pteropod oozes have only a limited occurrence on the ocean floor. There is also large scale carbonate production (~40% of the total carbonate production) in shelf seas via micro and macroalgae and also corals (Milliman, 1993).

The principal opal-secreting organisms in the marine environment are diatoms and radiolaria, with minor amounts of silica being produced by the silicoflagellates and the sponges. Diatoms are phytoplankton with frustules, which range in size from a few micrometres to around 2 mm. Radiolarians are grazers, with tests ranging in size from a few tens to a few hundred micrometres.

Both carbonate- and opal-secreting organisms contribute biogenous components to deep-sea sediments, calcareous oozes (which contain >30% skeletal carbonate) covering ~50% of the deep-sea floor and siliceous oozes (which contain >30% skeletal opal) covering ~15%. However, the oceanic distributions of the calcareous and siliceous oozes are very different. This is illustrated in Fig. 15.2 and Fig. 13.5, from which it can be seen that, whereas the calcareous oozes are found mainly on topographic highs in mid-ocean areas, the siliceous oozes tend to be restricted to regions underlying coastal and equatorial upwelling, that is, they are correlated with high primary production in the surface waters. There are a number of reasons for this overall difference in the distributions of calcareous and siliceous oozes, and these generally can be related to the factors that control the output (production) of the organisms and the dissolution of their skeletal remains.

### 15.2.3 Shell production

The production of shells depends on the primary productivity of the ocean,. The two most striking features in the global distribution of primary production are: (i) high productivity in coastal, upwelling and high latitude areas and (ii) a generally low productivity in the central gyres (see Section 9.1.5). Diatoms are particularly abundant within the high productivity areas with relative high surface water macro nutrient concentrations, except where iron limitation is important, while foraminifera and to some extend coccolithophores are more widely distributed; as discussed in Chapter 9.

### 15.2.4 Shell dissolution–preservation processes and their impact on global biogeochemical cycles

After sinking from the surface layer of the ocean both calcareous and opaline skeletal remains undergo dissolution processes in response to physicochemical parameters. Seawater varies in the extent to which it is undersaturated with respect to carbonate, but it is always undersaturated with respect to opaline silica. As a result, dissolution affects carbonate and siliceous organisms to varying extents, and tends to strengthen the difference produced by the primary productivity gradients. For this reason it is convenient to treat the dissolution of calcareous and siliceous skeletal remains separately.

#### 15.2.4.1 Carbonate shells

From the time scientists first became interested in the oceans it was realized that the carbonate content of deep-sea sediments varies considerably from one location to another. Further, it soon became apparent that this was related to the depth of water under which the deposits had accumulated, with the higher carbonate contents being found in sediments located on topographic highs (see Fig. 13.5 and 15.2). There is, therefore, a first-order depth control on the occurrence of carbonate sediments in the World Ocean. In addition, there is a general depth that forms a boundary between the deposition of carbonate and noncarbonate sediments (or at least those containing only a few percent of carbonate). This is termed the calcium carbonate compensation depth. Above this

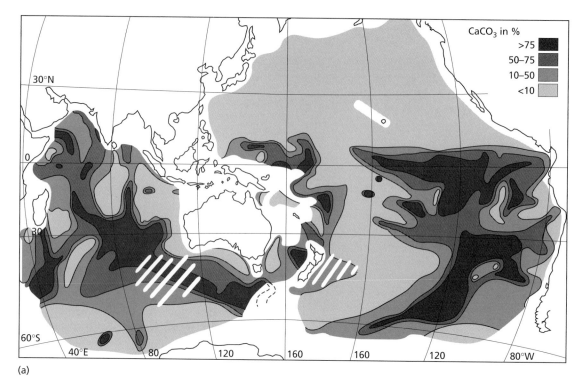

(a)

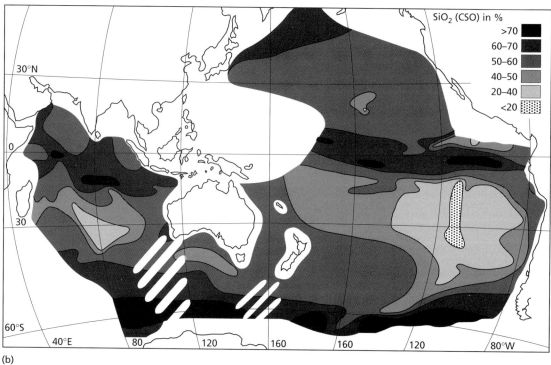

(b)

**Fig. 15.2** The distributions of calcium carbonate and opaline silica in deep-sea sediments. (a) The distribution of calcium carbonate in Indo-Pacific sediments (from Bostrom *et al.*, 1973). Reprinted with permission from Elsevier. (b) The distribution of opaline silica in Indo-Pacific sediments (from Bostrom *et al.*, 1973); data on a carbonate-, salt- and organic-matter-free basis. Reprinted with permission from Elsevier.

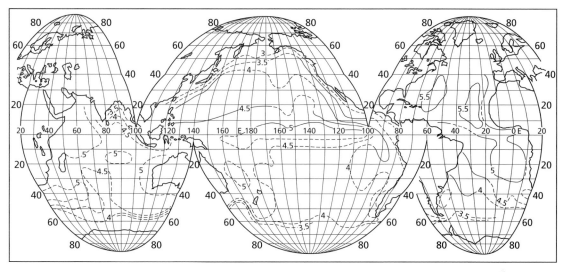

**Fig. 15.3** The distribution of the calcium carbonate compensation depth in the World Ocean (from Berger and Winterer, 1974). This is the depth (km) of the level at which the carbonate content of deep-sea sediments decreases to a few percent.

depth, carbonate, shells will accumulate in the sediments; at the depth itself, the rate of supply equals the rate of dissolution; at greater depths, dissolution exceeds supply and there is no *net* accumulation of significant amounts of carbonate material. The compensation depth differs for calcite and aragonite. For example, the calcite compensation depth (CCD) appears to be at ~4.5–5 km over much of the Atlantic, at ~5 km in the tropical Indian Ocean, and at ~3–5 km in the Pacific: see Fig. 15.3 and Fig. 15.4. In contrast, the aragonite compensation depth (ACD) is much shallower, ranging from ~3 km in the western tropical North Atlantic, to between ~1 and ~2 km in the western tropical Pacific, and as low as a few hundred metres in the tropical North Pacific (Berger, 1976). It is also important to recognize that in addition to the carbonate/non-carbonate deposition represented by the CCD boundary, there is dissolution of shell material above this depth in the sediments. This leads to variations in the relative degrees to which carbonate shells have undergone dissolution, and there is a horizon in the sediment that separates well-preserved from poorly preserved shell assemblages. The depth at which partial dissolution becomes evident is termed the lysocline.

The main feature in the distribution of calcium carbonate in deep-sea sediments therefore is its common occurrence on topographic highs. One of the keys necessary to understand how this depth

control operates lies in the fact that the degree to which seawater is undersaturated with respect to calcium carbonate varies down the water column. The upper waters are always supersaturated with respect to calcium carbonate but become undersaturated in intermediate and deep waters. It is this saturation–undersaturation depth pattern that can be regarded as being the driving force behind carbonate dissolution which is controlled by the physicochemistry of the ocean carbon system.

The chemistry of the ocean carbonate system is discussed in Section 9.5 and only the components of the system directly relevant to carbonate dissolution are repeated here. The fundamental equation for the dissolution of calcium carbonate is:

$$CaCO_{3(s)} \leftrightarrow Ca^{2+}_{(aq)} + CO^{2-}_{3(aq)} \tag{15.4}$$

and the solubility is then described by the solubility product Ks

$$K_{sp} = [Ca^{2+}][CO_3^{2-}] \tag{15.5}$$

The degree of saturation ($\Omega$) is then estimated by comparing the observed activities of calcium and carbonate to the solubility product.

$$\Omega = \frac{[Ca^{2+}][CO_3^{2-}]}{K_{sp}} \tag{15.6}$$

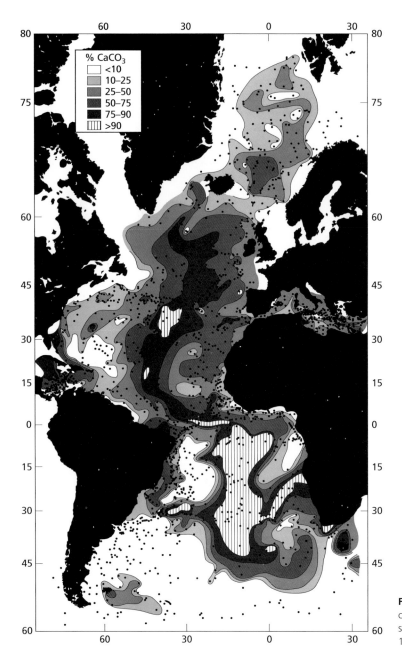

**Fig. 15.4** The present-day distribution of calcium carbonate in Atlantic Ocean surface sediments (from Biscaye *et al.*, 1976).

The solubility of both calcite and aragonite increases with increasing pressure (and so depth) and decreasing temperature, with the result that carbonate shells are susceptible to dissolution both during their transit down the water column and when they reach the bottom sediment. The pressure and depth effect is very much greater than the temperature effect. A change in temperature of 20°C, such as might occur between the surface ocean and 4 km depth, only produces a 4% change in solubility while the pressure effect of such a depth change produces a 200% change in solubility. This depth dependence of solubility therefore explains the broad pattern of calcium carbonate deposits in the open ocean, being absent from the deepest parts of the ocean floor while they are relatively abundant on shallower areas such as mid ocean ridges. This depth dependence also leads to much more effective preservation of carbonates on the shelf compared to the deep ocean although this is in part offset by higher organic matter fluxes which encourage dissolution in pore waters (Milliman, 1993).

The other key feature of the distribution of carbonate sediments in the ocean is the shallowing of the carbonate compensation depth from the Atlantic to the Pacific and Indian Oceans. This reflects the interaction of the sinking and decomposition of organic matter formed in surface waters and the global ocean circulation. The residence time of calcium in the oceans is very long and the distribution rather uniform, so the key driver of changes in solubility between the ocean basins is the carbonate ion concentrations. As discussed in Chapter 9, the global deep ocean circulation means that water that starts out at the surface in the Arctic Ocean makes its way through the deep oceans arriving in the northern Pacific and Indian Oceans many hundreds of years later. Throughout this time the water is isolated from exchange with the atmosphere and the products of the decay of sinking organic matter build up in the waters meaning that oxygen concentrations decrease and nutrients, alkalinity and $TCO_2$ increase. This increase in total alkalinity also drives a decrease in carbonate ion concentration and an increase in bicarbonate ion concentrations, thereby making calcium carbonate more soluble in the 'older' waters of the Pacific and Indian Oceans and raising the carbonate compensation depth to shallower levels (Broecker, 2004).

In theory it should be possible to calculate the exact depth of the point in the water column at which calcite or aragonite becomes undersaturated. However, our knowledge of the carbonate system is inadequate to do this and there are several confounding kinetic factors that also complicate such a calculation. The first is that large carbonate shells such as foraminifera dissolve relatively slowly in the modestly undersaturated deep ocean waters relative to their sinking rates. Hence carbonate material can certainly reach the sediments below the carbonate compensation depth. Furthermore, many carbonate skeletons are surrounded by an organic membrane and hence, not in direct contact with seawater, a situation that will again slow dissolution until such membranes are destroyed by bacterial action. If the sinking shells remain at the sediment surface below the carbonate compensation depth, they will dissolve. However, if they are stirred into the sediment by bioturbation they more into a different microenvironment in which organic matter decomposition may increase alkalinity and enhance calcium carbonate dissolution.

As noted earlier, the scale of fluxes through the carbonate system are very large and uncertain. Milliman (1993) estimated total carbonate production to be $5.3 \times 10^{15}$ g $CaCO_3$ yr$^{-1}$, with 46% of this on the shelf, 9% on the shelf slope and 45% on the oceans. A more recent budget (see Doney et al., 2009) suggests higher and rather uncertain open ocean carbonate production ($4$–$13 \times 10^{15}$ g $CaCO_3$ yr$^{-1}$ compared to $2.4 \times 10^{15}$ g $CaCO_3$ yr$^{-1}$ in the Milliman compilation) with only about $0.83 \times 10^{15}$ g $CaCO_3$ yr$^{-1}$ eventually being buried in the open ocean – as noted earlier most carbonate production on the shelf is preserved since the sediments are above the carbonate compensation depth. These estimates serve to illustrate the scale of the production and dissolution fluxes within the oceans, and also that our estimates of such processes are still uncertain.

Although the exact fluxes are uncertain, it is clear that the production, dissolution and sediment preservation patterns of plankton-produced calcium carbonate in the ocean system have important effects on the chemistry of seawater, and on the global $CO_2$ system (including climate). The formation of the skeletal carbonates, and their incorporation into deep-sea sediments, influences the marine budget of calcium and also Sr and Mg which can substitute for Ca in the calcium carbonate lattice. Overall, however, the carbonate skeletal material incorporated into deep-sea sediments tends to be relatively pure, and contains very low concentrations of trace metals other than Sr and Mg, although these can be associated with organic and sediment material trapped within the shells. Although the calcium carbonate skeletal material is relatively pure, the analysis of its isotopic (particularly carbon and oxygen isotopes) and trace element (including Mg and Cd) composition can provide a wealth of information on the water chemistry at the time of carbonate formation. Hence, the analysis of foraminifera shell and coral skeleton chemical composition along with foraminfera species composition (since different species are abundant in different ocean productivity and temperature regimes) forms the basis of most palaeooceanographic research and the reader is referred to a series of excellent chapters in Elderfield (2004) for further details of these palaeooceanography proxies.

In Chapter 9, the impact of increasing anthropogenic $CO_2$ emissions to the atmosphere on the oceans was noted. The result of increasing $CO_2$ emissions has been to lower surface water pH, and this water will slowly be transported into the deep ocean. Here it will effectively cause the carbonate compensation depth to rise and promote dissolution of the sediment calcium carbonate forming dissolved bicarbonate. There is a vast reservoir of carbonate within the oceans and on timescales of thousands of years this will effectively consume the anthropogenic $CO_2$ (Broecker, 2004). However, the timescales of this process are those of deep ocean circulation, that is, hundreds to thousands of years, and in the shorter term the predicted effects of concern are the shallowing of the aragonite compensation depth to the point at which it will approach the surface and may make it much more difficult for aragonite-forming organisms such as pteropods to survive (Doney *et al.*, 2009).

### 15.2.4.2 Opaline shells

The principal factors that control the distribution of opaline silica in deep-sea sediments can be summarized as follows. As seawater is universally undersaturated with respect to opaline silica, the extent to which this component is preserved in deep-sea sediments depends ultimately on the degree to which it has undergone, or escaped, dissolution. The preservation of opaline silica in deep-sea sediments has been reviewed by Ragueneau *et al.* (2000) who suggest this is influenced by the following factors.

1 The thermodynamic driving force, which is a function of the temperature, pressure and the $H_4SiO_4$ content of the bottom water. Temperature is particularly important here with rates of dissolution increasing by of the order of 20-fold over the range of ocean water temperatures.

2 The rate of production of opaline organisms in the overlying waters, that is, a steady-state situation develops in which the greater the amount of opal falling down the water column the greater the chance it will be preserved in the sediment. As a result, the highest concentrations of opal are found in the sediments under areas of coastal and equatorial upwelling.

3 Rates of dissolution are dependent on the exposed surface area. Thus rates differ between diatom (or radiolarian) species depending on their morphology. Diatoms also tend to produce thicker skeletons under iron limitation a process which will not alter their rate of dissolution, but will influence their preservation and their rate of sinking, and hence the depth within the water column at which they dissolve. Also opal skeletons are created surrounded by an organic sheath. This protects them from dissolution and after death until the microbial degradation of this sheath. Adsorption of organic matter onto opal will similarly slow dissolution.

4 The extent to which the opaline remains are diluted with non-opaline material, which acts to decrease the rate at which opal dissolves because this depends, among other factors, on the amount of reactive surface area exposed by the skeletal particles. This may also relate to a documented influence of aluminium in reducing opal solubility which may reflect a coating of skeletons with aluminosilicate formed in situ, or the effect of the substitution of Al in the opal lattice; for a review of this topic, see Van Cappellen and Qiu (1997a and b).

Opal dissolution continues on the sea bed and in the sediment–interstitial–water complex, which can lead to the diffusion of dissolved silica out of bottom sediments. The regeneration of silica from opaline shells transported to sediments depends on a complex interplay between sediment burial and mixing, dissolution of the shells, transport of silicic acid through the interstitial waters and the precipitation of authigenic silicate minerals.

The overall effect of the various fertility (production), dissolution and preservation processes is a very restricted, but highly distinctive, distribution of siliceous oozes in the oceans: see Fig. 15.2(b). This distribution correlates with the high primary productivity upwelling areas, and as a result the distribution of sediments rich in siliceous organisms is concentrated into three major bands (Fig. 13.5 and 15.2).

1 A Southern Hemisphere polar band, or ring, which nearly circles the globe. The sediments here are mainly diatomaceous oozes, which represents the main area of oceanic opal accumulation (DeMaster, 2002).
2 A band of opal rich sediments in the northern Pacific Ocean.
3 An equatorial belt in the Pacific Ocean, in which the sediments are relatively rich in radiolaria.

DeMaster (2002) also suggests there is significant opal burial on continental shelves where it will be incorporated within other terrigenous and biogenic sediments.

In addition to these three major bands, siliceous sediments are found in a number of coastal regions where upwelling occurs (e.g. in the Gulf of California and off the coast of Namibia). In general, diatoms are the dominant organisms in high-latitude and coastal siliceous sediments, whereas radiolarians predominate in the equatorial sediments.

Treguer et al. (1995) estimate global opal production as $240 \times 10^{12}$ mol Si yr$^{-1}$ with 50% of this exported below 200 m, a value consistent with the flux estimates of Sarmiento et al. (2007). Treguer et al. (1995) estimate opal burial as $7.1 \times 10^{12}$ mol Si yr$^{-1}$ (approximately balancing their estimate of input to the oceans which is dominated by rivers) implying a global burial efficiency of 3% which they suggest varies from almost zero in subtropical waters to >20% in the Southern Ocean. They suggest 90% of the dissolution occurs within the water column. Sarmiento et al. (2007) and Pondaven et al. (2000) suggest that preservation efficiencies are actually similar in all ocean basins and that it is the high production and efficient export of opal that creates the Southern Ocean diatomaceous oozes, rather than the variation in preservation efficiency, although in practice the two are closely linked.

### 15.2.5 Carbonate and opaline shell material in marine sediments: summary

The distribution of ocean productivity and the fact that diatoms are favoured in highly productive regions and coccoliths in less productive regions, combined with differences in the oceanic chemistries of the carbonate and silica systems, has led to distinct geographical patterns in the distributions of calcareous and siliceous oozes. Thus, in general, the calcare-

ous deposits are concentrated in relatively shallow open-ocean regions particularly in the Atlantic Ocean, whereas the siliceous deposits are found mainly at the edges of the oceans and particularly in the Southern Ocean.

## 15.3 'Hydrogenous' components: halmyrolysates and precipitates

### 15.3.1 Definition

There is some confusion in the literature over the terminology used to describe those components of marine sediments which have been formed predominantly inorganically from constituents dissolved in seawater.

For the present purposes an attempt will be made to classify the dissolved elements that take part in inorganic component-forming reactions by distinguishing between them on the basis of their sources. Thus, the following categories of elements are identified.
1 Direct seawater-derived elements, which are further subdivided into (i) hydrogenous elements, that is, those originating from the general background of elements dissolved in seawater, and (ii) hydrothermal elements, that is, those originating from the debouching of hydrothermal solutions at the ridge-crest spreading centres (see Section 5.1).
2 Oxic diagenetically derived elements, that is, those generated close to the sediment surface following their release on the oxidative destruction of organic carbon.
3 Suboxic diagenetically derived elements, that is, those originating from interstitial waters at some depth in the sediment following the destruction of organic carbon by secondary oxidants.
The sediment components formed from these various types of elements can then be classified into precipitates (primary) and halmyrolysates (secondary) as proposed by Elderfield (1976). This hydrogenous–hydrothermal–diagenetic trinity offers a convenient framework within which to describe those components of marine sediments that are formed in the oceanic environment by inorganic reactions involving dissolved elements, and it is this inorganic origin that distinguishes them from the biogenous components. As the individual components are described, an attempt will be made to identify any coupling

between the *processes* that utilize these genetically different elements.

The *precipitates*, or primary inorganic components, in marine sediments include oxyhydroxides, carbonates, phosphates, sulfides, sulfates and evaporite minerals; and the *halmyrolysates*, or secondary inorganic components, include glauconite, chamosite, palagonite, montmorillonite (smectite) and the zeolites. Thus, there is a wide range of halmyrolysates and precipitates in the marine environment. However, the influence they have on marine sedimentation varies from one component to another, both *quantitatively*, in their role as sediment-forming material, and *geochemically*, in the extent to which they influence chemical processes in the oceans. In both of these senses, the most important inorganically produced components found in marine sediments are (i) ferromanganese nodules, (ii) ferromanganese oxyhydroxides, and (iii) ferromanganese ridge-crest iron and manganese oxides. Because of their geochemical importance, these ferromanganese deposits will therefore serve as prime examples of process-orientated components formed inorganically in the marine environment.

## 15.3.2 Ferromanganese deposits in the oceans

Without assuming, at least at this stage, any genetic association in the sequence, the geochemical importance of ferromanganese deposits will be considered in the order, encrustations → ferromanganese nodules → sediment ferromanganese oxyhydroxides → hydrothermal precipitates. The sequence is set up in this way in order to consider first a hydrogenous end-member ferromanganese component (encrustations), then mixed hydrogenous–hydrothermal–diagenetic ferromanganese components (nodules), and finally a hydrothermal end-member ferromanganese component (hydrothermal precipitates). In this way it should be possible to identify how the pure hydrogenous and hydrothermal signals operate, and also how they combine with each other, and with the diagenetic signal, to form a variety of ferromanganese components.

## 15.3.3 Ferromanganese encrustations

Ferromanganese encrustations, or crusts, are relatively thin deposits on submarine rock outcrops, or on objects such as boulders or volcanic slabs. For example, on the Blake Plateau there is a manganese oxide crust 'pavement' that covers an area of ~5000 km$^2$ (Pratt and McFarlin, 1966). Further, these crusts can be the typical ferromanganese deposit in seamount provinces. One particularly interesting feature of the crusts is that those which grow on exposed rock surfaces have acquired the elements necessary for their growth directly from the overlying seawater (*hydrogenous* supply) without receiving an input from sediment sources. In this sense, it will be shown below that they can be thought of as representing an end-member in the oceanic ferromanganese depositional sequence. Because they are assumed to derive their composition from seawater, the analysis of these crusts has been used in studies of the past history of seawater composition (e.g. Goldstein and Hemming, 2003).

## 15.3.4 Ferromanganese nodules

### 15.3.4.1 Occurrence

The first detailed description of the widespread occurrence of ferromanganese nodules in the oceans is usually attributed to Murray and Renard (1891) from collections made during the *Challenger Expedition* (1873–76). The nodules are present throughout the sediment column, but the highest concentrations are found on the surface. Here, they can cover vast areas of the sea floor in some regions; for example, ~50% of the sediment surface in the western Pacific has a blanket of nodules. They have been reported in association with many types of sediment, and Cronan (1977) identified two principal factors that control their abundance.

1 Rate of accumulation of the host sediment. The highest number of nodules are found on those deep-sea sediments that are deposited at relatively slow rates of around a few millimetres per 1000 yr (mm $10^3$ yr$^{-1}$); these include pelagic clays and siliceous oozes. High concentrations of nodules also can occur where clay and biogenic sedimentation is inhibited as a result of current action, for example, on the tops of seamounts.

2 The availability of suitable nuclei for oxide accretion. Various materials (e.g. a fragment of pumice, a shark's tooth, a foram skeleton, a piece of consolidated clay) can be utilized for this purpose, but volcanic nuclei are especially common, and this may

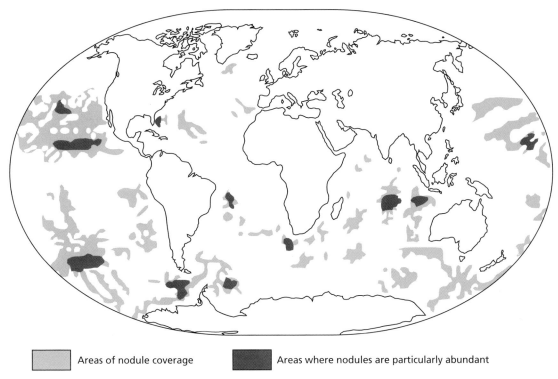

Areas of nodule coverage     Areas where nodules are particularly abundant

**Fig. 15.5** Generalized distribution of ferromanganese nodules in the World Ocean (from Cronan, 1980). Reprinted with permission from Elsevier.

explain the relatively high concentrations of nodules in areas of volcanism.

A generalized pattern of the distribution of ferro-manganese nodules is illustrated in Fig. 15.5. Nodules represent one of the largest manganese deposits on Earth and have attracted proposals to use them as a source of manganese and the many other metals that are concentrated within them. However, the difficulty of extracting the nodules from the depths of the ocean means that to date there has been no commercial activity.

### 15.3.4.2 Morphology

The nodules can have a wide variety of shapes and sizes. When attempting to assess the role they play in marine geochemistry it is useful to make a distinction between macronodules and micronodules. This is a purely arbitrary division in which macronodules have diameters in the centimetre (or greater) range, and micronodules have diameters in the millimetre (or less) range. However, the distinction has some meaning in the sense that, whereas the macronodules

may be considered to be foreign bodies on, or in, the sediment host, the micronodules are scattered throughout the deposits and are much more intimately associated with the other sediment-forming components, that is, they are part of the general sediment. Further, the micronodules often do not have a nucleus, and may in fact be composed of oxides that have sedimented directly from the water column.

Some typical macronodules are illustrated in Fig. 15.6. These macronodules have a wide variety of physical forms, and various schemes have been proposed for their morphological classification; for a detailed discussion of this topic,, see Raab and Meylan (1977). Perhaps the most widely used morphological framework is that in which the nodules are referred to as being some variant of a basic discoidal–ellipsoidal–spherical shape pattern. It is now recognized that the shape adopted by a macronodule is related to the manner in which it has grown, and in particular to whether it has received its component elements (mainly Mn and Fe) from overlying seawater or from underlying interstitial water. Manheim (1965) has described the two nodule

**Fig. 15.6** Typical morphologies of ferromanganese macro-nodules from the Atlantic and Indian Oceans (photographs supplied by D.S. Cronan and S.A. Moorby; (c), (e) and (f) have also been reproduced in Cronan, 1980). (a–c) *Shapes* include (a) and (b), semi-spherical shapes with 'knobbly' surface; (c) flattened nodule. (d,e) *Surface textures*: (d) smooth; (e) granular. (f) *A typical internal structure*, showing concentric series of light and dark bands (see text). Scale bars are in centimeters.

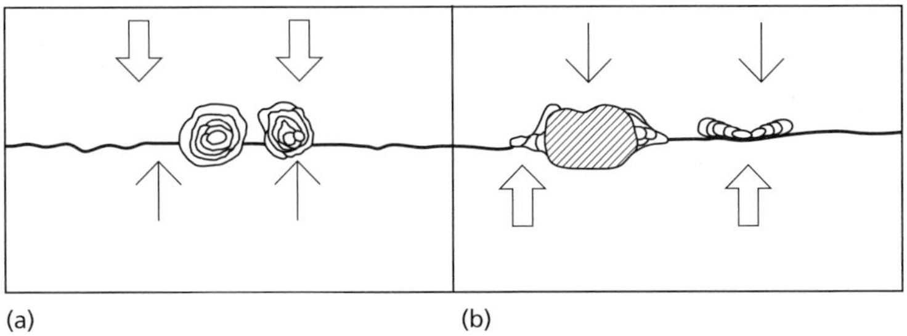

(a)                      (b)

**Fig. 15.7** Schematic representation of ferromanganese nodule end-member morphologies (from Manheim, 1965). The size of the arrows indicates the proportion and direction of metal supply. (a) This illustrates a typical situation in the open-ocean, with the nodules lying on an oxidized sediment substrate: here the dominant supply of metals is from the overlying seawater. (b) This illustrates a typical situation in nearshore and freshwater environments, with the nodules lying on a sediment substrate that is partly reducing in character: here the dominant supply of metals is via intersitial waters from below the substrate surface.

end-member morphologies that might arise from these different element supply mechanisms: see Fig. 15.7(a) and (b).

1 In Fig. 15.7(a), the nodule is formed on an oxidizing substrate and has received only a minor supply of dissolved material from the underlying sediment. Most of the elements therefore are derived from the overlying seawater and the nodule grows slowly, with occasional overturning, so that it acquires a generally spherical shape.

2 In Fig. 15.7(b), most of the elements are supplied from the interstitial waters below the sediment surface, following their suboxic release during the diagenetic sequence (see Section 14.1 and Section 15.3.4.6). They are then precipitated by oxidative processes at a suitable surface, in this case the nodule nucleus, which may be partly buried in the sediment. As the precipitation process continues, the sides nearest the source will take up the thickest layer of this bottom-derived metal input, whereas the upper sides will rely on a supply from the overlying seawater. As a result, nodules of this type will have a flattened discoidal shape.

These are, however, two extreme end-member morphologies, and a variety of intermediate nodule shapes can be found. In general, the extreme seawater end-member is really represented by encrustations on exposed rock surfaces, and the nodule (concretion) form closest to that of the seawater end-member is probably that found on the tops of current-washed seamounts. The extreme *diagenetic* end-member is best developed in nearshore and hemi-pelagic regions, where the diagenetic sequence can proceed to a late stage. Between these two extremes, individual nodules can have a supply of elements from both of the principal sources. Thus, the upper surfaces are supplied from the overlying seawater, whereas the lower surfaces are diagenetic in origin.

### 15.3.4.3 Growth rates

The measurement of natural series radionuclides such as $^{230}Th$ in the nodules allows their growth rates to be estimated. Such measurements reveal that these nodules and crusts grow very slowly. The rate at which a ferromanganese nodule grows is intimately related to the environment within which it forms, the slowest accretion rates of a few millimetres per million years (mm $10^6 yr^{-1}$) being found for the hydrogenous end-member, intermediate rates of a few hundred millimetres per million years being characteristic of the diagenetic end-member, and the highest rates of a few thousand millimetres per million years being shown by the hydrothermal end-member crusts. If it is accepted that at least some pelagic nodules do grow at an average rate of a few millimetres per million years, then this is considerably slower than the rates at which the host sediments accumulate, that is, a few millimetres per thousand years. This raises the question of, why do the manganese nodules not get buried in the sediments? This question has never really been satisfactorily answered, but the answer probably lies in some sort of periodic disturbance by biota or ocean currents to prevent burial.

### 15.3.4.4 Mineralogy

Much of our knowledge of the mineralogy of the manganese and iron phases in the nodules has come from the pioneering work of Buser and Grutter (1956). Subsequently, many other studies have been carried out on the mineralogy of the nodules, and for a detailed review of this topic the reader is referred to Burns and Burns (1977). The nodule mineralogy may be summarized as follows.

There are at least four mineral phases found in ferromanganese nodules and crusts: $\delta MnO_2$, 7 Å and 10 Å manganite, and iron hydroxide.

1 $\delta MnO_2$ forms aggregates of randomly orientated sheets.

2 7 Å manganite and 10 Å manganite have a double-layer structure consisting of ordered sheets of $MnO_2$ with disordered layers of metal ions, such as $Mn^{2+}$ and $Fe^{2+}$, coordinated with $O^{2-}$, $OH^-$ and $H_2O$, lying between them. In the 10 Å manganites the water and hydroxyls are probably present as discrete layers, whereas in the 7 Å manganites they form a single layer.

3 Amorphous iron hydroxide ($FeOOH \cdot nH_2O$), which may be converted to goethite ($\alpha$-$FeOOH$); other iron-rich phases, for example, lepidocrocite, can also be present.

The manganese phases have been related to known minerals, and in the most commonly adopted terminology 10 Å manganite is referred to as toderokite and 7 Å manganite is termed birnessite. Thus, the principal mineral phases in the nodules are $\delta MnO_2$, birnessite, toderokite and some form of hydrous iron oxide.

### 15.3.4.5 Chemistry

Ferromanganese nodules are particularly rich in manganese. However, they also contain relatively large concentrations of a variety of trace metals. Cronan (1976) assessed the magnitudes of the enrichments of a series of elements in ferromanganese nodules relative to their average crustal abundances, and on this basis he was able to identify three groups of elements.

1 The enriched elements, which were divided into a number of subgroups depending on the extent of their enrichment:

(a) Mn, Co, Mo and Th, which are enriched in the nodules by a factor of more than 100;

(b) Ni, Ag, Ir and Pb, which are enriched by a factor of between 50 and 100;

(c) B, Cu, Zn, Cd, Yb, W and Bi, which have enrichments between 10 and 50;

(d) P, V, Fe, Sr, Y, Zr, Ba, La and Hg, which are enriched by less than a factor of 10.

2 The non-enriched–non-depleted elements; these include Na, Mg, Ca, Sn, Ti, Ga, Pd and Au.

3 The depleted elements, which include Al, Si, K, Sc, Cr; however, the depletions are only slight.

A compilation of the average abundances of elements in ferromanganese nodules from the major oceans is given in Table 15.2, together with average elemental enrichment factors. The data in this table give an indication of how the composition of the nodules varies on an inter-oceanic basis. To some extent the variations in the chemical compositions of the nodules can be related to their mineralogy which, in turn, is a function of the depositional environment. Price and Calvert (1970) concluded that Pacific Ocean nodules have a continuous range of chemical and mineralogical compositions which extend between those of two end-members.

1 *Marginal end-member.* This has a disc-shaped morphology. Chemically, it is characterized by a high Mn:Fe ratio, and is relatively enriched in Mn, Cu, Ni and Zn. Mineralogically, it is composed of both $\delta MnO_2$ and toderokite, and some varieties also contain birnessite. This end-member is typically found in marginal and hemi-pelagic sediments, and is equivalent to the *diagenetic* nodule identified above.

2 *Open-ocean end-member.* This has a generally spherical shape. Chemically it is typified by a low Mn:Fe ratio, and is relatively enriched in Fe, Co and Pb. Mineralogically, it consists mainly, or even exclusively, of $\delta MnO_2$. This end-member is best represented by ferromanganese crusts and nodules from seamounts, and is equivalent to the seawater nodule identified above.

### 15.3.4.6 Formation

The fundamental geochemistry of Mn and Fe ultimately explains the formation of ferromanganese crusts, concretions and nodules (see Maynard, 2003). The key points are summarized below.

• Both Mn and Fe are abundant in the crust but in their oxidized forms, Mn(IV) and Fe(III), are very

**Table 15.2** Average abundances of elements in ferromanganese oxide deposits from each of the major oceans (in wt.%), and enrichment factors for each element (from Cronan, 1976) (Superscript numbers denote powers of ten, e.g.$^{-6}$ = ×10$^{-6}$).

| Element | Pacific Ocean | Atlantic Ocean | Indian Ocean | Southern Ocean | World Ocean average | Crustal abundance | Enrichment factor |
|---|---|---|---|---|---|---|---|
| B | 0.0277 | – | – | – | – | 0.0010 | 27.7 |
| Na | 2.054 | 1.88 | – | – | 1.9409 | 2.36 | 0.822 |
| Mg | 1.710 | 1.89 | – | – | 1.8234 | 2.33 | 0.782 |
| Al | 3.060 | 3.27 | 3.60 | – | 3.0981 | 8.23 | 0.376 |
| Si | 8.320 | 9.58 | 11.40 | – | 8.624 | 28.15 | 0.306 |
| P | 0.235 | 0.098 | – | – | 0.2244 | 0.105 | 2.13 |
| K | 0.753 | 0.567 | – | – | 0.6427 | 2.09 | 0.307 |
| Ca | 1.960 | 2.96 | 3.16 | – | 2.5348 | 4.15 | 0.610 |
| Sc | 0.00097 | – | – | – | – | 0.0022 | 0.441 |
| Ti | 0.674 | 0.421 | 0.629 | 0.640 | 0.6424 | 0.570 | 1.13 |
| V | 0.053 | 0.053 | 0.044 | 0.060 | 0.0558 | 0.0135 | 4.13 |
| Cr | 0.0013 | 0.007 | 0.0014 | – | 0.0014 | 0.01 | 0.14 |
| Mn | 19.78 | 15.78 | 15.12 | 11.69 | 16.174 | 0.095 | 170.25 |
| Fe | 11.96 | 20.78 | 13.30 | 15.78 | 15.608 | 5.63 | 2.77 |
| Co | 0.335 | 0.318 | 0.242 | 0.240 | 0.2987 | 0.0025 | 119.48 |
| Ni | 0.634 | 0.328 | 0.507 | 0.450 | 0.4888 | 0.0075 | 65.17 |
| Cu | 0.392 | 0.116 | 0.274 | 0.210 | 0.2561 | 0.0055 | 46.56 |
| Zn | 0.068 | 0.084 | 0.061 | 0.060 | 0.0710 | 0.007 | 10.14 |
| Ga | 0.001 | – | – | – | – | 0.0015 | 0.666 |
| Sr | 0.085 | 0.093 | 0.086 | 0.080 | 0.0825 | 0.0375 | 2.20 |
| Y | 0.031 | – | – | – | – | 0.0033 | 9.39 |
| Zr | 0.052 | – | – | 0.070 | 0.0648 | 0.0165 | 3.92 |
| Mo | 0.044 | 0.049 | 0.029 | 0.040 | 0.0412 | 0.00015 | 274.66 |
| Pd | 0.602$^{-6}$ | 0.574$^{-6}$ | 0.391$^{-6}$ | – | 0.553$^{-6}$ | 0.665$^{-6}$ | 0.832 |
| Ag | 0.0006 | – | – | – | – | 0.000007 | 85.71 |
| Cd | 0.0007 | 0.0011 | – | – | 0.00079 | 0.00002 | 39.50 |
| Sn | 0.00027 | – | – | – | – | 0.00002 | 13.50 |
| Te | 0.0050 | – | – | – | – | – | – |
| Ba | 0.276 | 0.498 | 0.182 | 0.100 | 0.2012 | 0.0425 | 4.73 |
| La | 0.016 | – | – | – | – | 0.0030 | 5.33 |
| Yb | 0.0031 | – | – | – | – | 0.0003 | 10.33 |
| W | 0.006 | – | – | – | – | 0.00015 | 40.00 |
| Ir | 0.939$^{-6}$ | 0.932$^{-6}$ | – | – | 0.935$^{-6}$ | 0.132$^{-7}$ | 70.83 |
| Au | 0.266$^{-6}$ | 0.302$^{-6}$ | 0.811$^{-7}$ | – | 0.248$^{-6}$ | 0.400$^{-6}$ | 0.62 |
| Hg | 0.82$^{-4}$ | 0.16$^{-4}$ | 0.15$^{-6}$ | – | 0.50$^{-4}$ | 0.80$^{-5}$ | 6.25 |
| Tl | 0.017 | 0.0077 | 0.010 | – | 0.0129 | 0.000045 | 286.66 |
| Pb | 0.0846 | 0.127 | 0.070 | – | 0.0867 | 0.00125 | 69.36 |
| Bi | 0.0006 | 0.0005 | 0.0014 | – | 0.0008 | 0.000017 | 47.05 |

insoluble in seawater while in their reduced form, Mn(II) and Fe(II), are more soluble. Hence, mobilization of the oxidized form under reducing (or acidic) conditions followed by reprecipitation under oxidizing and alkaline conditions is accepted to be at that heart of ferromanganese nodule formation.

- Mn(II) oxidation can be catalysed by Mn(IV) oxides (autocatalysis) which can act to focus further Mn precipitation and encourage nodule and concretion development.
- Iron and manganese oxides are also well known scavenging agents with high surface areas and a

strong tendency to adsorb other cations which will contribute to the accumulation of other metals within the nodules and concretions.

• Fe oxidation is faster than Mn oxidation and in the presence of sulfide reduced iron is insoluble as FeS, while Mn(II) is more soluble in the presence of sulfide. These properties can promote the separation of Mn and Fe during a reductive dissolution/oxic reprecipitation process.

The two end-member nodules described above are formed by two end-member processes, the marginal nodules acquiring most of their component metals from the interstitial waters of the host sediments (diagenetic nodules) and the open-ocean varieties taking most of their metals from the overlying water (seawater nodules). It must be stressed, however, that there is a continuous spectrum of oceanic ferromanganese nodules between these two extreme end-members. This is evidenced by the fact that many nodules from abyssal regions have both a seawater source (to their upper surfaces) and a diagenetic source (to their lower surfaces). Thus, many nodules are compositionally different on their seawater-facing and sediment-facing surfaces. Nonetheless, the basic differences are still related to the two different prime sources of metals to nodules.

According to Dymond *et al.* (1984), 'the compositions of ferromanganese nodules respond in a consistent manner to the sea floor environment in which they form'. In this respect the two end-member (i.e. seawater and diagenetic) nodule source concept provides a useful framework for a discussion of how the nodules formed. However, it was pointed out above that the components formed from both seawater and diagenetic sources can be subdivided into a number of genetic classes. For example, seawater sources can be either hydrogenous or hydrothermal, and diagenetic sources can be either oxic or suboxic in character. As a result, nodules having a seawater source can be subdivided into hydrogenous and hydrothermal types, and those having a diagenetic source can be subdivided into oxic and suboxic types. Hydrothermal processes will be considered separately in Section 15.3.6, and for the moment attention will be focused on the three nodule accretionary modes identified by Dymond *et al.* (1984). These are:

**1** that resulting from hydrogenous precipitation from seawater;

**2** that arising from oxic diagenetic processes;

**3** that associated with suboxic diagenetic processes.

Dymond *et al.* (1984) then tested their three-mode accretionary model with reference to the formation of ferromanganese nodules at three contrasting MANOP sites in the Pacific (sites H, S and R). This study will serve as a blueprint for a discussion of how the nodule-forming processes operate. To aid the discussion, model compositions of ferromanganese components arising from the three accretionary modes are given in Table 15.3.

*The seawater (or hydrogenous) nodule end-member.* This type of ferromanganese deposit is formed by the direct precipitation of colloidal metal oxides from seawater at the growth site, together with the accretion to the depositional surface of suspended Fe–Mn precipitates formed in the water column. The extreme example of a hydrogenous end-member ferromanganese deposit is provided by encrustations that are formed on exposed rock surfaces outside the direct influence of hydrothermal activity. Aplin and Cronan (1985a) described a series of such encrustations from the Line Islands (Central Pacific). The crusts have relatively low Mn : Fe ratios, and the principal minerals in them are $\delta MnO_2$ and amorphous $FeOOH \cdot nH_2O$; of these the $\delta MnO_2$ was the most important trace-metal-bearing phase, containing Co, Mo, Ni, Zn and Cd, with only Ba appearing to be associated specifically with the iron hydroxide phase. The authors concluded that the crusts had formed directly from the slow accumulation of trace-metal-enriched oxides, which had been deposited directly from the overlying water column, and that the trace-metal–mineral associations had resulted from the different scavenging behaviours of Mn and Fe oxides in the water.

The accretion of the hydrogenous ferromanganese deposits is constrained by the concentration of Mn and the kinetics of its oxidation, and Piper *et al.* (1984) have reviewed the processes involved in the formation of marine manganese minerals. According to these authors the Mn oxide that forms initially may be a metastable hausmannite, which undergoes ageing to $\delta MnO_2$ possibly by a disproportionation reaction as suggested by Hem (1978). Thus

$$3H_2O + \tfrac{1}{2}O_2 + 3Mn^{2+} = Mn_3O_4 + 6H^+ \qquad (15.7)$$

**Table 15.3** Chemical composition of 'end-member' ferromanganese nodules; units, $\mu g\,g^{-1}$.

| Element | Ferromanganese crusts | | Ferromanganese nodules | |
| | Hydrogenous | | Oxic | Suboxic |
| | Line Islands, Pacific* | MANOP Site H, Pacific† | MANOP Site H, Pacific† | MANOP Site H, Pacific† |
| --- | --- | --- | --- | --- |
| Mn | 204 000 | 222 000 | 316 500 | 480 000 |
| Fe | 170 000 | 190 000 | 44 500 | 4 900 |
| Co | 5 500 | 1 300 | 280 | 35 |
| Ni | 3 900 | 5 500 | 10 100 | 4 400 |
| Zn | 590 | 750 | 2 500 | 2 200 |
| Cu | – | 1 480 | 4 400 | 2 000 |
| Al | 16 000 | 11 800 | 27 100 | 7 500 |
| Ti | 12 000 | 5 300 | 1 700 | 365 |
| Si | – | 52 000 | 59 000 | 16 300 |
| Na | – | 10 400 | 16 100 | 32 800 |
| K | – | 4 900 | 8 200 | 6 200 |
| Mg | – | 10 400 | 23 000 | 13 800 |
| Ca | 26 000 | 26 000 | 15 200 | 12 500 |
| Mn:Fe | 1.2 | 1.2 | 7.1 | 98 |
| Co:Mn | 0.027 | 0.006 | 0.0009 | 0.00007 |
| Ni:Mn | 0.019 | 0.024 | 0.032 | 0.0092 |
| Cu:Mn | 0.0075 | 0.0022 | 0.020 | 0.0024 |
| Zn:Mn | 0.0029 | 0.0034 | 0.0079 | 0.0046 |

*Data from Aplin and Cronan (1985a).
†Data from Dymond et al. (1984).

and

$$Mn_3O_4 + 4H^+ = \delta MnO_2 + 2Mn^{2+} + 2H_2O \qquad (15.8)$$

Although the best end-member composition for the hydrogenous ferromanganese deposits is provided by crusts formed directly on to exposed rock surfaces, this type of accretion process also contributes to all nodule surfaces that are exposed to seawater, and it is evident that in addition to crusts, nodules having a large hydrogenous component can be found in the marine environment. Once a sediment substrate is introduced, however, the conditions for ferromanganese oxide formation change dramatically. In particular, the nodules can become partly buried and so can acquire a diagenetic input of dissolved metals. Although it is therefore convenient to treat diagenetic end-member nodules separately from the point of view of the principal processes involved in their generation, it must be stressed that in a strict sense most marine nodules will be mixed-source types.

*The diagenetic nodule end-member.* The seawater–diagenetic metal-supply system can result in nodules exhibiting top–bottom compositional differences. The extent to which these differences are found depends on how far the diagenetic sequence has proceeded. This, in turn, depends on the magnitude of the down-column carbon flux, which controls the sediment substrate depositional environment (see Chapter 14). In this respect, Dymond et al. (1984) distinguished between nodules formed under conditions of oxic and suboxic diagenesis.

1 *Oxic diagenesis.* In oxic diagenesis, degradable organic matter is destroyed by dissolved $O_2$, and at this stage particulate $MnO_2$ is not broken down for utilization as a secondary oxidant (see Chapter 14). There is no doubt that nodules formed on *oxic* substrates receive a proportion of their metals directly from seawater (hydrogenous supply). However, *direct* precipitation from seawater is not the sole, or even the dominant, route by which Mn and associated metals are supplied to nodules accreting on oxic sediment substrates, the reason being that the

sediments themselves can act as metal sources. A number of mechanisms have been proposed to describe the manner in which the sediment source operates under conditions of oxic diagenesis which include direct release of metals bound to organic matter which decomposes in sediments, desorption of metals bound to sediments and release from sediments (Fig. 15.8a) or release of metals bound within mineral phases in the sediments (Fig. 15.8b). (Callender and Bowser, 1980; Lyle *et al.*, 1984; Dymond *et al.*, 1984)

2 *Suboxic diagenesis.* In suboxic diagenesis, the diagenetic sequence is driven by the utilization of secondary oxidants for the degradation of organic carbon, usually at a relatively shallow depth in the sediment (see Chapter 14). Manganese dioxide can act as one of these secondary oxidants and in the process $Mn^{2+}$ (and other trace elements associated with the oxides) is released into the interstitial waters where it can migrate upwards under concentration gradients to undergo reprecipitation to $MnO_2$ in the upper oxic zone , that is, the elements reach the site of nodule formation from below. The depth at which this suboxic process takes place can be affected by the strength of the down-column carbon-mediated signal. For example, Dymond *et al.* (1984) reported the presence of suboxic diagenetic nodules at MANOP site H, but pointed out that Mn released from the utilization of $MnO_2$ was reprecipitated below the sediment surface on which the nodules were forming. To explain the formation of the suboxic diagenetic nodules, the authors therefore suggested that the site had received a supply of elements from pulses of biogenic debris, which had been sufficiently strong to decrease the depth at which suboxic diagenesis normally takes place in the sediments.

In oxic diagenesis, the destruction of labile material from the organic carriers results in the supply of relatively large amounts of metals, such as Cu, Ni and Zn, giving rise to the formation of Cu-, Ni- and Zn-rich toderokite. The suboxic diagenesis of Mn oxides from the organic matter pulses, however, does not release such large amounts of Cu and Ni, and the Mn-rich toderokite formed is less stable than the Ni-, Cu- and Zn-rich variety, and can collapse on dehydration to yield birnessite. Dymond *et al.* (1984) estimated the overall growth rates of the suboxic nodules to be of the order of ~100–200 mm $10^6$ yr$^{-1}$.

### 15.3.4.7 Ferromanganese nodules: summary

There are three general modes by which non-hydrothermal ferromanganese crusts and nodules accrete: these are the *seawater* mode, the *oxic diagenetic* mode and the *suboxic diagenetic* mode. The crusts represent an individual end-member formed by the seawater mode, but the nodules are usually top–bottom mixtures of the seawater mode and one, or both, of the diagenetic modes. The signatures associated with each of the three modes are summarized in the following.

1 Surface ferromanganese components formed by the seawater mode contain $\delta MnO_2$. They have low Mn:Fe ratios (~1), low Ni:Co ratios, and accrete at slow rates, which probably are in the order of ~1–2 mm $10^6$ yr$^{-1}$. Ferromanganese crusts on exposed rock surfaces and nodules formed on the tops of seamounts, and other swept sediment surfaces, are formed by this mode. In addition, the upper surfaces of all nodules have a component derived from the seawater accretion mode.

2 Sediment surface ferromanganese components formed by the oxic diagenetic mode contain a Cu-, Ni- and Zn-rich toderokite. The metals have been supplied by particulate matter via a variety of oxic diagenetic mechanisms, and the components have intermediate Mn:Fe ratios (e.g. ~5–10) and accrete at rates of a few tens of millimetres per million years. These components are found on the lower surfaces of nodules that grow on sediment substrates where the diagenetic sequence has not progressed beyond the oxic stage.

3 Sediment surface ferromanganese components formed by the suboxic diagenetic mode contain a Mn-rich toderokite, which may undergo transformation to birnessite. The metals necessary for the formation of these components are released from particulate matter during suboxic diagenesis. In some instances they may arise from the oxidative utilization of $MnO_2$ at depth in the sediment. If the $Mn^{2+}$ released in this manner is trapped below the sediment surface, however, a secondary source of suboxic $Mn^{2+}$ may be provided by pulses of biogenic material that are sufficiently strong to permit suboxic diagenesis to occur close to, or even at, the sediment surface. The components formed during suboxic diagenesis tend to have relatively high Mn:Fe ratios (often >20), and accrete episodically at rates that can

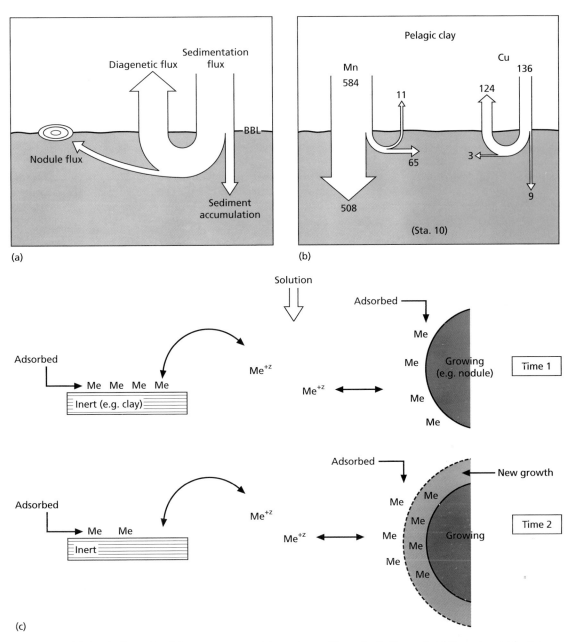

**Fig. 15.8** Models for the formation of ferromanganese nodules under conditions of oxic diagenesis. (a) General model for the early diagenetic release of trace metals (from Callender and Bowser, 1980). (b) Adsorption–bioturbation model for the transfer of loosely held elements to growing nodules (from Lyle et al., 1984); see text for details. Reprinted with permission from Elsevier.

be as high as several hundred millimetres per million years. Suboxic diagenetic components characterize the lower surfaces of nodules formed on marginal and hemi-pelagic sediments in which the diagenetic sequence has reached the $MnO_2$ oxidative utilization stage.

### 15.3.5 Sediment ferromanganese oxyhydroxides

In addition to macronodules, oceanic sediments contain micronodules and small-sized oxyhydroxides of a non-nodular origin. These are an intimate part of the sediment, and in the present section attention will be focused on the question, 'Do they form part of a crust → nodule → sediment oxyhydroxide continuum, or is there a major decoupling between the nodule–sediment system?'

Aplin and Cronan (1985a,b) gave data on the chemical composition of a series of ferromanganese crusts, sediment oxyhydroxides, $\delta MnO_2$-rich macronodules and toderokite-rich macronodules from the central Pacific. The authors used a selective leaching technique to determine the compositions of the sediment oxyhydroxides, and then compared these with those of the crusts and nodules from the same sample suite. The data are summarized in Table 15.4, and two principal conclusions can be drawn from this investigation.

1 The chemical compositions of the crusts, sediment oxyhydroxides and $\delta MnO_2$-rich nodules are generally similar. There are, however, some differences between the three phases; for example, the Mn:Fe, Cu:Mn and Co:Mn ratios varied, with those of the $\delta MnO_2$-rich nodules being intermediate between those of the crusts and the sediment oxyhydroxides. According to Aplin and Cronan (1985a,b) this can be explained in the following manner. The crusts derive their elements largely from seawater (hydrogenous supply). In contrast, the oxyhydroxides acquire most of their elements from processes occurring in the sediments, and in the region studied much of this elemental input arises from the decay of labile biologically transported material during early diagenesis (oxic *diagenetic* supply). The two phases therefore represent a seawater and an oxic diagenetic end-member, respectively. The $\delta MnO_2$-rich nodules, which grow at the sediment surface, receive elements from both the hydrogenous seawater and the oxic diagenetic sources; thus, they have a composition that is intermediate between those of the seawater and the oxic diagenetic end-members. In this crust → $\delta MnO_2$-rich nodule → sediment oxyhydroxide sequence it is therefore oxic diagenetic reactions that perturb the composition of the seawater end-member.

2 The chemical compositions of the toderokite-rich nodules were quite distinct from those of the members of the crust → $\delta MnO_2$-rich nodule → sediment oxyhydroxide sequence. For example, it can be seen from Table 15.4 that the Mn:Fe, Cu:Mn and Ni:Mn ratios are higher, and the Co:Mn ratio lower, in the toderokite-rich nodules than in the other components. These toderokite-rich nodules are indicative of more intense (i.e. suboxic) diagenesis, which involves the remobilization of $Mn^{2+}$ from the breakdown of Mn oxides, and Aplin and Cronan (1985b) suggested that in the area under investigation the

**Table 15.4** Elemental ratios in sediment oxyhydroxides, ferromanganese crusts, $\delta MnO_2$-rich nodules and toderokite-rich nodules from the central Pacific (data from Aplin and Cronan, 1985b).

| Elemental ratio | Component | | | |
| --- | --- | --- | --- | --- |
| | Sediment oxyhydroxide* | Fe–Mn crusts | $\delta MnO_2$-rich nodules | Toderokite-rich nodules |
| Mn:Fe | 0.99 | 1.15 | 0.97 | 2.6 |
| Cu:Mn | 0.020 | 0.009 | 0.013 | 0.032 |
| Ni:Mn | 0.019 | 0.020 | 0.020 | 0.040 |
| Co:Mn | 0.012 | 0.026 | 0.019 | 0.009 |
| Pb:Mn | 0.004 | 0.006 | 0.0048 | 0.002 |
| Zn:Mn | 0.007 | 0.003 | 0.003 | 0.004 |

*Leached fraction of sediment.

toderokite, which is a Cu-, Ni- and Zn-poor variety, had precipitated directly from the sediment interstitial waters.

It would appear, therefore, that although $\delta MnO_2$-rich nodules can have both a seawater and a diagenetic supply of trace metals, there is a general continuum of compositions in the sequence crust $\rightarrow$ $\delta MnO_2$-rich nodule $\rightarrow$ sediment oxyhydroxides. A major decoupling is found, however, in the sediment–nodule system, at the seawater–diagenetic boundary, between the $\delta MnO_2$-rich and the toderokite-rich nodules. This results because suboxic diagenesis is a recycling process, which can release elements already incorporated into the sediments, and depends largely on the extent to which the diagenetic sequence has proceeded. Oxic diagenesis, for example, when labile biologically transported elements are solubilized, can supply metals to both sediment oxyhydroxides and nodules. The decoupling break is strongest at the point where suboxic diagenesis releases $Mn^{2+}$, however, which can replace $Cu^{2+}$, $Ni^{2+}$ and $Zn^{2+}$ in toderokite and can result in the formation of a Cu-, Ni- and Zn-poor toderokite variety, which may subsequently form birnessite (see Section 15.3.4.6).

### 15.3.6 Hydrothermal ferromanganese deposits

The convection of seawater through freshly generated oceanic crust at the centres of sea-floor spreading is now recognized as being a process of global geochemical significance (see Sections 5.1 and 6.3). The elements that are solubilized during hydrothermal activity are subsequently removed from solution via solid phases, and the processes involved can be described in terms of the Red Sea system, a system in which the complete sequence of metal-rich precipitates can be found in sediments surrounding the discharge points of the mineralizing solutions.

In the Red Sea system hot saline brines become enriched in metals by reacting at high temperature with underlying evaporites, shales and volcanic rocks, and on discharge into seawater they gather in deeps in the central rift valley area. According to Cronan (1980) the Atlantis II Deep, an area of present-day metalliferous sediment deposition, can be used to illustrate the range of hydrothermal precipitates formed in the oceans. The general precipitation sequence from the hydrothermal solutions

occurs in the order: sulfides (e.g. sphalerite, pyrite, chalcopyrite, marcasite, galena) $\rightarrow$ iron silicates (e.g. smectite, chamosite, amorphous silicates) $\rightarrow$ iron oxides $\rightarrow$ manganese oxides. This fractionation of the precipitates manifests itself in their spatial separation from the venting source; for example, the iron and manganese oxides sometimes form distinct haloes at a distance from the discharge point. Thus, marine hydrothermal deposits can be thought of as representing various stages in the evolution of the mineralizing solutions. It was pointed out in Section 5.1, however, that the fractionation sequence may commence before the hydrothermal mineralizing solutions actually reach the venting outlet. Thus, in the white smokers the mixing of seawater and hydrothermal fluids occurs at depth in the system, with the result that sulfides have been removed within the vents before the fluids are debouched into seawater at a relatively low temperature. Here, therefore, the sulfide stage in the sequence has been reached under the sea floor. In the black smokers, however, hydrothermal fluids are vented directly on to the sea floor at high temperature, and the sulfides are precipitated directly on to the sea bed. Some of these sulfides are used to build the venting chimneys and some, together with other still dissolved hydrothermally derived metals, are dispersed to carry on the sequence around, and away from, the vents.

Hydrothermal deposits have been found, either at the sediment surface or close to the sediment–basalt basement interface, at many marine locations; for example, Lalou (1983) identified 70 such locations. The metal compositions of a series of hydrothermal deposits from a number of the more important hydrothermal study regions are given in Table 15.5.

Hydrothermal sulfides are generally restricted either to within the vents themselves or to the immediate vicinity of the venting system. In contrast, Fe and Mn dispersion haloes can cover a much greater area, and ferromanganese oxyhydroxides, nodules and crusts are the characteristic sedimentary components of much open-ocean hydrothermal activity. This occurs particularly along ridge axis, which often contains the water impacted by the mid ocean ridges, but the impact can extend hundreds of kilometres beyond the ridge axis as well (German and Von Damm, 2003). It is of interest, therefore, to compare the compositions of these end-member hydrothermal ferromanganese components with those that have a

**Table 15.5** Chemical compositions of hydrothermal deposits; units, $\mu g\,g^{-1}$.

|    | A* | B† | C‡ | D§ | E¶ | Fi | G** | H†† | I‡‡ | J§§ |
|----|------|------|------|------|------|------|------|------|------|------|
| Mn | 60000 | 46000 | 39043 | 430000 | 550000 | 470000 | 40000 | 279000 | 410000 | 380000 |
| Fe | 180000 | 141000 | 587 | 1600 | 2000 | 6600 | 232900 | 10500 | 8000 | 27000 |
| Al | 5000 | 23000 | – | 1800 | – | 2000 | 7900 | 12700 | 9000 | 6900 |
| Ti | 200 | – | – | – | – | 28 | – | – | 400 | 1060 |
| Co | 105 | 64 | 19 | 24 | 39 | 13 | 10 | 82 | 33 | 30 |
| Ni | 430 | 820 | 353 | 880 | 180 | 125 | 80 | 371 | 310 | 400 |
| Cu | 730 | 910 | 43 | 450 | 50 | 80 | 76 | 206 | 120 | 80 |
| Zn | 380 | 330 | – | 540 | 2020 | 90 | 35 | 83 | 400 | 310 |
| Mo | 30 | – | – | – | – | 540 | – | – | 900 | – |
| V | 450 | – | – | – | – | 110 | 152 | 214 | 110 | – |

\* Metal-rich sediment at crest of East Pacific Rise, data on a carbonate-free basis (Bostrom and Peterson, 1969).
† Metal-rich sediment, Bauer Deep, central Pacific, data on a carbonate-free basis (Dymond et al., 1973).
‡ Hydrothermal deposits, TAG area, Mid-Atlantic Ridge (Scott et al., 1974).
§ Hydrothermal deposits, TAG area, Mid-Atlantic Ridge (Toth, 1980).
¶ Hydrothermal deposits, Galapagos spreading centre (Moore and Vogt, 1976).
i Hydrothermal deposits, Galapagos mounds (Moorby and Cronan, 1983).
\*\* Hydrothermal clay-rich deposit, FAMOUS area, Mid-Atlantic Ridge (Hoffert et al., 1978).
†† Hydrothermal Fe-Mn concretions, FAMOUS area, Mid-Atlantic Ridge (Hoffert et al., 1978).
‡‡ Hydrothermal deposits, southwest Pacific island arc system (Moorby et al., 1984).
§§ Hydrothermal deposits, Gulf of Aden (Cann et al., 1977).

*hydrogenous* origin. For this purpose, representative chemical and mineralogical analyses, together with growth rate characteristics, for a hydrogenous crust (column A), an oxic nodule (column B), a suboxic nodule (column C) and a ridge-centre hydrothermal crust (column D) are given in Table 15.6. When hydrothermal crusts are considered it must be remembered, however, that the parent hydrothermal activity is not confined to active centres of sea-floor spreading, but that it can also occur in other regions associated with volcanic activity. One such region can be found in island arcs and their associated marginal basins. For example, Moorby et al. (1984) described ferromanganese crusts recovered from several sites on the Tonga–Kermadec Ridge in the southwest Pacific island arc. The average chemical and mineralogical composition of these crusts is also included in Table 15.6 (column E).

Although most crusts are to some extent mixtures of end-member sources, it is apparent from the various data in Table 15.6 that there are a number of important differences between the hydrothermal and hydrogenous crusts. These can be summarized as follows.

1 *Growth rates*. In general, the hydrogenous crusts appear to accrete at relatively slow rates, which are of the order of a few millilitres per million years. In contrast, the hydrothermal crusts grow at much faster rates, which are of the order of several hundred millimetres per million years. The principal reason for this difference is probably related to the fact that the hydrothermal crusts are formed in the vicinity of Mn-rich mineralizing solutions, whereas hydrogenous nodules have to rely on a background supply of Mn from normal seawater.

2 *Mineralogy*. Hydrogenous crusts are usually dominated by $\delta MnO_2$ phases. In contrast, the predominant minerals in the hydrothermal crusts are often toderokite or birnessite, that is, minerals with relatively high Mn:Fe ratios; however, a wide range of manganese minerals has in fact been reported in hydrothermal crusts.

3 *Chemistry*. There are two principal differences between the chemical compositions of the hydrothermal and hydrogenous crusts. First, the hydrothermal crusts generally are depleted in Co, Ni and Cu compared with their hydrogenous counterparts; this probably is a result of the uptake of these elements

**Table 15.6** Some general characteristics of oceanic Fe–Mn deposits; units, $\mu g\,g^{-1}$.

| | Chemical composition | | | | |
| --- | --- | --- | --- | --- | --- |
| | A* | B* | C* | D† | E‡ |
| | Hydrogenous crust | Oxic nodule | Suboxic nodule | Hydrothermal crust | Hydrothermal crust |
| Mn | 222 000 | 316 500 | 480 000 | 410 000 | 550 000 |
| Fe | 190 000 | 44 500 | 4900 | 8000 | 2000 |
| Co | 1300 | 280 | 35 | 33 | 39 |
| Ni | 5500 | 10 100 | 4400 | 310 | 180 |
| Cu | 1480 | 4400 | 2000 | 120 | 50 |
| Zn | 750 | 2500 | 2200 | 400 | 2020 |
| Mn : Fe | 1.2 | 7.1 | 98 | 51 | 275 |
| Principal mineralogy of Fe–Mn phases | $\delta MnO_2$ | $\delta MnO_2$, toderokite | Toderokite | Birnessite | Birnessite, toderokite |
| Approximate growth rates (mm $10^6\,yr^{-1}$) | 1–2 | 10–50 | 100–200 | 500 | 1000–2000 |

\* Data from Dymond *et al.* (1984), see Table 15.4.
† Data from Moorby *et al.* (1984); non-spreading centre hydrothermal deposit, southwest Pacific island arc.
‡ Data from Moore and Vogt (1976); spreading centre hydrothermal deposit, Galapagos region.

in sulfides early in the hydrothermal sequence. Secondly, the hydrothermal crusts exhibit an extreme fractionation of Mn from Fe, evidenced in their relatively very high Mn : Fe ratios, which probably result from the removal of Fe in the sulfide, silicate and Fe oxide phases that precipitate before the Mn oxides in the hydrothermal pulses. Thus, the crusts, in which the constituent elements are derived directly from seawater, are enriched in Mn relative to Fe in hydrothermal regions; however, metal-rich sediments, which are found around active ridge crest regions, do not show this fractionation of Mn from Fe because they contain a variety of hydrothermal components, including Fe oxides. The precipitation of hydrothermal ferromanganese deposits does result in the efficient scavenging of metals from seawater (see Chapter 5 and German and Von Damm, 2003) so even the hydrothermal crusts do not have one exclusive source of material. One of the principal reasons for the differences between hydrogenous and hydrothermal ferromanganese components, especially with regard to the faster growth rates of the latter, is that the hydrothermal activity involves the local introduction of Fe- and Mn-rich solutions into seawater at a relatively high rate. In contrast, hydrogenous components acquire their Fe and Mn from 'background' seawater. The two types of end-member ferromanganese deposits therefore have independent metal sources, for example, ridge-crest metalliferous material having a local hydrothermal source and abyssal plain nodules having a 'background' hydrogenous source.

Clearly, hydrothermal activity is an important source for Fe and Mn (and some other metals) to seawater. However, there is no doubt that end-member ferromanganese deposits have a number of individual characteristics, which result largely from differences between the local 'pulsed' (hydrothermal) and smoothed out seawater 'background' (hydrogenous) metal sources. The question of the ocean-wide importance of hydrothermal sources to ferromanganese deposits therefore must remain unresolved at present.

### 15.3.7 Ferromanganese deposits in the oceans: summary

1 Ferromanganese components in oceanic sediments consist of crusts, nodules and oxyhydroxides.
2 The formation of these ferromanganese components can be related to a hydrogenous–hydrothermal–diagenetic Mn supply trinity.

3 At the mid-ocean ridges the supply of Mn can be dominated by hydrothermal pulses. Away from the ridges, non-hydrothermal sources become more important, and there is an increasing contribution of elements to ferromanganese components from 'background' seawater, either directly (hydrogenous supply) or via diagenetic processes that involve a sediment substrate.

4 In open-ocean areas, oxic diagenesis releases elements close to the sediment surface, but in marginal regions suboxic diagenesis drives the processes involved in the liberation of elements following recycling at depth in the sediment column.

5 The ferromanganese components formed via this hydrogenous–hydrothermal–diagenetic trinity play an extremely important role in removing a variety of dissolved elements from seawater and also exert a major influence on the composition of deep-sea sediments.

## 15.4 Cosmogenous components

### 15.4.1 Definition

These are components that have been formed in outer space and have reached the surface of the Earth via the atmosphere. According to Chester and Aston (1976) the extraterrestrial material that has been identified in deep-sea sediments includes cosmic spherules and microtectites, together with cosmic-ray-produced radioactive and stable nuclides. However, these cosmic components form a very minor part of oceanic sediments, and for a detailed review of the topic the reader is referred to Brownlee (1981).

### 15.4.2 Cosmic spherules

Essentially, there are two kinds of spherules in deep-sea sediments, the iron and the stony types.

#### 15.4.2.1 The iron spherules

Small black magnetic spherules have been found in sediments from all the major oceans (see e.g. Brunn et al., 1955; Crozier, 1961; Millard and Finkelman, 1970). They were first described by Murray and Renard (1891), and it was these authors who suggested that they had a cosmic origin. This type of

spherule has a density of ~$6\,g\,cm^{-3}$, and consists of a metallic (Fe–Ni) nucleus surrounded by a shell of magnetite and wüstite, the latter being an iron oxide that is virtually non-existent on Earth but is found in meteorite fusion crusts (see e.g. Murrell et al., 1980). According to Murray and Renard (1891), this type of cosmic spherule originated as molten droplets ablated from a meteorite. The droplets then underwent oxidation, during which the outer magnetite shell was formed as they entered the Earth's atmosphere. In general, this theory is still accepted, and these spherules are considered to have a cosmic origin (see e.g. Parkin et al., 1967; Blanchard and Davies, 1978; Blanchard et al., 1980). Estimates of the influx, or accretion, rates of interplanetary dust particles (IDPs) to the surface of the Earth vary widely, ranging from >$10^6$ to <$10^3\,ton\,yr^{-1}$; in one estimate, Esser and Turekian (1988) used osmium isotope systematics to derive a rate of $4.9 \times 10^4\,ton\,yr^{-1}$ for the accretion rate of C-1 carbonaceous chondrite material to the Earth's surface. Interplanetary dust of the zodical cloud is inhomogeneously distributed in space, being localized in bands related to asteroid families, and is confined in rings locked to the Earth's orbit (see e.g. Dermott et al., 1994). It has been suggested that variations in the Earth's orbit induce changes in the rate at which IDPs are accreted, higher accretion occurring during periods of high inclination and/or eccentricity, and Muller and MacDonald (1995) proposed that the accretion varies with a 100 k yr periodicity in the Earth's orbital inclination, which is linked into glacial cycles. Farley and Patterson (1995) used the $^3$He extraterrestrial flux to the sea floor to confirm the 100 k yr periodicity in the Atlantic; however, Marcantonio et al. (1996) did not find $^3$He evidence to support the 100 k yr periodicity in sediments from the equatorial Pacific. It is thought by some workers that the accretion rates of the iron spherules may actually approach zero at certain times. For example, Parkin et al. (1967) sampled clean air at Barbados and reported that iron spherules with diameters >$20\,\mu m$ were reaching the surface of the Earth at ~$350\,ton\,yr^{-1}$, but that most of this could in fact be accounted for by contamination effects. These authors suggested that variations in the accretion rates of the magnetic spherules, as recorded in deep-sea sediments and ice cores, result from the spasmodic availability of suitable parent meteorites in outer space.

### 15.4.2.2 The stony spherules

These have a density of ~3 g cm$^{-3}$ and diameters in the range ~15–250 µm, although they often depart from a spherical shape. Essentially, they consist of fine-grained silicates (mainly olivine), magnetite and glass. It is thought generally that this type of spherule has originated from stony meteorites, although the actual mechanism involved in their formation is not clear. For example, Parkin *et al.* (1967) have suggested that they are in fact micrometeorites, which undergo relatively little physical alteration on passage through the atmosphere. Blanchard *et al.* (1980), however, showed that the stony spherules have similarities with chondrite fusion crusts, and concluded that they are products of the atmospheric heating of stony meteoroids, a theory that would account for the presence of the magnetite in them. On the basis of their deep-sea sediment storage, Murrell *et al.* (1980) have estimated the accretion rate of the stony spherules to be ~90 ton yr$^{-1}$. Some of the stony spherules identified in deep-sea sediments, however, may have a terrestrial origin (see e.g. Fredricksson and Martin, 1963).

### 15.4.3 Microtectites

These are small bodies of green-black glass found in large numbers over a few restricted regions of the Earth's surface. They have been identified in bands in deep-sea sediments from the Indian Ocean and the equatorial Atlantic, and probably originate from meteorite impacts (Glass, 1990).

### 15.5 Summary

We have now described the compositions and origins of the various components that are found in oceanic sediments. In the following chapter, we will attempt to show how the chemical signals associated with the formation of these components combine together to impose a control on the overall chemical composition of the sediments found in the oceans.

### References

Aplin, A.C. and Cronan, D.S. (1985a) Ferromanganese oxide deposits from the central Pacific Ocean, I. Encrustations from the Line Islands Archipelago. *Geochim. Cosmochim. Acta*, **40**, 427–436.

Aplin, A.C. and Cronan, D.S. (1985b) Ferromanganese oxide deposits from the central Pacific Ocean, II. Nodules and associated sediments. *Geochim. Cosmochim. Acta*, **49**, 437–451.

Berger, W.H. (1976) Biogenous deep sea sediments: production, preservation and interpretation, in *Chemical Oceanography*, J.P. Riley and R. Chester (eds), Vol. 5, pp.265–388. London: Academic Press.

Berger, W.H. and Winterer, E.L. (1974) Plate stratigraphy and the fluctuating carbonate line, in Pelagic Sediments on Land and under the Sea, K.J. Hsü and H.C. Jenkyns (eds), Oxford: Blackwell Scientific Publications, International Association of Sedimentologists, Special Publication 1, 11–48.

Biscaye, P.E. (1965) Mineralogy and sedimentation of Recent deep-sea clay in the Atlantic Ocean and adjacent seas and oceans. *Geol. Soc. Am. Bull.*, **76**, 830–832.

Biscaye, P.E., Kolla, V. and Turekian, K.K. (1976) Distribution of calcium carbonate in surface sediments of the Atlantic Ocean. *J. Geophys. Res.*, **81**, 2595–2603.

Blanchard, M.B. and Davies, A.S. (1978) Analysis of ablation debris from natural and artificial iron meteorites. *J. Geophys. Res.*, **83**, 1793–1808.

Blanchard, M.B., Brownlee, D.E., Bunch, T.E., Hodge, P.W. and Kyte, F.T. (1980) Meteor ablation spheres from deep-sea sediments. *Earth Planet. Sci. Lett.*, **46**, 178–190.

Bostrom, K. and Peterson, M.N.A (1969) The origin of aluminium-poor ferromanganoan sediments in areas of high heat flow on the East Pacific rise. *Mar. Geol.*, **7**, 427–447.

Bostrom, K., Kraemer, T. and Gartner, S. (1973) Provenance and accumulation rates of opaline silica, Al, Ti, Fe, Mn, Cu, Ni and Co in Pacific pelagic sediments. *Chem. Geol.*, **11**, 123–148.

Broecker, W.S. (2004) *The oceanic CaCO$_3$ cycle*, in *Treatise on Geochemistry Vol. 6, The Oceans and Marine Geochemistry*, H. Elderfield (ed.) H.D. Holland and K.K. Turekian (series eds), pp.529–549. London: Elsevier.

Brownlee, D.E. (1981) Extraterrestrial components, in The Sea, C. Emiliani (ed.), Vol. 7, pp.733–762. New York: John Wiley & Sons, Inc.

Brunn, A.F., Langer, E. and Pauly, H. (1955) Magnetic particles found by raking the deep sea bottom. *Deep Sea Res.*, **2**, 230–246.

Burns, R.G. and Burns, V.M. (1977) Mineralogy, in *Marine Manganese Deposits*, G.P. Glasby (ed.), pp.185–248. Amsterdam: Elsevier.

Buser, W. and Grutter, A. (1956) Uber die Natur der Mangankollen. *Schweiz. Mineral. Petrogr. Mitt.*, **36**, 49–62.

Callender, E. and Bowser, C.J. (1980) Manganese and copper geochemistry of interstitial fluids from manganese-rich sediments of the northeastern equatorial Pacific Ocean. *Am. J. Sci.*, **280**, 1063–1096.

Cann, J.R., Winter, C.K. and Pritchard, R.G. (1977) A hydrothermal deposit from the floor of the Gulf of Aden. *Mineral. Mag.*, **41**, 193–199.

Chester, R. and Aston, S.R. (1976) The geochemistry of deep-sea sediments, in *Chemical Oceanography*, J.P. Riley and R. Chester (eds), Vol. 6, pp.281–390. London: Academic Press.

Cronan, D.S (1976) Manganese nodules and other ferromanganese oxide deposits, in *Chemical Oceanography*, J.P. Riley and R. Chester (eds), Vol. 5, pp.217–263. London: Academic Press.

Cronan, D.S. (1977) Deep-sea nodules: distribution and geochemistry, in *Marine Manganese Deposits*, G.P. Glasby (ed.), pp.11–44. Amsterdam: Elsevier.

Cronan, D.S. (1980) *Underwater Minerals*. London: Academic Press.

Crozier, D.S. (1961) Micrometeorite measurements – satellite and ground-level data compared. *J. Geophys. Res.*, **66**, 2793–2795.

Dermott, S.F., Jayaraman, S., Xu, Y.L., Gustafson, B.A.S. and Lou, J.C. (1994) A circumsolar ring of asteroidal dust in resonant lock with the Earth. *Nature*, **369**, 719–723.

DeMaster, D.J. (2002) The accumulation and cycling of biogenic silica in the Southern Ocean: revisiting the marine silica budget. *Deep-Sea Research II*, **49**, 3155–3167.

Doney, S.C., Fabry, V.J., Feely, R.A. and Kleypass, J.A. (2009) Ocean acidification: the other $CO_2$ problem. *Ann. Rev. Mar. Sci.*, **1**, 169–192.

Dymond, J.M., Corliss, J.B., Heath, G.R., Field, C.W., Dasch, E.J. and Veeh, H.H. (1973) Origin of metalliferous sediments from the Pacific Ocean. *Geol. Soc. Am. Bull.*, **84**, 335–372.

Dymond, J.M., Lyle, M., Finney, B., *et al.* (1984) Ferromanganese nodules from MANOP Sites H, S and R – control of mineralogical and chemical composition by multiple accretionary processes. *Geochim. Cosmochim. Acta*, **48**, 931–949.

Elderfield, H. (1976) Hydrogenous material in marine sediments: excluding manganese nodules, in *Chemical Oceanography*, J.P. Riley and R. Chester (eds), Vol. 5, pp.137–215. London: Academic Press.

Elderfield, H. (2004) *Treatise on Geochemistry, Vol. 6, The Oceans and Marine Geochemistry*, H.D. Holland and K.K. Turekian (series eds). London: Elsevier.

Esser, B.K. and Turekian, K.K. (1988) Accretion rate of extraterrestrial particles determined from osmium isotope systematics of Pacific pelagic clay and manganese nodules. *Geochim. Cosmochim. Acta*, **52**, 1383–1388.

Farley, K.A. and Patterson, D.B. (1995) A 100-kyr periodicity in the flux of extraterrestrial $^3$He to the sea floor. *Nature*, **378**, 600–603.

Fredricksson, K. and Robbin Martin, L. (1963) The origin of black spherules in Pacific islands, deep-sea sediments and Antarctic ice. *Geochim. Cosmochim. Acta*, **27**(3), 241–251.

German, C.R. and Von Damm, K.L. (2003) Hydrothermal Processes, in *Treatise on Geochemistry*, H. Elderfield (ed.), Vol. 6, H.D. Holland and K.K. Turekian (series eds), pp.181–222. London: Elsevier.

Glass, B.P. (1990) Tektites and microtektites: key facts and inferences. *Tectonophysics*, **171**, 393–404.

Goldberg, E.D. (1954) Marine geochemistry. Chemical scavengers of the sea. *J. Geol.*, **62**, 249–255.

Goldstein, S.L. and Hemming, S.R. (2003) *Long-lived isotopic tracers in oceanography, paleooceanography and ice sheet dynamics*, in Treatise on Geochemistry, Vol. 6. H. Elderfield (ed.), The Oceans and Marine Geochemistry, , H.D. Holland and K.K. Turekian (series eds), pp.453–489. London: Elsevier.

Griffin, J.J. and Goldberg, E.D. (1963) *Clay-mineral distribution in the Pacific Ocean*, in The Sea, M.N. Hill (ed.), Vol. 3, pp.728–741. New York: John Wiley & Sons, Inc.

Griffin, J.J., Windom, H. and Goldberg, E.D. (1968) The distribution of clay minerals in the World Ocean. *Deep Sea Res.*, **15**, 433–459.

Hem, J.D. (1978) Redox processes at surfaces of manganese oxide and their effects on aqueous metal ions. *Chem. Geol.*, **21**, 199–218.

Hoffert, M., Perseil, A., Hekinian, R., *et al.* (1978) Hydrothermal deposits sampled by diving saucer in Transform Fault 'A' near 37°N on the Mid-Atlantic Ridge, FAMOUS area. *Oceanol. Acta*, **1**, 73–86.

Lalou, C. (1983) Genesis of ferromanganese deposits: hydrothermal origin, in *Hydorthermal Processes at Seafloor Spreading Centres*, P.A. Rona, K. Bostrom, L. Laubier and K.L. Smith (eds), pp.503–534. New York: Plenum.

Lyle, M., Heath, G.R. and Robbins, J.M. (1984) Transport and release of transition elements during early diagenesis: sequential leaching of sediments from MANOP sites M and H, Part I. pH5 acetic acid leach. *Geochim. Cosmochim. Acta*, **48**, 1705–1715.

Mackenzie, F.T. and Garrels, R.M. (1966) Chemical mass balance between rivers and oceans. *Am. J. Sci.*, **264**, 507–525.

Manheim, F.T. (1965) Manganese–iron accumulations in the shallow marine environment, in *Symposium on Marine Geochemistry*, D.R. Shrink and J.T. Corless (eds), pp.217–276. Narragansett, RI: Narragansett Marine Laboratory, University of Rhode Island, Occasional Publication, no. 3.

Marcantonio, F., Anderson, R.F., Stute, M., Kumar, N., Scholosser, P. and Mix, A. (1996) Extraterrestrial $^3$He as a tracer of marine sediment transport and accumulation. *Nature*, **383**, 705–707.

Maynard, J.B. (2003) Manganiferous Sediments, Rocks, and Ores, in *Treatise on Geochemistry*, F.T. Mackenzie (ed.) Vol. 7, H.D. Holland and K.K. Turekian (series eds), pp.289–308. London: Elsevier.

Millard, H.T. and Finkelman, R.B. (1970) Chemical and mineralogical compositions of cosmic and terrestrial spherules from a marine sediment. *J. Geophys. Res.*, **75**, 2125–2134.

Milliman, J.D. (1993) Production and accumulation of calcium carbonate in the ocean: budget of a nonsteady state. *Global Biogeochem. Cycl.*, 7, 927–957.

Moorby, S.A. and Cronan, D.S. (1983) The geochemistry of hydrothermal and pelagic sediments from the Galapagos Hydrothermal Mounds DSDP Leg 70. *Mineral. Mag.*, 47, 291–300.

Moorby, S.A., Cronan, D.S. and Clasby, G.P. (1984) Geochemistry of hydrothermal Mn-oxide deposits from the S.W. Pacific island arc. *Geochim. Cosmochim. Acta*, 48, 433–441.

Moore, W.S. and Vogt, P.R. (1976) Hydrothermal manganese crust from two sites near the Galapagos spreading axis. *Earth Planet. Sci. Lett.*, 29, 349–356.

Muller, R.A. and MacDonald, G.J. (1995) Glacial cycles and orbital inclination. *Nature*, 377, 107–108.

Murray, J. and Renard, A.F. (1891) *Deep-Sea Deposits. London: Scientific Report of the Challenger Expedition, no. 3*. London: HMSO.

Murrell, M.T., Davies, P.A., Nishizumi, K. and Millard, H.T. (1980) Deep-sea spherules from Pacific clay: mass distribution and influx rate. *Geochim. Cosmochim. Acta*, 44, 2067–2074.

Parkin, D.W., Delany, A.C. and Delany, A.C. (1967) A search for airborne cosmic dust on Barbados. *Geochim. Cosmochim. Acta*, 31, 1311–1320.

Pondaven, P., Ragueneau, O., Treguer, P., Hauvespre, A., Dezileau, L. and Reyss, J.L. (2000) Resolving the 'opal paradox' in the Southern ocean. *Nature*, 405, 168–172.

Piper, D.Z., Basler, J.R. and Bischoff, J.L. (1984) Oxidation state of marine manganese nodules. *Geochim. Cosmochim. Acta*, 48, 2347–2355.

Pratt, R.M. and McFarlin, P.F. (1966) Manganese pavements on the Blake Plateau. *Science*, 151, 1080–1082.

Price, N.B. and Calvert, S.E. (1970) Compositional variation in Pacific Ocean ferromanganese nodules and its relationship to sediment accumulation rates. *Mar. Geol.*, 9, 145–171.

Raab, W.J. and Meylan, M.A. (1977) Morphology, in *Marine Manganese Deposits*, G.P. Glasby (ed.), pp.109–146. Amsterdam: Elsevier.

Ragueneau, O. *et al.* (2000) A review of the Si cycle in the modern ocean: recent progress and missing gaps in the application of biogenic opal as a paleoproductivity proxy. *Global and Planetary Change*, 26, 317–365.

Riley, J.P. and Chester, R. (1971) *Introduction to Marine Chemistry*. London: Academic Press.

Sarmiento, J.L., Simeon, J., Gnanadesikan, A., Gruber, N., Key, R.M. and Schlitzer, R. (2007) Deep ocean biogeochemistry of silicic acid and nitrate. *Global Biogeochem. Cycl.*, 21, doi: 10.1029/2006GB002720.

Sayles, F.L. and Mangelsdorf, P.C. (1977) The equilibration of clay minerals with seawater: exchange reactions. *Geochim. Cosmochim. Acta*, 41, 951–960.

Scott, M.R., Scott, R.B., Rona, P.A., Butler, L.W. and Nalwak, A.J. (1974) Rapidly accumulating manganese deposit from the median valley of the Mid-Atlantic Ridge. *Geophys. Res. Lett.*, 1, 355–358.

Sillen, L.G. (1961) The physical chemistry of sea water, in *Oceanography*, M. Sears (ed.), pp.549–581. American Association for the Advancement of Science, Publication, 67.

Sillen, L.G. (1963) How has sea water got its present composition? *Sven. Kem. Tidskr.*, 75, 161–177.

Sillen, L.G. (1965) Oxidation state of earth's ocean and atmosphere. I. A model calculation on earlier states. The myth of the 'probiotic soup'. *Ark. Kemi*, 24, 431–444.

Sillen, L.G. (1967) Gibbs phase rule and marine sediments, in *Equilibrium Concepts in Natural Water Systems*, W. Stumm (ed.), pp.57–69. Washington, DC: American Chemical Society, Advances in Chemistry Series, No. 67.

Toth, J.R. (1980) Deposition of submarine crusts rich in manganese and iron. *Geol. Soc. Am. Bull.*, 91, 44–54.

Treguer, P., Nelson, D.M., Van Bennekom, A.J., DeMaster, D.J., Leynaert, A. and Queguiner, B. (1995) The Silica Balance in the World Ocean: A Reestimate. *Science*, 268, 375–379.

Van Cappellen, P. and Qiu, L. (1997a) Biogenic silica dissolution in sediments of the Southern Ocean. I. Solubility. *Deep Sea Res.* II, 44, 1109–1128.

Van Cappellen, P. and Qiu, L. (1997b) Biogenic silica dissolution in sediments of the Southern Ocean. II. Kinetics. *Deep Sea Res.* II, 44, 1129–1149.

# 16     Unscrambling the sediment-forming chemical signals

We are now attempting to understand the factors that control the chemical composition of marine sediments, and from the point of view of the present volume, attention will be focused largely on the upper few metres of the sediment column, because it is here that material reacts directly with seawater. In this context, therefore, the oceanic water column can be viewed as a medium through which chemical signals, or fluxes, are transmitted from above to the upper portions of the bottom sediments. In addition, signals are also transmitted to this part of the sediment column from below via interstitial waters. This chapter will focus particularly on trace metals which can neither be created nor destroyed. The nutrient components and organic carbon are generally intensively cycled throughout the water and sediment column making it much more difficult to follow their individual behaviour.

A characteristic feature of deep-sea clays is that, relative to nearshore muds, they contain enhanced concentrations of some trace elements (e.g. Mn, Cu, Ni, Co, Pb). These 'excess' trace elements are further enhanced in pelagic sediments laid down at slow rates in areas remote from the land masses, and are even more enhanced in ferromanganese nodules and metalliferous active-ridge deposits. It was pointed out in Section 13.5 that some kind of sequential enhancement, or fractionation, is therefore occurring for the 'excess' trace elements within marine sediments. A central issue of marine geochemistry is to identify the processes by which such elemental enhancements and fractionations have arisen.

A number of authors have suggested that the overall chemical compositions of marine sediments can be considered to result from the contributions made by individual sediment fractions (see e.g.

Krishnaswami, 1976; Bacon and Rosholt, 1982). However, each of the fractions contains a number of individual sediment-forming components. Because of this, an approach in which the formation of individual components viewed in terms of chemical signals offers a potentially attractive insight into the factors that control the overall composition of marine sediments.

The major sediment-forming components have been described in Chapter 15, in which it became apparent that the processes involved in the generation of some components cannot be related to single chemical signals, but rather are the result of signal coupling. For example, ferromanganese nodules can acquire elements transmitted by signals associated with hydrogenous, hydrothermal and both oxic and suboxic diagenetic processes. To assess the chemical composition of marine sediments in terms of chemical signals it is therefore necessary to make a distinction between individual signals that are potentially able to give rise to individual components but that more often combine together to form multisource components. In the present approach, therefore, an attempt will be made to relate the chemistry of the sediments to chemical signals that are associated with the processes involved in the generation of the sediment-forming components. This is a purely artificial approach, but is adopted in order to provide a convenient framework within which to describe the factors that constrain the chemical composition of the sediments.

## 16.1 Definition of terminology

To simplify the often confusing terminology found in the literature, the chemical signals identified here will

---

*Marine Geochemistry*, Third Edition. Roy Chester and Tim Jickells.
© 2012 by Roy Chester and Tim Jickells. Published 2012 by Blackwell Publishing Ltd.

be defined on the basis of the processes that have been shown to be operative in the genesis of the sediment-forming components, that is, a process-oriented approach will be adopted.

Initially, a distinction will be made between detrital (sometimes also termed lithogenous or refractory) and non-detrital (sometimes termed non-lithogenous, or non-refractory) components (see also Section 3.1.4). This two-component classification involves a fundamental geochemical division between two different types of elements.

1 Detrital elements are part of the crystalline mineral matrix, usually in lattice-held associations, and have been carried through the mobilization–transportation cycle in a solid form. Thus, the detrital components, such as clay minerals, take part in particulate ↔ dissolved reactions, but the detrital matrix-associated elements themselves do not usually undergo exchange between the solid and dissolved states.

2 Non-detrital elements are not part of the mineral matrix, but have been removed from solution in association with either inorganic or organic hosts. Thus, these are the elements that take part in exchanges between the particulate and dissolved states.

It should be stressed that this classification takes no account of non-lattice → lattice transformations that can affect some elements, for example, during diagenesis or reverse weathering (see Chapter 14).

In the simple twofold classification, a single signal is involved in the formation of the non-detrital components. In reality, the non-detrital signal can be resolved into a number of individual signals associated with both the supply and removal of elements from solution. The nature of these individual non-detrital signals can be related to the processes involved in the generation of the sediment-forming components. In Chapter 15 it was shown that elements that were classified originally as 'hydrogenous' were related to a number of different processes, which were classified as being: (i) hydrogenous; (ii) hydrothermal; (iii) oxic diagenetic; and (iv) suboxic diagenetic in origin. It was also shown that there is a degree of coupling between some of the signals involved in the formation of the components; for example, that between hydrogenous ferromanganese crusts, sediment ferromanganese oxyhydroxides and $\delta MnO_2$-rich ferromanganese nodules, all of which receive their elements from overlying seawater. However, it was also shown that there was a major decoupling between these three components and toderokite-rich nodules, which were supplied mainly from interstitial waters following suboxic diagenesis at depth in the sediment column. Thus, a distinction could be made between elements supplied from the water column above, and those supplied from the interstitial water column below. In terms of the identification of chemical signals that result in the formation of sediment-forming components, it is necessary to take account of this decoupling break between seawater and interstitial-water sources. This is an extremely important distinction because hydrogenous, hydrothermal and oxic diagenetic processes, which release elements at or near the sediment surface and represent a primary source to the sediments, whereas suboxic mobilization at depth can involve a recycled supply of elements associated with the secondary oxidants that are utilized in suboxic diagenesis.

In the process-orientated approach used here for the classification of the chemical signals transmitted to marine sediments, the following signal types will therefore be identified.

1 The detrital signal. This is a background signal, which transmits elements carried in the crystalline matrix of lithogenous minerals.

2 The non-detrital signal. This transmits elements that have been removed from solution at some stage during their history. It is sub-divided into a number of categories.

(a) The biogenous signal. In the present context, this refers mainly to the signal associated with biological shell material, and excludes the organic carbon involved in both the down-column transport of trace metals to the sediment surface and subsequent diagenetic processes.

(b) The authigenic signal. The term authigenic will be used here to describe the primary 'background' signal transmitted through seawater. In an ocean-wide context this signal receives contributions from all sources that supply elements to seawater, but in the present definition elements from these sources are regarded as being smoothed out into a general background signal, that is, one that excludes the effects of localized inputs, such as those arising from hydrothermal sources. The removal of elements from 'background' seawater involves processes associated with both the hydrogenous and oxic diagenetic signals. The hydrogenous signal can result in the inorganic formation of components via the removal of dissolved elements from seawater without the necessity of a sediment substrate, for example, on to rock surfaces (see Section 15.3.4.6). In contrast, the oxic diagenetic signal requires the initial removal of a dissolved element from seawater on to a carrier phase and the down-column transport of the carrier, via the large organic aggregates (CPM + associated FPM) of the global carbon flux, to a sediment surface. Following this, some elements (e.g. 'nutrient-type') are released to become available for incorporation into the sediment as the labile fraction of the carriers is destroyed during oxic diagenesis. Other elements (e.g. 'scavenged-type') may be directly incorporated into the sediment in association with the FPM carriers; although the latter can be released and redistributed within the sediment (see e.g., Fig. 15.8c). The principal difference between the hydrogenous and the oxic diagenetic signals, however, is that the latter requires diagenesis to be initiated on a sediment substrate. However, the hydrogenous and the oxic diagenetic signals are interlinked in that they both acquire their elements from the same general source, that is, 'background' seawater, and reach the sediment surface directly from the overlying water column (a primary seawater source). In the present context, therefore, the hydrogenous and oxic diagenetic signals are combined into the authigenic signal, which may be defined here as 'that giving rise to components formed inorganically from elements originating in the overlying background seawater.

(c) The diagenetic, or sediment-recycled, signal. As used in the present context, this describes the signal transmitted through the interstitial water column following suboxic diagenetic mobilization; that is, this recycled signal largely involves redox-sensitive elements (e.g. Mn, Fe, Cu).

(d) The hydrothermal signal. This applies, in the sense used here, to the seawater signal resulting from high-temperature solutions at the ridge crests.

(e) The contaminant signal. This arises from the introduction of anthropogenic material into the oceans.

The detrital, authigenic and biogenous signals may be considered to operate on an ocean-wide basis and to transmit elements that have a primary 'background' seawater source. To establish a theoretical framework within which to unscramble the sediment-forming signals it therefore is convenient to envisage marine sediments as being formed by components originating from mixtures of these 'background' signals, upon which are superimposed perturbation spikes from more localized hydrothermal and contaminant seawater signals, and from the more widely occurring interstitial-water suboxic diagenetic signal. The elements associated with the suboxic diagenetic signal can have had a variety of primary sources, but the important point is that they have been recycled within the sediment complex itself. In the present context, therefore, any signal that is not associated directly with background seawater is regarded as a spike. The question that must now be considered is 'how can these various signals be unscrambled?' A variety of techniques have been used for this purpose, and these include the following.

1 Interpretation of total sediment chemical analyses. These can be used, for example, to assess the relative amounts of major components such as clays, carbonates and opaline silica in oceanic sediments (Fig. 13.5).

2 Spatial mapping of elemental concentrations. This can be used, for example, to establish elemental source–transport patterns (see e.g. Turekian and Imbrie, 1966).

3 Elemental accumulation rate comparisons. These can be used to determine the rates at which different components are formed (see e.g. Bender et al., 1971).

4 Factor analysis (see e.g. Krishnaswami, 1976), isotope analysis (see e.g. Thompson et al., 1984) and chemical leaching techniques (see e.g. Chester and Hughes, 1967). These can be utilized to assess the partitioning of elements between individual sediment phases.

Techniques such as those identified above will be introduced into the text as each of the individual chemical sediment-forming signals is described in the following sections.

## 16.2 The biogenous signal

The deposition of calcareous and siliceous shells in marine sediments, and the degree to which they are preserved, plays a dominant role in controlling the major element chemistry of the deposits. Siliceous sediments generally have a restricted distribution and, with a few important exceptions, are confined to the edges of the oceans. Carbonate sediments, however, are much more widespread and, in fact, the bulk composition of many deep-sea deposits may be considered to be made up of mixtures of carbonate and lithogenous, mainly clay, end-members.

The extent to which carbonate and siliceous shell materials affect the overall major element composition of deep-sea sediments can be estimated from the relative proportions of particular tracers present in them; Al (lithogenous), Ca (carbonate) and Si (siliceous), see for example, the major element analyses listed in Table 13.3. Thus, the chemical signals associated with the deposition of biogenous carbonate and siliceous shells can be unscrambled by relating them to the major element composition of the sediments. However, it was pointed out in Section 13.5 that, with the exception of Mg and Sr, which are found in the carbonate lattice, both carbonate and siliceous sediments are generally impoverished in those excess trace metals that are concentrated in deep-sea clays. Further, neither carbonate nor opaline shell phases act as significant carriers for the down-column transport of trace metals, although their organic matter may be a carrier. It may be concluded, therefore, that the biogenous shells are not important trace-metal hosts and act rather to dilute metals associated with other sediment components. Indeed, it is common practice to express the concentrations of trace metals in sediments on a 'carbonate-free' basis in order to overcome the dilution effects of the shell matrix.

## 16.3 The detrital and authigenic signals

With respect to the transport of trace metals to the sediment surface from background seawater, and the incorporation of the excess trace-metal fractions into deep-sea clays, attention therefore can be focused largely on the detrital and authigenic signals. In a 'normal' deep-sea clay, that is, one that has not been perturbed by major hydrothermal, contaminant or suboxic diagenetic spikes, the total concentration ($C_t$) of an element therefore may be expressed in terms of the sum of the contributions from detrital and authigenic components, thus:

$$C_t = C_d + C_a \qquad (16.1)$$

where $C$ is the concentration of an element and the suffixes t, d and a refer to the total, detrital and authigenic concentrations, respectively (see e.g. Krishnaswami, 1976). In terms of this two-component framework, an attempt will now be made to unscramble the detrital and authigenic signals.

### 16.3.1 The detrital signal

In the definition used here, the detrital signal transmits elements carried in the crystalline matrix of lithogenous minerals. Crust-derived weathering products, together with continental and submarine volcanic debris, make up most of the lithogenous material transported to the oceans over geological time. This lithogenous material comprises the bulk of deep-sea clays; for example, the oxides of aluminium and silicon alone account for ~80% of the total sediment. This crust-derived material is brought to the open ocean mainly along the fluvial and atmospheric routes, but it is important to remember that the material carried by these forms of transport consists of both detrital and non-detrital components and some non-detrital components may readily dissolve or exchange on reaching seawater. Clearly, therefore, the bulk composition of river and atmospheric particulates cannot be used to assess the composition of the detrital signal ($C_d$) transmitted to the oceans.

### 16.3.2 The authigenic signal

The geochemically important primary authigenic material in marine sediments consists mainly of ferromanganese crusts, nodules and sediment oxyhydroxides, which are formed by hydrogenous and oxic diagenetic processes, together with the population of elements associated with a wide variety of sediment components in clay intersheet and surface-adsorbed positions. The detailed chemistry of these primary authigenic components has been described in Section 15.3, and the average compositions of the crusts and nodules are summarized in Table 16.1. It can be seen from this table that these authigenic components are rich in those elements that are present in deep-sea clays in excess concentrations (see also Chapter 13.5). Some of the authigenic

**Table 16.1** Typical compositions of primary authigenic components in deep-sea sediments; units, µg g-1.

| Elements | Hydrogenous crusts* | Oxic nodules* | Suboxic nodules* | Pelagic clay† |
|---|---|---|---|---|
| Mn | 222 000 | 316 500 | 480 000 | 65 000 |
| Fe | 190 000 | 44 500 | 4 900 | 12 500 |
| Co | 1 300 | 280 | 35 | 116 |
| Ni | 5 500 | 10 100 | 4 400 | 293 |
| Cu | 1 480 | 4 400 | 2 000 | 570 |
| Zn | 750 | 2 500 | 2 200 | – |
| Al | 11 800 | 27 100 | 7 500 | 83 000 |

* Data from Table 15.6.
† Data from Table 13.4.

components, such as micronodules and oxyhydrox-ides, are an intimate part of the sediment complex, and in certain regions they can have a major influ-ence on the composition of the total sediments with respect to the excess trace elements.

## 16.4 Unscrambling the detrital and authigenic signals

The bulk of the non-biogenic sediment-forming components in deep-sea clays are detrital in origin. Authigenic components are quantitatively far less important, but they are enriched in some trace metals. However, the proportions in which the detri-tal and authigenic components are present in deep-sea sediments vary from one suite of deposits to another, and to evaluate their relative importance it is necessary to unscramble the signals associated with the two components. There are several ways in which this can be done, and some of these are dis-cussed in the following.

*The direct approach using chemical leaching proce-dures to determine elemental partitioning.* In this approach, the detrital and authigenic fractions of a sediment are actually separated from each other and analysed individually. This approach reflects the fact that metals bound to different chemical phases in the sediments will be dissolved using different chemical reagents. For example, Chester and Hughes (1967) used a chemical leaching technique to establish the partitioning of elements in a North Pacific pelagic clay core, and Chester and Messiha-Hanna (1970) applied a similar technique to a series of North Atlantic deep-sea surface sediments. The data for the

chemical compositions of the detrital and authigenic fractions of the sediments obtained in this way are given in Table 16.2.

*The indirect approach.* Several routes can be used in this approach.

*1 Background correction: detrital fraction.* When this route is adopted it is assumed that the chemical composition of the detrital fraction of deep-sea sedi-ments can be estimated with respect to some refer-ence material.

(a) Krishnaswami (1976) used the compositions of (i) continental shale; and (ii) nearshore sedi-ments as being representative of the crust-derived (lithogenous) material transported to the marine environment; that is, the baseline material upon which the 'excess' authigenic trace elements are superimposed. In terms of Equation 16.1, values of $C_t$ were obtained by direct analysis, and $C_d$ was assumed to be the same as that of either continen-tal shales or nearshore muds. Thus, the composi-tion of $C_a$ could be obtained from $C_t \sim C_d$. The compositions of the detrital fractions based on those of continental shales and nearshore muds are listed in Table 16.2(a) Col. 3, and the 'best estimate' for the derived authigenic fraction is given in Table 16.2(b) Col. 6.

(b) A similar approach using a different kind of background correction for the estimation of the detrital fraction was applied by Thompson *et al.* (1984) to a suite of deep-sea clays from the Nares Abyssal Plain in the northeast Atlantic. Two types of clay were found in this area; a red clay and a grey clay. On the basis of their total sediment analysis it was evident that the grey clays had only

**Table 16.2** Estimated chemical compositions of the detrital and authigenic fractions of deep-sea clays; units, $\mu g\,g^{-1}$.

(a) Detrital fraction.

| Element | Nares Abyssal Plain red clays* | Pacific pelagic clays† | Nares Abyssal Plain red clays‡ | Bermuda Rise carbonates§ | Atlantic deep-sea sediments¶ | Pacific pelagic clayi | Average nearshore mud (Wedepohl, 1960) | Average shale (Wedepohl, 1968) |
|---|---|---|---|---|---|---|---|---|
| Mn | 578 | 2087 | 770 | 605 | 550 | 740 | 850 | 850 |
| Fe | 42 280 | 43 300 | 51 800 | 51 240 | – | 51 800 | 69 900 | 47 000 |
| Cu | 51 | 244 | 45 | 36 | 67 | 212 | 48 | 45 |
| Co | 24 | 48 | 24 | 23 | 12 | 16 | 13 | 19 |
| Ni | 65 | 92 | 56 | 65 | 63 | 46 | 55 | 68 |
| Zn | 111 | – | 117 | 124 | – | – | 95 | 95 |
| V | 158 | – | 153 | – | 120 | 92 | 130 | 130 |
| Cr | – | – | – | – | 72 | 91 | 100 | 90 |

* Estimated using background correction with respect to local grey clays (Thompson et al., 1984).
† Estimated using background correction with respect to continental shales (Krishnaswami, 1976).
‡ Estimated using graphical procedure (Thompson et al., 1984).
§ Estimated using graphical procedure (Bacon and Rosholt, 1982).
¶ Estimated using chemical leaching technique (Chester and Messiha-Hanna, 1970).
i Estimated using chemical leaching technique (Chester and Hughes, 1967).

(b) Authigenic fraction.

| Element clays¶ | Bermuda Rise carbonates* | Pacific pelagic clay† | Atlantic deep-sea sediments‡ | Atlantic deep-sea sediments§ | Pacific pelagic |
|---|---|---|---|---|---|
| Mn | 4400 | 4380 | 3871 | 3112 | 7000 |
| Fe | – | – | – | – | 7500 |
| Cu | 110 | 101 | 110 | 128 | 225 |
| Co | 6 | 73 | 60 | – | 80 |
| Ni | 61 | 147 | 112 | 62 | 175 |
| Zn | 40 | – | – | – | – |

* Estimated using graphical procedure (Bacon and Rosholt, 1982).
† Estimated using chemical leaching technique (Chester and Hughes, 1967).
‡ Estimated using chemical leaching technique (Chester and Messiha-Hanna, 1970).
§ Estimated using chemical leaching technique (Thomas, 1987).
¶ Best estimate using a variety of techniques (Krishnaswami, 1976).

a negligible authigenic fraction, and it was assumed that for these sediments $C_t \sim C_d$. The data for the composition of the grey clays, which are therefore representative of the detrital fraction of deep-sea sediments, are given in Table 16.2(a) Col. 2.

*2 Graphical procedures: detrital and authigenic fractions.* Bacon and Rosholt (1982) and Thompson *et al.* (1984) made assumptions about the inter-relationships between detrital, biogenic and authigenic components within sediment cores and then extrapolated plots of concentrations against biogenic components and sedimentation rates respectively to estimate an end member authigenic component concentrations as shown in Table 16.2. These authors therefore assumed that the total trace metal content of the sediments could be expressed in terms of a two component terrigenous (detrital) and pelagic (authigenic) system by the equation:

$$C_{total} = f(C_{pelagic} - C_{terrigenous}) + C_{terrigenous} \quad (16.2)$$

Where $C_{total}$, $C_{pelagic}$ and $C_{terrigenous}$ are the metal concentrations in the total sediment, the pelagic and the terrigenous components respectively and f is the proportion of pelagic component in the sediment.

*3 Manganese nodule model: authigenic fraction.* Krishnaswami (1976) used this model, which is based on the assumption that the rates of precipitation for authigenic trace metals are the same in pelagic clays and ferromanganese nodules. Using compositional data obtained from nodules, Krishnaswami (1976) was then able to estimate the concentrations of a series of elements in the authigenic fractions of Pacific pelagic clays. These data are given in Table 16.2(b) Col. 6. Several different estimates therefore are available for the compositions of the detrital and authigenic components present in deep-sea sediments (see Table 16.2). Perhaps the most striking feature to emerge from these data is that there is a generally good agreement between the estimates, especially for the composition of the detrital fraction in deep-sea sediments. Thus, it is possible, at least to a first-order approximation, to unscramble the detrital from the authigenic signal if the total metal concentration of a total deep-sea sediment is known or if the total metal concentration can be chemically separated into its component parts. The signal strengths will, of course, be dependent on the relative rates at which the components associated

with the detrital and authigenic signals accumulate. The manner in which the proportions of the two components vary in oceanic sediments can be illustrated with respect to the Atlantic Ocean.

### 16.4.1 Regional variations in the relative authigenic and detrital component

Chester and Messiha-Hanna (1970) and Thomas (1987) used chemical leaching procedures to investigate the partitioning of a series of elements in samples from the present-day Atlantic sediment surface. A compilation of their data, expressed in terms of the two-component detrital–authigenic classification, is given in Table 16.3, and the data are illustrated in Fig. 16.1. From Table 16.3 it can be seen that the average partitioning signatures of the elements vary considerably; for example, Al is ~80% detrital, whereas Mn is ~70% authigenic in character. The predominant detrital nature of aluminium validates its use as a tracer of the detrital component noted earlier.

These proportions are ocean-wide averages, however, and the partitioning signatures of some elements vary with the environment of deposition of the host sediment. This can be illustrated with data from Thomas (1987) and Chester *et al.* (1988), who analysed sediments using the chemical leaching technique to directly differentiate between the various sediment phases for a range of elements in a North Atlantic marginal (hemi-pelagic) sediment and Mid-Atlantic Ridge sediment. The data are given in

**Table 16.3** Average detrital–authigenic partitioning of elements in North Atlantic deep-sea sediments (data from Chester and Messiha-Hanna, 1970) and Thomas, 1987).

| Element | Percentage of total concentration | |
| --- | --- | --- |
| | Detrital | Authigenic |
| Al | 81 | 19 |
| Fe | 82 | 18 |
| V | 71 | 29 |
| Cr | 70 | 30 |
| Ni | 55 | 45 |
| Cu | 44 | 56 |
| Co | 42 | 48 |
| Mn | 32 | 68 |

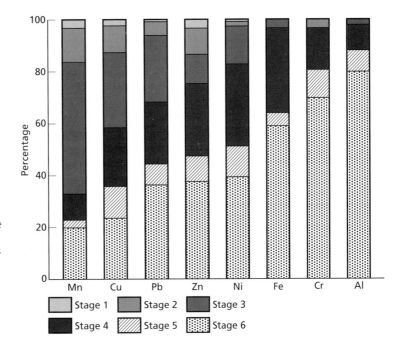

**Fig. 16.1** The average partitioning of some elements in North Atlantic surface deep-sea sediments (data from Chester and Messiha-Hanna, 1970) and Thomas (1987). The diagram is intended to highlight the detrital–authigenic partitioning of the elements; stage 6 represents the detrital fraction and stages 1–5 represent individual authigenic host fractions.

**Table 16.4** Variations in the detrital–authigenic partitioning of elements in North Atlantic deep-sea sediments (data from Thomas, 1987).

| Element | Marginal sediment: percentage of total concentration | | Mid-Atlantic Ridge sediment: percentage of total concentration | |
|---|---|---|---|---|
| | Detrital | Authigenic | Detrital | Authigenic |
| Al | 93 | 7 | 76 | 14 |
| Fe | 77 | 23 | 68 | 32 |
| Mn | 18 | 82 | 14 | 86 |
| Cu | 55 | 45 | 23 | 67 |
| Ni | 64 | 36 | 17 | 83 |
| Cr | 63 | 37 | 88 | 12 |
| Pb | 85 | 15 | 38 | 62 |

Table 16.4, and show that, whereas Al, Fe and Cr retain their detrital character (and Mn remains strongly authigenic) in sediments from both environments of deposition, Cu, Ni and Pb are considerably more authigenic in nature in the ridge than in the marginal sediments. Figure 16.2 illustrates the results

of an east–west, marginal → open-ocean → marginal, sediment transect. The principal partitioning trends can be summarized as follows.

**1** There is a general decrease in the proportion of detrital Cu (stage 6) away from the margins toward the mid-ocean regions, as the influence of continentally derived material decreases.

**2** The highest proportion of organically associated authigenic Cu (stage 5) is found in marginal sediments under regions of high surface-water productivity. In these regions there is a high sedimentation rate and an enhanced down-column flux. Under these conditions relatively high concentrations of organic matter, together with organically associated Cu, can be preserved in the sediments. This is an important finding because it suggests that in some marginal regions significant amounts of authigenic Cu can be stored in a relatively immobile form as the organic carriers escape oxic diagenesis at the sediment surface.

**3** There is a clear trend in the distribution of Cu associated with the easily reducible oxides of stage 3 (mainly new manganese oxides), with the highest contributions being found in mid-ocean deposits around the Mid-Atlantic Ridge and ridge flanks and

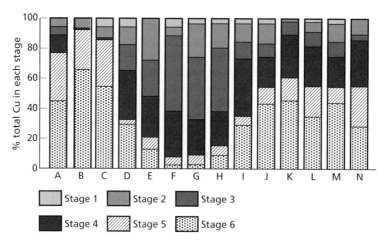

**Fig. 16.2** The partitioning signatures for total copper ($\Sigma$Cu) in Atlantic deep-sea sediments (from Chester *et al.*, 1988). The samples were collected on an east–west equatorial transect: A–D, eastern margins; E–H, Mid-Atlantic Ridge and ridge flanks; I–N, western margins. The Cu host associations are as follows: stage 1, loosely held or exchangeable associations; stage 2, carbonate and surface oxide associations; stage 3, easily reducible associations, mainly with (new) oxides and oxyhydroxides of manganese and amorphous iron oxides; stage 4, moderately reducible associations, mainly with (aged) manganese oxides and crystalline iron oxides; stage 5, organic association; stage 6, detrital, or residual, associations. Reprinted with permission from Elsevier.

the lowest in sediments from the marginal areas. These mid-ocean sediments also contain high concentrations of total Cu due to effective scavenging of water column copper during their formation (see Section 14.6).

This type of spatial variation in elemental partitioning signatures is extremely important in relation to understanding the processes that control the chemical compositions of deep-sea sediments, and can be explored further by considering Atlantic sediments in more detail and then using them to illustrate how the detrital and authigenic signals operate on an ocean-wide basis.

In a classic paper, Turekian and Imbrie (1966) provided data on the spatial distributions of Mn, Co, Cu, Ni and Cr in surface sediments from the North and South Atlantic. When the concentrations of Mn, Co, Cu and Ni were expressed on a carbonate-free basis there were clear patterns in their spatial distributions, the highest values being found in sediments from remote mid-ocean areas of low clay accumulation, and the lowest in deposits from the continentally adjacent abyssal plains: this is illustrated for Ni in Fig. 16.3(a). In contrast, there were no clear systematic large patterns in the spatial distribution of Cr. Thus, the investigation identified a geographic fractionation between those elements that have excess concentrations in pelagic clays (e.g. Mn, Co, Cu and Ni) and those that do not (e.g. Cr).

Chester and Messiha-Hanna (1970) used a two-stage sequential leaching technique to investigate the partitioning of a series of elements between the detrital and authigenic fractions of surface deep-sea sediments from the North Atlantic. A number of important conclusions can be drawn from this study, and these are summarized in the following.

1 Elements that have similar concentrations in pelagic clays and nearshore muds (see Table 13.4) are associated mainly with the detrital fraction of the sediments. For example, Fe, Cr, V and Al all have >70% of their total concentrations in the detrital fractions.

2 Elements such as Mn, Ni, Co and Cu, which are present in excess concentrations in pelagic clays, have a much higher proportion of their total concentrations in the authigenic associations. For example, >70% of the total Mn is present in the authigenic fractions (see previously). It may be concluded, therefore, that the elements that are enhanced in pelagic clays are authigenic in origin and have been

removed from solution at some time during their history.

3 There were clear trends in the spatial distributions of the authigenic, or excess, elements. This is illustrated for Ni in Fig. 16.3(b), from which it is evident that the proportion of detrital Ni decreases from ~60% near to the continents to <30% in the mid-ocean areas. Thus, the highest proportions of authigenic (non-detrital) Ni are found in sediment from the remote regions. It may be concluded, therefore, that the enhanced total sediment concentrations of Ni reported by Turekian and Imbrie (1966) for the mid-ocean areas (Fig. 16.3a) are the result of an increased contribution from authigenic Ni.

4 The detrital fractions of the North Atlantic deep-sea sediments have a similar composition to those of the total samples of nearshore muds and continental shales: see Table 13.4. According to Chester and Aston (1976) these nearshore muds may be considered to represent an early stage in the adjustment of land-derived material to those processes that operate to produce trace-metal enhancements in the marine environment. This therefore supports their use as indicators of the composition of the detrital fraction of deep-sea sediments, for example, in the background correction method (see previously).

## 16.4.2 Controls on the authigenic inputs to sediments

A number of techniques have been described that can be used to unscramble the detrital and authigenic sediment-forming signals, and in terms of the twofold background-signal classification it has been demonstrated that the concentrations of the excess trace metals in deep-sea sediments result from the authigenic signal. It is now necessary to consider both the nature of the driving force behind this authigenic signal and the manner in which the signal itself operates in the oceans. Up to this point, spatial elemental distributions, together with a variety of unscrambling techniques, have been used to distinguish between the detrital and authigenic sediment components. In order to describe the various theories that have been proposed to explain how the authigenic signal itself operates, it is necessary to introduce another parameter into the system, and this is the rate at which the host sediments accumulate. The lithogenous, or clay, fraction in deep-sea sediments

has a relatively low accumulation rate, generally in the range of a few millimetres per thousand years (see Section 13.1). Within this low accumulation pattern, however, there are considerable variations between sediments deposited in different oceanic areas. In the two-component detrital–authigenic classification, the detrital material is equivalent to the clay fraction, and the rate at which it accumulates obviously will affect the authigenic signal, as the latter can be swamped out at relatively fast detrital accumulation rates.

Originally, two general theories were proposed to explain how the 'detrital–authigenic' sediment-generating system operated on an ocean-wide basis.

*The trace-element-veil theory.* This theory was proposed by Wedepohl (1960), and the rationale underlying it is that the excess trace elements in deep-sea clays have a *constant supply* to all parts of the World Ocean, which is independent from, but superimposed upon, a variable clay accumulation rate. Thus, in the constant-flux model there is a veil of trace elements homogeneously distributed throughout the oceanic water column, from which the elements are removed by sediment components at rates that vary according to how fast the total deposits accumulate; that is, the longer the components are in contact with seawater, the higher the authigenic trace-metal content of the sediment.

Turekian (1967) criticized the trace-element-veil theory on a number of grounds, the major criticism being directed at the fundamental underlying concept that the independent removal of trace metals from solution is essentially constant throughout the World Ocean. He pointed out that this implies that there is a homogeneity in the trace-metal composition of seawater, whereas in fact variations occur in the geographical distributions of the metals in both surface and deep waters (see Chapter 11).

Turekian (1967) developed an alternative theory to explain the incorporation of the excess trace elements in deep-sea sediments, and this is described in the following.

*The differential transport theory.* Turekian (1967) used spatial trace-metal-distribution data for Atlantic surface sediments to develop this theory. It was shown above that the essential features in the spatial distributions of the *excess* trace metals in the Atlantic

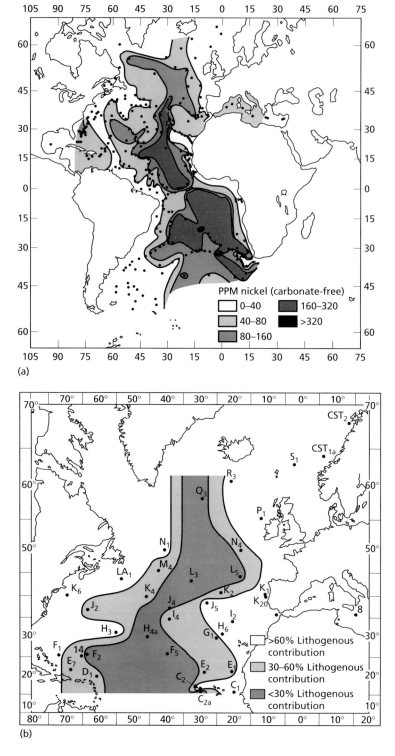

(a)

(b)

**Fig. 16.3** The distribution of Ni in Atlantic deep-sea sediments. (a) The distribution of total Ni (from Turekian and Imbrie, 1966). Reprinted with permission from Elsevier. (b) The distribution of lithogenous (detrital) Ni (from Chester and Messiha-Hanna, 1970). Reprinted with permission from Elsevier.

deep-sea sediments are that they have their highest concentrations in open-ocean areas and their lowest concentrations on the continentally adjacent abyssal plains. Turekian (1967) related these distribution patterns to the processes that govern sediment transport on a global-ocean scale. To do this, he distinguished between two principal transport mechanisms, which operate on material that escapes the coastal zone; (i) lateral off-shelf movement along the sea bed by bottom transport, and (ii) movement down the water column by vertical transport; (see also Section 13.2), In the differential transport theory it is proposed that each of the two transport mechanisms is associated with a specific particulate trace-metal population.

**1** Relatively large-sized particulate material that escapes the estuarine environment is often dumped on the continental shelves, from which it can be further transported to the deep-sea floor by processes including turbidity current transport as well as smaller scale more continuous transport processes (see Section 13.2). These shelf-sediment particles contain relatively high concentrations of lithogenous components, such as quartz, and generally have low trace-metal concentrations, most of which are located in detrital material. Thus, Turekian (1967) identified a trace-metal-poor particulate fraction, which is initially deposited on the shelves and may be transported to hemi-pelagic areas by bottom transport.

**2** The small-sized particles that escape the estuarine and coastal zones can be transported out into the open ocean by surface currents, where they are joined by particles transported directly via the atmospheric flux and augmented by biogenic material formed in the oceans. Here, the particles enter the vertical flux and settle out down the water column to form pelagic deposits on remote topographic highs and other mid-ocean areas. These particles are composed of material such as manganese and iron oxyhydroxides, together with clay minerals that have oxide, and/or organic matter, coatings. As they have relatively large specific surface areas, the particles are actively involved in the scavenging of trace metals from solution as they are transported along the river → estuarine → open-ocean → water-column → sediment, or atmospheric → open-ocean → water-column → sediment, pathways. Thus, they form a trace-metal-rich particulate fraction.

Turekian (1967) therefore suggested that the deposition of the large-sized trace-metal-poor particles around the continents by bottom processes, and the deposition of the small-sized trace-metal-rich particles via vertical water-column settling in mid-ocean areas, was mainly responsible for the fractionation of the excess trace elements between hemi-pelagic and pelagic deep-sea sediments in the Atlantic.

One of the major differences between the trace-element-veil and the differential transport theories revolves around the rates at which the excess elements accumulate in deep-sea sediments.

Turekian (1967) wrote a general equation for the accumulation rate of a trace element in a deep-sea sediment. Elderfield (1976) modified this equation into a form that, using the terminology adopted in this volume, can be expressed in the following way:

$$\Sigma F_E = F_E + [E]_D F_D \qquad (16.3)$$

where $\Sigma F_E$ is the total accumulation rate of an element in a sediment, $F_E$ is the constant rate of addition of $E$ from solution in seawater to the sediment, and $[E]_D$ is the concentration of the element in the detrital component of the sediment, which accumulates at a rate $F_D$. Thus, a further step has been taken in defining the two-component detrital–authigenic concept for the distribution of trace elements in deep-sea sediments by linking the signals involved to sediment accumulation rates. Originally, in terms of the trace-element-veil theory, it was thought that the coupling between the two signals involved a variable detrital accumulation rate and a constant authigenic accumulation rate. This constant-flux model for the accumulation of authigenic trace metals is based on the assumption of a uniform authigenic deposition rate over the entire ocean floor. If the rate of authigenic deposition is in fact uniform, and the detrital accumulation variable, then there will be a negative correlation between the authigenic concentration ($C_a$) and the sediment accumulation rate ($S$), because in clay sediments the latter is dependent mainly on the deposition of detrital material. Krishnaswami (1976) has expressed this relationship as

$$C_a = \frac{K}{S_{\rho_s}} \qquad (16.4)$$

and from this the total sediment concentration would be

$$C_t = C_a + C_d = \frac{K}{S_{\rho_s}} + C_d \qquad (16.5)$$

where $K$ is the authigenic deposition rate ($g\,cm^{-2}\,yr^{-1}$), $\rho_s$ is the in situ density of the sediment ($g\,cm^{-3}$) and $S$ is the sedimentation rate ($cm\,yr^{-1}$). This constant-flux model will therefore apply to elements that have (i) a homogeneous distribution in the oceans and (ii) residence times equal to or greater than the oceanic circulation times.

Krishnaswami (1976) used a variety of techniques to demonstrate that ~90% of the Mn, ~80% of the Ni and Co, and ~50% of the Cu in Pacific pelagic clays are authigenic in origin. In contrast, >90% of

the Sc, Ti and Th is detrital in character. The author then applied the constant-flux equations to his elemental concentration and sediment accumulation rate data and found, as predicted by the model, that there was a negative correlation between the detrital, or clay, sediment accumulation rates and the concentrations of Mn, Co, Ni, Fe and Cu in the Pacific pelagic sediments: see Fig. 16.4. These negative correlations applied to a series of different sediments, and Krishnaswami (1976) concluded that this was consistent with the constant-flux model of a uniform authigenic deposition superimposed on to a variable background detrital input. In contrast, he found that the concentrations of the detrital elements Sc, Ti and

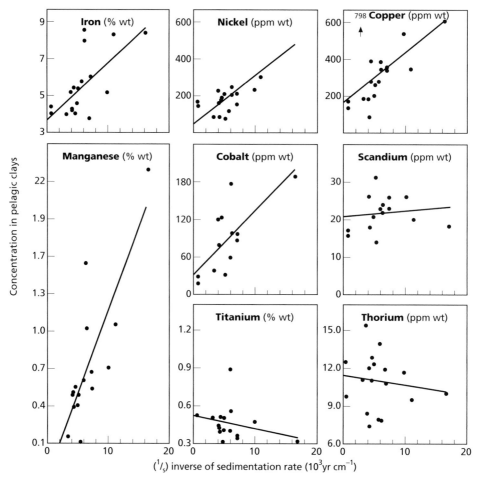

**Fig. 16.4** Scatter plots of elemental concentrations in Pacific pelagic clays against inverse sedimentation rates (from Krishnaswami, 1976). Reprinted with permission from Elsevier.

**Table 16.5** Estimates of the accumulation rates of authigenic elements in sediments from the World Ocean; units, $\mu g\, cm^{-2}\, 10^3\, yr^{-1}$.

| Element | Estimated 'global' authigenic flux to deep sea (Krishnaswami, 1976) | Estimated authigenic flux to Bermuda Rise sediments (Bacon and Rosholt, 1982) | Estimated authigenic flux to Nares Abyssal Plain red clays* | Estimated authigenic flux to Nares Abyssal Plain red clays† |
|---|---|---|---|---|
| Mn | 500 | 4300 ± 1100 | 1220 | 1264 ± 110 |
| Fe | 800 | – | 1560 | 2737 ± 503 |
| Co | 5 | 7.2 ± 5.7 | 14 | 13 ± 2.9 |
| Ni | 10 | 46 ± 16 | 20 | 17 ± 3.5 |
| Cu | 8 | 76 ± 26 | 28 | 26 ± 2.5 |
| Zn | – | 17 ± 20 | 5.1 | 7.2 ± 1.3 |
| V | – | – | 6.2 | 6.1 ± 2.5 |

*Estimated using detrital background correction method (Thompson *et al.*, 1984).
†Estimated using constant-flux method (Thompson *et al.*, 1984).

Th were independent of sediment accumulation rates: see Fig. 16.4. Krishnaswami (1976) also concluded that when the detrital sedimentation rate is high, that is, $\geq 10\, mm\; 10^3\, yr^{-1}$, the authigenic concentration will be small and $C_t \sim C_d$. When the detrital sedimentation rate is small, however, a pure authigenic fraction will be formed and $C_t \sim C_a$; for example, the extreme case for this would be found in hydrogenous ferromanganese crusts (see Section 15.3.3). From his data, Krishnaswami (1976) derived a best estimate of the authigenic deposition rate of a series of elements in the World Ocean, and these are given in Table 16.5.

Krishnaswami's data appeared to confirm the constant-flux model for the removal of authigenic trace metals from solution in seawater into pelagic clays, that is, the trace-element-veil theory was approximately confirmed. However, subsequent studies have challenged this. Two investigations will serve to illustrate this.

*Sediments from the Bermuda Rise.* Bacon and Rosholt (1982) carried out a study of the accumulation rates of a series of radionuclides and trace metals in carbonate-rich sediments from the Bermuda Rise (North Atlantic), an area in which the sediment regime is characterized by a rapid deposition of material transported by abyssal currents. The authors used a two-component sediment matrix, consisting of a terrigenous clay (detrital) component and a pelagic (authigenic) component, for the interpretation of their radionuclide and trace-metal data. In terms of this matrix, it was assumed that the total metal content of the sediments was given by Equation 16.2, and the authors were able to show that the sediments in the region were formed by the deposition of (i) a terrigenous component transported to the North American Basin from continental sources (detrital), added to which was (ii) a rain of biogenic debris produced in the surface waters, which had an associated flux of trace metals scavenged from the water column (pelagic). The sediments therefore were a mixture of these two end-member components. The magnitudes of the pelagic, or authigenic, fluxes to the sediments were estimated by relating the trace-metal data to the *excess* $^{230}Th$ in the sediments. This $^{230}Th$ excess is defined as 'unsupported' $^{230}Th$, and is obtained by subtracting the uranium-activity equivalent from the total $^{230}Th$ activity to derive the $^{230}Th$ scavenged from the water column (see Section 11.5) and use this to track other scavenged metal inputs. A summary of the trace-metal fluxes determined in this way by Bacon and Rosholt (1982) is given in Table 16.5, together with other authigenic flux estimates.

It is apparent from the data given in Table 16.5 that although the authigenic accumulation rate of Co on the Bermuda Rise is reasonably similar to that predicted as normal for the World Ocean by Krishnaswami (1976), those of Mn, Ni and Cu are considerably higher, thus throwing doubt on the constant-flux model. Bacon and Rosholt (1982) were unable to identify the reasons for the higher than expected authigenic fluxes of Mn, Ni and Cu on the Bermuda Rise, but pointed out that they may have been affected by, among other factors, an enhanced trace-metal scavenging by resuspended TSM, which is at a high level in this region of the North Atlantic (see Section 10.2). The authors also considered the important relationship between the residence time of an element and the magnitude of its authigenic flux. They pointed out that $^{230}$Th has such a short oceanic residence time that it must be deposited almost entirely in the basin within which it is generated. Trace metals having longer residence times, however, might be expected to migrate greater distances from their source of supply, and perhaps accumulate preferentially in areas, such as the Bermuda Rise, that are particularly efficient scavenging sinks. Bacon and Rosholt (1982) suggested that this might explain the relatively high authigenic fluxes for Ni and Cu in their survey area, as both elements have residence times of several thousand years (see Section 11.2), and so could undergo large-scale oceanic migration. They concluded, however, that an inter-oceanic migratory hypothesis of this kind is unlikely to explain the differences in the magnitude of the Atlantic and Pacific authigenic Mn signals because this element has a relatively short oceanic residence time, which is similar to that of $^{230}$Th. We now know that this reflects the higher input of Mn to the Atlantic Ocean compared to the Pacific, which, if the metal is mainly removed in the ocean to which it is initially introduced, would result in higher authigenic fluxes to some Atlantic deep-sea clays.

On the basis of down-core data, Bacon and Rosholt (1982) concluded that the concentrations of trace metals in the sediments are controlled mainly by dilution of the authigenic fraction by varying inputs of the terrigenous (detrital) end-member; for example, the clay flux varied by a factor of around four. However, the authigenic fluxes of the metals and radionuclides scavenged from the water column are not sensitive to this variation in the clay flux, and in fact they have remained almost constant from glacial to interglacial periods; that is, in the region itself the constant-flux model was approximated over time. Thus, the authigenic fluxes are not associated with the deposition of clay material, but are probably related to the deposition of biogenic carri-

ers, that is, in association with the down-column TSM fluxes: see Chapters 10, 11 and 12.

Three important overall conclusions therefore can be drawn from the study carried out by Bacon and Rosholt (1982).

1 Regions that are large accumulators of sediment can also act as accumulators of scavenged trace metals.

2 Although the authigenic fluxes of Mn, Ni and Cu on the Bermuda Rise differ from those in the Pacific, the constant-flux model appears to be approximated on a regional scale.

3 The trace-metal scavenging in the water column is controlled by biogenic rather than terrigenous phases.

*Sediments from the Nares Abyssal Plain.* Thompson *et al.* (1984) carried out an investigation into trace metal accumulation rates in a series of grey (excess thorium, that is, $^{230}$Th$_{ex,}$ -poor) and red ($^{230}$Th$_{ex}$-rich) clays from the Nares Abyssal Plain (NAP) in the North Atlantic; the techniques they used to derive the composition of the detrital fractions of the clays have been described above. The sedimentation regime in the region was related to detrital material that had been deposited rapidly from distal turbidity currents to form the grey clays, and slowly deposited from nepheloid layers to give rise to the red clays. On the basis of the total sediment geochemistry the red clays were strongly enriched in Mn, Co, Cu, Ni, Zn and V relative to the grey clays. Despite their colour differences, however, Sr isotope evidence showed that the clay fractions of both sediments had the same terrigenous origin, and the trace-metal enhancement in the red clays was attributed to their scavenging from the water column; that is, the slowly deposited red clays had received an additional trace-metal supply from the overlying seawater. Thus, the chemical composition of the clays is controlled by the mixing of the detrital and authigenic components, which, in turn, is controlled by their relative accumulation rates. As a result, in the grey clays $C_t \sim C_d$. This was the condition predicted by Krishnaswami (1976) when the detrital material accumulates at a sufficiently fast rate. For the red clays, however, the total metal flux ($F_t$) is the sum of the fluxes of the authigenic ($F_a$) and the detrital ($F_d$) components:

$$F_t = F_d + F_a \qquad (16.6)$$

$F_a$ is assumed to vary independently of $F_d$. Both $F_t$ and $F_d$ are the products of the total mean sediment accumulation rate ($S$) and the concentration of the metals in, respectively, the total sediment ($C_t$) and the detrital component ($C_d$). The net removal rate of the authigenic trace metals was then calculated by two routes:

1

$$F_a = (C_t - C_d)S \qquad (16.7)$$

where $C_d$ is assumed to be equal to the total metal content of the grey clays, that is, $C_t = C_d$ (the background correction method: see previously);

2

$$F_t = F_a + C_d S \qquad (16.8)$$

where a graphical procedure is used in which the total metal fluxes ($F_t = C_t S$) are plotted against sedimentation rates to give a regression line that allows both $F_a$ (the $S = 0$ intercept) and $C_d$ (the gradient) to be evaluated. This approach does not assume a knowledge of $C_d$, but does assume that the authigenic flux is constant for each metal throughout the survey area, that is, a constant-flux regional model.

The linearity of the plots confirmed the latter assumption, and the composition of the detrital fraction obtained from the gradients generally was similar to that obtained by assuming that it was equal to the total concentration of the grey clays. The form of the plots suggested that it is biogenic material and not terrigenous (detrital) clays that control the authigenic fluxes, a conclusion similar to that reached by Bacon and Rosholt (1982) for the Bermuda Rise. These biogenic trace-metal carriers have been destroyed at the sediment surface by oxic diagenesis, thus highlighting the importance of including oxic diagenetic processes in the authigenic signal.

The authigenic fluxes for the NAP derived from the two equations showed good agreement for Mn, Cu, Co, Ni and Zn: see Table 16.5. Although these authigenic fluxes are reasonably constant over the NAP survey area, comparison with data for other regions shows that the magnitudes of the fluxes do vary, both within ocean basins (e.g. those for sediments of the Bermuda Rise, Atlantic) and between

ocean basins (compared to those for the Pacific clays).

It was therefore becoming increasingly apparent that although the constant-flux model for the accumulation of authigenic elements in sediments can be approximated regionally, it does not apply on an ocean-wide basis. Despite this, however, Thompson *et al.* (1984) pointed out that the relative magnitudes of the authigenic fluxes generally are the same in different oceanic areas, and that there is a tendency for them to decrease in the order observed for the NAP, that is,

Fe > Mn > Cu ~ Ni > Co ~ V ~ Zn

Thompson *et al.* (1984) identified two factors that might be expected to control the relative magnitudes of the authigenic metal fluxes. These were (i) the relative values of their input fluxes to the oceans, that is, their geochemical abundances, and (ii) the relative efficiency of the scavenging processes that remove them from seawater. On the assumption that the oceanic inputs of the metals are dominated by fluvial sources, Thompson *et al.* (1984) used the data provided by Martin and Whitfield (1983) to show that the relative dissolved concentrations of the metals in river water are

Fe > Mn > Cu > Ni > Co

This is the same order as that of their relative authigenic fluxes, although distortions by factors such as hydrothermal sources were not considered by the authors in their model. Thompson *et al.* (1984) concluded, therefore, that although there are large variations in the absolute magnitudes of the authigenic metal fluxes from one oceanic region to another, these do not fractionate the elements to an extent that removes the underlying pattern imposed by their geochemical abundances.

Transition metals that have relatively long residence times in seawater, for example, Cu and Ni, will be expected to be transported further from their source areas before being removed from solution than would metals with shorter residence times, for example, Fe and Mn. Despite the geochemical abundance control the authors were therefore able to identify differences in the magnitudes of the authigenic fluxes that do, in fact, reflect oceanic reactivities of the individual elements. To illustrate this, Thompson *et al.* (1984) made a direct comparison

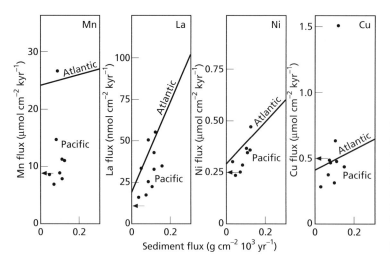

**Fig. 16.5** Comparison between the authigenic fluxes of elements in red clays from the Nares Abyssal Plain (Atlantic) and those of Pacific pelagic red clays. The ratios of the Atlantic to the average Pacific red clays are Mn = 2.5, La = 1.7, Ni = 1.2 and Cu = 0.75 (from Thompson *et al.*, 1984). Reprinted with permission from Elsevier.

between the authigenic fluxes they derived for the NAP red clays and those estimated by Krishnaswami (1976) for a series of Pacific clays. To do this, the data for Mn, Cu and Ni from the two oceans were plotted in a graphical form: see Fig. 16.5. The plots reveal that there is a wide range in the authigenic accumulation rates for Mn despite the narrow range in sediment accumulation rates. In contrast, the ranges in the authigenic accumulation rates for Cu and Ni are much smaller. Further, when derived from the data as displayed in Fig. 16.5, the authigenic fluxes for Cu and Ni in the Pacific clays are similar to those in the Atlantic clays at the NAP site, the ratios of the NAP : average Pacific fluxes being 0.75 and 1.2 for Cu and Ni, respectively. In contrast, the fluxes for Mn are lower for the Pacific clays, the ratio of the NAP : average Pacific fluxes being 2.5. The authors concluded that these features are consistent with the oceanic reactivities of the elements, Mn having a short oceanic residence time (<100 yr) resulting from rapid surface-water scavenging, and Cu and Ni having longer residence times (>1000 yr) owing to their lower scavenging and their involvement in the biogeochemical assimilation–regeneration cycle and release from bottom sediments. Thus, it would be expected that because of their longer residence times, inter-oceanic concentration variations for Cu and Ni will be smaller than those for Mn, that is, source-strength effects will be smoothed out to a much greater degree for Cu and Ni than for Mn, so allowing the Cu and Ni to be redistributed in

both seawater and the authigenic fractions of deep-sea sediments.

It may be concluded, therefore, that the geochemical abundances of the metals are the principal factor governing their authigenic sediment fluxes, but that fractionation occurs in accordance with their oceanic reactivities. This results in the reactive elements (e.g. Mn) having higher fluxes in Atlantic relative to Pacific clays because the Atlantic has a greater fluvial input than the Pacific, and in the less reactive elements (e.g. Cu, Ni) having generally similar fluxes in the two oceans.

Bacon and Rosholt (1982) and Thompson *et al.* (1984) concluded that the settling of biogenic carriers in the form of organic aggregates is the main driving force behind the down-column transport of trace metals to the sediments on both the Bermuda Rise and the NAP. The carrier phases are destroyed by oxic diagenesis at the sediment surface and some of the associated metals are taken into authigenic components. The ocean-wide importance of these carriers has been demonstrated by other workers. For example, Aplin and Cronan (1985) demonstrate a close similarity between the down-column fluxes and sediment accumulation rates for Ni, Co, Ti, Mn and Fe (Fig. 16.6a). By contrast Cu is supplied in excess of its accumulation rate, which is in agreement with its release back into seawater across the sediment–water interface (see Section 14.4.3). It is apparent, therefore, that down-column transport via biogenic carriers is the principal route by which

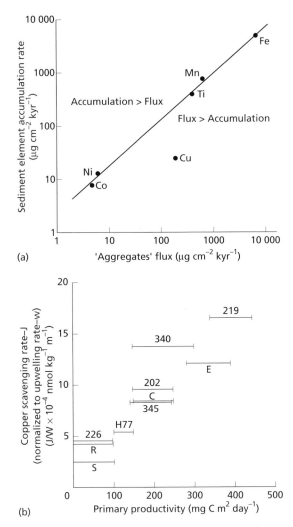

(a)

(b)

**Fig. 16.6** The rates of water-column removal processes for trace elements. (a) Comparison between biogenic carrier down-column fluxes and accumulation rates of elements in sediments from the southwest Pacific (from Aplin and Cronan, 1985). Reprinted with permission from Elsevier. (b) Estimates of the primary production of organic carbon in surface waters ($^{14}$C uptake) versus the relative scavenging rate of Cu in the deep ocean; numbers and letters indicate original data sources (from Collier and Edmond, 1984). Reprinted with permission from Elsevier.

controlled mainly by geochemical abundances, but metal reactivities in the water column are superimposed on input source strengths. In terms of their oceanic reactivities, the scavenging-type elements have relatively short residence times and so undergo restricted transport between the major ocean basins. The nutrient-type elements have longer residence times, and so have a greater potential for inter-ocean transfer. The down-column fluxes of these nutrient-type elements vary considerably throughout the oceans, however, depending largely on the magnitude of the down-column carbon-driven flux. For example, Collier and Edmond (1984) found a striking correlation between the scavenging rate of Cu in the water column and primary productivity in surface waters: see Fig. 16.6(b). The authors concluded that this relationship emphasizes the importance of regions of high primary productivity in driving the vertical transport of elements in the oceans; that is, these regions have relatively high down-column TSM fluxes (see Section 12.1) which influence scavenging rates (Section 11.5). As a result, nutrient-type elements can be removed from the water column on time-scales that prevent their inter-oceanic transport, and this can lead to patches of high concentrations of these elements in sediments deposited under regions of intense primary production.

It may therefore be concluded that the authigenic signal does not operate in a manner that yields an ocean-wide constant flux for those authigenic elements which have relatively short residence times in seawater. Authigenic elements that have longer residence times, however, can have their dissolved concentrations smoothed out between the major oceans, and therefore may approach the constant-flux model as originally proposed by Wedepohl (1960) and supported by data from Krishnaswami (1976). Thus, we have come full circle, and the distributions of the elements in deep-sea sediments can be linked to the processes controlling their oceanic residence times, and their general seawater chemistries, discussed in Chapter 11.

## 16.5 Signal spikes

### 16.5.1 Introduction

In the previous sections, the sources of the excess trace metals found in deep-sea clays were assessed in terms of their background transport via the

authigenic trace metals reach the sediment surface. In view of this, it is not surprising that although the authigenic processes provide an ocean-wide background for the supply of non-detrital metals to sediments, the magnitudes of the authigenic fluxes of some elements vary both between and within ocean basins (see Section 12.1). These fluxes appear to be detrital and authigenic signals. This involved a two-component system in which

$$C_t = C_d + C_a \qquad (16.1)$$

Within the ocean system, however, a series of more localized signals can impose perturbation spikes on this background, such that

$$C_t = C_d + C_a + C_s \qquad (16.9)$$

where $C_s$ refers to the concentration of an element arising from the perturbation spikes. These spikes are transmitted through the water column mainly by the hydrothermal and contaminant signals, and through interstitial waters by the diagenetic signal.

### 16.5.2 The hydrothermal signal spike

The concepts underlying hydrothermal activity have been discussed in Chapter 5, and the processes involved in the generation of hydrothermal sediment-forming components have been described in Section 15.3.6. Attention now will be directed to considering the question, 'How do these chemically specialized components affect the compositions of deep-sea sediments?' The hydrothermal emanations occur at the ridge-crest spreading centres and give rise to a variety of hydrothermal components, but as far as deep-sea sediments in general are concerned it is the ferromanganese precipitates that are the principal manifestations of hydrothermal activity.

It has been known for many years that metal-rich sediments are found in association with the mid-ocean ridge system. For example, Murray and Renard (1891) and Revelle (1944) reported the presence of such deposits in the eastern Pacific, and El Wakeel and Riley (1961) gave details of Mn- and Fe-rich calcareous ooze from the vicinity of the East Pacific Rise (EPR). Such deposits are now known to be associated with all ridge spreading centres (see Chapter 5 also German and Von Damm, 2003). Compositional data for some of these metal-rich sediments are given in Table 15.5 and Table 16.6.

Metal-rich deposits are formed on newly generated crust at the ridge spreading centres. During the process of sea-floor spreading this crust moves away from the ridges as the ocean basins are opened up, and the basement is covered by a blanket of normal deep-sea sediments. On the shallower ridge crests above the carbonate compensation depth (see Section 15.2) this is often carbonate ooze. Further away from the ridge crest in deeper water below the carbonate compensation depth pelagic clays will accumulate. As a result, although surface outcrops of

**Table 16.6** Chemical composition of some modern and ancient metalliferous sediments; units, $\mu g\,g^{-1}$, carbonate-free basis.

| Element | East Pacific Rise: crest* | East Pacific Rise: flanks* | Pacific: Northwest Nazca Plate† | | | | East Pacific Mid-Atlantic Ridge§ |
| | | | East Pacific | Bauer Deep | Central Basin | Rise: basal sediments‡ | |
|---|---|---|---|---|---|---|---|
| Fe | 180 000 | 105 000 | 302 000 | 158 300 | 121 100 | 200 700 | 76 000 |
| Mn | 60 000 | 30 000 | 99 200 | 57 500 | 39 600 | 60 600 | 4 100 |
| Cu | 730 | 960 | 1 450 | 1 171 | 985 | 790 | – |
| Ni | 430 | 675 | 642 | 1 066 | 1 307 | 460 | – |
| Co | 105 | 230 | – | – | – | 82 | – |
| Pb | – | – | – | – | – | 100 | – |
| Zn | 380 | 290 | 594 | 413 | 311 | 470 | – |
| V | 450 | 240 | – | – | – | – | – |
| Hg | – | – | – | – | – | – | 414 |
| As | 145 | 65 | – | – | – | – | 174 |
| Mo | 30 | 113 | – | – | – | – | – |
| Cr | 55 | 32 | – | – | – | – | – |
| Al | 5 000 | 46 300 | 5 100 | 32 400 | 67 400 | 27 300 | 57 900 |

*Bostrom and Peterson (1969). †Heath and Dymond (1977). ‡Cronan (1976). §Cronan (1972).

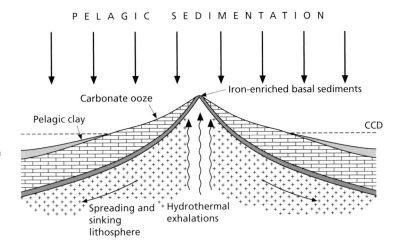

P E L A G I C     S E D I M E N T A T I O N

Carbonate ooze

Iron-enriched basal sediments

Pelagic clay

CCD

**Fig. 16.7** Schematic representation of a model for sediment accumulation at oceanic spreading ridge centres (from Davies and Gorsline, 1976; after Broecker, 1974; and Bostrom and Peterson, 1969). Reprinted with permission from Elsevier.

Spreading and sinking lithosphere

Hydrothermal exhalations

metal-rich sediments are usually found mainly at the ridge crests, Bostrom and Peterson (1969) predicted that they should also form a layer on top of the oceanic basement at all locations. A general model illustrating this process is shown in Fig. 16.7. Confirmation that metalliferous sediments are the first material deposited on the basaltic basement was provided by cores obtained during the Deep Sea Drilling Project (DSDP), which showed that Mn- and Fe-rich EPR analogue deposits were indeed found at the base of the sediment column at many oceanic locations (see e.g. von der Borch and Rex, 1970; Cronan *et al.*, 1972; Cronan, 1973; Dymond *et al.*, 1973; Cronan and Garrett, 1973; Horder and Cronan, 1981).

It may be concluded, therefore, that Fe- and Mn-rich metalliferous deposits are usually the first type of sediment to be laid down on the oceanic crust at the spreading centres, thus forming a bottom layer on to which other types of deep-sea sediments accumulate. Although the hydrothermal precipitates have their source at the ridge-crest venting systems, the sediments formed there are rarely composed of pure hydrothermal material, but usually are mixtures of the metal-rich end-member precipitates and other sediment-forming components: mainly carbonate shells, which are preserved on the topographic highs. Marchig and Grundlach (1982) identified a prototype undifferentiated end-member formed at an initial stage in the evolution of hydrothermal mate-

**Table 16.7** Chemical composition of a prototype undifferentiated hydrothermal precipitate*; units, $\mu g\,g^{-1}$, carbonate-free basis.

| Element | Concentration | Element | Concentration |
|---------|---------------|---------|---------------|
| Si | 36 000 | Mo | 134 |
| Al | 3 600 | Zr | 73 |
| Mn | 91 800 | Nb | 2 |
| Fe | 281 000 | Rb | 9.3 |
| Ti | 300 | Ce | 13 |
| Cu | 1 300 | Th | 61 |
| Zn | 393 | V | 1923 |
| Co | 111 | W | 21 |
| Cr | 30 | Y | 101 |
| Ni | 600 | La | 98 |

*Data for the <63 μm fractions of sediment cores from the East Pacific Rise (Marchig and Grundlach, 1982).

rial from the EPR. From the data given by these authors an estimate is therefore available of the composition of the Fe- and Mn-rich end-member that is precipitated from hydrothermal fluids on mixing with seawater (see Table 16.7). Some authors have suggested that hydrothermal fluids can be dispersed on an ocean-wide basis (see e.g. Lalou, 1983), and it is clear that hydrothermal manganese signals can be transmitted for long distances via the global mid-depth water circulation patterns (see Section 15.3.6). The question that arises, therefore, is

'How widespread is the Fe- and Mn-rich hydrothermal end-member that is generated from these pulses?' To answer this question, it is necessary to unscramble the effects of the hydrothermal pulse signal from those of other sediment-forming signals.

Bostrom *et al.* (1969) addressed the problem of unscrambling the hydrothermal signal on an oceanic-wide basis. Various authors have identified a series of metal-rich deep-sea sediments associated with areas of high heat flow on the EPR. Relative to normal deep-sea sediments these deposits are enriched in elements such as Fe, Mn, Cu, Ni and Pb, which are associated with the colloidal-sized hydrothermal particles. Another important characteristic of the metalliferous sediments is that they are depleted in lithogenous elements such as Al and Ti: see Table 16.6. This enrichment–depletion pattern was utilized by Bostrom *et al.* (1969) to characterize a hydrothermal signature. To quantify the signature, the authors used the ratio $Al:(Al + Fe + Mn)$ as an indicator of metalliferous sedimentation. In this way,

the hydrothermal end-member was characterized as having $Al:(Al + Fe + Mn)$ ratios of $<10 \times 10^2$, and the normal deep-sea sediment end-member as having ratios $>60 \times 10^2$. Sediments that are mixtures of the two end-members have intermediate ratios. The authors then plotted values of the hydrothermal ratio for deep-sea sediments, and showed that the metalliferous deposits are indeed characteristic of the mid-ocean ridge system throughout the World Ocean: see Fig. 16.8. The lowest ratios, and the greatest coverage of hydrothermal sedimentation, were found in the Pacific Ocean around the EPR, and the influence of the hydrothermal material on surrounding sediments decreased in the order Pacific Ocean ridge system > Indian Ocean ridge system > Atlantic Ocean ridge system. This rank order correlates well with the relative rates of ocean-floor spreading, which can indicate the general degree to which a ridge is active; thus, the spreading rate in the Pacific is $\sim 2.0$–$6.0\,\mathrm{cm\,yr^{-1}}$, in the South Atlantic and Indian Oceans it is $\sim 1.5$–$3.0\,\mathrm{cm\,yr^{-1}}$, and in the North Atlantic it is $\sim 1.0$–$1.4\,\mathrm{cm\,yr^{-1}}$.

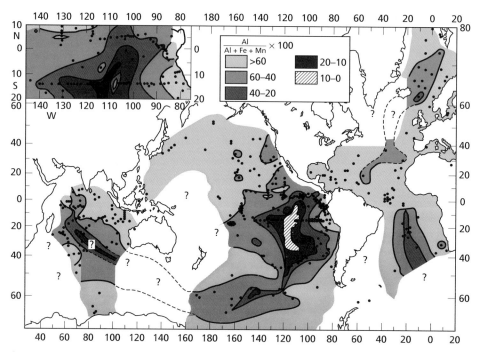

**Fig. 16.8** The distribution of the hydrothermal ratio, $Al:(Al + Fe + Mn)$, in sediments from the World Ocean (from Bostrom *et al.*, 1969). Note that low ratios are associated with spreading ridges. Insert is a detailed map for the region of the tropical equatorial Pacific. Reprinted with permission from Elsevier.

Bostrom *et al.* (1969) therefore were able to identify the widespread occurrence of metalliferous deposits on the mid-ocean ridge system by unscrambling the hydrothermal signal in terms of the sediment Al : (Al + Fe + Mn) ratio. Thus, hydrothermal deposits have relatively high concentrations of Fe and Mn. In addition, the hydrothermal deposits can be differentiated from 'normal' deep-sea sediments on the basis of the rates at which the Fe and Mn in them accumulate. This was considered in Section 15.3.6 with respect to ferromanganese nodules and crusts, and now will be discussed in terms of deep-sea sediments. It was shown in Section 16.4 that, although there is not a constant ocean-wide authigenic deposition rate for Mn in deep-sea sediments, the rates do appear to be in the range ~1000–5000 $\mu g\,cm^{-2}\,10^3\,yr^{-1}$ for the Atlantic, and somewhat lower for the Pacific. For example, Krishnaswami (1976) estimated an authigenic deposition rate for Mn of ~500 $\mu g\,cm^{-2}\,10^3\,yr^{-1}$ for Pacific pelagic clays, and Bender *et al.* (1971) calculated an average total deposition rate for Mn of ~1300 $\mu g\,cm^{-2}\,10^3\,yr^{-1}$ for Pacific deep-sea sediments from the same oceanic region. In contrast, Bender *et al.* (1971) used radiochemical data to estimate that the accumulation rate of Mn in a core from the crest of the EPR was ~35 000 $\mu g\,cm^{-2}\,10^3\,yr^{-1}$, that is, this Mn was accumulating almost 25 times faster than in normal deep-sea sediments. Dymond and Veeh (1975) gave data on the accumulation rates of Mn, Fe and Al in sediments on a transect extending west to east across the EPR, and clearly demonstrated enhanced accumulation rates for Mn and Fe in sediments deposited on the ridge crest. The Mn accumulation rates on the crest reached values of ~28 000 $\mu g\,cm^{-2}\,10^3\,yr^{-1}$, that is, the same order of magnitude as those reported by Bender *et al.* (1971). The maximum rate for the accumulation of Fe on the EPR was ~82 000 $\mu g\,cm^{-2}\,10^3\,yr^{-1}$. According to Dymond and Veeh (1975), the authigenic accumulation rate for Fe in normal deep-sea sediments from this region is ~200 $\mu g\,cm^{-2}\,10^3\,yr^{-1}$, and Krishnaswami (1976) estimated an average rate of ~800 $\mu g\,cm^{-2}\,10^3\,yr^{-1}$ for accumulation of Fe in Pacific pelagic clays. Clearly, therefore, the accumulation of Fe on the crest of the EPR is of the order of 100–400 times faster than it is in normal Pacific deep-sea sediments.

It may be concluded, therefore, that authigenic deposition cannot explain the enhanced accumulation rates of Mn and Fe on the crestal regions of the EPR. Rather, it would appear that the high accumulation rates are the result of an enhanced supply of the elements from a hydrothermal signal. However, as they form, hydrothermal precipitates can remove dissolved elements directly from seawater, so that not all the elements that accumulate at fast rates around the ridge crests necessarily have a hydrothermal origin.

Another way of characterizing the hydrothermal source of an element is by making a direct comparison between the compositions of hydrothermal solutions and those of metalliferous sediments deposited in the same area. Such an approach was adopted by Von Damm *et al.* (1985). These authors determined the concentrations of Mn, Fe, Ni, Cu, Zn, Co, Cd, Ag and Pb in hydrothermal fluids vented from the high-temperature system at 21°N on the EPR, and compared the elemental ratios in the fluids with those in associated metalliferous sediments using Fe as a tracer of hydrothermal sources. The results may be summarized as follows.

1 For Co and Ni (which have an insignificant hydrothermal source), the ratios in the venting solutions were lower than in the sediments, implying that additional amounts of these elements are scavenged from seawater by the hydrothermal Fe–Mn precipitates.

2 For Mn, Zn, Cu and Ag, the ratios in the hydrothermal solutions were greater than, or equal to, those in the sediments, indicating that although the hydrothermal input is the major source of the elements a proportion of them may be lost to seawater, that is, they may escape the immediate venting area.

3 The Pb : Fe ratios were very similar in both the venting solutions and the metalliferous sediments, implying a predominantly hydrothermal origin for the Pb.

The hydrothermal signal results from the emanation of mineralizing solutions at the ridge-crest spreading centres. These metal-rich emanations impose spikes on the general background of biogenous, detrital and authigenic components that combine to form normal deep-sea sediments. The components derived from the hydrothermal sources form an important class of metal-rich sediments. By unscrambling the hydrothermal signal, however, it has been demonstrated that the metalliferous deposits themselves have only a relatively small areal extent and, with respect to the sediment surface, they are largely confined to regions around the mid-ocean ridge system.

### 16.5.3 The contaminant signal spike

The changes in the inputs of metals and nutrients to the oceans were discussed in Chapters 3–6 and the impact on metal and nutrient concentrations within the oceans were considered in Chapter 9 and 11. It is of interest at this stage, however, to demonstrate how contaminant spikes can influence the elemental chemistry of marine sediments.

#### 16.5.3.1 Nearshore sediments

Contaminant, or anthropogenically-generated, elements are brought to the ocean reservoir by the same pathways that transport the material released naturally during crustal mobilization. Both spatial and temporal 'fingerprints' resulting from the inputs of contaminant elements have been recorded in near-shore sediments, which accumulate at relatively fast rates close to the sources of pollution. Various techniques can be used to identify elements transmitted by contaminant signals, and to unscramble them from natural background inputs. Two examples will serve to illustrate how such techniques can assess the effects that contaminant spikes can have on coastal deposits.

*The spatial distribution of contaminant elements in coastal sediments.* A large proportion of many of the elements associated with contaminant spikes have been introduced into seawater in a dissolved state *and have subsequently been removed from solution*; thus, they form part of the non-detrital sediment fraction. To separate this fraction from the crystalline mineral matrix, Chester and Voutsinou (1981) applied a chemical leaching technique to sediments from two Greek gulfs, one of which had received pollutant inputs and the other that was reasonably pristine in character. By using the sediments of the unpolluted gulf as a background baseline, the authors were able to (i) identify contaminant trace metals; and (ii) map their spatial distributions in sediments from the polluted gulf (Thermaikos Gulf). To illustrate this, the spatial distribution of Zn in Thermaikos Gulf is shown in Fig. 16.9. There is a heavily industrialized region at the head of this gulf, and it

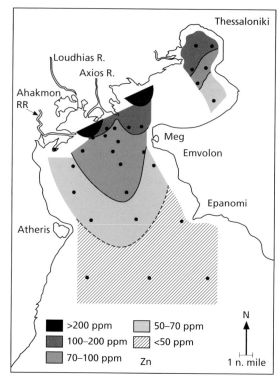

**Fig. 16.9** Distribution of non-detrital Zn in surface sediments from Thermaikos Gulf, Greece (from Chester and Voutsinou, 1981). Reprinted with permission from Elsevier.

is evident from Fig. 16.9 that the non-detrital concentrations of Zn are highest in sediments in this region and decrease out into the central gulf; thus, local inputs have imposed a fingerprint on the distribution of this trace metal in the surficial sediments of the polluted gulf.

*The temporal distribution of contaminant elements in coastal sediments.* Because of their relatively fast rates of deposition coastal sediments can sometimes provide a historical record of contaminant inputs. This was demonstrated by Chow *et al.* (1973) with respect to the deposition of Pb in coastal sediments off southern California. The authors carried out a survey of the distribution of Pb in dated sediment columns from a number of basins in the area. To aid their interpretation of the data, Al (which has a mainly natural origin) was used as a normalizing element to establish the background levels of Pb in the sediments, and isotopic ratios were used to identify the sources of the Pb itself. Sediments were sampled from the following localities.

1 The Soledad Basin. This is remote from the prevailing winds blowing off southern California and is free from waste discharges, and therefore was used as a control area.

2 The San Pedro, Santa Monica and Santa Barbara Basins. These are inner basins in the Los Angeles area.

3 White Point. This is close to a waste outfall and provided sediments containing industrial and domestic wastes; however, the sediments here had accumulated so rapidly that only a recent age could be assigned to them over the interval studied.

The Pb concentrations, and Pb:Al ratios, in the sediments from the various basins are illustrated in Fig. 16.10, and the results of the study can be summarized as follows.

1 The Pb concentrations in the sediments of the Soledad Basin are generally invariant and are less than ~20 µg g$^{-1}$. Further, the surface deposits contain Pb which had an isotopic composition similar to that of weathered material from the Baja California province. The sediments in this basin therefore were used as a baseline against which to compare the Pb distributions in sediments from the other basins.

2 Relative to those of the Soledad Basin, the sediments of the inner basins contain higher concentrations of Pb in their upper portions, which decreased towards baseline levels at depths of ~8 cm. Below this depth, the isotopic ratios of the Pb were similar to those that represent the pre-pollution supply. The sediments in these basins are anoxic and had not suffered bioturbation, and it could be shown that the rates of contaminant Pb accumulation had started to increase in the 1940s. The increased inputs are clearly reflected in contaminant spikes in the down-column Pb profiles in the inner basin sediments.

3 The sediments from White Point have the highest Pb concentrations of several hundreds of micrograms per gram (µg g$^{-1}$), and Pb isotopic ratios that are similar to those in petrol (gasoline) then sold in southern California.

### 16.5.3.2 Deep-sea sediments

At present, slowly accumulating deep-sea sediments have not yet recorded major inputs of most anthropogenically-generated elements to the extent that concentration spikes are present. Evidence is now beginning to appear, however, which suggests that the situation is changing; for example, artificial

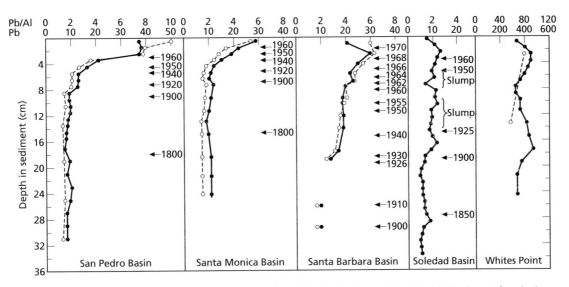

**Fig. 16.10** The down-column concentrations of Pb (solid circles), and Pb:Al ratios (open circles), in dated sediments from basins off the coast of southern California (from Chow *et al.*, 1973); concentrations in µg g$^{-1}$ and sediment ages marked as year of deposition.

radionuclides have been reported in the upper portions of deep-sea sediments (see e.g. Lapicque *et al.*, 1987).

It is elements that have a strong anthropogenic signal to the oceans, and a relatively short residence time in the water column, that have the greatest potential for being recorded in deep-sea sediments. Lead is a prime example of such an element:

1 it has a strong atmospheric anthropogenic signal to open-ocean surface waters (see Section 6.4);

2 it is a scavenging-type element that is rapidly removed from surface waters within a few years (see Section 11.3); and

3 it has a relatively short overall residence time in the oceans of around 100 yr (see Section 11.2 and 3).

Veron *et al.* (1987) have reported data that do indeed offer evidence of recent Pb pollution in northeast Atlantic deep-sea sediments. These authors showed that Pb concentrations in the surficial sediments (21 and 15 μg g$^{-1}$ in the top 1 cm) of two short cores were higher than those in the underlying 10 cm (6.0 and 2.8 μg g$^{-1}$: see Fig. 16.11). The amount of

Pb stored in the surficial sediments was of the same order as the amount of Pb in the overlying water column, and the authors concluded that the surficial Pb spike was derived from anthropogenic sources.

It is apparent, therefore, that contaminant spikes are beginning to be identified in surficial deep-sea sediments, but it will be a considerable time before such spikes are recorded for elements that have relatively long oceanic residence times.

### 16.5.4 The diagenetic signal spike

Trace metals are transported down the water column from primary seawater sources in association with large-sized particulate organic carrier phases. Once they are deposited at the sediment surface diagenesis is initiated, and the carriers are either destroyed during oxic diagenesis, or buried and subsequently at least partially destroyed during suboxic diagenesis. Many of the reactions that occur in the upper sediment, and the benthic boundary layer, are driven by oxic diagenesis. Because this affects elements that have been removed from seawater and subsequently transported down the water column, oxic diagenesis has been classified as part of the authigenic signal. Some of the processes associated with oxic diagenesis can lead to the transport of elements downwards into the sediments through interstitial waters; this was illustrated in Section 14.4.3 with respect to Cu. The point is, however, that these elements were originally transported down the seawater column and released during oxic diagenesis. The authigenic signal therefore is the basic background control on the spatial distribution of the chemical elements in surface deep-sea sediments, although the distribution can be perturbed by hydrothermal and contaminant spikes. The vertical distribution of elements added to the sediment column from seawater sources can, however, be affected by remobilization and transport through interstitial waters associated with suboxic diagenesis. As defined in the present context, therefore, the diagenetic signal refers to that transmitted through the interstitial water column following suboxic diagenesis, and can affect both natural and contaminant elements.

During suboxic diagenesis dissolved trace metals are released from both organic carriers and secondary oxidants, and can reach concentrations in the interstitial waters in excess of those in the overlying

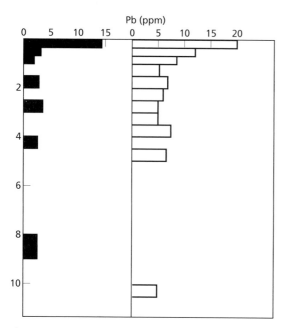

**Fig. 16.11** Anthropogenic effects on lead in the marine environment. Depth distributions of lead in two deep-sea sediment cores from the northeast Atlantic (from Veron *et al.*, 1987); units in μg g$^{-1}$. The surficial samples in both cores contained excess Pb, which was attributed to an anthropogenic source.

appears to be non-detrital in origin (see e.g. Krishnaswami, 1976); that is, the element has a partitioning signature that is intermediate between those of $\Sigma Fe$ and $\Sigma Mn$. A number of features can be seen in the accumulation rate pattern of $\Sigma Cu$ in Pacific deep-sea sediments.

**1** There is a tendency for $\Sigma Cu$ to accumulate faster around the continents, which presumably reflects the near-source strength of the detrital Cu signal.

**2** In the North Pacific there is a general decrease in the accumulation rates of $\Sigma Cu$ towards mid-ocean areas as the influence of the detrital signal decreases. Here, $\Sigma Cu$ accumulates at rates of $<25 \mu g\,cm^{-2}$ $10^3\,yr^{-1}$. From the data provided by Krishnaswami (1976), authigenic Cu accumulates in normal Pacific pelagic clays at $\sim8 \mu g\,cm^{-2}$ $10^3\,yr^{-1}$, so that on this basis around one-third of the $\Sigma Cu$ arises from the authigenic signal; however, this is a minimum estimate, and the figure may well reach the 50% authigenic partitioning reported by Krishnaswami (1976).

**3** In the South Pacific, the accumulation of $\Sigma Cu$, like those of $\Sigma Mn$ and $\Sigma Fe$, is dominated by the hydrothermal signal, and around the EPR $\Sigma Cu$ accumulates at rates of $>200 \mu g\,cm^{-2}$ $10^3\,yr^{-1}$, which is similar to the rate of accumulation associated with the detrital input from the US mainland in the North Pacific.

**4** There is a band of relatively high $\Sigma Cu$ accumulation around the Equator, which is considerably better developed than those for either $\Sigma Mn$ or $\Sigma Fe$. This band corresponds to a zone of high primary productivity in the surface waters, and the enhanced accumulation of $\Sigma Cu$ probably results from an accelerated down-column authigenic flux from the carbon-rich overlying waters (see Figs 9.5 and 16.6b). Bostrom *et al.* (1973) reported that the distribution and accumulation rate patterns of $\Sigma Ni$ and $\Sigma Co$ in the Indo-Pacific generally were similar to those for $\Sigma Cu$.

It may be concluded, therefore, that the distributions and accumulation rate patterns of elements in surface sediments from the Indo-Pacific can be used to illustrate how sediment-forming signals interact on a global scale. The main features in the distribution and accumulation rates of opaline silica in the sediments can be related to the biogenic signal, and those for $\Sigma Al$ can be explained in terms of the predominance of the detrital signal. Patterns in the distributions and accumulation rates of $\Sigma Fe$, $\Sigma Mn$ and $\Sigma Cu$ can be interpreted mainly in terms of the global

interaction of the detrital, authigenic and hydrothermal signals. It must be remembered, however, that within the sediment column the suboxic diagenetic signal can redistribute redox-sensitive elements such as manganese, the overall effect decreasing in the order: shelf > hemi-pelagic > pelagic sediments (see Section 14.1.3).

## 16.7 Unscrambling the sediment-forming chemical signals: summary

**1** The chemical composition of marine sediments can be interpreted within a framework in which the elements are envisaged as being transmitted to the deposits via a series of signals associated with a variety of geochemical processes. The components making up the sediments are formed by elements that are transmitted either by individual signals or, more usually, by signal couplings.

**2** Detrital, biogenous and authigenic signals operate on an ocean-wide scale. These are the background signals, which carry the elements having a direct seawater source.

**3** A series of perturbation spikes from more localized hydrothermal and contaminant seawater signals, and from interstitial-water diagenetic signals, are superimposed on the background signals.

**4** Sediments from different oceanic environments record varying signal strengths. For example:

(a) authigenic signals have their greatest influence on pelagic sediments deposited in regions away from the spreading ridges;

(b) diagenetic signals are strongest in hemi-pelagic sediments, in which suboxic diagenesis takes place;

(c) hydrothermal signals can dominate in areas in the vicinity of spreading ridges, where metal-rich sediments can be formed;

(d) effects of contaminant signals are confined mainly to coastal deposits, but are starting to appear in deep-sea sediments.

## References

Aplin, A.C. and Cronan, D.S. (1985) Ferromanganese oxide deposits in the central Pacific Ocean, II. Nodules and associated elements. *Geochim. Cosmochim. Acta*, **49**, 437–451.

Bacon, M.P. and Rosholt, J.N. (1982) Accumulation rates of Th-230, Pa-231, and some transition metals on the Bermuda Rise. *Geochim. Cosmochim. Acta*, **46**, 651–666.

Bender, M.L. (1971) Does upward diffusion supply the excess in manganese in sediments? *J. Geophys. Res.*, **76**, 4212–4215.

Bender, M.L., Broecker, W., Gornitz, V., *et al.* (1971) Geochemistry of three cores from the East Pacific Rise. *Earth Planet. Sci. Lett.*, **12**, 425–433.

Bostrom, K. and Peterson, M.N.A. (1969) The origin of aluminium-poor ferromanganoan sediments in areas of high heat flow on the East Pacific Rise. *Mar. Geol.*, **7**, 427–447.

Bostrom, K., Peterson, M.N.A., Joensuu, O. and Fisher, D.E. (1969) Aluminium-poor ferromanganoan sediments on active oceanic ridges. *J. Geophys. Res.*, **74**, 3261–3270.

Bostrom, K., Kraemer, T. and Gartner, S. (1973) Provenance and accumulation rates of opaline silica, Al, Ti, Fe, Mn, Cu, Ni and Co in Pacific pelagic sediments. *Chem. Geol.*, **11**, 123–148.

Broecker, W.S. (1974) *Chemical Oceanography*. New York: Harcourt Brace Jovanovich.

Chester, R. and Aston, S.R. (1976) The geochemistry of deep-sea sediments, in *Chemical Oceanography*, J.P. Riley and R. Chester (eds), Vol. 6, pp.281–390. London: Academic Press.

Chester, R. and Hughes, M.J. (1967) A chemical technique for the separation of ferrogmanganese minerals, carbonate minerals and adsorbed trace elements from pelagic sediments. *Chem. Geol.*, **3**, 199–212.

Chester, R. and Hughes, M.J. (1969) The trace element geochemistry of a North Pacific pelagic clay core. *Deep Sea Res.*, **13**, 627–634.

Chester, R. and Messiha-Hanna, R.G. (1970) Trace element partition patterns in North Atlantic deep-sea sediments. *Geochim. Cosmochim. Acta*, **34**, 1121–1128.

Chester, R. and Voutsinou, F.G. (1981) The initial assessment of trace metal pollution in coastal sediments. *Mar. Pollut. Bull.*, **12**, 84–91.

Chester, R., Thomas, A., Lin, F.J., Basaham, A.S. and Jacinto, G. (1988) The solid state speciation of copper in surface water particulates and oceanic sediments. *Mar. Chem.*, **24**, 261–292.

Chow, T.J., Bruland, K.W., Bertine, K., Soutar, A., Koide, M. and Goldberg, E.D. (1973) Lead pollution: records in Southern California coastal sediments. *Science*, **181**, 551–552.

Collier, R. and Edmond, J.M. (1984) The trace element geochemistry of marine biogenic particulate matter. *Prog. Oceanogr.*, **13**, 113–199.

Cronan, D.S. (1972) The Mid-Atlantic Ridge near 45°N, XVII: Al, As, Hg and Mn in ferruginous sediments from the median valley. *Can. J. Earth Sci.*, **9**, 319–323.

Cronan, D.S. (1973) Basal ferruginous sediments cored during Leg 16, Deep Sea Drilling Project, in *Initial Reports of the Deep-Sea Drilling Project*, Vol. XVI, pp.601–604. Washington, DC: US Government Printing Office.

Cronan, D.S. (1976) Basal metalliferous sediments from the eastern Pacific. *Geol. Soc. Am. Bull.*, **87**, 929–934.

Cronan, D.S. and Garrett, D.E. (1973) Distribution of elements in metalliferous Pacific sediments collected during the Deep Sea Drilling Project. *Nature*, **242**, 88–89.

Cronan, D.S., van Andel, Tj.H., Heath, G.R. *et al.* (1972) Iron-rich basal sediments from the eastern equatorial Pacific: Leg 16, Deep Sea Drilling Project. *Science*, **175**, 61–63.

Davies, T.A. and Gorsline, D.S. (1976) Oceanic sediments and sedimentary processes, in *Chemical Oceanography*, J.P. Riley and R. Chester (eds), Vol. 5, pp.1–80. London: Academic Press.

Dymond, J. and Veeh, H.H. (1975) Metal accumulation rates in the southeast Pacific and the origin of metalliferous sediments. *Earth Planet. Sci. Lett.*, **28**, 13–22.

Dymond, J., Corless, J.B., Heath, G.R., Field, C.W., Dasch, E.J. and Veeh, H.H. (1973) Origin of metalliferous sediments from the Pacific Ocean. *Geol. Soc. Am. Bull.*, **84**, 3355–3372.

Elderfield, H. (1976) Manganese fluxes to the oceans. *Mar. Chem.*, **4**, 103–132.

El Wakeel, S.K. and Riley, J.P. (1961) Chemical and mineralogical studies of deep-sea sediments. *Geochim. Cosmochim. Acta*, **25**, 110–146.

German, C.R. and Von Damm, K.L. (2003) Hydrothermal Processes, in *Treatise on Geochemistry*, H. Elderfield (ed.), Vol. 6, H.D. Holland and K.K. Turekian (series eds), pp.181–222. London: Elsevier.

Heath, G.R. and Dymond, J. (1977) Genesis and transformation of metalliferous sediments from the East Pacific Rise, Bauer Deep and Central Basin, northwest Nazca Plate. *Geol. Soc. Am. Bull.*, **88**, 723–733.

Horder, M.F. and Cronan, D.S. (1981) The geochemistry of some basal sediments from the western Indian Ocean. *Oceanol. Acta*, **4**, 213–221.

Krishnaswami, S. (1976) Authigenic transition elements in Pacific pelagic clays. *Geochim. Cosmochim. Acta*, **40**, 425–434.

Lalou, C. (1983) Genesis of ferromanganese deposits: hydrothermal origin, in *Hydrothermal Processes at Seafloor Spreading Centres*, P.A. Rona, K. Bostrom, L. Laubier and K.L. Smith (eds), pp.503–534. New York: Plenum.

Lapicque, G., Livingston, H.D., Lambert, C.E., Bard, E. and Labeyrie, L.D. (1987) Interpretation of 239,240Pu in Atlantic sediments with a non-steady state input model. *Deep Sea Res.*, **34**, 1841–1850.

Marchig, V. and Grundlach, H. (1982) Iron-rich metalliferous sediments on the East Pacific Rise: prototype of undifferentiated metalliferous sediments on divergent plate boundaries. *Earth Planet. Sci. Lett.*, **58**, 361–382.

Martin, J.M. and Whitfield, M. (1983) The significance of the river input of chemical elements to the oceans, in *Trace Metals in Seawater*, C.S. Wong, E. Boyle, K.W. Bruland, J.D. Burton and E.D. Goldberg (eds), pp.256–296. New York: Plenum.

Murray, J. and Renard, A.F. (1891) *Deep-sea Deposits*. London: Scientific Report, *Challenger* Expedition, no. 3. London: HMSO.

Revelle, R.R. (1944) Scientific results of the cruise VII of the 'Carnegie'. *Publ. Carnegie Inst.*, **556**, 1–180.

Sawlan, J.J. and Murray, J.W. (1983) Trace metal remobilization in the interstitial waters of red clay and hemipelagic marine sediments. *Earth Planet. Sci. Lett.*, **64**, 213–230.

Thomas, A.R. (1987) *Glacial–interglacial variations in the geochemistry of North Atlantic deep-sea deposits*. PhD Thesis, University of Liverpool.

Thompson, J., Carpenter, M.S.N., Colley, S., Wilson, T.R.S., Elderfield, H. and Kennedy, H. (1984) Metal accumulation in northwest Atlantic pelagic sediments. *Geochim. Cosmochim. Acta*, **48**, 1935–1948.

Turekian, K.K. (1967) Estimates of the average Pacific deep-sea clay accumulation rate from material balance considerations. *Prog. Oceanogr.*, **4**, 226–244.

Turekian, K.K. and Imbrie, J. (1966) The distribution of trace elements in deep-sea sediments of the Atlantic Ocean. *Earth Planet. Sci. Lett.*, **1**, 161–168.

Veron, A., Lambert, C.E., Isley, A., Linet, P. and Grousset, F. (1987) Evidence of recent lead pollution in deep-north east Atlantic sediments. *Nature*, **326**, 278–281.

Von Damm, K.L., Edmond, J.M., Grant, B., Measures, C.J., Walden, B. and Weiss, R.F. (1985) Chemistry of submarine hydrothermal solutions at 21°N, East Pacific Rise. *Geochim. Cosmochim. Acta*, **49**, 2197–2220.

Von der Borch, C.C. and Rex, R.W. (1970) Amorphous iron oxide precipitates in sediments cored during Leg 5, Deep Sea Drilling Project, in *Initial Reports of the Deep Sea Drilling Project*, Vol. 5, pp.541–544. Washington, DC: US Government Printing Office.

Wedepohl, K.H. (1960) Spureanalytische Untersuchungen an Tiefseetonen aus dem Atlantik. *Geochim. Cosmochim. Acta*, **18**, 200–231.

Wedepohl, K.H. (1968) Chemical fractionation in the sedimentary environment, in *Origin Distribution of the Elements*, L.H. Ahrens (ed.), pp.999–1016. Oxford: Pergamon.

# Part IV
# The Global Journey: Synthesis

# 17    Marine geochemistry: an overview

The 'global journey', which traced material from its sources, through the ocean reservoir, and to the sediment sink, has now been completed. In the present chapter, the various strands will be brought together in an overview of the present state-of-the-art in marine geochemistry.

## 17.1 How the system works

The question that was posed in the Introduction was 'How do the oceans work as a chemical system?' The route that was selected in an attempt to answer this question involved:

1 identifying the pathways followed by the material that entered the ocean reservoir from both external and internal sources, and quantifying the magnitudes of the fluxes associated with them;

2 describing the physical, biological and chemical processes that occur within the water column, and relating them to the fluxes that carry material to the sediment (and rock) sink; and

3 outlining the various processes that interact to control the composition of the sediments themselves.

There are a wide variety of inorganic and organic components in seawater, but it was pointed out in the *Introduction* that in order to follow the route outlined above particular attention would be paid to a selected number of components that trace particular processes. These will include some trace and major components of seawater.

The primary global-scale sources of material to the oceans are river run-off, atmospheric deposition and hydrothermal exhalations, all of which supply both dissolved and particulate material to the ocean system. The system therefore is dominated by the large fluxes of material that enter it across boundaries, and it has become apparent that the key to understanding the driving force behind many of the processes that operate in the water column lies in the particulate ↔ dissolved interactions which take place during the throughput of material from its sources to the sediment sink. During this source → sink journey the dissolved constituents encounter regions of relatively high particle concentrations, for example, in estuaries (especially those having a turbidity maximum) and river plumes, in the sea-surface microlayer, under conditions of high primary production, in the regions of hydrothermal venting systems and in nepheloid layers generated by boundary currents at the western edges of the ocean basins. All these high-particle regions became zones of enhanced dissolved ↔ particulate reactivity. In addition, there is a background microcosm of particles dispersed throughout all the water column. The overall effect is that the ocean may be considered to be a particle-dominated system even though the actual particle concentrations is low, and the composition of the seawater phase is controlled to a large extent by the 'great particle conspiracy'. It is this conspiracy that is the key to Forchhammer's 'facility with which the elements in seawater are made insoluble'.

Essentially, the ocean system consists of two layers of water: a thin, warm, less dense surface layer, which caps a more dense, thick, cold, deep-water layer. The two layers are separated by the thermocline and pycnocline mixing barriers. Primary production in the oceans is confined to the upper layer of approximately 100 m where light can penetrate. This primary production actively takes up many elements and also creates the particulate matter that scavenges other elements and ultimately fuels the biogeochemistry (and indeed most of the life) within the oceans. Transport of nutrients to the surface layer from external sources or by mixing with deep waters is a key control on ocean productivity, along with the supply of sunlight.

Once they have reached the seawater system the dissolved and particulate elements are subjected to a complex series of transport–removal processes. Non-reactive elements will tend to behave in a generally conservative manner. Their distributions will be controlled by physical processes, such as water mass mixing, their residence times in the ocean will be relatively long (millions of years) compared to more reactive elements and to ocean water mixing times. Their distribution within the ocean will be relatively uniform. As the degree of reactivity of an element increases it becomes progressively influenced by bio-geochemical processes, and begins to behave in a non-conservative manner showing gradients of concentrations within the oceans with both depth and location. The degree of reactivity exhibited by an element in seawater therefore exerts a basic control on how it moves through the ocean system. River run-off and atmospheric deposition both deliver dissolved and particulate material to the surface ocean via exchange across the estuarine–sea and air–sea interfaces. This surface ocean is a zone of relatively high particle concentration, the externally delivered particles being swamped in most areas by internally produced biological particles, the products of primary and secondary production. Large-sized, biologically produced aggregates, which consist of a significant fraction of faecal material, sink from the surface layer. At relatively shallow depths a large fraction of the aggregates, usually >90%, undergoes destruction with the components regenerated as dissolved species within the water column. However, it is the fraction that escapes destruction and falls to the sea bed that drives the principal transport of material to the sediment surface via the global carbon flux. The total oceanic particle population therefore consists of a wide particle-size spectrum. However, it is convenient to divide it into fine particulate matter (FPM), that is, the suspended population, and coarse particulate matter (CPM), that is, the sinking population. The production and coupling between the populations involves a continuous series of aggregation–disaggregation processes, the end-point of which is to produce a fine particle population suspended in the water column, and the down-column transport which links the two populations and involves a piggy-back type of reversible FPM association with the large, fast-sinking CPM aggregates. Both the FPM and CPM particulate populations take part in a series of complex biologically and chemically mediated reactions, both in the surface waters and at depth in the water column. These particle-driven reactions are the principal control on the chemical composition of seawater, which is regulated by a balance between the rate of addition of a dissolved component and its rate of removal via sinking particulate material to the sediment sink.

Many elements in seawater have an oceanic residence time that is relatively short compared with both the rate at which they have been added to seawater over geological time and the holding time of the oceans. The controls on the short residence times of these dissolved elements are imposed by uptake reactions with the oceanic particle population. A number of process-orientated trace elements were used in the text as examples of reactive oceanic components in order to illustrate how this 'great particle conspiracy' operates. Although it is difficult to distinguish between biologically-dominated and inorganically-dominated controls on dissolved trace elements in the water column, two general particle-association removal routes were distinguished. Both of these routes are ultimately driven by the global carbon flux, either directly, with carbon-associated carriers (nutrient-type trace metal removal reactions), or indirectly, via a coupling between the small-sized inorganic (FPM) and the large-sized carbon particles (CPM) (scavenging-type removal reactions). Although some trace metals do in fact exhibit mixed removal processes, the distinction does underline a fundamental dichotomy in the oceanic behaviour of a number of trace metals. Three broad types of behaviour can be distinguished.

1 Conservative-type behaviour where the concentrations of elements vary little with depth or location and elements have oceanic residence times of millions of years.

2 Scavenging-type trace metals, which show a surface-enrichment–depth-depletion profile, undergo extensive reactions with the fine (FPM) fraction of oceanic TSM, and although these reactions often involve reversible equilibria, in which there is continuous exchange between the dissolved and particulate states, their residence times in the oceans tend to

be relatively short, that is, of the order of a few hundreds of years or even less. These trace metals reach the sediment surface in association with their inorganic or organic host particles.

3 Nutrient-type elements become involved in the major oceanic biogeochemical cycles and undergo a surface-depletion–subsurface-regeneration enrichment at depth in the water column. This applies to major components of living matter such as the macronutrients; for example nitrogen and phosphorus, essential trace metals such as Zn and some other metals that have limited biological role. In these cases, for example that of cadmium, the nutrient like behaviour may reflect a biological requirement or uptake via biological pathways designed for other elements which are not able to completely discriminate between elements of similar charge and size. Uptake in surface waters and recycling throughout the water column results in the nutrient-type elements having residence times that are relatively long compared with those of the scavenging-type elements and with the oceanic mixing time. For example, Bruland (1980) set a lower limit residence time for the nutrient-type elements of about 5000 yr (see also Table 11.1 which implies that some elements with slightly shorter residence times may also show nutrient-type behaviour). The nutrient elements are carried to the sediment reservoir in association with the carbon fraction of the large-sized (CPM) oceanic TSM population. The fate of these elements depends on the diagenetic environment in the upper part of the sediment column.

(a) Under oxic conditions the carriers are rapidly destroyed to release their associated elements at the sediment surface.

(b) The carrier phases alternatively may be buried and broken down at depth within the sediments under oxic or suboxic conditions, and the associated elements released at depth in the sediments into the interstitial waters from which they may be recycled to the water column or retained in the sediments.

In addition to particle reactivity, water transport acts as a control on the distributions of dissolved elements in the ocean system. This has different effects on the nutrient-type and the scavenging-type elements. The trace-element particulate carriers sink to deeper water where the nutrient-type elements are released back into solution, and this imposes a fundamental control on their oceanic distributions. The oceanic deep-water 'grand tour' transports water from the main deep-water sources in the Atlantic through the Indian Ocean and finally to the North Pacific. As a result, components such as the nutrients and the nutrient-type trace elements, which have a deep-water recycling stage, build up in concentration in the deep-waters of the Pacific relative to those of the Atlantic. In contrast, the scavenging-type elements reflect more local processes and tend to decrease in deep water concentrations along the flow path from the deep Atlantic to the deep Pacific. Thus, the differences between the conservative, scavenging-type and nutrient-type elements are reflected not only in their distribution in the water column and manner in which they are taken out of solution, but also in their water-column residence times.

Elements are introduced into the oceans from both continental and oceanic crustal sources, but for most elements the continental source dominates. The geochemical abundances of the elements in the continental crust source material will therefore exert a fundamental control on their oceanic distributions. However, because they are removed relatively rapidly from the water column the scavenging-type elements tend to be fractionated from the nutrient-type and conservative-type elements. In this way, therefore, the concentration of a specific dissolved element in seawater will depend in part on its geochemical abundance in the crust, and in part on the efficiency with which it is removed to the solid phase, that is, its oceanic reactivity.

The down-column transport of components from the surface ocean is dominated by the carbon flux, which varies in relation to the extent of primary production in the surface waters; thus, the strength of the flux is greatest under the regions of intense productivity, which are located mainly at the edges of the ocean basins. Lateral mid-water and bottom flows can also transport material in the oceans. These flows include both the large scale ocean circulation and processes that resuspend sediments within the oceans and at the ocean margins. The throughput of material to the sea bed involves a coupling between the vertical flux and a series of these lateral fluxes. It is the removal of dissolved components via association with the particles transported by this flux combination that controls the elemental composition of seawater by delivering material across the benthic boundary layer to the sediment sink and so taking it out of the water column.

The composition of oceanic sediments is therefore determined by the nature of the components transported by the various down-column and lateral fluxes, and by the relative strengths of the individual fluxes themselves. The overall result of these various factors is to set up an ocean particle flux regime in which the following general trends can be identified.

1 During descent down the water column a large fraction of the organic matter from the surface layer is destroyed in the upper waters, but that which reaches the sediment surface is related to primary production so that the most organic-rich deposits are found fringing the continents under the areas of intense production.

2 The major sediment-forming biogenic components are opaline silica and carbonate shell material. There is a fundamental biogeographical dichotomy in the distribution of planktonic populations that is evidenced in a coastal high-productivity regime, which is characterized by silica-secreting diatoms, and an oceanic low-productivity regime, which is characterized by smaller phytoplankton and includes carbonate-secreting coccoliths. The extent to which seawater is undersaturated with respect to calcium carbonate increases with depth so that away from the coastal regime its preservation is restricted mainly to mid-ocean topographic highs located above the carbonate compensation depth. In contrast, seawater is everywhere undersaturated with respect to silica and siliceous shells can accumulate only in those regions where the supply rate exceeds the rate of dissolution, that is, under areas of high primary production at the ocean margins. The overall result of these dissolution processes is that siliceous deposits are generally found at the edges of the oceans, whereas carbonates are concentrated in mid-ocean ridge areas. Thus, dissolution constraints on the preservation of shell material and differential controls on dissolution control the distribution of carbonate and opaline material on the ocean floor.

3 In open-ocean areas below the carbonate compensation depth the sediments are dominated by inorganic components particularly clays, reflecting inputs from land, plus hydrogenous phases formed in situ. These accumulate very slowly and dominate in these areas only because other biogenic components are not preserved in the sediments.

A series of chemical signals is transmitted through the ocean system, and these combine together to control the composition of the bottom sediments. The background transport of material to sediments operates on a global-ocean scale, and the signals involved can be subdivided into biogenous, detrital and authigenic types. The authigenic signal is the primary background signal, which carries elements derived from solution in seawater to the sediment surface. For elements that have relatively short residence times in seawater, the authigenic signal does not operate on a constant-flux basis. Dissolved elements that have a longer residence time, however, can have their concentrations smoothed out between the major oceans, and their deposition may be relatively uniform across the oceans. As a result, because they are removed relatively rapidly from the water column the scavenging-type elements tend to be fractionated from the nutrient-type elements in the sediments as well as in the water column (see previously). Thus, Mn (a scavenging-type element) has an oceanic reactivity pattern imposed on its removal into sediments, with the result that the magnitude of its authigenic flux appears to be related directly to that of its input flux; for example, authigenic Mn is accumulating faster in the Atlantic, which has stronger fluvial and atmospheric fluxes, than in the Pacific. In contrast, less reactive elements, such as the nutrient-type Ni and the mixed-type Cu, generally have similar authigenic fluxes in the two oceans, that is, the geochemical abundance (Mn) versus oceanic reactivity (Ni, Cu) control. In addition to the background signals, more localized signals, such as those associated with hydrothermal activity and anthropogenic effects, can transport components to the sediment surface from the overlying seawater.

Even when the components are actually incorporated into the sediments they have not simply been locked away in a static reservoir, but have in fact entered a diagenetically active and biogeochemically dynamic environment. The diagenetic reactions are controlled by the manner in which processes in the sedimentary environment destroy organic carbon; that is, even the small fraction of organic matter that survives the journey down the water column is subjected to further degradation in the sediment reservoir. The intensity of diagenesis is controlled by the amount of organic carbon that reaches the sediment surface, and there is a redox-driven diagenetic sequence in the sediments, in which a variety of oxidants are switched on as the previous one is exhausted, in the general order: oxygen then nitrate~manganese oxides then iron oxides then sulfate. The extent to which this sequence progresses depends on the amount of organic matter reaching the sediment surface. This organic matter supply is related to the extent of primary production in the surface waters, the delivery to the sediments (supply rate) and the rate at which the sediments accumulate (burial rate). Within the sediments the organic matter supply creates a lateral diagenetic sequence, in which the diagenetic intensity decreases in the order: nearshore to hemi-pelagic to pelagic sediments. Components released in the diagenetic reactions are transported through the sediment interstitial waters. Some of these components are trapped in the solid phases and so can impose a diagenetic spike on the sediments. Others can escape back into seawater, however, thus providing a secondary, that is, recycled, oceanic source. Thus, rather than acting as a static sink, sediments can recycle some elements, either (i) retaining but redistributing them in the sediment column; or (ii) losing them back to seawater. Despite these recycling processes, the sediment reservoir remains the ultimate sink for particulate material that flows through the ocean system, and so also acts as the major sink for particle-reactive elements that are removed from the dissolved phase.

It is apparent, therefore, that the major process that controls the dissolved-element composition of seawater is a balance between the rates at which the elements are added to the system and the rates at which they are removed by the throughput of particulate material that delivers them to the sediment sink. During their residence time in seawater, dissolved elements are transported by physical circulation, and undergo a series of biogeochemical reactions by which ultimately they are taken up by particulate matter, which also is transported by vertical and horizontal movements. Overall, therefore, it may be concluded that the oceanic chemical system is driven by a physical–chemical–biological process trinity. This process trinity operates on both the particulate and dissolved material introduced into the ocean reservoir and controls their passage through the system to the sediment sink. It also continues to influence the fate of the elements within the sediment sink itself; for example, physical processes resuspend sediments into the water column, and biogeochemical processes are active in diagenesis.

## 17.2 Balancing the books

A process-orientated approach to marine geochemistry has been adopted in an attempt to elucidate the oceanic cycles of various constituents, and one possible way of answering the question 'To what extent have we understood how the oceans work as a chemical system?' is to establish accounting procedures that can be used to assess the quantitative aspect of the cycles, for example, by attempting to construct mass balances for the system. This can be done on a variety of scales, but here we will focus on whole ocean budgets which have the advantage of clearly defined boundaries. A variety of physical and biogeochemical processes control both the removal of a dissolved element from seawater and its transfer to the deposition sink(s). Although these processes work differently for different elements, if the oceans are assumed to be in a steady state, then there should be a balance for any constituent between its input and its output rates, that is, if sufficient data are available it should be possible to construct a mass balance for the constituent. This exercise in geochemical accountancy therefore serves to both synthesise and test our understanding of the geochemical system. Mass budgets also serve to contextualize recent perturbations of the global system, such as have occurred for lead and nitrogen and the associated estimates of residence times provide a context for understanding the rates of change in ocean concentrations created by these changed inputs. In the case of some elements such as carbon (see Fig. 9.18), the development and management of the global cycle is now a societal priority in the face of the impact of rising $CO_2$ emissions on climate.

Various authors have attempted to produce overall mass balances for the dissolved constituents found in seawater. However, a few examples will serve to illustrate how the mass-balance approach has evolved in terms of recent advances in our understanding of how the ocean system works. It should always be remembered that these budgets are necessarily uncertain and based on considerable extrapolations. Furthermore different budgets will often contain rather different flux estimates to other budgets. The budgets here are taken from particular examples in the literature. The individual fluxes used by these authors will not agree exactly with those presented in, for example, Chapter 6 although in most cases they are of similar magnitude.

## 17.2.1 Major Ions

A long standing goal in oceanography has been to understand the sources and sinks for the major ions in ocean waters and ultimately answer the question, 'Why is the sea salty?' A budget for the sources and sinks of the seven major ions in seawater is presented in Table 17.1. This budget only considers riverine inputs, ignoring the transport of atmospheric dust for example, but this is a good approximation of the total input for these soluble major ions. The budget considers removal by the formation of evaporite minerals, which are dominated by sodium chloride and calcium sulfates and carbonates, ion exchange of clays as they enter the ocean (see Chapter 5), reactions at mid-ocean ridges and the precipitation of calcium carbonate. The consumption of sulfate during suboxic diagenesis is not considered in this particular budget and is probably a significant addition sulfate sink. The major ions in ocean waters all have conservative-type distributions and very long residence times, with the exception of alkalinity which is somewhat shorter and does show minor surface water depletions in concentrations. Under such conditions validating the steady state assumption is particularly problematic because there is very little information available with which to test if concentrations of a component in seawater have remained essentially constant over multi-million year timescales. For sodium and chloride it is thought

likely that the precipitation of evaporite deposits in ocean margin environments is the main sink. Although there are vast evaporite deposits known from the geological record, there are no deposits forming at the required magnitude at the present time. However, given the very long residence time of these ions in the ocean, the steady state assumption only requires that such sinks balance outputs over million year timescales and on those time-scales, changes in climate and geological environments have the potential to create such evaporite basins.

Based on the budget in Table 17.1, ultimately the sea is salty and its composition dominated by sodium chloride because these ions are relatively abundant in the crust, but are only very slowly removed from seawater compared to other ions and have much longer residence time than the ocean water itself. Hence the sodium chloride concentrations build up and come to dominate the major ion composition. In the case of magnesium, the main sink is uptake by basalts within hydrothermal systems, and prior to the discovery of the importance of these systems, geochemists had struggled to balance the magnesium budgets. The sinks for calcium and carbonate by contrast are biogenic calcium carbonate deposits, and the river supply is significantly augmented by release of calcium at mid-ocean ridges. The budget of Milliman (1993; see Chapter 15) suggests that only half of the total calcium carbonate formed in the surface ocean is buried, the rest being redissolved

**Table 17.1** Geochemical budget for major seawater ions and alkalinity (based on McDuff and Morel, 1980). All fluxes as $10^{12}\,mol\,yr^{-1}$.
Residence times in millions of years.
Positive numbers are inputs, negative numbers are sinks.
Note that for chloride the balance of inputs and removal is assumed, hence the perfect balance.
No numbers in a box reflect fluxes that are thought to be insignificant and '?' represent fluxes whose significance is not know.

| Ion | River Input | Evaporites + air-sea exchange | Ion Exchange | Hydrothermal Activity | Low Temperature Basalt Weathering | Carbonate Deposition | Net Balance | Residence Time $10^6$ yr |
|---|---|---|---|---|---|---|---|---|
| $Cl^-$ | +10 | −10 | | ? | | | 0 | 71 |
| $Na^+$ | +11.8 | −9.3 | −1.9 | ? | | | 0.6 | 52 |
| $Mg^{2+}$ | +8 | −0.5 | −1.2 | −7.8 | | | −1.5 | 9 |
| $SO_4^{2-}$ | +3.7 | −0.5 | | −3.8 | | | −0.6 | 10 |
| $K^+$ | +3.2 | −0.1 | −0.4 | +1.3 | −4 | | 0 | 4 |
| $Ca^{2+}$ | +17.1 | −0.1 | +2.6 | +3.1 | +2 | −24.7 | 0 | +0.65 |
| Alkalinity | +47.8 | | +0.5 | −0.4 | | −49.4 | −1.5 | +0.06 |

at depth. This illustrates that for many elements the internal cycles within the ocean are large compared to the external inputs.

## 17.2.2 Nutrients

*Phosphorus.* Broecker and Peng (1982) present a very simple phosphorus budget in which they estimate the total P input to the oceans to be $3.2 \times 10^{10} \, \text{mol} \, \text{yr}^{-1}$ which is exactly balanced by the sedimentary sink. Their estimate of the flux of phosphorus from the surface ocean to the deep is 10 times larger than the input or burial flux, emphasizing again the importance of the internal cycle with phosphorus atoms being recycled within the oceans many times before their ultimate burial. Recently Wallmann (2010) has revisited the ocean phosphorus budget. His budget considers in particular the extensive recycling within sediments with almost 80% of the phosphorus reaching oceanic sediments being recycled and >95% of that reaching continental margin sediments. Wallman's analysis suggests that the oceans are not currently in phosphorus steady state with net loss rates of $11.6 \times 10^{10} \, \text{mol} \, \text{yr}^{-1}$, which are small compared to the large scale internal cycle and phosphorus inventory. Wallman speculates that the imbalance may reflect the fact that the ocean composition is still adjusting to the end of the last glaciation, during which time weathering inputs may have been higher and, with lower sea level, continental margin sedimentation would have been greatly reduced. Such a response time is consistent with a phosphorus residence time of the order of 16 000 to 38 000 years (Ruttenberg, 1993).

*Silicon.* The importance of internal cycles is illustrated further by the silicon cycle in Table 17.2 which is based on a budget by Treguer *et al.* (1995). These authors estimate the overall uncertainties in their input and output estimates as 33% and 25% respectively and within these uncertainties conclude that the budget is approximately balanced.

The input budget for Si is dominated by river inputs but aeolian and inputs associated with direct hydrothermal inputs, and subsequent seafloor weathering of basalts, are significant. The main sink is the burial of biogenic opal, principally in deep ocean siliceous oozes, mostly around Antactica. Treguer

**Table 17.2** Budget of Si in the world ocean $10^{12} \, \text{mol} \, \text{yr}^{-1}$ (Treguer *et al.*, 1995).

| Flux Source | Inputs |
|---|---|
| River input | 5 |
| Eolian | 0.5 |
| Seafloor weathering | 0.4 |
| Hydrothermal | 0.2 |
| Total | 6.1 |

| | Outputs |
|---|---|
| Biogenic opal burial coastal | 1.2 |
| Biogenic opal burial oceanic | 5.9 |
| Total | 7.1 |

*et al.* (1995) estimate the total surface water production of opal to be about $240 \times 10^{12} \, \text{mol} \, \text{yr}^{-1}$, with 60% of this regenerated within the euphotic zone (approximately the upper 100 m), and in total 88% of the Si regenerated within the ocean water column. They estimate that in total about $30 \times 10^{12} \, \text{mol} \, \text{yr}^{-1}$ Si reach the sediments and of that only $7 \times 10^{12} \, \text{mol} \, \text{yr}^{-1}$ is finally buried. Thus the internal cycling of Si again dwarfs the external inputs.

*Nitrogen.* This consideration excludes the role of $N_2$ gas except in terms of its biological fixation and focuses on all other forms of nitrogen termed here 'fixed' nitrogen.

As discussed in Chapter 9, the internal cycling of nitrogen and phosphorus within the oceans are closely tied to one another as revealed by the close correlations between the dissolved concentrations of the two with a near constant N/P ratio – the Redfield ratio. This reflect the essential requirement for both elements within biological systems and hence their link to the carbon cycle. However, despite the closely coupled internal cycling of these elements within the oceans, their inputs to the oceans are very different. The nitrogen cycle has been the subject of a great deal of recent study that has served to emphasise its complexity. New processes have been identified (e.g. Anammox) and the estimates of the quantitative significance of others such as marine biological nitrogen fixation have been revised considerably (see Chapter

**Table 17.3** Oceanographic budget for fixed nitrogen (Brandes *et al.*, 2007).

| | | |
|---|---|---|
| Inputs | Nitrogen fixation | $5.7–10.7 \times 10^{12}\,mol\,yr^{-1}$ |
| | Atmospheric + fluvial inputs | $7.1–10.7 \times 10^{12}\,mol\,yr^{-1}$ |
| Outputs | Water column denitrification | $5.7–7.1 \times 10^{12}\,mol\,yr^{-1}$ |
| | Sedimentary denitrification | $14.3–17.9 \times 10^{12}\,mol\,yr^{-1}$ |
| | Burial | $1.8 \times 10^{12}\,mol\,yr^{-1}$ |

9) in the light of new information. A recent budget prepared by Brandes *et al.* (2007) is shown in Table 17.3 and the input fluxes are consistent with those presented in Chapter 6.

As noted in Chapter 9, nitrogen fixation, fluvial and atmospheric inputs are all important sources of nitrogen to the oceans and the latter two have probably been increased significantly by anthropogenic activity. Burial of nitrogen associated with organic matter is a small sink compared with denitrification and other nitrogen reduction processes where nitrate is used as an alternative electron acceptor in suboxic processes in the sediment or in the low oxygen zones of the oceans which are found principally in the Indian and Pacific oceans. These rapid rates of loss via denitrification reduce the ocean nitrogen residence time to 1500–3200 years (depending on whether the input or removal term is used in the calculation), much shorter than that of phosphorus.

The other notable feature is that the budget is probably not in balance. The nitrogen cycle has been particularly intensively studied cycle. Recent research has tended to exacerbate rather than reduce the discrepancies between inputs and losses. Atmospheric and fluvial inputs have been increased by human activity and the system will not yet have adjusted to that input so it might be anticipated that inputs would be greater than losses, but the problem is the reverse with outputs greater than inputs. Denitrification rates may be underestimated since the full impact of newly discovered nitrogen loss processes such as Anammox are still being evaluated, and if this is the case the imbalance could be even greater. Brandes *et al.* (2007) consider the option that the system is not in balance and is responding to the post glacial sea level rise and the attendant impacts on denitrification, but evidence on changes in the ocean nitrogen cycle from the isotopic composition of ocean sediment cores suggest this is unlikely. Brandes

*et al.* (2007) therefore suggest that the budgetary imbalance reflect our underestimation of input processes, particularly nitrogen fixation. Further research will be needed to better describe the ocean nitrogen cycle and given the central importance of this cycle for life in the ocean and the scale of human disturbance of the nitrogen cycle, such research is clearly a priority.

*Trace Metals.* The budgets of the major ions and silicon are dominated by removal by specific well defined processes. By contrast, the removal of trace metals is dominated by adsorption or uptake as minor components in other phases. Fluxes associated with detrital components of clay material are augmented by authigenic fluxes derived from water column sources. Theses authigenic fluxes have been estimated using a variety of approaches and these are presented in Table 16.5. The difficulties of making accurate measurements of many of the trace metals, and of identifying the sinks, complicates accurate budgeting. In addition the assumption of steady state may not be appropriate if contemporary estimates of river and atmospheric fluxes are enhanced by human activity. Despite these caveats, the attempts to construct global budgets can still be instructive.

Collier and Edmond (1984) developed a budget for cadmium. Their estimate of inputs is substantially lower than that presented in Chapter 6 and this may reflect the inclusion of estuarine cadmium desorption in the fluxes in Chapter 6. The sediment burial rate of Collier and Edmond is rather lower than that of Simpson (1981; see Table 17.4) and both estimates are lower than the input estimates. This may reflect limitations of the data or enhancements of contemporary river and atmospheric fluxes by anthropogenic activity. Regardless of these limitations the most striking conclusion of the flux estimates is that, as for the nutrients discussed above, that the internal cycling rates are much greater than the fluxes in and

**Table 17.4** Oceanographic budget for cadmium.

| Input (atmosphere + river) | 21–68 × 10⁶ mol yr⁻¹ (a) |
|---|---|
| | 165 × 10⁶ mol yr⁻¹ (b) |
| Flux out of the euphotic zone | 1200–2600 × 10⁶ mol yr⁻¹ (a) |
| Deep ocean regeneration | 72–1800 × 10⁶ mol yr⁻¹ (a) |
| Deep ocean sedimentation | 2.2 × 10⁶ mol yr⁻¹ (a) |
| | 18 × 10⁶ mol yr⁻¹ (c) |

(a) Collier and Edmond (1984), (b) Table 6.13 (c) Simpson (1981).

**Table 17.5** Oceanographic budget for copper.

| Input (atmospheric + river) | 1116–1620 × 10⁶ mol yr⁻¹ (a) |
|---|---|
| | 1150–1700 × 10⁶ mol yr⁻¹ (b) |
| Flux to deep ocean biogenic + scavenging | 4320–5400 × 10⁶ mol yr⁻¹ (a) |
| | 1080–6120 × 10⁶ mol yr⁻¹ (a) |
| Regeneration from sediments | 3600–21 600 × 10⁶ mol yr⁻¹ (a) |
| Burial | 453–4302 × 10⁶ mol yr⁻¹ (c) |

(a) Collier and Edmond (1984), (b) Table 6.13, (c) Table 16.3.

**Table 17.6** Oceanographic budget for uranium (Klinkhammer and Palmer, 1991).

| Input | Rivers | 36 × 10⁶ mol yr⁻¹ |
|---|---|---|
| Removal | Margin sediments | 28 × 10⁶ mol yr⁻¹ |
| | Hydrothermal systems | 16 × 10⁶ mol yr⁻¹ |

out of the ocean system. This reflects the nutrient like cycling of cadmium.

Collier and Edmond also developed a model for copper (see Table 17.5). In the case of copper their estimates of inputs are very similar to those in Chapter 6 and within the range of sediment burial estimates, suggesting approximate geochemical balance. Again as for cadmium and the nutrients, there is intense cycling in the water column involving both transport associated with primary production and in the case of copper deep water scavenging, plus intense regeneration at the sediment water interface, consistent with the observed behaviour of copper (see Section 11.3).

Uranium is a trace element within the oceans, but one with a very long residence time ($5 \times 10^5$ yr) and approximately conservative behaviour because when in the oxidized U(VI) form it is present as a simple cation $UO_2^{2+}$. However, under reducing conditions during suboxic diagenesis principally in ocean margin sediments, or associated with reactions with basalt in hydrothermal systems, it forms U(IV) which is highly particle reactive. The uranium budget developed by Klinkhammer and Palmer (1991) is shown in Table 17.6.

## 17.3 The role of the ocean margins

The mass balances described above involved the surface- and deep-water reservoirs. In addition, it is

necessary to distinguish between two sub-reservoirs, that is, the ocean margins and the open-ocean (see Section 6.5 and 11.4). The ocean margins, which include estuaries and the continental shelf, make up only ~7% of the global ocean surface, but account for a significant fraction of oceanic primary production (probably around 20%). They are also impacted by large terrigenous particulate and dissolved fluvial fluxes: for a discussion of the role of the ocean margins in global processes the reader is referred to the volume edited by Mantoura et al. (1991).

Martin and Thomas (1994) attempted to assess the fluxes and fates of the dissolved nutrient-type trace metals Cd, Cu, Ni and Zn in the ocean margins. The authors used a variety of approaches to estimate the input fluxes of the trace metals to the ocean margins from (i) the external 'telluric' source, that is, fluvial and atmospheric; and (ii) the internal oceanic source, derived from upwelled water. It was not possible to quantify the potential additional source of trace-metal mobilization from shelf sediments, and as a result the overall mass balance calculations could not be closed. Despite this, two important conclusions could be drawn from the study.

**1** A comparison of the data indicated that when the export and input fluxes are compared it appears that the standing crop of trace metals in the margin regions cannot be sustained by the telluric sources alone, but is highly dependent on dissolved inputs from the deep ocean.

**2** Residence times of the trace metals in the margins were estimated by dividing the dissolved stock by the sum of the telluric inputs. The values derived in this way were strikingly low, the margin residence times ranging from 0.3–0.9 yr for Cd to 0.7–2.7 yr for Cu, Ni and Zn. On the basis of these data it may be concluded that the margin and open-ocean subsystems exchange trace metals more intensively than was at first thought. Such exchange is also very important for nutrients and is responsible in part for driving the high primary productivity fo the coastal oceans (see Section 6.5).

## 17.4 Conclusions

The last few decades have seen a quantum leap in our understanding of how the ocean works as a chemical system. As a result, marine geochemists are now beginning to have at least a first-order understanding of many of the processes that drive oceanic chemistry. The present volume has described the manner in which these processes operate within a global ocean source–sink framework. The next step is to evaluate the chemistry of the ocean system in relation to planetary geochemistry. In this context it is of interest to summarize the concepts developed by Whitfield and his co-workers who have attempted to find a rationale for the composition of seawater that is based on fundamental chemical principles.

The concentration of an element in seawater is controlled by a combination of its input strengths (geochemical abundance) and its output strengths (oceanic reactivity). Many of the elements present in seawater have been released from crustal rocks, transported to the oceans, and then become incorporated into marine sediments. These sediments, however, are eventually recycled back to the continents during the processes of sea-floor spreading and

mountain building; these processes are illustrated in Fig. 17.1, in which the dynamic nature of the structural and spatial relationships between large-scale geological phenomena is shown. Because of these large-scale recycling processes Whitfield and Turner (1979) concluded that the same material has been cycled through the ocean system several times during the history of the Earth, and suggested that it therefore should be possible to use the partitioning of the elements between the ocean and the crustal rocks to gain an insight into the nature of seawater itself. The ocean system is assumed to be in a steady state, such that the rate of input of an element balances its rate of removal. The time an element resides in seawater can be expressed in terms of its mean oceanic residence time (MORT), $\overline{t}_Y$ (Section 11.2), which may be defined as

$$\overline{t}_Y = Y_S^0 / \overline{J}_Y^0 \qquad (17.1)$$

where $Y_S^0$ is the total mass of the element $Y$ dissolved in the ocean reservoir and $\overline{J}_Y^0$ is the mean flux of $Y$ through the reservoir in unit time. The superscript zero emphasizes that the values refer to a system in steady state. The MORT is a measure of the reactivity of an element in the ocean system. This is because elements that are highly reactive will have low MORT values and a rapid throughput, whereas those that are unreactive will have high MORT

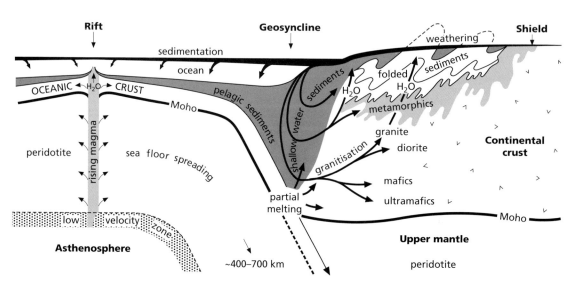

**Fig. 17.1** The recycling of oceanic sediments during the processes of sea-floor spreading and mountain building (from Degens and Mopper, 1976). Reprinted with permission from Elsevier.

values and will tend to accumulate in the system. The MORT values therefore are essential parameters that describe the steady-state composition of seawater, and Whitfield (1979) has suggested that Forchhammer's 'facility with which the elements in seawater are made insoluble' has found a quantitative expression in the MORT concept. Thus, a MORT value is a direct measure of the ease with which an element is removed to the sediment sink by incorporation into the solid phase. This affinity of an element for the solid phase can be described by a coefficient that expresses its partitioning between the water and rock phases (partitioning coefficient, $K_Y$), which is calculated as the ratio between the mean concentration of the element in natural water to that in crustal rock. Whitfield and Turner (1979) found a linear relationship between the seawater–crustal-rock partition coefficient ($\log K_Y(SW)$) and the MORT ($\log t_Y$) values of elements (see Fig. 17.2a). It was therefore demonstrated that the MORT value of an element is related directly to its partitioning between the oceans and crustal rocks. Thus, a relationship was established between the reactivity of an element in the oceans (MORT) and its long-term recycling through the global reservoir system ($K_Y$).

Whitfield and Turner (1979) suggested that the partitioning of elements between solid and liquid phases in seawater (and river water) could be rationalized using a simple electrostatic model, in which the fundamental chemical control on the solid–liquid partitioning coefficients is related to the electronegativity of an element. This can be quantified by an electronegativity function ($Q_{YO}$), which is a measure of the attraction that an oxide-based mineral lattice will exert on the element. The authors then showed that the electronegativity function ($Q_{YO}$) of an element can be related directly to its partition coefficient ($K_Y$). The relationship between crustal-rock/ seawater partition coefficients and the electronegativity functions of the elements is illustrated in Fig. 17.2(b), from which it can be seen that the elements that are more strongly bound to the solid phase (high $Q_{YO}$ values) have small partition coefficients (low $K_Y$ values). It is apparent, therefore, that the manner in which an element is partitioned between crustal rock and seawater is dependent on the extent to which it is attracted to the oxide-based mineral lattice. The correlations between $Q_{YO}$ and $K_Y$ thus offer a theoretical explanation for variations in the partition

coefficients of the elements, which is based ultimately on differences in their electronic structures, which themselves are a function of their atomic number and so their chemical periodicity. Whitfield and Turner (1983) drew attention to the fact that this chemical periodicity, which involves a link between electronic structure and chemical behaviour, provides a rationalization of the inorganic chemistry of all the elements. There is therefore a fundamental regularity in the organization of the elements, which, as the authors pointed out, is sometimes forgotten when attempts are made to assess their behaviour in natural systems. For this reason, the correlation between the partition coefficients of the elements and their electronegativities represents an important step forward in our understanding of the chemistry of seawater. Further, the MORT–partition-coefficient– electronegativity-function relationships permit a number of the basic aspects of oceanic chemistry to be predicted on the basis of theoretical chemical concepts. For example, Whitfield (1979) derived a general equation relating to the MORT value and the electronegativity function of an element, and showed that MORT values derived from the electronegativity functions agreed with observed values within an order of magnitude; that is, MORT values can be predicted reasonably well from a knowledge of the electrochemical properties of the elements. A second equation was proposed, which related the electronegativity function of an element to the global mean value of its river input, and this was used to estimate the composition of seawater. For most elements the estimated mean global composition of seawater again agreed with the observed values within an order of magnitude, even though the concentrations themselves range over 12 orders of magnitude; the predicted–observed seawater composition comparison is illustrated in Fig. 17.2(c).

The MORT–partition-coefficient–electronegativity -function relationships developed by Whitfield and co-workers therefore provide a series of theoretical chemical concepts which suggest that the overall composition of seawater is controlled by geological processes that are governed by relatively simple geochemical rules. As a result, the concentration of an element in seawater is controlled by its abundance in the crust (geochemical abundance) and by the ease with which it can be taken into solid sedimentary phases (oceanic reactivity). Ultimately the same

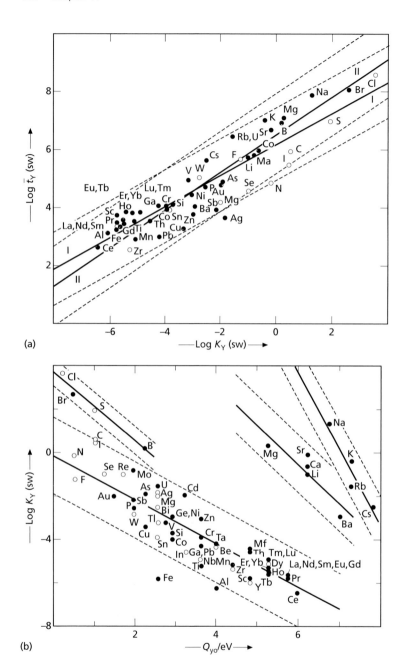

**Fig. 17.2** Relationships in the 'Whitfield ocean'. (a) The relationship between the mean oceanic residence time (MORT, $t_Y$), and the seawater–crustal-rock partition coefficient, $K_Y$ (SW) (from Turner et al., 1980). Reprinted with permission from Elsevier. (b) The relationship between the seawater–crustal-rock partition coefficient, $K_Y$ (SW), and the electronegativity function, $Q_{YO}$ (from Turner et al., 1980). Reprinted with permission from Elsevier. (c) Comparison between the observed, $Y_s$ (obs), and calculated, $Y_s$ (calc), compositions of seawater (from Whitfield, 1979). Reprinted with permission from Elsevier.

processes also control the ease with which most elements are weathered from the crust as well (see Fig. 3.2) and scavenged in the oceans see Section 11.5). Thus, we now have a wider theoretical framework within which to interpret the factors that control the chemical composition of seawater.

Overlain on these geochemical controls are the role of biological processes which particularly impact C, N, P and Si cycling, but also a range of key trace elements such as zinc. The production of particulate matter by biological processes also plays a key role in providing most of the particulate matter, that

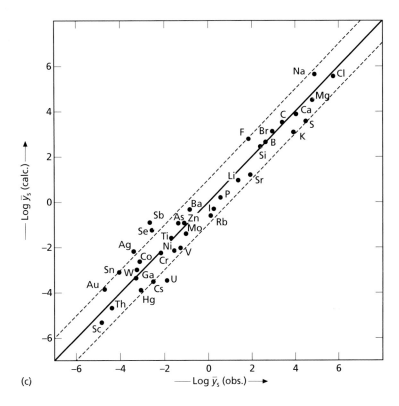

**Fig. 17.2** *Continued*   (c)

cycles through the ocean. The creation of this organic matter, predominantly by phytoplankton, and its decomposition or consumption by zooplankton, bacteria and viruses creates the vast internal cycling system within the oceans that cycles so many of these elements between the surface and deep oceans many times before their burial. Recent advances in our understanding of the key nutrient cycling, such as the description of new nitrogen loss processes and the considerable increase in our estimates of nitrogen fixation rates, demonstrate the limits of our understanding of these key biological cycles, even as human perturbation of these and the closely related carbon cycles become issue of major societal interest, Iron also plays a key role in biological systems (Boyd and Ellwood, 2010), but differs from the other key biological elements in also being highly particle reactive, creating a situation in which it is much more rapidly removed from the oceans to the sediments than other nutrients. This creates a situation of iron limitation in areas of the ocean where external inputs are inadequate to sustain productivity.

To come full circle, therefore, it may be concluded that the overall *composition* of seawater is controlled by relatively simple geochemical rules. The *distribution* of the elements within seawater, however, is dependent on the physical–biological–chemical process trinity that drives the ocean system. The present volume has been concerned with the manner in which the processes involved in this process trinity operate on the throughput of material in the ocean system. The ultimate aim of marine geochemistry must be to produce a rationale for oceanic chemistry based on fundamental chemical principles, which can therefore provide a set of general rules enabling the concentration and behaviour of elements in seawater to be described within a coherent pattern. It is apparent that such an approach is already gaining ground. The models produced will be refined in the future as marine geochemists struggle towards an explanation of Forchhammer's 'facility with which the elements in seawater are made insoluble'.

This future promises to be exciting.

# References

Boyd, P.W. and Ellwood, M.J. (2010) The biogeochemical cycle of iron in the ocean. *Nature Geosciences*, **3**, 675–682.

Brandes, J.A., Devol, A.H. and Deutsch, C. (2007) New developments in the marine nitrogen cycle. *Chem. Rev.*, **107**, 577–589.

Broecker, W.S. and Peng, T-H (1982) *Tracers in the Sea*. Palisades: Lamont-Doherty Geological Observatory.

Bruland, K.W. (1980) Oceanographic distributions of cadmium, zinc, nickel and copper in the North Pacific. *Earth Planet. Sci. Lett.*, **47**, 176–198.

Collier, R.W. and Edmond, J.M. (1984) The trace element geochemistry of marine biogenic particulate matter. *Prog. Oceanogr.*, **13**, 113–199.

Degens, E.T. and Mopper, K. (1976) Factors controlling the distribution and early diagenesis of organic matter in marine sediments, in *Chemical Oceanography*, J.P. Riley and R. Chester (eds), Vol. 5, pp.59–113. London: Academic Press.

Klinkhammer, G.P. and Palmer, M.R. (1991) Uranium in eth oceans: where it goes and why. *Geochim. Cosmochim. Acta.*, **55**, 1799–1806.

Mantoura, R.F.C., Martin, J.M. and Wollast, R. (eds) (1991) *Ocean Margin Processes in Global Change*. Chichester: John Wiley & Sons, Ltd.

Martin, J.M. and Thomas, A.J. (1994) The global insignificance of telluric input of dissolved trace metals (Cd, Cu, Ni and Zn) to ocean margins. *Mar. Chem.*, **46**, 165–178.

McDuff, R.E. and Morel, F.M.M. (1980) The geochemical control of seawater (Sillen revisited) *Environ. Sci. Tech.*, **14**, 1182–1186.

Milliman, J.D. (1993) Production and accumulation of calcium carbonate in the ocean: budget of a nonsteady state. *Global Biogeochem. Cycl.*, **7**, 927–957.

Ruttenberg, K.C. (1993) Reassessment of the oceanic residence time of phosphorus. *Chem. Geol.*, **107**, 405–409.

Simpson, W.R. (1981) A critical review of cadmium in the marine environment. *Prog. Oceanogr.*, **10**, 1–70.

Treguer, P., Nelson, D.M., Van Bennekom, A.J., DeMaster, D.J., Leynaert, A. and Queguiner, B. (1995) The Silica Balance in the World Ocean: A Reestimate *Science*, **268**, 375–379.

Turner, D.R., Dickson, A.G. and Whitfield, M. (1980) Water–rock partition coefficients and the composition of natural waters: a reassessment. *Mar. Chem.*, **9**, 211–218.

Wallmann, K. (2010) Phosphorus imbalance in the global ocean. *Global Biogeochem. Cycl.*, **24**, doi:10.1029/2009GB003643.

Whitfield, M. (1979). The mean oceanic residence time (MORT) concept—a rationalization. *Mar. Chem.*, **8**, 101–123.

Whitfield, M. and Turner, D.R. (1979) Water–rock partition coefficients and the composition of river and seawater. *Nature*, **278**, 132–136.

Whitfield, M. and Turner, D.R. (1983) Chemical periodicity and the speciation and cycling of the elements, in *Trace Metals in Seawater*, C.S. Wong, E. Boyle, K.W. Bruland, J.D. Burton and E.D. Goldberg (eds), pp.719–750. New York: Plenum.

# Index